HALF-DISCRETE
HILBERT-TYPE
INEQUALITIES

HALF-DISCRETE HILBERT-TYPE INEQUALITIES

$$\sum_{n=1}^{\infty}\sum_{m=1}^{\infty}\frac{a_m b_n}{m+n} < \pi\left\{\sum_{m=1}^{\infty}a_m^2\sum_{n=1}^{\infty}b_n^2\right\}^{\frac{1}{2}}$$

$$\int_0^{\infty}\int_0^{\infty}\frac{f(x)g(y)}{x+y}\,dx\,dy < \pi\left\{\int_0^{\infty}f^2(x)\,dx\int_0^{\infty}g^2(y)\,dy\right\}^{\frac{1}{2}}$$

Bicheng Yang
Guangdong University of Education, China

Lokenath Debnath
University of Texas-Pan American, USA

World Scientific

NEW JERSEY · LONDON · SINGAPORE · BEIJING · SHANGHAI · HONG KONG · TAIPEI · CHENNAI

Published by

World Scientific Publishing Co. Pte. Ltd.

5 Toh Tuck Link, Singapore 596224

USA office: 27 Warren Street, Suite 401-402, Hackensack, NJ 07601

UK office: 57 Shelton Street, Covent Garden, London WC2H 9HE

Library of Congress Cataloging-in-Publication Data
Yang, Bicheng, author.
  Half-discrete Hilbert-type inequalities / by Bicheng Yang (Guangdong University of Education,
China) & Lokenath Debnath (University of Texas- Pan American, USA).
    pages cm
  Includes bibliographical references and index.
  ISBN 978-981-4504-97-3 (hardcover : alk. paper)
  1. Inequalities (Mathematics) 2. Mathematical analysis. I. Debnath, Lokenath, author. II. Title.
  QA295.Y36 2014
  515'.46--dc23

                                                        2013036829

British Library Cataloguing-in-Publication Data
A catalogue record for this book is available from the British Library.

Printed in Singapore

Preface

Historically, mathematical analysis has been the major and significant branch of mathematics for the last three centuries. Indeed, inequalities became the heart of mathematical analysis. Many great mathematicians have made significant contributions to numerous new developments on the subject which led to the discovery of many new inequalities with proofs and useful applications in the fields of mathematical physics, pure and applied mathematics. Indeed, mathematical inequalities became an important branch of modern mathematics in the twentieth century through the pioneering work entitled *Inequalities* by G.H. Hardy, J.E. Littlewood and G. Polya which was first published as a treatise in 1934. This unique publication represents a paradigm of precise logic, full of elegant inequalities with rigorous proofs and useful applications in mathematics.

During the twentieth century, discrete and integral inequalities played a fundamental role in mathematics and have a wide variety of applications in many areas of pure and applied mathematics. In particular, David Hilbert (1862-1943) first proved Hilbert's double series inequality without exact determination of the constant in his lectures on integral equations. Herman Weyl (1885-1955) published a proof of the Hilbert double series inequality in 1908. Subsequently, Isaac Schur (1875-1941) gave a new proof of the Hilbert double series inequality in 1911 with the best possible sharp constant, and also discovered the integral analogue of the Hilbert double series inequality which became known as the Hilbert integral inequality. In 1925, G.H. Hardy (1877-1947) provided the best extension of it by introducing a pair of conjugate exponent which became known as the Hardy-Hilbert inequality. The Hilbert-type inequalities are a more wider class of analytic inequalities with bilinear kernels which include the Hardy-Hilbert inequality as a particular case.

The mathematical theory of inequalities, in general, and the Hilbert-type inequalities, in particular, and their applications has grown considerably during the last one century. With the advent of new ideas and proofs, new results and applications, studies are continually being added to the major subject of mathematical inequalities, where these are themselves developing and coalescing. It is becoming more and more desirable for mathematicians to study the Hilbert-type inequalities as a whole. Yet it is increasingly difficult for them to do so since major and important articles often appears in many different journals.

The difficulty can be alleviated if a single research monograph containing a coherent account of recent developments, especially, if written to be accessible to both graduate students and research professionals. It is our hope that this monograph will first interest, then prepare readers to undertake research projects on the Hilbert-type inequalities and their applications, by providing that background of fundamental ideas, methods, proofs, and results essential to understanding the specialized literature of this vast area. Many ideas, major results, methods, proofs, and examples presented in this volume are either motivated by, or borrowed from works cited in the Bibliography. We wish to express our gratitude to the authors of their works.

The writing of this monograph was also greatly influenced by the famous quotations of David Hilbert and G.H. Hardy as follows:

> As long as a branch of knowledge offers an abundance of problems, it is full of vitality.

David Hilbert

> ... we have always found with most inequalities, that we have a little new to add.

> ... in a subject (inequalities) like this, which has applications in every part of mathematics but never been developed systematically.

G.H. Hardy

So, this monograph deals with an extensive account of the theory and applications with numerous examples of any half-discrete Hilbert-type inequalities in a self-contained and rigorous manner using the methods of real and functional analysis, operator theory, the way of weight functions, and special functions. Throughout this book, special attention is given to proofs of the best constant factors of all inequalities.

The first chapter deals with recent developments of the Hilbert-type discrete and integral inequalities by introducing kernels, weight functions, and multi-parameters. Included are numerous examples and applications, many extensions, generalizations, refinements of Hilbert-type inequalities involving special functions such as beta, gamma, logarithm, trigonometric, hyperbolic, Bernoulli's numbers and functions, Euler's constant, zeta functions and hypergeometric functions. Special attention is given to many equivalent inequalities and to conditions under which the constant factors involved in inequalities are the best possible.

Chapter 2 contains some improvements of the celebrated Euler-Maclaurin summation formula, optimization of methods of estimating the series and the weight functions. Included are many useful theorems, corollaries, and inequalities with applications involving new inequalities on the Hurwitz zeta function, the Riemann zeta function and the extended Stirling formula.

Chapter 3 is devoted to the half-discrete Hilbert-type inequalities with general homogeneous kernels and their many extensions. Included are several equivalent inequalities, operator expressions, and the reverses of the Hilbert-type inequalities with many generalizations, applications and particular examples.

The main objective of Chapter 4 is to derive half-discrete Hilbert-type inequalities with a general non-homogeneous kernel, and their extensions. Many equivalent inequalities, and their operator expressions, two classes of reverse inequality, many extensions and particular examples are included in this chapter.

Chapter 5 contains two kinds of multi-dimensional half-discrete Hilbert-type inequalities with the best possible constant factors. These inequalities are extensions of the two-dimensional cases studied in Chapters 3 and 4. Included are equivalent forms, operator expressions, reverses, proofs of many important theorems with corollaries, and many particular examples.

The final Chapter 6 deals with two-kinds of multiple half-discrete Hilbert-type inequalities with the best possible constant factors. Included are equivalent inequalities, operator expressions, the reverses, proofs of major theorems and corollaries, and numerous examples with particular kernels. It is shown that theorems and corollaries of this chapter reduce to the corresponding results of Chapters 3 and 4 as special cases.

Basically, this monograph is designed as a modern source of many new half-discrete Hilbert-type inequalities with numerous examples and applications. It also provides new results and information that put the reader at the forefront of current research. A large number of research papers and books have been included in the bibliography to stimulate new interest in future advanced study and research. A short index of this book is added in the end to include a wide variety of terms and topics so that they are useful for the reader.

Our special thanks to Ms. Veronica Chavarria who cheerfully typed the manuscript with constant changes and revisions. In spite of the best efforts of everyone involved, some typographical errors will doubtlessly remain. We wish to express thanks to Ms. Lai Fun Kwong, Dr. S.C. Lim and the Production Department of the World Scientific Publishing company for their help and cooperation.

Bicheng Yang   Lokenath Debnath
Guang Zhou, Guangdong   Edinburg, Texas
P.R. China   U.S.A.

Acknowledgments

This work is supported by The National Natural Science Foundation of China (No. 61370186), and 2012 Knowledge Construction Special Foundation Item of Guangdong Institution of Higher Learning College and University (No. 2012KJCX0079, in China).

Contents

Chapter 1

Recent Developments of Hilbert-Type Inequalities with Applications

"As long as a branch of knowledge offers an abundance of problems, it is full of vitality."

"The organic unity of mathematics is inherent in the nature of this science, for mathematics is the foundation of all exact knowledge of natural phenomena."

David Hilbert

1.1 Introduction

This chapter deals with some recent developments of Hilbert-type discrete and integral inequalities by introducing kernels, weight functions, and multi-parameters. Included are numerous generalizations, extensions and refinements of Hilbert-type inequalities involving many special functions such as beta, gamma, logarithm, trigonometric, hyperbolic, Bernoulli's functions and Bernoulli's numbers, Euler's constant, zeta function and hypergeometric functions with many applications. Special attention is given to many equivalent inequalities and to conditions under which the constant factors involved in inequalities are the best possible. Many particular cases of Hilbert-type inequalities are presented with numerous applications. A large number of major books and research papers published in recent years are included to stimulate new interest in future study, especially in research on half-discrete Hilbert-type inequalities and their applications.

1

1.2   Hilbert's Inequality and Hilbert's Operator

1.2.1   *Hilbert's Discrete and Integral Inequalities*

Historically, mathematical analysis has been the major and significant branch of mathematics for the last three centuries. Indeed, inequalities became the heart of mathematical analysis. Many great mathematicians have made significant contributions to many new developments of the subject which led to the discovery of many new inequalities with proofs and useful applications in many fields of mathematical physics, pure and applied mathematics. Indeed, mathematical inequalities became an important branch of modern mathematics in the twentieth century through the pioneering work entitled *Inequalities* by G.H. Hardy, J.E. Littlewood and G. Pòlya [21] which was first published as a treatise in 1934. This unique publication represents a paradigm of precise logic, full of elegant inequalities with rigorous proofs and useful applications in mathematics.

It is appropriate to mention a delighted quotation of Anthony Zygmund (1900-1992) "Hardy, Littlewood and Pòlya's book has been one of the most important books in analysis in the last few decades. It had an impact on the trend of research and is still influencing it. In looking through the book now one realises how little one would like to change the existing text."

During the twentieth century, discrete and integral inequalities played a fundamental role in mathematics and have a wide variety of applications in many areas of pure and applied mathematics. In particular, David Hilbert (1862-1943) first proved Hilbert's double series inequality without exact determination of the constant in his lectures on integral equations. If $\{a_m\}$ and $\{b_n\}$ are two real sequences such that $0 < \sum_{m=1}^{\infty} a_m^2 < \infty$ and $0 < \sum_{n=1}^{\infty} b_n^2 < \infty$, then the *Hilbert's double series inequality* is given by

$$\sum_{n=1}^{\infty} \sum_{m=1}^{\infty} \frac{a_m b_n}{m+n} < \pi \left\{ \sum_{m=1}^{\infty} a_m^2 \sum_{n=1}^{\infty} b_n^2 \right\}^{\frac{1}{2}}. \qquad (1.1)$$

This famous inequality was proved by Hilbert in the early 1900s with the constant 2π in place of π. Several years after Hilbert's proof, Issai Schur (1875-1941) [66] gave a new proof in 1911 which revealed that (1.1) actually holds for constant π which is the best possible sharp constant.

In 1908, Weyl [77] published a proof of Hilbert's inequality (1.1), and Schur [66] also discovered the integral analogue of (1.1), which became known as the *Hilbert's integral inequality* in the form

$$\int_0^{\infty} \int_0^{\infty} \frac{f(x) g(y)}{x+y} dx dy < \pi \left\{ \int_0^{\infty} f^2(x) \, dx \int_0^{\infty} g^2(y) \, dy \right\}^{\frac{1}{2}}, \qquad (1.2)$$

where f and g are measurable functions such that $0 < \int_0^\infty f^2(x)\,dx < \infty$ and $0 < \int_0^\infty g^2(y)\,dy < \infty$, and π in (1.2) is still the best possible constant factor. A large number of generalizations, extensions and refinements of both (1.1) and (1.2) are available in literature in Hardy *et al.* [21], Mitrinović *et al.* [57], Kuang [47] and Hu [34].

Considerable attention has been given to the well-known classical Hardy-Littlewood-Sobolev (HLS) inequality (see Hardy *et al.* [21]) in the form

$$\int_{\mathbf{R}^n} \int_{\mathbf{R}^n} \frac{f(x)\,g(y)}{|x-y|^\lambda} dx\,dy \leq C_{n,\lambda,p} \|f\|_p \|g\|_q, \tag{1.3}$$

for every $f \in L^p(\mathbf{R}^n)$ and $g \in L^q(\mathbf{R}^n)$, where $0 < \lambda < n$, $1 < p, q < \infty$ such that $\frac{1}{p} + \frac{1}{q} + \frac{\lambda}{n} = 2$, and $\|f\|_p$ is the $L^p(\mathbf{R}^n)$ norm of the function f. For arbitrary p and q, an estimate of the upper bound of the constant $C_{n,\lambda,p}$ was given by Hardy, Littlewood and Sobolev, but no sharp value is known still now. However, for the special case, $p = q = 2n/(2n-\lambda)$, the sharp value of the constant was found as

$$C_{p,\lambda,n} = C_{n,\lambda} = \pi^{\frac{\lambda}{2}} \frac{\Gamma\left(\frac{n-\lambda}{2}\right)}{\Gamma\left(n-\frac{\lambda}{2}\right)} \left[\frac{\Gamma\left(\frac{n}{2}\right)}{\Gamma(n)}\right]^{\frac{\lambda}{n}-1}, \tag{1.4}$$

and the equality in (1.3) holds if and only if $g(x) = c_1 f(x)$ and $f(x) = c_2 h\left(\frac{x}{\mu^2} - a\right)$, where $h(x) = \left(1 + |x|^2\right)^{-d}$, $2d = (\lambda - 2n)$, $a \in \mathbf{R}^n$, c_1, c_2, $\mu \in \mathbf{R}/\{0\}$.

In 1958, Stein and Weiss [67] generalized the double weighted inequality of Hardy and Littlewood in the form with the same notation as in (1.3):

$$\left| \int_{\mathbf{R}^n} \int_{\mathbf{R}^n} \frac{f(x)\,g(x)}{|x|^\alpha |x-y|^\lambda |y|^\beta} dx\,dy \right| \leq C_{\alpha,\beta,n,\lambda,p} \|f\|_p \|g\|_q, \tag{1.5}$$

where $\alpha + \beta \geq 0$, and the powers of α and β of the weights satisfy the following conditions $1 - \frac{1}{p} - \frac{\lambda}{n} < \frac{\alpha}{n} < 1 - \frac{1}{p}$, and $\frac{1}{p} + \frac{1}{q} + \frac{1}{n}(\lambda + \alpha + \beta) = 2$. Inequality (1.5) and its proof given by Stein and Weiss [67] represent some major contribution to the subject.

On the other hand, Chen *et al.* [8] used weighted Hardy-Littlewood-Sobolev inequalities (1.3) and (1.5) to solve systems of integral equations. In 2011, Khotyakov [42] suggested two proofs of the sharp version of the HLS inequality (1.3). The first proof is based on the invariance property of the inequality (1.3), and the second proof uses some properties of the fast diffusion equation with the conditions $\lambda = n - 2$, $n \geq 3$ on the sharp HLS inequality (1.3).

1.2.2   *Operator Formulation of Hilbert's Inequality*

Suppose that $\mathbf{R}$ is the set of real numbers and $\mathbf{R}_+^m = \overbrace{(0,\infty) \times \cdots \times (0,\infty)}^{m}$, for $p > 1$, $l^p := \{a = \{a_n\}_{n=1}^{\infty} |||a||_p = \{\sum_{n=1}^{\infty} |a_n|^p\}^{\frac{1}{p}} < \infty\}$ and $L^p(\mathbf{R}_+) := \{f|||f||_p = \{\int_0^{\infty} |f(x)|^p dx\}^{1/p} < \infty\}$ are real normal spaces with the norms $||a||_p$ and $||f||_p$. We express inequality (1.1) using the form of operator as follows: $T : l^2 \to l^2$ is a linear operator, for any $a = \{a_m\}_{m=1}^{\infty} \in l^2$, there exists a sequence $c = \{c_n\}_{n=1}^{\infty} \in l^2$, satisfying

$$c_n = (Ta)(n) = \sum_{m=1}^{\infty} \frac{a_m}{m+n}, \qquad n \in \mathbf{N}, \tag{1.6}$$

where $\mathbf{N}$ is the set of positive integers. Hence, for any sequence $b = \{b_n\}_{n=1}^{\infty} \in l^2$, we define the inner product of Ta and b as follows

$$(Ta, b) = (c, b) = \sum_{n=1}^{\infty} \left(\sum_{m=1}^{\infty} \frac{a_m}{m+n} \right) b_n = \sum_{n=1}^{\infty} \sum_{m=1}^{\infty} \frac{a_m b_n}{m+n}. \tag{1.7}$$

Using (1.7), inequality (1.1) can be rewritten in the operator form

$$(Ta, b) < \pi \, ||a||_2 \, ||b||_2 \,, \tag{1.8}$$

where $||a||_2$, and $||b||_2 > 0$. It follows from Wilhelm [78] that T is a bounded operator and the norm $||T|| = \pi$ and T is called *Hilbert's operator* with the kernel $\frac{1}{m+n}$. For $||a||_2 > 0$, the equivalent form of (1.8) is given as $||Ta||_2 < \pi \, ||a||_2$, that is,

$$\sum_{n=1}^{\infty} \left(\sum_{m=1}^{\infty} \frac{a_m}{m+n} \right)^2 < \pi^2 \sum_{n=1}^{\infty} a_n^2, \tag{1.9}$$

where the constant factor π^2 is still the best possible. Obviously, inequality (1.9) and (1.1) are equivalent (see Hardy *et al.* [21]).

We may define Hilbert's integral operator as follows: $\widetilde{T} : L^2(\mathbf{R}_+) \to L^2(\mathbf{R}_+)$, for any $f \in L^2(\mathbf{R}_+)$, there exists a function, $h = \widetilde{T}f \in L^2(\mathbf{R}_+)$, satisfying

$$\left(\widetilde{T}f \right)(y) = h(y) = \int_0^{\infty} \frac{f(x)}{x+y} dx, \qquad y \in (0, \infty). \tag{1.10}$$

Hence, for any $g \in L^2(\mathbf{R}_+)$, we may still define the inner product of $\widetilde{T}f$ and g as follows:

$$\left(\widetilde{T}f, g \right) = \int_0^{\infty} \left(\int_0^{\infty} \frac{f(x)}{x+y} dx \right) g(y) \, dy = \int_0^{\infty} \int_0^{\infty} \frac{f(x) g(y)}{x+y} dx dy. \tag{1.11}$$

Setting the norm of f as $\|f\|_2 = \left\{\int_0^\infty f^2(x)\,dx\right\}^{\frac{1}{2}}$, if $\|f\|_2$, and $\|g\|_2 > 0$, then (1.2) my be rewritten in the operator form

$$\left(\tilde{T}f, g\right) < \pi \|f\|_2 \|g\|_2. \tag{1.12}$$

It follows that $\|\tilde{T}\| = \pi$ (see Carleman [5]), and we have the equivalent form of (1.2) as $\left\|\tilde{T}f\right\|_2 < \pi \|f\|_2$ (see Hardy et *al* [21]), that is,

$$\int_0^\infty \left(\int_0^\infty \frac{f(x)}{x+y}\,dx\right)^2 dy < \pi^2 \int_0^\infty f^2(x)\,dx, \tag{1.13}$$

where the constant factor π^2 is still the best possible. It is obvious that inequality (1.13) is the integral analogue of (1.9).

1.2.3 *A More Accurate Discrete Hilbert's Inequality*

If we set the subscripts m, n of the double series from zero to infinity, then, we may rewrite inequality (1.1) equivalently in the following form:

$$\sum_{n=0}^\infty \sum_{m=0}^\infty \frac{a_m b_n}{m+n+2} < \pi \left\{\sum_{n=0}^\infty a_n^2 \sum_{n=0}^\infty b_n^2\right\}^{\frac{1}{2}}, \tag{1.14}$$

where the constant factor π is still the best possible. Obviously, we may raise the following question: Is there a positive constant $\alpha\,(< 2)$, that makes inequality still valid as we replace 2 by α in the kernel $\frac{1}{m+n+2}$? The answer is positive. That is the following more accurate Hilbert's inequality (for short, Hilbert's inequality) (see Hardy *et al.* [21]):

$$\sum_{n=0}^\infty \sum_{m=0}^\infty \frac{a_m b_n}{m+n+1} < \pi \left\{\sum_{n=0}^\infty a_n^2 \sum_{n=0}^\infty b_n^2\right\}^{\frac{1}{2}}, \tag{1.15}$$

where the constant factor π is the best possible.

Since for a_m, $b_n \geq 0$, $\alpha \geq 1$,

$$\sum_{n=0}^\infty \sum_{m=0}^\infty \frac{a_m b_n}{m+n+\alpha} \leq \sum_{n=0}^\infty \sum_{m=0}^\infty \frac{a_m b_n}{m+n+1},$$

then, by (1.15) and for $\alpha \geq 1$, we obtain

$$\sum_{n=0}^\infty \sum_{m=0}^\infty \frac{a_m b_n}{m+n+\alpha} < \pi \left\{\sum_{n=0}^\infty a_n^2 \sum_{n=0}^\infty b_n^2\right\}^{\frac{1}{2}}. \tag{1.16}$$

For $1 \leq \alpha < 2$, inequality (1.16) is a refinement of (1.14). Obviously, we have a refinement of (1.9), which is equivalent to (1.16) as follows:

$$\sum_{n=0}^{\infty} \left(\sum_{m=0}^{\infty} \frac{a_m}{m+n+\alpha} \right)^2 < \pi^2 \sum_{n=0}^{\infty} a_n^2 \qquad (1 \leq \alpha < 2). \tag{1.17}$$

For $0 < \alpha < 1$, in 1936, Ingham [39] proved: If $\alpha \geq \frac{1}{2}$, then

$$\sum_{n=0}^{\infty} \sum_{m=0}^{\infty} \frac{a_m a_n}{m+n+\alpha} \leq \pi \sum_{n=0}^{\infty} a_n^2; \tag{1.18}$$

and if $0 < \alpha < \frac{1}{2}$, then

$$\sum_{n=0}^{\infty} \sum_{m=0}^{\infty} \frac{a_m a_n}{m+n+\alpha} \leq \frac{\pi}{\sin(\alpha\pi)} \sum_{n=0}^{\infty} a_n^2. \tag{1.19}$$

Note. If we put $x = X + \frac{\alpha}{2}$, $y = Y + \frac{\alpha}{2}$, $F(X) = f\left(X + \frac{\alpha}{2}\right)$ and $G(Y) = g\left(Y + \frac{\alpha}{2}\right)$ $(\alpha \in \mathbf{R})$ in (1.2), then we obtain

$$\int_{-\frac{\alpha}{2}}^{\infty} \int_{-\frac{\alpha}{2}}^{\infty} \frac{F(X)G(Y)}{X+Y+\alpha} dX dY < \pi \left\{ \int_{-\frac{\alpha}{2}}^{\infty} F^2(X) dX \int_{-\frac{\alpha}{2}}^{\infty} G^2(Y) dY \right\}^{\frac{1}{2}}. \tag{1.20}$$

For $\alpha \geq \frac{1}{2}$, inequality (1.20) is an integral analogue of (1.18) with $G = F$. However, if $0 < \alpha < \frac{1}{2}$, inequality (1.20) is not an integral analogue of (1.19), because two constant factors are different.

Using the improved version of the Euler-Maclaurin summation formula and introducing new parameters, several authors including Yang ([111], [174]) recently obtained several more accurate Hilbert-type inequalities and some new Hardy-Hilbert inequality with applications.

1.2.4 *Hilbert's Inequality with One Pair of Conjugate Exponents*

In 1925, by introducing one pair of conjugate exponents (p, q) with $\frac{1}{p} + \frac{1}{q} = 1$, Hardy [20] gave an extension of (1.1) as follows:

If $p > 1$, a_m, $b_n \geq 0$, such that $0 < \sum_{m=1}^{\infty} a_m^p < \infty$ and $0 < \sum_{n=1}^{\infty} b_n^q < \infty$, then

$$\sum_{n=1}^{\infty} \sum_{m=1}^{\infty} \frac{a_m b_n}{m+n} < \frac{\pi}{\sin\left(\frac{\pi}{p}\right)} \left\{ \sum_{m=1}^{\infty} a_m^p \right\}^{\frac{1}{p}} \left\{ \sum_{n=1}^{\infty} b_n^q \right\}^{\frac{1}{q}}, \tag{1.21}$$

where the constant factor $\frac{\pi}{\sin(\pi/p)}$ is the best possible. The equivalent discrete form of (1.21) is as follows:

$$\sum_{n=1}^{\infty} \left(\sum_{m=1}^{\infty} \frac{a_m}{m+n} \right)^p < \left[\frac{\pi}{\sin\left(\frac{\pi}{p}\right)} \right]^p \sum_{n=1}^{\infty} a_n^p, \qquad (1.22)$$

where the constant factor $\left[\frac{\pi}{\sin(\pi/p)} \right]^p$ is still the best possible. Similarly, inequalities (1.15) and (1.17) (for $\alpha = 1$) may be extended to the following equivalent forms (see Hardy *et al.* [21]):

$$\sum_{n=0}^{\infty} \sum_{m=0}^{\infty} \frac{a_m b_n}{m+n+1} < \frac{\pi}{\sin\left(\frac{\pi}{p}\right)} \left\{ \sum_{m=0}^{\infty} a_m^p \right\}^{\frac{1}{p}} \left\{ \sum_{n=0}^{\infty} b_n^q \right\}^{\frac{1}{q}}, \quad (1.23)$$

$$\sum_{n=0}^{\infty} \left(\sum_{m=0}^{\infty} \frac{a_m}{m+n+1} \right)^p < \left[\frac{\pi}{\sin\left(\frac{\pi}{p}\right)} \right]^p \sum_{n=0}^{\infty} a_n^p, \qquad (1.24)$$

where the constant factors $\frac{\pi}{\sin(\pi/p)}$ and $\left[\frac{\pi}{\sin(\pi/p)} \right]^p$ are the best possible. The equivalent integral analogues of (1.21) and (1.22) are given as follows:

$$\int_0^{\infty} \int_0^{\infty} \frac{f(x)\,g(y)}{x+y} dx dy < \frac{\pi}{\sin\left(\frac{\pi}{p}\right)} \left\{ \int_0^{\infty} f^p(x)\,dx \right\}^{\frac{1}{p}} \left\{ \int_0^{\infty} g^q(y)\,dy \right\}^{\frac{1}{q}},$$

$$(1.25)$$

$$\int_0^{\infty} \left(\int_0^{\infty} \frac{f(x)}{x+y} dx \right)^p dy < \left[\frac{\pi}{\sin\left(\frac{\pi}{p}\right)} \right]^p \int_0^{\infty} f^p(x)\,dx. \qquad (1.26)$$

We call (1.21) and (1.23) as Hardy-Hilbert's inequality and call (1.25) as *Hardy-Hilbert's integral inequality*.

Inequality (1.23) may be expressed in the form of operator as follows: $T_p : l^p \rightarrow l^p$ is a linear operator, such that for any non-negative sequence $a = \{a_m\}_{m=1}^{\infty} \in l^p$, there exists $T_p a = c = \{c_n\}_{n=1}^{\infty} \in l^p$, satisfying

$$c_n = (T_p a)(n) = \sum_{m=0}^{\infty} \frac{a_m}{m+n+1}, \qquad n \in \mathbf{N}_0 = \mathbf{N} \cup \{0\}. \qquad (1.27)$$

And for any non-negative sequence $b = \{b_n\}_{n=1}^{\infty} \in l^q$, we can define the formal inner product of $T_p a$ and b as follows:

$$(T_p a, b) = \sum_{n=0}^{\infty} \left(\sum_{m=0}^{\infty} \frac{a_m}{m+n+1} \right) b_n = \sum_{n=0}^{\infty} \sum_{m=0}^{\infty} \frac{a_m b_n}{m+n+1}. \qquad (1.28)$$

Then inequality (1.23) may be rewritten in the operator form

$$\left(T_p a, b\right) < \frac{\pi}{\sin\left(\frac{\pi}{p}\right)} \|a\|_p \|b\|_q, \tag{1.29}$$

where $\|a\|_p, \|b\|_q > 0$. The operator T_p is called *Hardy-Hilbert's operator.*

Similarly, we define the following *Hardy-Hilbert's integral operator* $\widetilde{T}_p$: $L^p\left(\mathbf{R}_+\right) \to L^p\left(\mathbf{R}_+\right)$ as follows: For any $f\left(\geq 0\right) \in L^p\left(\mathbf{R}_+\right)$, there exists a $h = \widetilde{T}_p f \in L^p\left(\mathbf{R}_+\right)$, defined by

$$\left(\widetilde{T}_p f\right)(y) = h\left(y\right) = \int_0^\infty \frac{f\left(x\right)}{x+y} dx, \qquad y \in \mathbf{R}_+. \tag{1.30}$$

And for any $g\left(\geq 0\right) \in L^q\left(\mathbf{R}_+\right)$, we can define the formal inner product of $\widetilde{T}_p f$ and g as follows:

$$\left(\widetilde{T}_p f, g\right) = \int_0^\infty \int_0^\infty \frac{f\left(x\right) g\left(y\right)}{x+y} dx dy. \tag{1.31}$$

Then inequality (1.25) may be rewritten in the operator form as follows:

$$\left(\widetilde{T}_p f, g\right) < \frac{\pi}{\sin\left(\frac{\pi}{p}\right)} \|f\|_p \|g\|_q. \tag{1.32}$$

On the other hand, if (p, q) is not a pair of conjugate exponents, then we have the following results (see Hardy *et al.* [21]):

If $p > 1$, $q > 1$, $\frac{1}{p} + \frac{1}{q} \geq 1$, $0 < \lambda = 2 - \left(\frac{1}{p} + \frac{1}{q}\right) \leq 1$, then

$$\sum_{n=1}^\infty \sum_{m=1}^\infty \frac{a_m b_n}{(m+n)^\lambda} \leq K \left\{\sum_{m=1}^\infty a_m^p\right\}^{\frac{1}{p}} \left\{\sum_{n=1}^\infty b_n^q\right\}^{\frac{1}{q}}, \tag{1.33}$$

where $K = K\left(p, q\right)$ relates to p, q, only for $\frac{1}{p} + \frac{1}{q} = 1$, $\lambda = 2 - \left(\frac{1}{p} + \frac{1}{q}\right) = 1$, the constant factor K is the best possible. The integral analogue of (1.33) is given by

$$\int_0^\infty \int_0^\infty \frac{f\left(x\right) g\left(y\right)}{\left(x+y\right)^\lambda} dx dy \leq K \left\{\int_0^\infty f^p\left(x\right) dx\right\}^{\frac{1}{p}} \left\{\int_0^\infty g^q\left(y\right) dy\right\}^{\frac{1}{q}}. \tag{1.34}$$

We also find an extension of (1.34) as follows (see Mitrinović *et al* [57]):

If $p > 1$, $q > 1$, $\frac{1}{p} + \frac{1}{q} > 1$, $0 < \lambda = 2 - \left(\frac{1}{p} + \frac{1}{q}\right) < 1$, then

$$\int_{-\infty}^\infty \int_{-\infty}^\infty \frac{f\left(x\right) g\left(y\right)}{|x+y|^\lambda} dx dy \leq k\left(p, q\right) \left\{\int_{-\infty}^\infty f^p(x) dx\right\}^{\frac{1}{p}} \left\{\int_{-\infty}^\infty g^q(x) dx\right\}^{\frac{1}{q}}. \tag{1.35}$$

For $f\left(x\right) = g\left(x\right) = 0$, $x \in \left(-\infty, 0\right]$, inequality (1.35) reduces to (1.34). Leven [52] also studied the expression forms of the constant factors in (1.33) and (1.34). But he did not prove their best possible property. In 1951, Bonsall [2] considered the case of (1.34) for the general kernel.

1.2.5 A Hilbert-type Inequality with the General Homogeneous Kernel of Degree -1

If $\alpha \in \mathbf{R}$, the function $k(x, y)$ is measurable in $\mathbf{R}_+^2$, satisfying for any x, y, $u > 0$, $k(ux, uy) = u^\alpha k(x, y)$, then $k(x, y)$ is called the *homogeneous function* of degree α. In 1934, Hardy et *al* [21] published the following theorem: Suppose that $p > 1$, $\frac{1}{p} + \frac{1}{q} = 1$, $k_1(x, y) (\geq 0)$ is a homogeneous function of degree -1 in $\mathbf{R}_+^2$. If $f(x), g(y) \geq 0, f \in L^p(\mathbf{R}_+), g \in L^q(\mathbf{R}_+)$, $k = \int_0^\infty k_1(u, 1) u^{-\frac{1}{p}} du$ is finite, then we have $k = \int_0^\infty k_1(1, u) u^{-\frac{1}{q}} du$ and the following equivalent integral inequalities:

$$\int_0^\infty \int_0^\infty k_1(x, y) f(x) g(y) \, dx dy \leq k \left\{ \int_0^\infty f^p(x) dx \right\}^{\frac{1}{p}} \left\{ \int_0^\infty g^q(y) dy \right\}^{\frac{1}{q}},$$
(1.36)

$$\int_0^\infty \left(\int_0^\infty k_1(x, y) f(x) \, dx \right)^p dy \leq k^p \int_0^\infty f^p(x) \, dx,$$
(1.37)

where the constant factor k is the best possible. Moreover, if $a_m, b_n \geq 0, a = \{a_m\}_{m=1}^\infty \in l^p, b = \{b_n\}_{n=1}^\infty \in l^q$, both $k_1(u, 1) u^{\frac{-1}{p}}$ and $k_1(1, u) u^{\frac{-1}{q}}$ are decreasing in $\mathbf{R}_+$, then we have the following equivalent discrete forms:

$$\sum_{n=1}^\infty \sum_{m=1}^\infty k_1(m, n) a_m b_n \leq k \left\{ \sum_{m=1}^\infty a_m^p \right\}^{\frac{1}{p}} \left\{ \sum_{n=1}^\infty b_n^q \right\}^{\frac{1}{q}},$$
(1.38)

$$\sum_{n=1}^\infty \left(\sum_{m=1}^\infty k_1(m, n) a_m \right)^p \leq k^p \sum_{n=1}^\infty a_n^p.$$
(1.39)

For $0 < p < 1$, if $k = \int_0^\infty k_1(u, 1) u^{-\frac{1}{p}} du$ is finite, then we have the reverses of (1.36) and (1.37).

Note. We have not seen any proof of (1.36)–(1.39) and the reverse examples in [21].

We call $k_1(x, y)$ the kernel of (1.36) and (1.37). If all the integrals and series in the right hand side of inequalities (1.36)–(1.39) are positive, then we can obtain the following particular examples (see Hardy *et al.* [21]):

(1). For $k_1(x, y) = \frac{1}{x+y}$ in (1.36)–(1.39), they reduce to (1.25), (1.26), (1.21) and (1.22);

(2). If $k_1(x, y) = \frac{1}{\max\{x, y\}}$ in (1.36)–(1.39), they reduce the following two pairs of equivalent forms:

$$\int_0^\infty \int_0^\infty \frac{f(x)\,g(y)}{\max\{x,y\}}dxdy < pq\left\{\int_0^\infty f^p(x)\,dx\right\}^{\frac{1}{p}}\left\{\int_0^\infty g^q(y)\,dy\right\}^{\frac{1}{q}},$$
$$(1.40)$$

$$\int_0^\infty\left(\int_0^\infty \frac{f(x)}{\max\{x,y\}}dx\right)^p dy < (pq)^p \int_0^\infty f^p(x)\,dx; \qquad (1.41)$$

$$\sum_{n=1}^\infty \sum_{m=1}^\infty \frac{a_m b_n}{\max\{m,n\}} < pq\left\{\sum_{m=1}^\infty a_m^p\right\}^{\frac{1}{p}}\left\{\sum_{n=1}^\infty b_n^q\right\}^{\frac{1}{q}}, \qquad (1.42)$$

$$\sum_{n=1}^\infty\left(\sum_{m=1}^\infty \frac{a_m}{\max\{m,n\}}\right)^p < (pq)^p \sum_{n=1}^\infty a_n^p; \qquad (1.43)$$

(3). If $k_1(x,y) = \frac{\ln(x/y)}{x-y}$ in (1.36)–(1.39), they reduce to the following two pairs of equivalent inequalities:

$$\int_0^\infty \int_0^\infty \frac{\ln\left(\frac{x}{y}\right) f(x)\,g(y)}{x-y}dxdy$$
$$< \left[\frac{\pi}{\sin\left(\frac{\pi}{p}\right)}\right]^2\left\{\int_0^\infty f^p(x)\,dx\right\}^{\frac{1}{p}}\left\{\int_0^\infty g^q(y)\,dy\right\}^{\frac{1}{q}}, \quad (1.44)$$

$$\int_0^\infty\left(\int_0^\infty \frac{\ln\left(\frac{x}{y}\right) f(x)}{x-y}dx\right)^p dy < \left[\frac{\pi}{\sin\left(\frac{\pi}{p}\right)}\right]^{2p}\int_0^\infty f^p(x)dx; \qquad (1.45)$$

$$\sum_{n=1}^\infty \sum_{m=1}^\infty \frac{\ln\left(\frac{m}{n}\right) a_m b_n}{m-n} < \left[\frac{\pi}{\sin\left(\frac{\pi}{p}\right)}\right]^2\left\{\sum_{m=1}^\infty a_m^p\right\}^{\frac{1}{p}}\left\{\sum_{n=1}^\infty b_n^q\right\}^{\frac{1}{q}},$$
$$(1.46)$$

$$\sum_{n=1}^\infty\left(\sum_{m=1}^\infty \frac{\ln\left(\frac{m}{n}\right) a_m}{m-n}\right)^p < \left[\frac{\pi}{\sin\left(\frac{\pi}{p}\right)}\right]^{2p}\sum_{n=1}^\infty a_n^p. \qquad (1.47)$$

Note. The constant factors in the above inequalities are all the best possible. We call (1.42) and (1.46) *Hardy-Littlewood-Pòlya's inequalities* (or *H-L-P inequalities*). We find that the kernels in the above inequalities are all decreasing functions. But this is not necessary. For example, we find the following two pairs of equivalent forms with the non-decreasing kernel (see Yang [121]):

$$\int_0^\infty \int_0^\infty \frac{\left|\ln\left(\frac{x}{y}\right)\right| f(x)\, g(y)}{\max\{x,y\}} dx dy$$

$$< (p^2 + q^2) \left\{\int_0^\infty f^p(x)\, dx\right\}^{\frac{1}{p}} \left\{\int_0^\infty g^q(y)\, dy\right\}^{\frac{1}{q}}, \quad (1.48)$$

$$\int_0^\infty \left(\int_0^\infty \frac{\left|\ln\left(\frac{x}{y}\right)\right| f(x)}{\max\{x,y\}} dx\right)^p dy < (p^2 + q^2)^p \int_0^\infty f^p(x)\, dx, \quad (1.49)$$

$$\sum_{n=1}^\infty \sum_{m=1}^\infty \frac{\left|\ln\left(\frac{m}{n}\right)\right| a_m b_n}{\max\{m,n\}} < (p^2 + q^2) \left\{\sum_{m=1}^\infty a_m^p\right\}^{\frac{1}{p}} \left\{\sum_{n=1}^\infty b_n^q\right\}^{\frac{1}{q}},$$

$$(1.50)$$

$$\sum_{n=1}^\infty \left(\sum_{m=1}^\infty \frac{\left|\ln\left(\frac{m}{n}\right)\right| a_m}{\max\{m,n\}}\right)^p < (p^2 + q^2)^p \sum_{n=1}^\infty a_n^p, \quad (1.51)$$

where the constant factors $(p^2 + q^2)$ and $(p^2 + q^2)^p$ are the best possible. Other types of inequalities with the best constant factors are as follows (see Xin and Yang [88]):

$$\int_0^\infty \int_0^\infty \frac{\left|\ln\left(\frac{x}{y}\right)\right| f(x)\, g(y)}{x + y} dx dy$$

$$< c_0(p) \left\{\int_0^\infty f^p(x)\, dx\right\}^{\frac{1}{p}} \left\{\int_0^\infty g^p(y)\, dy\right\}^{\frac{1}{q}}, \quad (1.52)$$

$$\int_0^\infty \left(\int_0^\infty \frac{\left|\ln\left(\frac{x}{y}\right)\right| f(x)}{x + y} dx\right)^p dy < c_0^p(p) \int_0^\infty f^p(x)\, dx; \quad (1.53)$$

$$\sum_{n=1}^\infty \sum_{m=1}^\infty \frac{\left|\ln\left(\frac{m}{n}\right)\right| a_m b_n}{m + n} < c_0(2) \left\{\sum_{m=1}^\infty a_m^2 \sum_{n=1}^\infty b_n^2\right\}^{\frac{1}{2}}, (1.54)$$

$$\sum_{n=1}^\infty \left(\sum_{m=1}^\infty \frac{\left|\ln\left(\frac{m}{n}\right)\right| a_m}{m + n}\right)^2 < c_0^2(2) \sum_{n=1}^\infty a_n^2, \quad (1.55)$$

where the constant factor $c_0(p) = 2 \sum_{n=1}^\infty (-1)^{n-1} \left[\frac{1}{\left(n - \frac{1}{p}\right)^2} - \frac{1}{\left(n - \frac{1}{q}\right)^2}\right]$.

1.2.6 *Two Multiple Hilbert-type Inequalities with the Homogeneous Kernels of Degree* $(-n+1)$

Suppose that $n \in \mathbf{N} \backslash \{1\}$, n numbers $p, q, \cdots, r$ satisfying $p, q, \cdots, r > 1$, $p^{-1} + q^{-1} + \cdots + r^{-1} = 1$, $k(x, y, \cdots, z) \geq 0$ is a homogeneous function of degree $(-n+1)$. If

$$k = \int_0^\infty \int_0^\infty \cdots \int_0^\infty k(1, y, \cdots, z) y^{-\frac{1}{q}} \cdots z^{-\frac{1}{r}} dy \cdots dz$$

is a finite number, $f, g, \cdots, h$ are non-negative measurable functions in $\mathbf{R}_+$, then, we have the following multiple Hilbert-type integral inequality (see Hardy *et al.* [21]):

$$\int_0^\infty \int_0^\infty \cdots \int_0^\infty k(x, y, \cdots, z) f(x) g(y) \cdots h(z) dx dy \cdots dz$$

$$\leq k \left(\int_0^\infty f^p(x) dx \right)^{\frac{1}{p}} \left(\int_0^\infty g^q(y) dy \right)^{\frac{1}{q}} \cdots \left(\int_0^\infty h^r(z) dz \right)^{\frac{1}{r}}. \quad (1.56)$$

Moreover, if $a_m, b_n, \cdots, c_s \geq 0$, $k(1, y, \cdots, z) x^0 y^{-\frac{1}{q}} \cdots z^{-\frac{1}{r}}$, $k(x, 1, \cdots, z) x^{-\frac{1}{p}} y^0 \cdots z^{-\frac{1}{r}}, \cdots, k(x, y, \cdots, 1) x^{-\frac{1}{p}} y^{-\frac{1}{q}} \cdots z^0$ are all decreasing functions with respect to any single variable in $\mathbf{R}_+$, then, we have

$$\sum_{s=1}^\infty \cdots \sum_{n=1}^\infty \sum_{m=1}^\infty k(m, n, \cdots, s) a_m b_n \cdots c_s$$

$$\leq k \left(\sum_{m=1}^\infty a_m^p \right)^{\frac{1}{p}} \left(\sum_{n=1}^\infty b_n^q \right)^{\frac{1}{q}} \cdots \left(\sum_{s=1}^\infty c_s^r \right)^{\frac{1}{r}}. \quad (1.57)$$

Note. The authors did not write and prove that the constant factor k in the above inequalities is the best possible. For two numbers p and q $(n = 2)$, inequalities (1.56) and (1.57) reduce respectively to (1.36) and (1.38).

1.3 Modern Research for Hilbert-type Inequalities

1.3.1 *Modern Research for Hilbert's Integral Inequality*

In 1979, based on an improvement of Hölder's inequality, Hu [33] proved a refinement of (1.2) (for $f = g$) as follows:

$$\int_0^\infty \int_0^\infty \frac{f(x) f(y)}{x + y} dx dy$$

$$< \pi \left\{ \left(\int_0^\infty f^2(x) dx \right)^2 - \frac{1}{4} \left(\int_0^\infty f^2(x) \cos \sqrt{x} dx \right)^2 \right\}^{\frac{1}{2}}. \quad (1.58)$$

Since then, Hu [34] published many interesting results similar to (1.58).

In 1998, Pachpatte [58] gave an inequality similar to (1.2) as follows: For $a,\ b > 0$,

$$\int_0^a \int_0^b \frac{f(x)\,g(y)}{x+y} dx dy$$
$$< \frac{\sqrt{ab}}{2} \left\{ \int_0^a (a-x)\, f'^2(x)\, dx \int_0^b (b-x)\, g'^2(y)\, dy \right\}^{\frac{1}{2}}. \quad (1.59)$$

Some improvements and extensions have been made by Zhao and Debnath [173], Lu [55] and He and Li [23]. We can also refer to other works of Pachpatte in [59].

In 1998, by introducing parameters $\lambda \in (0,1]$ and $a,\ b \in \mathbf{R}_+\ (a < b)$, Yang [96] gave an extension of (1.2) as follows:

$$\int_a^b \int_a^b \frac{f(x)\,g(y)}{(x+y)^\lambda} dx dy$$
$$< B\left(\frac{\lambda}{2}, \frac{\lambda}{2}\right) \left[1 - \left(\frac{a}{b}\right)^{\frac{\lambda}{4}}\right] \left\{ \int_a^b x^{1-\lambda} f^2(x)\, dx \int_a^b x^{1-\lambda} g^2(y)\, dy \right\}^{\frac{1}{2}},$$
$$\quad (1.60)$$

where $B(u,v)$ is the beta function.

In 1999, Kuang [46] gave another extension of (1.2) as follows: For $\lambda \in \left(\frac{1}{2}, 1\right]$,

$$\int_0^\infty \int_0^\infty \frac{f(x)\,f(y)}{x^\lambda + y^\lambda} dx dy < \frac{\pi}{\lambda \sin\left(\frac{\pi}{2\lambda}\right)} \left\{ \int_0^\infty f^2(x)\, dx \int_0^\infty g^2(y)\, dy \right\}^{\frac{1}{2}}. \quad (1.61)$$

We can refer to the other works of Kuang in [47] and [48].

In 1999, using the methods of algebra and analysis, Gao [15] proved an improvement of (1.2) as follows:

$$\int_0^\infty \int_0^\infty \frac{f(x)\,f(y)}{x+y} dx dy < \pi\sqrt{1-R} \left\{ \int_0^\infty f^2(x)\, dx \int_0^\infty g^2(y)\, dy \right\}^{\frac{1}{2}}, \quad (1.62)$$

where $\|f\| = \int_0^\infty f^2(x) dx^{1/2}$, $R = \frac{1}{\pi}\left(\frac{u}{\|g\|} - \frac{v}{\|f\|}\right)^2$, $u = \sqrt{\frac{2}{\pi}}\,(g,e)$, $v = \sqrt{2\pi}\,(f, e^{-x})$, $e(y) = \int_0^\infty \frac{e^x}{x+y} dx$. We also refer to works of Gao and Hsu in [17].

In 2002, using the operator theory, Zhang [172] gave an improvement of (1.2) as follows:

$$\int_0^\infty \int_0^\infty \frac{f(x)\,f(y)}{x+y}\,dxdy$$

$$\leq \frac{\pi}{\sqrt{2}}\left\{\int_0^\infty f^2(x)\,dx \int_0^\infty g^2(x)\,dx + \left(\int_0^\infty f(x)\,g(x)\,dx\right)^2\right\}^{\frac{1}{2}}.$$

$$(1.63)$$

1.3.2 On the Way of Weight Coefficient for Giving a Strengthened Version of Hilbert's Inequality

In 1991, for making an improvement of (1.1), Hsu and Wang [32] raised the way of weight coefficient as follows: At first, using Cauchy's inequality in the left hand side of (1.1), it follows that

$$I = \sum_{n=1}^{\infty}\sum_{m=1}^{\infty}\frac{a_m b_n}{m+n} = \sum_{n=1}^{\infty}\sum_{m=1}^{\infty}\frac{1}{m+n}\left[\left(\frac{m}{n}\right)^{\frac{1}{4}}a_m\right]\left[\left(\frac{n}{m}\right)^{\frac{1}{4}}b_n\right]$$

$$\leq \left\{\sum_{m=1}^{\infty}\left[\sum_{n=1}^{\infty}\frac{1}{m+n}\left(\frac{m}{n}\right)^{\frac{1}{2}}\right]a_m^2 \sum_{n=1}^{\infty}\left[\sum_{m=1}^{\infty}\frac{1}{m+n}\left(\frac{m}{n}\right)^{\frac{1}{2}}\right]b_n^2\right\}^{\frac{1}{2}}.$$

$$(1.64)$$

Then, we define the weight coefficient

$$\omega(n) = \sum_{m=1}^{\infty}\frac{1}{m+n}\left(\frac{m}{n}\right)^{\frac{1}{2}},\qquad n\in\mathbf{N},\qquad(1.65)$$

and rewrite (1.64) as follows:

$$I \leq \left\{\sum_{m=1}^{\infty}\omega(m)\,a_m^2 \sum_{n=1}^{\infty}\omega(n)\,b_n^2\right\}^{\frac{1}{2}}.\qquad(1.66)$$

Setting

$$\omega(n) = \pi - \frac{\theta(n)}{n^{1/2}},\qquad n\in\mathbf{N},\qquad(1.67)$$

where, $\theta(n) = (\pi - \omega(n))\,n^{1/2}$, and estimating the series of $\theta(n)$, it follows that

$$\theta(n) = \left[\pi - \sum_{m=1}^{\infty}\frac{1}{m+n}\left(\frac{m}{n}\right)^{\frac{1}{2}}\right]n^{1/2} > \theta = 1.1213^{+}.\qquad(1.68)$$

Thus, result (1.67) yields

$$\omega\left(n\right) < \pi - \frac{\theta}{n^{1/2}}, \qquad n \in \mathbf{N}, \quad \theta = 1.1213^{+}. \tag{1.69}$$

In view of (1.66), a strengthened version of (1.1) is given by

$$I < \left\{\sum_{n=1}^{\infty}\left(\pi - \frac{\theta}{n^{1/2}}\right)a_n^2 \sum_{n=1}^{\infty}\left(\pi - \frac{\theta}{n^{1/2}}\right)b_n^2\right\}^{\frac{1}{2}}. \tag{1.70}$$

Hsu and Wang [32] also raised an open question how to obtain the best value θ of (1.70). In 1992, Gao [14] gave the best value $\theta = \theta_0 = 1.281669^{+}$.

Xu and Gao [91] proved a strengthened version of (1.11) given by

$$I < \left\{\sum_{n=1}^{\infty}\left[\frac{\pi}{\sin\left(\frac{\pi}{p}\right)} - \frac{p-1}{n^{1/p} + n^{-1/q}}\right]a_n^p\right\}^{\frac{1}{p}}$$

$$\times \left\{\sum_{n=1}^{\infty}\left[\frac{\pi}{\sin\left(\frac{\pi}{p}\right)} - \frac{q-1}{n^{1/q} + n^{-1/p}}\right]b_n^q\right\}^{\frac{1}{q}}. \tag{1.71}$$

In 1997, using the way of weight coefficient and the improved Euler-Maclaurin's summation formula, Yang and Gao ([144], [18]) showed that

$$I < \left\{\sum_{n=1}^{\infty}\left[\frac{\pi}{\sin\left(\frac{\pi}{p}\right)} - \frac{1-\gamma}{n^{1/p}}\right]a_n^p\right\}^{\frac{1}{p}}$$

$$\times \left\{\sum_{n=1}^{\infty}\left[\frac{\pi}{\sin\left(\frac{\pi}{p}\right)} - \frac{1-\gamma}{n^{1/q}}\right]b_n^q\right\}^{\frac{1}{q}}, \tag{1.72}$$

where $1 - \gamma = 0.42278433^{+}$ (γ is the Euler constant).

In 1998, Yang and Debnath [140] gave another strengthened version of (1.11), which is an improvement of (1.71). We can also refer to some strengthened versions of (1.15) and (1.23) in papers of Yang [97], Yang and Debnath [143].

1.3.3 *Hilbert's Inequality with Independent Parameters*

In 1998, using the optimized weight coefficients and introducing an independent parameter $\lambda \in (0, 1]$, Yang [96] provided an extension of (1.2) as follows:

If $0 < \int_0^\infty x^{1-\lambda} f^2(x)\,dx < \infty$ and $0 < \int_0^\infty x^{1-\lambda} g^2(x)\,dx < \infty$, then

$$\int_0^\infty \int_0^\infty \frac{f(x)g(y)}{(x+y)^\lambda}\,dxdy$$

$$< B\left(\frac{\lambda}{2}, \frac{\lambda}{2}\right) \left\{ \int_0^\infty x^{1-\lambda} f^2(x)\,dx \int_0^\infty x^{1-\lambda} g^2(x)\,dx \right\}^{\frac{1}{2}}, \quad (1.73)$$

where the constant factor $B\left(\frac{\lambda}{2}, \frac{\lambda}{2}\right)$ is the best possible. The proof of the best possible property of the constant factor was given by Yang [95], and the expressions of the beta function $B(u,v)$ are given in Wang and Guo [73]:

$$B(u,v) = \int_0^\infty \frac{t^{u-1}dt}{(1+t)^{u+v}} = \int_0^1 (1-t)^{u-1} t^{v-1} dt$$

$$= \int_1^\infty \frac{(t-1)^{u-1}\,dt}{t^{u+v}} \qquad (u, v > 0). \quad (1.74)$$

Some extensions of (1.21), (1.23) and (1.25) were given by Yang and Debnath ([98], [141], [142]) as follows: If $\lambda > 2 - \min\{p, q\}$, then

$$\int_0^\infty \int_0^\infty \frac{f(x)g(y)}{(x+y)^\lambda}\,dxdy < B\left(\frac{p+\lambda-2}{p}, \frac{q+\lambda-2}{q}\right)$$

$$\times \left\{ \int_0^\infty x^{1-\lambda} f^p(x)\,dx \right\}^{\frac{1}{p}} \left\{ \int_0^\infty x^{1-\lambda} g^q(x)\,dx \right\}^{\frac{1}{q}}.$$

$$(1.75)$$

If $2 - \min\{p, q\} < \lambda \leq 2$, then

$$\sum_{n=1}^\infty \sum_{m=1}^\infty \frac{a_m b_n}{(m+n)^\lambda} < B\left(\frac{p+\lambda-2}{p}, \frac{q+\lambda-2}{q}\right)$$

$$\times \left\{ \sum_{n=1}^\infty n^{1-\lambda} a_n^p \right\}^{\frac{1}{p}} \left\{ \sum_{n=1}^\infty n^{1-\lambda} b_n^q \right\}^{\frac{1}{q}}, \quad (1.76)$$

$$\sum_{n=0}^\infty \sum_{m=0}^\infty \frac{a_m b_n}{(m+n+1)^\lambda} < B\left(\frac{p+\lambda-2}{p}, \frac{q+\lambda-2}{q}\right)$$

$$\times \left\{ \sum_{n=0}^\infty \left(n+\frac{1}{2}\right)^{1-\lambda} a_n^p \right\}^{\frac{1}{p}} \left\{ \sum_{n=0}^\infty \left(n+\frac{1}{2}\right)^{1-\lambda} b_n^q \right\}^{\frac{1}{q}}, \quad (1.77)$$

where the constant factor $B\left(\frac{p+\lambda-2}{p}, \frac{q+\lambda-2}{q}\right)$ is the best possible. Yang [99] also proved that (1.76) is valid for $p = 2$ and $\lambda \in (0, 4]$. Yang ([100],

[115]) gave another extensions of (1.21) and (1.23) as follows: If $0 < \lambda \leq \min\{p, q\}$, then

$$\sum_{n=1}^{\infty} \sum_{m=1}^{\infty} \frac{a_m b_n}{m^\lambda + n^\lambda}$$

$$< \frac{\pi}{\lambda \sin\left(\frac{\pi}{p}\right)} \left\{ \sum_{n=1}^{\infty} n^{(p-1)(1-\lambda)} a_n^p \right\}^{\frac{1}{p}} \left\{ \sum_{n=1}^{\infty} n^{(q-1)(1-\lambda)} b_n^q \right\}^{\frac{1}{q}}, \quad (1.78)$$

and if $0 < \lambda \leq 1$, then

$$\sum_{n=0}^{\infty} \sum_{m=0}^{\infty} \frac{a_m b_n}{\left(m + \frac{1}{2}\right)^\lambda + (n + \frac{1}{2})^\lambda}$$

$$< \frac{\pi}{\lambda \sin\left(\frac{\pi}{p}\right)} \left\{ \sum_{n=0}^{\infty} \left(n + \frac{1}{2}\right)^{p-1-\lambda} a_n^p \right\}^{\frac{1}{p}} \left\{ \sum_{n=0}^{\infty} \left(n + \frac{1}{2}\right)^{q-1-\lambda} b_n^q \right\}^{\frac{1}{q}}.$$

$$(1.79)$$

In 2004, Yang [104] proved the following dual form of (1.21):

$$\sum_{n=1}^{\infty} \sum_{m=1}^{\infty} \frac{a_m b_n}{m + n} < \frac{\pi}{\sin\left(\frac{\pi}{p}\right)} \left\{ \sum_{n=1}^{\infty} n^{p-2} a_n^p \right\}^{\frac{1}{p}} \left\{ \sum_{n=1}^{\infty} n^{q-2} b_n^q \right\}^{\frac{1}{q}}. \quad (1.80)$$

Inequality (1.80) reduces to (1.21) when $p = q = 2$. For $\lambda = 1$, (1.79) reduces to the dual form of (1.23) as follows:

$$\sum_{n=0}^{\infty} \sum_{m=0}^{\infty} \frac{a_m b_n}{m + n + 1}$$

$$< \frac{\pi}{\sin\left(\frac{\pi}{p}\right)} \left\{ \sum_{n=0}^{\infty} \left(n + \frac{1}{2}\right)^{p-2} a_n^p \right\}^{\frac{1}{p}} \left\{ \sum_{n=0}^{\infty} \left(n + \frac{1}{2}\right)^{q-2} b_n^q \right\}^{\frac{1}{q}}. \quad (1.81)$$

We can find some extensions of the H-L-P inequalities with the best constant factors such as (1.40)-(1.51) (see Wang and Yang [75], Yang [102]–[103]) by introducing some independent parameters.

In 2001, by introducing some parameters, Hong [29] gave a multiple integral inequality, which is an extension of (1.21). He et al [28] gave a similar result for particular conjugate exponents. For making an improvement of their works, Yang [101] gave the following inequality, which is a best extension of (1.21): If $n \in \mathbf{N} \setminus \{1\}$, $p_i > 1$, $\sum_{i=1}^{n} \frac{1}{p_i} = 1$, $\lambda > n - \min_{1 \leq i \leq n} \{p_i\}$,

$f_i(t) \geq 0$ and $0 < \int_0^\infty t^{n-1-\lambda} f_i^{p_i}(t)\, dt < \infty$: $(i = 1, 2, \cdots, n)$, then, we have

$$\int_0^\infty \cdots \int_0^\infty \frac{\prod_{i=1}^n f_i(x_i)}{\left(\sum_{i=1}^n x_i\right)^\lambda} dx_1 \cdots dx_n$$

$$< \frac{1}{\Gamma(\lambda)} \prod_{i=1}^n \Gamma\left(\frac{p_i + \lambda - n}{p_i}\right) \left\{ \int_0^\infty t^{n-1-\lambda} f_i^{p_i}(t)\, dt \right\}^{\frac{1}{p_i}}, \qquad (1.82)$$

where the constant factor $\frac{1}{\Gamma(\lambda)} \prod_{i=1}^n \Gamma\left(\frac{p_i + \lambda - n}{p_i}\right)$ is the best possible. In particular, for $\lambda = n - 1$, it follows that

$$\int_0^\infty \cdots \int_0^\infty \frac{\prod_{i=1}^n f_i(x_i)}{\left(\sum_{i=1}^n x_i\right)^{n-1}} dx_1 \cdots dx_n$$

$$< \frac{1}{\Gamma(\lambda)} \prod_{i=1}^n \Gamma\left(1 - \frac{1}{p_i}\right) \left\{ \int_0^\infty f_i^{p_i}(t) dt \right\}^{\frac{1}{p_i}}. \qquad (1.83)$$

In 2003, Yang and Rassias [147] introduced the way of weight coefficient and considered its applications to Hilbert-type inequalities. They summarized how to use the way of weight coefficients to obtain some new improvements and generalizations of the Hilbert-type inequalities. Since then, a number of authors discussed this problem (see Brnetić *et al.* [3]– [4], Chen and Xu [6], Gao [16], Gao *et al.* [19], He *et al.* [26]–[27], Jia *et al.* [40]–[41], Krnić *et al.* [43]–[45], Laith [50], Lu [56], Salem [65], Sulaiman [68]–[69], Wang and Xin [74], Xie and Lu [80]–[81], Xu [91], and Yang [110]). But how to give an uniform extension of inequalities (1.80) and (1.21) with a best possible constant factor, this was solved in 2004 by introducing two pairs of conjugate exponents (see Yang [105]).

1.3.4   *Hilbert-type Inequalities with Multi-parameters*

In 2004, by introducing an independent parameter $\lambda > 0$ and two pairs of conjugate exponents (p, q) and (r, s) with $\frac{1}{p} + \frac{1}{q} = \frac{1}{r} + \frac{1}{s} = 1$, Yang [105] gave an extension of (1.2) as follows: If $p, r > 1$, and the integrals of the right hand side are positive, then

$$\int_0^\infty \int_0^\infty \frac{f(x)\, g(y)}{x^\lambda + y^\lambda} dx dy$$

$$< \frac{\pi}{\lambda \sin\left(\frac{\pi}{r}\right)} \left\{ \int_0^\infty x^{p\left(1 - \frac{\lambda}{r}\right) - 1} f^p(x)\, dx \right\}^{\frac{1}{p}} \left\{ \int_0^\infty x^{q\left(1 - \frac{\lambda}{s}\right) - 1} g^q(x)\, dx \right\}^{\frac{1}{q}},$$

$$(1.84)$$

where the constant factor $\frac{\pi}{\lambda \sin(\frac{\pi}{r})}$ is the best possible.

For $\lambda = 1$, $r = q$, $s = p$, inequality (1.84) reduces to (1.25); for $\lambda = 1$, $r = p$, $s = q$, inequality (1.84) reduces to the dual form of (1.25) as follows:

$$\int_0^\infty \int_0^\infty \frac{f(x)g(y)}{x+y}dxdy$$
$$< \frac{\pi}{\sin\left(\frac{\pi}{p}\right)} \left\{\int_0^\infty x^{p-2}f^p(x)\,dx\right\}^{\frac{1}{p}} \left\{\int_0^\infty x^{q-2}g^q(x)\,dx\right\}^{\frac{1}{q}}. \quad (1.85)$$

In 2005, by introducing an independent parameter $\lambda > 0$, and two pairs of generalized conjugate exponents $(p_1, p_2, \cdots, p_n)$ and $(r_1, r_2, \cdots, r_n)$ with $\sum_{i=1}^n \frac{1}{p_i} = \sum_{i=1}^n \frac{1}{r_i} = 1$, Yang *et al.* [149] gave a multiple integral inequality as follows:

For $p_i, r_i > 1$ $(i = 1, 2, \cdots, n)$,

$$\int_0^\infty \cdots \int_0^\infty \frac{\prod_{i=1}^n f_i(x_i)}{\left(\sum_{i=1}^n x_i\right)^\lambda}dx_1 \cdots dx_n$$
$$< \frac{1}{\Gamma(\lambda)} \prod_{i=1}^n \Gamma\left(\frac{\lambda}{r_i}\right) \left\{\int_0^\infty t^{p_i\left(1-\frac{\lambda}{r_i}\right)-1}f_i^{p_i}(t)\,dt\right\}^{\frac{1}{p_i}}, \quad (1.86)$$

where the constant factor $\frac{1}{\Gamma(\lambda)} \prod_{i=1}^n \Gamma\left(\frac{\lambda}{r_i}\right)$ is the best possible. For $n = 2$, $p_1 = p$, $p_2 = q$, $r_1 = r$ and $r_2 = s$, inequality (1.86) reduces to the following:

$$\int_0^\infty \int_0^\infty \frac{f(x)g(y)}{(x+y)^\lambda}dxdy$$
$$< B\left(\frac{\lambda}{r}, \frac{\lambda}{s}\right) \left\{\int_0^\infty x^{p\left(1-\frac{\lambda}{r}\right)-1}f^p(x)\,dx\right\}^{\frac{1}{p}} \left\{\int_0^\infty x^{q\left(1-\frac{\lambda}{s}\right)-1}g^q(x)\,dx\right\}^{\frac{1}{q}}.$$
$$(1.87)$$

It is obvious that inequality (1.87) is another best extension of (1.25).

In 2006, using two pairs of conjugate exponents (p, q) and (r, s) with p, $r > 1$, Hong [31] gave a multi-variable integral inequality as follows:

If $\mathbf{R}_+^n = \{x = (x_1, x_2, \cdots, x_n); x_i > 0, i = 1, 2, \cdots, n\}$, α, β, $\lambda > 0$,

$\|x\|_\alpha = \left(\sum_{i=1}^n x_i^\alpha\right)^{\frac{1}{\alpha}}$, $f, g \geq 0$, $0 < \int_{\mathbf{R}_+^n} \|x\|_\alpha^{p\left(n-\frac{\beta\lambda}{r}\right)-n} f^p(x)\,dx < \infty$ and

$0 < \int_{\mathbf{R}_+^n} \|x\|_\alpha^{q\left(n - \frac{\beta\lambda}{s}\right) - n} g^q(x)\, dx < \infty$, then

$$\int_{\mathbf{R}_+^n} \int_{\mathbf{R}_+^n} \frac{f(x)\, g(y)\, dx dy}{\left(\|x\|_\alpha^\beta + \|y\|_\alpha^\beta\right)^\lambda} < \frac{\Gamma^n\left(\frac{1}{\alpha}\right)}{\beta \alpha^{n-1} \Gamma\left(\frac{n}{\alpha}\right)} B\left(\frac{\lambda}{r}, \frac{\lambda}{s}\right)$$

$$\times \left\{\int_{\mathbf{R}_+^n} \|x\|_\alpha^{p\left(n - \frac{\beta\lambda}{r}\right) - n} f^p(x)\, dx\right\}^{\frac{1}{p}} \left\{\int_{\mathbf{R}_+^n} \|x\|_\alpha^{q\left(n - \frac{\beta\lambda}{s}\right) - n} g^q(x)\, dx\right\}^{\frac{1}{q}},$$

(1.88)

where the constant factor $\frac{\Gamma^n\left(\frac{1}{\alpha}\right)}{\beta \alpha^{n-1} \Gamma\left(\frac{n}{\alpha}\right)} B\left(\frac{\lambda}{r}, \frac{\lambda}{s}\right)$ is the best possible. In particular, for $n = 1$, (1.88) reduces to Hong's work in [30]; for $n = \beta = 1$, (1.88) reduces to (1.87). In 2007, Zhong and Yang [181] generalized (1.88) to a general homogeneous kernel and proposed the reversion.

We can find another inequality with two parameters as follows (see Yang [108]):

$$\sum_{n=1}^\infty \sum_{m=1}^\infty \frac{a_m b_n}{(m^\alpha + n^\alpha)^\lambda}$$

$$< \frac{1}{\alpha} B\left(\frac{\lambda}{r}, \frac{\lambda}{s}\right) \left\{\sum_{n=1}^\infty n^{p\left(1 - \frac{\alpha\lambda}{r}\right) - 1} a_n^p\right\}^{\frac{1}{p}} \left\{\sum_{n=1}^\infty n^{q\left(1 - \frac{\alpha\lambda}{s}\right) - 1} b_n^q\right\}^{\frac{1}{q}}, \quad (1.89)$$

where $\alpha, \lambda > 0$, $\alpha\lambda \le \min\{r, s\}$. In particular, for $\alpha = 1$, we have

$$\sum_{n=1}^\infty \sum_{m=1}^\infty \frac{a_m b_n}{(m + n)^\lambda}$$

$$< B\left(\frac{\lambda}{r}, \frac{\lambda}{s}\right) \left\{\sum_{n=1}^\infty n^{p\left(1 - \frac{\lambda}{r}\right) - 1} a_n^p\right\}^{\frac{1}{p}} \left\{\sum_{n=1}^\infty n^{q\left(1 - \frac{\lambda}{s}\right) - 1} b_n^q\right\}^{\frac{1}{q}}. \quad (1.90)$$

For $\lambda = 1$, $r = q$, inequality (1.90) reduces to (1.21), and for $\lambda = 1$, $r = p$, (1.90) reduces to (1.80). Also we can obtain the reverse form as follows (see Yang [109]):

$$\sum_{n=0}^\infty \sum_{m=0}^\infty \frac{a_m b_n}{(m + n + 1)^2}$$

$$> 2 \left\{\sum_{n=0}^\infty \left[1 - \frac{1}{4(n + 1)^2}\right] \frac{a_n^p}{2n + 1}\right\}^{\frac{1}{p}} \left\{\sum_{n=0}^\infty \frac{b_n^q}{2n + 1}\right\}^{\frac{1}{q}}, \quad (1.91)$$

where $0 < p < 1$, $\frac{1}{p} + \frac{1}{q} = 1$. The other results on the reverse of the Hilbert-type inequalities are found in Xi [79] and Yang [112].

In 2006, Xin [87] gave a best extension of H-L-P integral inequality (1.45) as follows:

$$\int_0^\infty \int_0^\infty \frac{\ln\left(\frac{x}{y}\right) f(x) g(y)}{x^\lambda - y^\lambda} dx dy < \left[\frac{\pi}{\sin\left(\frac{\pi}{r}\right)}\right]^2$$

$$\times \left\{\int_0^\infty x^{p\left(1-\frac{\lambda}{r}\right)-1} f^p(x)\, dx\right\}^{\frac{1}{p}} \left\{\int_0^\infty x^{q\left(1-\frac{\lambda}{s}\right)-1} g^q(x)\, dx\right\}^{\frac{1}{q}}. \quad (1.92)$$

In 2007, Zhong and Yang [176] gave an extension of another H-L-P integral inequality (1.40) as follows:

$$\int_0^\infty \int_0^\infty \frac{f(x) g(y)}{\max\{x^\lambda, y^\lambda\}} dx dy$$

$$< \frac{rs}{\lambda} \left\{\int_0^\infty x^{p\left(1-\frac{\lambda}{r}\right)-1} f^p(x)\, dx\right\}^{\frac{1}{p}} \left\{\int_0^\infty x^{q\left(1-\frac{\lambda}{s}\right)-1} g^q(x)\, dx\right\}^{\frac{1}{q}}. \quad (1.93)$$

Zhong and Yang [177] also gave the reverse form of (1.93).

Considering a particular kernel, Yang [118] proved

$$\sum_{n=1}^\infty \sum_{m=1}^\infty \frac{a_m b_n}{(\sqrt{m} + \sqrt{n}) \sqrt{\max\{m, n\}}}$$

$$< 4\ln 2 \left\{\sum_{n=1}^\infty n^{\frac{p}{2}-1} a_n^p\right\}^{\frac{1}{p}} \left\{\sum_{n=1}^\infty n^{\frac{q}{2}-1} b_n^q\right\}^{\frac{1}{q}}. \quad (1.94)$$

Yang [114] also proved that

$$\sum_{n=1}^\infty \sum_{m=1}^\infty \frac{a_m b_n}{(m + an)^2 + n^2}$$

$$< \left(\frac{\pi}{2} - \arctan a\right) \left\{\sum_{n=1}^\infty \frac{a_n^p}{n}\right\}^{\frac{1}{p}} \left\{\sum_{n=1}^\infty \frac{b_n^q}{n}\right\}^{\frac{1}{q}} \quad (a \geq 0). \quad (1.95)$$

Using the residue theory, Yang [123] obtained the following inequality

$$\int_0^\infty \int_0^\infty \frac{f(x) g(y)}{(x + ay)(x + by)(x + cy)} dx dy$$

$$< k \left\{\int_0^\infty x^{-\frac{p}{2}-1} f^p(x)\, dx\right\}^{\frac{1}{p}} \left\{\int_0^\infty x^{-\frac{q}{2}-1} g^q(x)\, dx\right\}^{\frac{1}{q}}, \quad (1.96)$$

where $k = \frac{1}{(\sqrt{a}+\sqrt{b})(\sqrt{b}+\sqrt{c})(\sqrt{a}+\sqrt{c})}$ $(a, b, c > 0)$. The constant factors in the above new inequalities are all the best possible. Some other new results are proved by several authors including He *et al.* [27], Li and He [53], Xie [81]–[84].

1.4   Some New Applications for Hilbert-type Inequalities

1.4.1   *Operator Expressions of Hilbert-type Inequalities*

Suppose that H is a separable Hilbert space and $T : H \to H$ is a bounded self-adjoint semi-positive definite operator. In 2002, Zhang [172] proved the following inequality:

$$(a, Tb)^2 \leq \frac{\|T\|^2}{2} \left(\|a\|^2 \|b\|^2 + (a, b)^2 \right) \qquad (a, b \in H), \qquad (1.97)$$

where (a, b) is the inner product of a and b, and $\|a\| = \sqrt{(a, a)}$ is the norm of a. Since the Hilbert integral operator $\widetilde{T}$ defined by (1.10) satisfies the condition of (1.97) with $\left\|\widetilde{T}\right\| = \pi$, then inequality (1.2) may be improved as (1.63). Since the operator T_p defined by (1.27) (for $p = q = 2$) satisfies the condition of (1.97) (see Wilhelm [78]), we may improve (1.15) to the following form:

$$\sum_{n=0}^{\infty} \sum_{m=0}^{\infty} \frac{a_m b_n}{m + n + 1} < \frac{\pi}{\sqrt{2}} \left\{ \sum_{n=0}^{\infty} a_n^2 \sum_{n=0}^{\infty} b_n^2 + \left(\sum_{n=0}^{\infty} a_n b_n \right)^2 \right\}^{\frac{1}{2}}. \qquad (1.98)$$

The key of applying (1.97) is to obtain the norm of the operator and to show the semi-definite property. Now, we consider the concept and the properties of Hilbert-type integral operator as follows:

Suppose that $p > 1$, $\frac{1}{p} + \frac{1}{q} = 1$, $L^r(\mathbf{R}_+)$ $(r = p, q)$ are real normal linear spaces and $k(x, y)$ is a non-negative symmetric measurable function in $\mathbf{R}_+^2$ satisfying

$$\int_0^{\infty} k(x, t) \left(\frac{x}{t} \right)^{\frac{1}{r}} dt = k_0(p) \in \mathbf{R} \qquad (x > 0). \qquad (1.99)$$

We define an integral operator as

$$T : L^r(\mathbf{R}_+) \to L^r(\mathbf{R}_+) \ (r = p, q),$$

for any $f (\geq 0) \in L^p(\mathbf{R}_+)$, there exists $h = Tf \in L^p(\mathbf{R}_+)$, such that

$$(Tf)(y) = h(y) := \int_0^{\infty} k(x, y) f(x) \, dx \qquad (y > 0). \qquad (1.100)$$

Or, for any $g (\geq 0) \in L^q(\mathbf{R}_+)$, there exists $\widetilde{h} = Tg \in L^q(\mathbf{R}_+)$, such that

$$(Tg)(x) = \widetilde{h}(x) := \int_0^{\infty} k(x, y) g(y) \, dy \qquad (x > 0). \qquad (1.101)$$

In 2006, Yang [113] proved that the operator T defined by (1.100) or (1.101) are bounded with $\|T\| \leq k_0(p)$. The following are some results of [113]: If $\varepsilon > 0$, is small enough and the integral $\int_0^\infty k(x,t)\left(\frac{x}{t}\right)^{\frac{1+\varepsilon}{r}} dt\,(r = p, q; x > 0)$ is convergent to a constant $k_\varepsilon(p)$ independent of x satisfying $k_\varepsilon(p) = k_0(p) + o(1)\,(\varepsilon \to 0^+)$, then $\|T\| = k_0(p)$. If $\|T\| > 0$, $f \in L^p(\mathbf{R}_+)$, $g \in L^q(\mathbf{R}_+)$, $\|f\|_p, \|g\|_q > 0$, then we have the following equivalent inequalities:

$$(Tf, g) < \|T\| \cdot \|f\|_p \|g\|_q, \tag{1.102}$$

$$\|Tf\|_p < \|T\| \cdot \|f\|_p. \tag{1.103}$$

Some particular cases are considered in this paper [113].

Yang [119] also considered some properties of Hilbert-type integral operator (for $p = q = 2$). For the homogeneous kernel of degree -1, Yang [125] found some sufficient conditions to obtain $\|T\| = k_0(p)$. We can see some properties of the discrete Hilbert-type operator in the discrete space in Yang ([116], [120], [122], [124]). Recently, Bényi and Choonghong [1] proved some new results concerning best constants for certain multi-linear integral operators. In 2009, Yang [129] summarized the above part results. Some other works about Hilbert-type operators and inequalities with the general homogeneous kernel and multi-parameters were provided by several other authors (see Huang [35], Liu and Yang [54], Wang and Yang [76], Xin and Yang [89], Yang [114] and [133], Yang and Krnic [145] and Yang and Rassias [148]).

1.4.2 *Some Basic Hilbert-type Inequalities*

If the Hilbert-type inequality relates to a single symmetric homogeneous kernel of degree -1 (such as $\frac{1}{x+y}$, or $\frac{|ln(x/y)|}{x+y}$) and the best constant factor is a more brief form, which does not relate to any conjugate exponents (such as (1.2)), then we call it basic Hilbert-type integral inequality. Its series analogue (if exists) is also called basic Hilbert-type inequality. If the simple homogeneous kernel is of degree $-\lambda\,(\lambda > 0)$ with a parameter λ and the inequality cannot be obtained by a simple transform to a basic Hilbert-type integral inequality, then we call it a basic Hilbert-type integral inequality with a parameter.

For examples, we call the following integral inequality that is, (1.2) as

$$\int_0^\infty \int_0^\infty \frac{f(x)g(y)}{x+y}dxdy < \pi \left\{ \int_0^\infty f^2(x)\,dx \int_0^\infty g^2(x)\,dx \right\}^{\frac{1}{2}}, \tag{1.104}$$

and the following H-L-P inequalities (for $p = 2$ in (1.40) and (1.45)):

$$\int_0^\infty \int_0^\infty \frac{f(x)g(y)}{\max\{x,y\}}dxdy < 4 \left\{ \int_0^\infty f^2(x)\,dx \int_0^\infty g^2(x)\,dx \right\}^{\frac{1}{2}}, \tag{1.105}$$

$$\int_0^\infty \int_0^\infty \frac{\ln\left(\frac{x}{y}\right) f(x) g(y)}{x - y} dx dy < \pi^2 \left\{\int_0^\infty f^2(x) dx \int_0^\infty g^2(x) dx\right\}^{\frac{1}{2}}$$

(1.106)

basic Hilbert-type integral inequalities. In 2005, Yang [106] gave the following basic Hilbert-type integral inequality:

$$\int_0^\infty \int_0^\infty \frac{\left|\ln\left(\frac{x}{y}\right)\right| f(x) g(y)}{\max\{x, y\}} dx dy < 8 \left\{\int_0^\infty f^2(x) dx \int_0^\infty g^2(x) dx\right\}^{\frac{1}{2}};$$

(1.107)

In 2011, Yang [134] gave the following basic Hilbert-type integral inequalities:

$$\int_0^\infty \int_0^\infty \frac{\left|\ln\left(\frac{x}{y}\right)\right| f(x) g(y)}{x + y} dx dy < c_0 \left\{\int_0^\infty f^2(x) dx \int_0^\infty g^2(x) dx\right\}^{\frac{1}{2}},$$

(1.108)

$$\int_0^\infty \int_0^\infty \frac{\arctan\sqrt{\frac{x}{y}}}{x + y} f(x) g(y) dx dy$$

$$< \frac{\pi^2}{4} \left\{\int_0^\infty f^2(x) dx \int_0^\infty g^2(x) dx\right\}^{\frac{1}{2}},$$

(1.109)

where, $c_0 = 8 \sum_{n=1}^\infty \frac{(-1)^n}{(2n-1)^2} = 7.3277^+$.

Yang ([106], [126], [128], [130], [146]) also gave a basic Hilbert-type integral inequality with a parameter $\lambda \in (0, 1)$:

$$\int_0^\infty \int_0^\infty \frac{f(x) g(y)}{|x - y|^\lambda} dx dy$$

$$< 2B\left(1 - \lambda, \frac{\lambda}{2}\right) \left\{\int_0^\infty x^{1-\lambda} f^2(x) dx \int_0^\infty x^{1-\lambda} g^2(x) dx\right\}^{\frac{1}{2}}.$$

(1.110)

Similar to discrete inequality (1.19), the following integral inequality

$$\int_0^\infty \int_0^\infty \frac{f(x) g(y)}{(x + y)^\lambda} dx dy$$

$$< B\left(\frac{\lambda}{2}, \frac{\lambda}{2}\right) \left\{\int_0^\infty x^{1-\lambda} f^2(x) dx \int_0^\infty x^{1-\lambda} g^2(x) dx\right\}^{\frac{1}{2}},$$

(1.111)

is called *basic Hilbert-type integral inequality with a parameter* $\lambda \in (0, \infty)$.

Also we find the following basic Hilbert-type inequalities:

$$\sum_{n=1}^{\infty}\sum_{m=1}^{\infty} \frac{a_m b_n}{m+n} < \pi \left\{ \sum_{n=1}^{\infty} a_n^2 \sum_{n=1}^{\infty} b_n^2 \right\}^{\frac{1}{2}}, \tag{1.112}$$

$$\sum_{n=1}^{\infty}\sum_{m=1}^{\infty} \frac{a_m b_n}{\max\{m,n\}} < 4 \left\{ \sum_{n=1}^{\infty} a_n^2 \sum_{n=1}^{\infty} b_n^2 \right\}^{\frac{1}{2}}, \tag{1.113}$$

$$\sum_{n=1}^{\infty}\sum_{m=1}^{\infty} \frac{\ln\left(\frac{m}{n}\right) a_m b_n}{m-n} < \pi^2 \left\{ \sum_{n=1}^{\infty} a_n^2 \sum_{n=1}^{\infty} b_n^2 \right\}^{\frac{1}{2}}. \tag{1.114}$$

It follows from (1.50) with $p = q = 2$ that

$$\sum_{n=1}^{\infty}\sum_{m=1}^{\infty} \frac{\left|\ln\left(\frac{m}{n}\right)\right| a_m b_n}{\max\{m,n\}} < 8 \left\{ \sum_{n=1}^{\infty} a_n^2 \sum_{n=1}^{\infty} b_n^2 \right\}^{\frac{1}{2}}. \tag{1.115}$$

In 2010, Xin and Yang [88] proved the following inequality

$$\sum_{n=1}^{\infty}\sum_{m=1}^{\infty} \frac{\left|\ln\left(\frac{m}{n}\right)\right| a_m b_n}{m+n} < c_0 \left\{ \sum_{n=1}^{\infty} a_n^2 \sum_{n=1}^{\infty} b_n^2 \right\}^{\frac{1}{2}}, \tag{1.116}$$

where, $c_0 = 8 \sum_{n=1}^{\infty} \frac{(-1)^n}{(2n-1)^2} = 7.3277^+$. Inequalities (1.115) and (1.116) are new basic Hilbert-type inequalities. We still have a basic Hilbert-type inequality with a parameter $\lambda \in (0, 4]$ as follows (see Yang [100]):

$$\sum_{n=1}^{\infty}\sum_{m=1}^{\infty} \frac{a_m b_n}{(m+n)^\lambda} < B\left(\frac{\lambda}{2}, \frac{\lambda}{2}\right) \left\{ \sum_{n=1}^{\infty} n^{1-\lambda} a_n^2 \sum_{n=1}^{\infty} n^{1-\lambda} b_n^2 \right\}^{\frac{1}{2}}. \tag{1.117}$$

1.4.3 *Some Applications to Half-discrete Hilbert-type Inequalities*

In recent years, Dračic *et al.* [11]–[13], Pogany [60]–[61] made some new important contributions to discrete Hilbert-type inequalities with nonhomogeneous kernels using special functions. On the other hand, in 2006-2011, Xie *et al.* ([81]–[86]) have investigated many Hilbert-type integral inequalities with the particular kernels such as $|x + y|^{-\lambda}$ similar to inequality (1.35). In 2010-2012, Yang ([131]–[132]) considered the compositions of two discrete Hilbert-type operators with two conjugate exponents and kernels $(m + n)^{-\lambda}$ and in 2010, Liu and Yang [54] also studied the compositions of two Hilbert-type integral operators with the general homogeneous kernel of negative degree and obtained some new results with applications.

Hardy *et al.* [21] proved some results of half-discrete inequalities with the general non-homogeneous kernel in Theorem 351 without any proof of the constant factors as best possible. Yang [107] introduced an interval variable to give a half-discrete inequality with the particular non-homogeneous kernel as $\frac{1}{(1+u(x)u(n))^\lambda}$, and proved that the constant factor is the best possible. In 2011, Yang [135] gave the following half-discrete Hilbert-type inequalities with the best possible constant factor $B(\lambda_1, \lambda_2)$:

$$\int_0^\infty f(x) \left(\sum_{n=1}^\infty \frac{a_n}{(x+n)^\lambda} \right) dx < B(\lambda_1, \lambda_2) ||f||_{p,\phi}||a||_{q,\psi}, \qquad (1.118)$$

where $\lambda_1 \lambda_2 > 0$, $0 \le \lambda_2 \le 1$, $\lambda_1 + \lambda_2 = \lambda$,

$$||f||_{p,\phi} = \left\{ \int_0^\infty \varphi(x) f^p(x) dx \right\}^{1/p} > 0,$$

$$||a||_{q,\psi} = \left\{ \sum_{n=1}^\infty a_n^q \right\}^{1/q} > 0, \qquad \varphi(x) = x^{p(1-\lambda_1)-1},$$

$\psi(n) = n^{q(1-\lambda_2)-1}$. Zhong *et al.* ([174]–[177]) has investigated several half-discrete Hilbert-type inequalities with particular kernels.

Using the way of weight functions and the techniques of discrete and integral Hilbert-type inequalities with some additional conditions on the kernel, a half-discrete Hilbert-type inequality with a general homogeneous kernel of degree $-\lambda \in \mathbf{R}$ is obtained as follows:

$$\int_0^\infty f(x) \sum_{n=1}^\infty k_\lambda(x,n) a_n dx < k(\lambda_1)||f||_{p,\phi}||a||_{q,\psi}, \qquad (1.119)$$

where $k(\lambda_1) = \int_0^\infty k_\lambda(t,1) t^{\lambda_1-1} dt \in \mathbf{R}_+$. This is an extension of the above particular result with the best constant factor $k(\lambda_1)$ (see Yang and Chen [138]).

If the corresponding integral inequality of a half-discrete inequality is a basic Hilbert-type integral inequality, then we call it the basic half-discrete Hilbert-type inequality. substituting some particular kernels in the main result found in [138] leads to some basic half-discrete Hilbert-type inequalities as follows:

$$\int_0^\infty f(x) \sum_{n=1}^\infty \frac{a_n}{x+n} dx < \frac{\pi}{2} ||f||_2 ||a||_2, \tag{1.120}$$

$$\int_0^\infty f(x) \sum_{n=1}^\infty \frac{\ln(x/n)a_n}{x-n} dx < \left(\frac{\pi}{2}\right)^2 ||f||_2 ||a||_2, \tag{1.121}$$

$$\int_0^\infty f(x) \sum_{n=1}^\infty \frac{a_n}{\max\{x,n\}} dx < 4||f||_2 ||a||_2, \tag{1.122}$$

$$\int_1^\infty f(x) \sum_{n=1}^\infty \frac{|\ln(x/n)|a_n}{\max\{x,n\}} dx < 16||f||_2 ||a||_2, \tag{1.123}$$

$$\int_1^\infty f(x) \sum_{n=1}^\infty \frac{|\ln(x/n)|a_n}{x+n} dx < c_0 ||f||_2 ||a||_2, \tag{1.124}$$

where $c_0 = 8 \sum_{n=1}^\infty \frac{(-1)^n}{(2n-1)^2} = 7.3277^+$.

1.5 Concluding Remarks

(1) Many different kinds of Hilbert-type discrete and integral inequalities with applications are presented in this chapter. Special attention is given to new results proved in recent years. Included are many generalizations, extensions and refinements of Hilbert-type discrete and integral inequalities involving many special functions such as beta, gamma, hypergeometric, trigonometric, hyperbolic, zeta, Bernoulli's functions and Bernoulli's numbers and Euler's constant.

(2) For more information about Hilbert-type discrete and integral inequalities, the reader is referred to three recent research monograph by Yang ([128], [130], and [134]) who presented many new results on integral and discrete-type operators with general homogeneous kernels of degree of real numbers and two pairs of conjugate exponents as well as the related inequalities. These research monographs and a recent paper by Debnath and Yang [10] also contained recent developments of both discrete and integral types of operators and inequalities with proofs, examples and applications.

Chapter 2

Improvements of the Euler-Maclaurin Summation Formula and Applications

"Since a general solution must be judged impossible from want of analysis, we must be content with the knowledge of some special cases, and that all the more, since the development of various cases seems to be the only way to bringing us at least to a more perfect knowledge."

Leonhard Euler

2.1 Introduction

In 1732, Leonhard Euler (1707-1783) stated *the Euler-Maclaurin summation formula* which was independently discovered by Euler and Colin Maclaurin (1698-1746) in the period of 1732-1742. This chapter deals with some preliminary results to improve the Euler-Maclaurin summation formula for optimizing the methods of estimation of series and the weight coefficients. Many useful theorems, corollaries and inequalities are discussed. Included are some applications involving new inequalities on Hurwitz's zeta function restricted to the real axis, the Riemann zeta function and the extended Stirling formula.

2.2 Some Special Functions Relating Euler-Maclaurin's Summation Formula

2.2.1 *Bernoulli's Numbers*

We define a function $G(x)$ as follows:

$$G(x) = \frac{x}{e^x - 1}, \quad x \in \mathbf{R} \quad \left(G(0) = \lim_{x \to 0} G(x) = 1 \right).$$

It is obvious that the power series of $G(x)$ at $x = 0$ possesses a positive convergence radius. Assuming that the sequence $\{B_n\}_{n=0}^{\infty}$ is defined by the exponent creation function of $G(x)$, and

$$G(x) = \sum_{n=0}^{\infty} B_n \frac{x^n}{n!},$$

we obtain

$$1 = \frac{e^x - 1}{x} \sum_{n=0}^{\infty} B_n \frac{x^n}{n!}$$

$$= \sum_{n=0}^{\infty} \frac{1}{(n+1)!} x^n \sum_{n=0}^{\infty} \frac{B_n}{n!} x^n = \sum_{n=0}^{\infty} \left[\sum_{k=0}^{n} \frac{B_k}{k!(n-k+1)!} \right] x^n.$$

Comparing to the coefficients of x^n on two sides of the above equality, it follows that

$$B_0 = 1, \quad \text{and} \quad \sum_{k=0}^{n} \frac{B_k}{k!(n-k+1)!} = 0 \quad (n \in \mathbf{N}).$$

We can obtain the recursion formula of $\{B_n\}_{n=0}^{\infty}$ as follows:

$$B_0 = 1, B_n = -n! \sum_{k=0}^{n-1} \frac{B_k}{k!(n-k+1)!}, \quad n \in \mathbf{N}. \tag{2.1}$$

Since we have

$$G(-x) := \frac{-x}{e^{-x} - 1} = \frac{x(e^x - 1 + 1)}{e^x - 1} = x + G(x),$$

then, in view of

$$G(x) = \sum_{n=0}^{\infty} B_n \frac{x^n}{n!},$$

it follows that

$$\sum_{n=0}^{\infty} B_n \left[(-1)^n - 1 \right] \frac{x^n}{n!} = x.$$

Comparing the coefficients of $\frac{x^n}{n!}$ on two sides of the above equality, we obtain

$$B_1 = -\frac{1}{2}, \quad B_{2k+1} = 0, \quad k \in \mathbf{N}. \tag{2.2}$$

By (2.1) and (2.2), we can find the constants of $\{B_{2n}\}_{n=1}^{\infty}$ step by step as follows:

$$B_2 = \frac{1}{6}, \quad B_4 = -\frac{1}{30}, \quad B_6 = \frac{1}{42}, \quad B_8 = -\frac{1}{30}, \quad \cdots .$$

We call B_n ($n \in \mathbf{N}_0 = \mathbf{N} \cup \{0\}$) the Bernoulli numbers. In a general way, we also obtain the following formula for the Bernoulli numbers (see Xu and Wang [92]):

$$B_{2k} = (-1)^{k+1} \frac{(2k)!}{2^{2k-1}\pi^{2k}} \sum_{n=1}^{\infty} \frac{1}{n^{2k}}, \quad k \in \mathbf{N}. \tag{2.3}$$

2.2.2 **Bernoulli's Polynomials**

Suppose that the function $B_n(t)$ is defined by the following exponent creation function:

$$e^{tx}\, G(x) = \sum_{n=0}^{\infty} B_n(t)\, \frac{x^n}{n!}. \tag{2.4}$$

In the convergence interval of (2.4), we have

$$\sum_{n=0}^{\infty} B_n(t)\frac{x^n}{n!} = e^{tx}G(x) = \sum_{n=0}^{\infty} \frac{t^n}{n!}\, x^n \sum_{n=0}^{\infty} \frac{B_n}{n!}\, x^n$$

$$= \sum_{n=0}^{\infty} \left[\sum_{k=0}^{n} \binom{n}{k} B_k\, t^{n-k}\right] \frac{x^n}{n!}. \tag{2.5}$$

Comparing to the coefficients of the term $\frac{x^n}{n!}$ on two sides of (2.5), we get the following formula:

$$B_n(t) = \sum_{k=0}^{n} \binom{n}{k} B_k\, t^{n-k}, \quad n \in \mathbf{N}_0. \tag{2.6}$$

We call $B_n(t)$ $(n \in \mathbf{N}_0)$ Bernoulli polynomials. In particular, we obtain

$$B_0(t) = 1, \quad B_1(t) = t - \frac{1}{2},$$

$$B_2(t) = t^2 - t + \frac{1}{6},$$

$$B_3(t) = t^3 - \frac{3}{2}t^2 + \frac{1}{2}t,$$

$$B_4(t) = t^4 - 2t^3 + t^2 - \frac{1}{30}, \cdots,$$

and prove the equations that (see Xu and Wang [92])

$$B_n'(t) = nB_{n-1}(t), \quad \int_0^1 B_n(t)dt = 0, \quad B_{n-1}(0) = B_{n-1}, \quad n \in \mathbf{N}.$$

On the other hand, in view of $B_0(t) = 1, B_n'(t) = nB_{n-1}(t)$ and $B_{n-1}(0) = B_{n-1}$, $n \in \mathbf{N}$, by integration, we find

$$B_n(t) = B_n + n \int_0^1 B_{n-1}(t)dt, \quad n \in \mathbf{N}. \tag{2.7}$$

We may use (2.7) to define the Bernoulli polynomials in (2.6). In fact, by mathematical induction, for $n = 0$, we have $B_0(t) = 1$. Assuming that (2.6) is valid for n, then, for $n + 1$, it follows that

$$B_{n+1}(t) = B_{n+1} + (n+1) \int_0^1 B_n(t)dt$$

$$= B_{n+1} + (n+1) \sum_{k=0}^{n} \binom{n}{k} B_k \int_0^1 t^{n-k} dt$$

$$= \sum_{k=0}^{n+1} \binom{n+1}{k} B_k \, t^{n+1-k}.$$

Thus, the function (2.6) is the root of functional equation (2.7).

2.2.3 *Bernoulli's Functions*

For any $t \in \mathbf{R}$, we denote $[t]$ the maximal integer that does not exceed t and $\{t\} = t - [t]$. Then it follows that Bernoulli functions $P_k(t) = B_k(\{t\})(k \in \mathbf{N})$ are periodic functions with the least positive periodic 1. It is obvious that $P_k(t)$ are of bounded variation in any finite interval. $P_1(t) = \{t\} - \frac{1}{2}$ is not continuous in the integers, but it is differentiable in the other points. $P_2(t)$ is continuous in $\mathbf{R}$ and differentiable in the set of all the non-integers. For $k \geq 3$, $P_k(t)$ are continuous and differentiable in $\mathbf{R}$.

We can prove that $P_{2k}(t)$ are even functions and $P_{2k-1}(t)$ are odd functions. In fact, setting $H(t, x) = e^{tx} G(x)$, since

$$e^{(1-t)x} \frac{x}{e^x - 1} = e^{-tx} \frac{xe^x}{e^x - 1} = e^{t(-x)} \frac{-x}{e^{-x} - 1},$$

we obtain $H(1 - t, x) = H(t, -x)$, and by (2.4), it follows that

$$\sum_{n=0}^{\infty} B_n(1-t) \frac{x^n}{n!} = \sum_{n=0}^{\infty} B_n(t) \frac{(-x)^n}{n!} = \sum_{n=0}^{\infty} (-1)^n B_n(t) \frac{x^n}{n!}.$$

Hence, comparing the coefficients of $\frac{x^n}{n!}$ on two sides of the above equality, we obtain

$$B_n(1 - t) = (-1)^n B_n(t) \quad (n \in \mathbf{N}_0),$$

and then, it follows that

$$P_n(-t) = P_n(1 - t) = (-1)^n P_n(t), \quad n \in \mathbf{N}.$$

Equivalently, we may define the Bernoulli functions $P_n(t)$ $(n \in \mathbf{N})$ by the following recursion functional equations:

$$P_1(t) = \{t\} - \frac{1}{2},$$

$$P_{n+1}(t) = B_{n+1} + (n+1) \int_0^t P_n(t)dt, \quad n \in \mathbf{N}. \tag{2.8}$$

Since $P_1(t)$ is not continuous at any integer t, then, in view of the integral properties, $P_2(t)$ is continuous at any t, but it is not differentiable at integer t. And then $P_n(t)$ $(n \geq 3)$ are differentiable on the real axis.

Considering the value of $P_2(k)$ at any integer k, it is obvious that $P_2(k) \left(= B_2 = \frac{1}{6}\right)$ is the maximum value, and at $t = k + \frac{1}{2}$, since $P_2' \left(k + \frac{1}{2}\right) = 0$, then the minimum value of $P_2(t)$ is $P_2 \left(k + \frac{1}{2}\right)$ expressed by

$$P_2 \left(k + \frac{1}{2}\right) = P_2 \left(\frac{1}{2}\right) = B_2 + 2 \int_0^{\frac{1}{2}} \left(t - \frac{1}{2}\right) dt = -\frac{1}{12}.$$

The roots of $P_3(t)$ are divided into two classes. One is the set of integers k, that make $P_3(k) = B_3 = 0$. The other class is the set of all $k + \frac{1}{2}$. In fact, by the definition of Bernoulli polynomial, it follows that

$$P_3 \left(k + \frac{1}{2}\right) = P_3 \left(\frac{1}{2}\right) = 3 \int_0^{\frac{1}{2}} P_2(t) dt$$

$$= 3 \int_0^{\frac{1}{2}} \left(t^2 - t + \frac{1}{6}\right) dt = 0.$$

Since the points that make the value of $P_4(t)$ are maximum or minimum at the roots of $P_3(t)$, these points are divided into two classes. In a general way, by the Rabbe formula (see Xu and Wang [92], Proposition 98), we have

$$B_n(kt) = k^{n-1} \sum_{i=0}^{k-1} B_n \left(t + \frac{i}{k}\right), k, \quad n \in \mathbf{N}. \tag{2.9}$$

Setting $k = 2, t = 0$, we find

$$B_n \left(\frac{1}{2}\right) = (2^{1-n} - 1) B_n(0) = (2^{1-n} - 1) B_n. \tag{2.10}$$

Hence, both $P_{2n+1}(t)$ and $P_3(t)$ possess the same two classes roots (see Xu and Wang [92], Proposition 93). Since the positions of the points that make the value of $P_{2n+2}(t)$ maximum or minimum are the positions of the points that make $P_{2n+2}'(t) = 0$, then both $P_{2n+2}(t)$ and $P_2(t)$ possess the same points that make the value maximum or minimum. Since $P_{2n+2}(k) = P_{2n+2}(0) = B_{2n+2}$, then, by (2.10), we obtain

$$P_{2n+2} \left(k + \frac{1}{2}\right) = P_{2n+2} \left(\frac{1}{2}\right) = - \left(1 - \frac{1}{2^{2n+1}}\right) B_{2n+2}, \tag{2.11}$$

$$\left| P_{2n+2} \left(k + \frac{1}{2}\right) \right| < |B_{2n+2}|, \quad n \in \mathbf{N}_0. \tag{2.12}$$

In view of the above formulas, we discover that

$$(\max P_{2n+2}(t)) (\min P_{2n+2}(t)) < 0,$$

$$\max P_{2n+2}(t) \neq - \min P_{2n+2}(t), \quad n \in \mathbf{N}_0..$$

2.2.4 *The Euler-Maclaurin Summation Formula*

Assuming that $m,\, n \in \mathbf{N}_0$, $m < n$, $f(t)$ is a piecewise smooth continuous function in $[m, n]$, then we have the following formula:

$$\sum_{k=m+1}^{n} f(k) = \int_{m}^{n} f(t)dt + \frac{1}{2}f(t)|_{m}^{n} + \int_{m}^{n} P_1(t)f'(t)dt. \qquad (2.13)$$

In fact, since $P_1(t) = t - k - \frac{1}{2}$, $t \in [k, k+1)$, integration by parts gives

$$\int_{m}^{n} P_1(t)f'(t)dt = \sum_{k=m}^{n-1} \int_{k}^{k+1} \left(t - k - \frac{1}{2}\right) df(t)$$

$$= \sum_{k=m}^{n-1} \left[\left(t - k - \frac{1}{2}\right) f(t)|_{k}^{k+1} - \int_{k}^{k+1} f(t)dt \right]$$

$$= \frac{1}{2} \sum_{k=m}^{n-1} (f(k+1) + f(k)) - \int_{m}^{n} f(t)dt$$

$$= \frac{1}{2} (f(m) - f(n)) + \sum_{k=m+1}^{n} f(k) - \int_{m}^{n} f(t)dt.$$

Hence, we obtain (2.13).

In a general way, we have the following Euler-Maclaurin summation formula (see Knopp [51]):

Theorem 2.1. *Assuming that $m,\, n \in \mathbf{N}_0$, $m < n$, $f(t) \in C^q[m, n](q \in \mathbf{N})$, then, we have*

$$\sum_{k=m+1}^{n} f(k) = \int_{m}^{n} f(t)dt + \sum_{k=1}^{q} \frac{(-1)^k B_k}{k!} f^{(k-1)}(t)|_{m}^{n}$$

$$+ \frac{(-1)^{q+1}}{q!} \int_{m}^{n} P_q(t)\, f^{(q)}(t)dt. \qquad (2.14)$$

Proof. We prove (2.14) by mathematical induction. For $q = 1$ (indicating $f^{(0)}(t) = f(t)$), since $B_1 = -\frac{1}{2}$, by (2.13), we have (2.14). Suppose that (2.14) is valid for $q(\in \mathbf{N})$. Then, for $q + 1$, since $P'_{q+1}(t) = (q + 1)P_q(t)$, integration by parts, we find

$$\int_{m}^{n} P_q(t)f^{(q)}(t)\, dt = \frac{1}{q+1} \int_{m}^{n} f^{(q)}(t)\, dP_{q+1}(t)$$

$$= \frac{B_{q+1}}{q+1} f^{(q)}(t)|_{m}^{n} - \frac{1}{q+1} \int_{m}^{n} P_{q+1}(t)f^{(q+1)}(t)dt. \qquad (2.15)$$

Then substitution of (2.15) in (2.14), it follows that (2.14) is valid for $q + 1$. $\square$

Note. If $f^q(t)$ is a constant function, then it follows

$$\int_m^n P_q(t)f^{(q)}(t)dt = 0.$$

Since $B_{2q+1} = 0$ $(q \in \mathbf{N})$, we may reduce (2.14) in the following form:

Corollary 2.1. *Assuming that* m, n, $q \in \mathbf{N}_0, m < n, f(t) \in C^{2q+1}[m,n]$, *we have*

$$\sum_{k=m}^n f(k) = \int_m^n f(t)dt + \frac{1}{2}(f(n) + f(m))$$

$$+ \sum_{k=1}^q \frac{B_{2k}}{(2k)!} f^{(2k-1)}(t)|_m^n + \delta_q(m,n), \qquad (2.16)$$

where

$$\delta_q(m,n) = \frac{1}{(2q+1)!} \int_m^n P_{2q+1}(t)f^{(2q+1)}(t)\, dt, \qquad (2.17)$$

and $\delta_q(m,n)$ *is called the residue term of* $(2q+1)th$ *order in (2.16).*

Note. *For* $q = 0$, *we define that the series on the right hand side of (2.16) is 0. In particular, for* $q = 0$, *we have*

$$\sum_{k=m}^n f(k) = \int_m^n f(t)dt + \frac{1}{2}(f(n) + f(m)) + \int_m^n P_1(t)f'(t)dt.$$

If $f(t) \in C^3[m,n]$, *then, we still have*

$$\int_m^n P_1(t)f'(t)dt = \frac{1}{12}f'(t)|_m^n + \frac{1}{6}\int_m^n P_3(t)f'''(t)dt.$$

For any $r \in \mathbf{R}$ and $k \in \mathbf{N}_0$, we define the combination number $\binom{r}{k}$ as follows (see Qu [64]):

$$\binom{r}{k} = \begin{cases} 1, & k = 0, \\ \frac{r(r-1)\cdots(r-k+1)}{k!}, & k \in \mathbf{N}. \end{cases}$$

Example 2.1. Suppose that $f(t) = (t+a)^l$ $(l \in \mathbf{N}, 0 < a \leq 1; t \in [0,\infty))$. Then $f^{(i)}(t) = i! \binom{l}{i} (t+a)^{l-i}$, for $i < l$; $f^{(i)}(t) = $ constant, for $i \geq l$. In (2.14), for $2q = l$ or $2q + 1 = l$, we have $q = \left[\frac{l}{2}\right]$ and $\delta_q(m,n) = 0$. By (2.16) (setting $m = 0$), we find

$$\sum_{k=0}^n (k+a)^l = \frac{(n+a)^{l+1}}{l+1} + \frac{1}{2}(n+a)^l + \sum_{k=1}^{\left[\frac{l}{2}\right]} \frac{B_{2k}}{2k} \binom{l}{2k-1} (n+a)^{l-2k+1}$$

$$- \left[\frac{1}{l+1} a^{l+1} - \frac{1}{2} a^l + \sum_{k=1}^{\left[\frac{l}{2}\right]} \frac{B_{2k}}{2k} \binom{l}{2k-1} a^{l-2k+1} \right]. \quad (2.18)$$

For $a = \frac{c}{b}$ $(0 < c < b)$ in (2.18), we may deduce a summation formula of an arithmetical sequence with the power of non-negative integer (see Yang [150]), and for $a = 1$, replacing n by $n-1$ in (2.18), we still have (see Yang [151])

$$\sum_{k=1}^{n} k^l = \frac{n^{l+1}}{l+1} + \frac{1}{2}n^l + \sum_{k=1}^{[\frac{l}{2}]} \frac{B_{2k}}{2k} \binom{l}{2k-1} n^{l-2k+1}. \qquad (2.19)$$

2.3   Estimations of the Residue Term about a Class Series

In (2.17), if $f^{(2q-1)}(t)$ is a bounded variation function, then we find the following estimation (see Cheng [9]):

$$|\delta_q(m,n)| = \frac{1}{(2q+1)!} \left| \int_m^n P_{2q+1}(t) df^{(2q)}(t) \right|$$

$$= \frac{1}{(2q+1)!} \left| P_{2q+1}(t) f^{(2q)}(t)|_m^n - (2q+1) \int_m^n P_{2q}(t) f^{(2q)}(t) dt \right|$$

$$= \frac{1}{(2q)!} \left| \int_m^n P_{2q}(t) df^{(2q-1)}(t) \right| \le \frac{1}{(2q)!} |B_{2q}| V_m^n \left(f^{(2q-1)} \right). \quad (2.20)$$

We next refine (2.20) in the following theorem by adding some conditions.

2.3.1   *An Estimation under the More Fortified Conditions*

Theorem 2.2. *Assuming that $\mathbf{Z}$ is the set of all integers, m, $n \in \mathbf{Z}$, $q \in \mathbf{N}_0$, $m < n$, $g(t) \in C^3[m,n]$, $g^{(k)}(t) \le 0 \, (\ge 0)$, $t \in [m,n](k=1,3)$, if there exist two intervals $I_k \subset [m,n]$, such that $g^{(k)}(t) < 0 (> 0)$, $t \in I_k(k=1,3)$, then, we have the following estimation (see Yang and Zhu [152]):*

$$\tilde{\delta}_q(m,n) = \frac{1}{(2q+1)!} \int_m^n P_{2q+1}(t) \, g(t) \, dt$$

$$= \varepsilon_q \frac{B_{2q+2}}{(2q+2)!} \, g(t)|_m^n, \qquad 0 < \varepsilon_q < 1. \qquad (2.21)$$

Setting $n = \infty$, in addition, $g(\infty) = g'''(\infty) = 0$, then, we have

$$\tilde{\delta}_q(m,\infty) = \frac{1}{(2q+1)!} \int_m^{\infty} P_{2q+1}(t) \, g(t) \, dt$$

$$= \varepsilon_q \frac{-B_{2q+2}}{(2q+2)!} \, g(m), \qquad 0 < \varepsilon_q < 1. \qquad (2.22)$$

In particular, if $q = 0$, it follows that

$$\tilde{\delta}_0(m, \infty) = \int_m^\infty P_1(t)\, g(t)\, dt = -\frac{\varepsilon_0}{12}\, g(m), \quad 0 < \varepsilon_0 < 1. \tag{2.23}$$

Proof. In view of (2.12), we have

$$\max_{t \in \mathbf{R}} |P_{2q+2}(t)| = |P_{2q+2}(m)| = |B_{2q+2}|.$$

Since $|P_{2q+2}(t)|$ is a non-constant continuous function, $|g'(t)| > 0$, $t \in I_1 \subset [m, n]$, and $g'(t) \le 0 \ (\ge 0)$ in $[m, n]$, then we find

$$\left| \int_m^n P_{2q+2}(t) g'(t) dt \right| \le \int_m^n |P_{2q+2}(t)|\, |g'(t)|\, dt < |B_{2q+2}| \int_m^n |g'(t)| dt$$

$$= \left| B_{2q+2} \int_m^n g'(t) dt \right| = |B_{2q+2} g(t)|_m^n|.$$

Hence, there exists a constant $\varepsilon_q \in (0, 2)$ such that

$$\int_m^n P_{2q+2}(t) g'(t) dt = B_{2q+2}\, g(t)|_m^n\, (1 - \varepsilon_q). \tag{2.24}$$

Integration by parts and in view of (2.24), we find

$$\tilde{\delta}_q(m, n) = \frac{1}{(2q+1)!} \int_m^n P_{2q+1}(t)\, g(t)\, dt$$

$$= \frac{1}{(2q+2)!} \int_m^n g(t)\, dP_{2q+2}(t)$$

$$= \frac{1}{(2q+2)!} \left[g(t) P_{2q+2}(t)|_m^n - \int_m^n P_{2q+2}(t) g'(t) dt \right] \tag{2.25}$$

$$= \frac{1}{(2q+2)!} [B_{2q+2}\, g(t)|_m^n - B_{2q+2}\, g(t)|_m^n\, (1 - \varepsilon_q)]$$

$$= \varepsilon_q \frac{B_{2q+2}}{(2q+2)!}\, g(t)|_m^n, \quad 0 < \varepsilon_q < 2. \tag{2.26}$$

In the following, we show that $0 < \varepsilon_q < 1$ in (2.26). It is obvious that both the integral $\int_m^n P_{2q+1}(t)\, g(t)\, dt$ and the term $B_{2q+2} g(t)|_m^n$ keep the same sign by (2.26). Since $g'''(t)$ and $g'(t)$ possess the same property of the sign, by the same way of obtaining (2.26), it follows that both $\int_m^n P_{2q+3}(t)\, g''(t) dt$ and $B_{2q+4}\, g''(t)|_m^n$ also keep the same sign.

Integration by parts, we obtain

$$\int_m^n P_{2q+2}(t) g'(t)\, dt = \frac{1}{2q+3} \int_m^n g'(t) dP_{2q+3}(t)$$

$$= \frac{1}{2q+3} \left[g'(t) P_{2q+3}(t)|_m^n - \int_m^n P_{2q+3}(t)\, g''(t) dt \right]$$

$$= \frac{-1}{2q+3} \int_m^n P_{2q+3}(t)\, g''(t) dt. \tag{2.27}$$

By (2.3), we have $B_{2q+4}B_{2q+2} < 0$ $(q \in \mathbf{N}_0)$. In view of $g''(t)|_m^n g(t)|_m^n > 0$, it is obvious that $(B_{2q+4}\, g''(t)|_m^n)\,(B_{2q+2}\, g(t)|_m^n) < 0$ and then $\int_m^n P_{2q+3}(t)g''(t)dt$ and $B_{2q+2}g(t)|_m^n$ keep the different sign. Hence, by (2.27), $\int_m^n P_{2q+2}(t)g'(t)dt$ and $B_{2q+2}g(t)|_m^n$ keep the same sign. Thus, by virtue of (2.24), we find $\varepsilon_q \in (0,1)$, and then (2.21) follows.

For $n = \infty$, in the same way, we obtain (2.22).   □

Note. In Theorem 2.2, (i) if $g'(t) < 0$ (> 0) replaces to $g'(t) = 0$ $(t \in [m,n])$, then, we have

$$\widetilde{\delta}_q(m,n) = \frac{1}{(2q+1)!} \int_m^n P_{2q+1}(t)\, g(t)\, dt$$

$$= \frac{B_{2q+2}}{(2q+2)!}\, g(t)|_m^n = 0; \tag{2.28}$$

(ii) if $g^{(3)}(t) < 0$ (> 0) replaces to $g^{(3)}(t) = 0$ $(t \in [m,n])$, then we have

$$\widetilde{\delta}_q(m,n) = \frac{1}{(2q+1)!} \int_m^n P_{2q+1}(t)\, g(t)dt = \frac{B_{2q+2}}{(2q+2)!}\, g(t)|_m^n. \tag{2.29}$$

In fact, in view of $\int_m^n P_{2q+3}(t)\, g''(t)dt = 0$, by (2.27), it follows that $\int_m^n P_{2q+2}(t)\, g'(t)dt = 0$ and then, by (2.25), we have (2.29).

Example 2.2. We can show that the function $g(t) = \frac{\ln t}{t-1}$, $t \in \mathbf{R}_+$ $(g(1) = 1)$ satisfies the conditions of $(-1)^k g^{(k)}(t) > 0$ $(k = 0,1,2,3)$. In fact, we have

$$g(t) = \frac{\ln[1+(t-1)]}{t-1} = \sum_{k=0}^{\infty}(-1)^k \frac{(t-1)^k}{k+1}$$

$$= \sum_{k=0}^{\infty} \frac{(-1)^k k!}{k+1}\, \frac{(t-1)^k}{k!}, \quad -1 < t-1 \le 1, \tag{2.30}$$

and then

$$g^{(k)}(1) = \frac{(-1)^k}{k+1}\, k! \quad (k \in \mathbf{N}_0).$$

In particular, we obtain

$$g^{(0)}(1) = 1, \quad g'(1) = -\frac{1}{2}, \quad g''(1) = \frac{2}{3}, \quad g'''(1) = -\frac{3}{2}, \cdots.$$

It is obvious that $g(t) > 0$.

We find

$$g'(t) = \frac{h(t)}{t(t-1)^2}, \qquad h(t) = t - 1 - t\ln t.$$

Since $h'(t) = -\ln t > 0$, $0 < t < 1$; $h'(t) < 0$, $t > 1$, then $\max h(t) = h(1) = 0$ and $g'(t) < 0$ $(t \neq 1)$. In view of $g'(1) = -\frac{1}{2}$, then, we find $g'(t) < 0$ $(t > 0)$.

We also obtain

$$g''(t) = \frac{J(t)}{t^2(t-1)^3}, \quad J(t) = -(t-1)^2 - 2t(t-1) + 2t^2 \ln t.$$

Since $J'(t) = -4(t-1) + 4t \ln t$ and $J''(t) = 4 \ln t < 0$, $0 < t < 1$; $J''(t) > 0$, $t > 1$, then, $\min J'(t) = J'(1) = 0$ and $J'(t) > 0$ $(t \neq 1)$; $J'(1) = 0$. Hence, $J(t)$ is strictly increasing with $J(1) = 0$ and then, $J(t) < 0$, $0 < t < 1$; $J(t) > 0$, $t > 1$. It follows that $g''(t) > 0$ $(t \neq 1)$. Since $g''(1) = \frac{2}{3}$, then, we find $g''(t) > 0$ $(t > 0)$.

We have

$$g'''(t) = \frac{L(t)}{t^3(t-1)^4},$$

$$L(t) = 2(t-1)^3 + 3t(t-1)^2 + 6t^2(t-1) - 6t^3 \ln t,$$

$$L'(t) = 9(t-1)^2 + 18t(t-1) - 18t^2 \ln t,$$

$$L''(t) = 36(t-1) - 36t \ln t, \quad L'''(t) = -36 \ln t.$$

Then $L'''(t) > 0$, $0 < t < 1$; $L'''(t) < 0$, $t > 1$; $L''(1) = 0$ and $\max L''(t) = L''(1) = 0$. Hence, $L''(t) < 0$ $(t \neq 1)$ and $L'(t)$ is strictly decreasing with $L'(1) = 0$ and then, $L'(t) > 0$, $0 < t < 1$; $L'(t) < 0$, $t > 1$. Therefore, $\max L(t) = L(1) = 0$ and $L(t) < 0$ $(t \neq 1)$. It follows that $g'''(t) < 0 (t \neq 1)$. Since $g'''(1) = -\frac{3}{2}$, then, we find $g'''(t) < 0$ $(t > 0)$.

In view of the above results and (2.21), we have

$$\tilde{\delta}_q(m,n) = \frac{1}{(2q+1)!} \int_m^n P_{2q+1}(t) \frac{\ln t}{t-1} dt$$

$$= \varepsilon_q \frac{B_{2q+2}}{(2q+2)!} \left(\frac{\ln n}{n-1} - \frac{\ln m}{m-1} \right), \quad 0 < \varepsilon_q < 1. \quad (2.31)$$

Corollary 2.2. *Assuming that $m_i \in \mathbf{Z}$, $q \in \mathbf{N_0}$, $m_i < m_{i+1}$, $g(t) \in C^3[m_i, m_{i+1}]$, $g^{(k)}(t) \leq 0$ (≥ 0), $t \in [m_i, m_{i+1}]$ $(k = 1,3; i = 1,2,\cdots,s)$, $m = m_1$, $n = m_{s+1}$, if there exist two intervals $I_k \subset [m,n]$ such that $g^{(k)}(t) < 0$ (>0), $t \in I_k (k = 1,3)$, then, we still have (2.21). Moreover, if $g^{(k)}(\infty) = 0$ $(k = 0,2)$, then, we have (2.22).*

Proof. Without loss of generality, assuming that $g^{(k)}(t) \leq 0$, $t \in [m_i, m_{i+1}]$ $(k = 1,3; i = 1,2,\cdots,s)$, $g^{(k)}(t) < 0, t \in I_k$ $(k = 1,3)$, and $B_{2q+2} < 0$, by (2.21), it follows that

$$0 \leq \int_{m_i}^{m_{i+1}} P_{2q+1}(t) \, g(t) \, dt \leq \frac{B_{2q+2}}{2q+2} \, g(t)|_{m_i}^{m_{i+1}} \quad (i = 1,\cdots,s).$$

There exists an i, such that the above inequalities keep the strict signs. Hence, we find

$$0 < \int_m^n P_{2q+1}(t)\, g(t)\, dt$$

$$= \sum_{i=1}^s \int_{m_i}^{m_{i+1}} P_{2q+1}(t)\, g(t)\, dt < \frac{B_{2q+2}}{2q+2} g(t)|_m^n.$$

Therefore, we have (2.21), and then, the other results hold. □

By (2.27), we still have

$$\int_m^n P_{2q}(t)\, g(t)\, dt = \frac{-1}{2q+1} \int_m^n P_{2q+1}(t) g'(t) dt \quad (q \in \mathbf{N}).$$

Then, by Theorem 2.2, it follows that

Corollary 2.3. *Assuming that m, $n \in \mathbf{Z}$, $q \in \mathbf{N}$, $m < n$, $g(t) \in C^4[m,n]$, $g^{(k)}(t) \leq 0\ (\geq 0)$, $t \in [m,n](k = 2,4)$, if there exist two intervals $I_k \subset [m,n]$, such that $g^{(k)}(t) < 0\ (> 0)$, $t \in I_k(k = 2,4)$, then we have the following estimation:*

$$\int_m^n P_{2q}(t)\, g(t)\, dt = \frac{-B_{2q+2}\, \varepsilon_q}{(2q+1)(2q+2)} g'(t)|_m^n, \quad 0 < \varepsilon_q < 1. \tag{2.32}$$

Setting $n = \infty$, if in addition, $g'(\infty) = g^{(4)}(\infty) = 0$, then we have

$$\int_m^\infty P_{2q}(t)\, g(t)\, dt = \frac{B_{2q+2}\, \varepsilon_q}{(2q+1)(2q+2)} g'(m), \quad 0 < \varepsilon_q < 1. \tag{2.33}$$

2.3.2 *Some Estimations under the More Imperfect Conditions*

Corollary 2.4. *Assuming that m, $n \in \mathbf{Z}$, $q \in \mathbf{N}_0$, $m < n$, $g(t) \in C^1[m,n]$, $g'(t) \leq 0\ (\geq 0)$, $t \in [m,n]$, if there exists an interval $I_1 \subset [m,n]$ such that $g'(t) < 0\ (> 0)$, $t \in I_1$, then we have the following estimation:*

$$\widetilde{\delta}_q(m,n) = \frac{1}{(2q+1)!} \int_m^n P_{2q+1}(t)\, g(t)\, dt$$

$$= \widetilde{\varepsilon}_q \frac{2B_{2q+2}}{(2q+2)!} g(t)|_m^n, \quad 0 < \widetilde{\varepsilon}_q < 1. \tag{2.34}$$

Setting $n = \infty$, and $g(\infty) = 0$, then, we have

$$\widetilde{\delta}_q(m,\infty) = \frac{1}{(2q+1)!} \int_m^\infty P_{2q+1}(t)\, g(t)\, dt$$

$$= \widetilde{\varepsilon}_q \frac{-2B_{2q+2}}{(2q+2)!} g(m), \quad 0 < \widetilde{\varepsilon}_q < 1. \tag{2.35}$$

In particular, for $q = 0$, it follows that

$$\tilde{\delta}_0(m, \infty) = \int_m^\infty P_1(t)\, g(t)\, dt = \frac{-\tilde{\varepsilon}_0}{6}\, g(m), \quad 0 < \tilde{\varepsilon}_0 < 1. \tag{2.36}$$

Proof. It is obvious that in the conditions of the corollary, we obtain (2.24). Then (2.21) and (2.22) are valid for $\varepsilon_q \in (0, 2)$. Setting $\varepsilon_q = 2\tilde{\varepsilon}_q$, we have (2.34) and (2.35). ∎

We can refine Corollary 2.4 in the following theorem:

Theorem 2.3. *Assuming that $m, n \in \mathbf{Z}$, $q \in \mathbf{N}_0$, $m < n$, $g(t)$ is a monotone piecewise smooth continuous function in $[m, n]$, then we have the following estimation:*

$$\tilde{\delta}_q(m, n) = \frac{1}{(2q+1)!} \int_m^n P_{2q+1}(t)\, g(t)\, dt$$

$$= \tilde{\varepsilon}_q \frac{2B_{2q+2}}{(2q+2)!} \left(1 - \frac{1}{2^{2q+2}}\right) g(t)|_m^n, \quad 0 < \tilde{\varepsilon}_q < 1. \tag{2.37}$$

Setting $n = \infty$ and $g(\infty) = 0$, then, we have

$$\tilde{\delta}_q(m, \infty) = \frac{1}{(2q+1)!} \int_m^\infty P_{2q+1}(t)\, g(t)\, dt$$

$$= \tilde{\varepsilon}_q \frac{-2B_{2q+2}}{(2q+2)!} \left(1 - \frac{1}{2^{2q+2}}\right) g(m), \quad 0 < \tilde{\varepsilon}_q < 1. \tag{2.38}$$

In particular, for $q = 0$, it follows that

$$\tilde{\delta}_0(m, \infty) = \int_m^\infty P_1(t)\, g(t)\, dt = -\frac{\tilde{\varepsilon}_0}{8}\, g(m), \quad 0 < \tilde{\varepsilon}_0 < 1. \tag{2.39}$$

Proof. If $g(t)$ is constant, then, two sides of (2.37) and (2.38) equal to zero (in (2.36), $g(t) = 0$). In the following, we assume that $g(t)$ is a non-constant function in $[m, n]$ with $g'(t) \le 0$.

(i) If $B_{2q+2} < 0$, since $g(n) < g(m)$, $\int_{m+k-1}^{m+k} P_{2q+1}(t) dt = 0$ $(k \in \mathbf{N})$, then, by (2.26), we find

$$\int_m^n P_{2q+1}(t)\, g(t)\, dt = \varepsilon_q \frac{B_{2q+2}}{2(q+1)}\, g(t)|_m^n \quad (0 < \varepsilon_q < 2),$$

$$0 < \int_m^n P_{2q+1}(t)\, g(t)\, dt = \sum_{k=1}^{n-m} \int_{m+k-1}^{m+k} P_{2q+1}(t)[g(t) - g(m+k)]dt$$

$$= \sum_{k=1}^{n-m} \left\{ \int_{m+k-1}^{m+k-\frac{1}{2}} P_{2q+1}(t)\, [g(t) - g(m+k)]dt \right.$$

$$\left. + \int_{m+k-\frac{1}{2}}^{m+k} P_{2q+1}(t)\, [g(t) - g(m+k)]dt \right\}$$

$$= \sum_{k=1}^{n-m} \left\{ [g(m+k-1) - g(m+k)] \int_{m+k-1}^{m+k-\frac{1}{2}} P_{2q+1}(t)\, dt \right\} + \sum_{k=1}^{n-m} \alpha_k,$$

$$\tag{2.40}$$

where α_k is defined by

$$\alpha_k = \int_{m+k-1}^{m+k-\frac{1}{2}} P_{2q+1}(t)\, [g(t) - g(m+k-1)]\, dt$$

$$+ \int_{m+k-\frac{1}{2}}^{m+k} P_{2q+1}(t)\, [g(t) - g(m+k)]\, dt.$$

Since $g(t)$ is decreasing, then, it follows that

$$g(t) - g(m+k-1) \leq 0, \quad t \in \left(m+k-1, \quad m+k-\frac{1}{2} \right),$$

$$g(t) - g(m+k) \geq 0, \quad t \in \left(m+k-\frac{1}{2}, \quad m+k \right).$$

In view of $P_{2q+2}(m+k-1) = B_{2q+2} < 0$, by (2.11), it follows that $P_{2q+2}\left(m+k-\frac{1}{2}\right) > 0$. Hence, $P_{2q+2}(t)$ is strictly increasing in $\left(m+k-1, \quad m+k-\frac{1}{2}\right)$ with $P'_{2q+2}(t) > 0$ and

$$P_{2q+1}(t) = \frac{1}{2q+2} P'_{2q+2}(t) > 0, \qquad t \in \left(m+k-1, \quad m+k-\frac{1}{2} \right).$$

Similarly, we have

$$P_{2q+1}(t) = \frac{1}{2q+2} P'_{2q+2}(t) < 0, \quad t \in \left(m+k-\frac{1}{2}, \quad m+k \right).$$

Hence, we find

$$P_{2q+1}(t)\, [g(t) - g(m+k-1)] \leq 0, \quad t \in \left(m+k-1, \quad m+k-\frac{1}{2} \right),$$

$$P_{2q+1}(t)\, [g(t) - g(m+k)] \leq 0, \quad t \in \left(m+k-\frac{1}{2}, \quad m+k \right),$$

and then $\alpha_k \leq 0$. Since $g(t)$ is of non-constant, there exists a positive integer k_0 ($\leq n - m$), such that $\alpha_{k_0} < 0$, and then, $\sum_{k=1}^{n-m} \alpha_k < 0$.

By (2.11), we still have

$$\sum_{k=1}^{n-m} [g(m+k-1) - g(m+k)] \int_{m+k-1}^{m+k-\frac{1}{2}} P_{2q+1}(t) \, dt$$

$$= -\int_{m-1}^{m-\frac{1}{2}} P_{2q+1}(t) \, dt \sum_{k=1}^{n-m} [g(m+k) - g(m+k-1)]$$

$$= -\frac{1}{2q+2} \left[P_{2q+2}\left(m - \frac{1}{2}\right) - P_{2q+2}(m-1) \right] g(t)|_m^n$$

$$= -\frac{1}{2q+2} \left[\left(\frac{1}{2^{2q+1}} - 1\right) B_{2q+2} - B_{2q+2} \right] g(t)|_m^n$$

$$= \frac{2}{2q+2} \left(1 - \frac{1}{2^{2q+2}}\right) B_{2q+2} \, g(t)|_m^n.$$

In view of (2.40), we find

$$0 < \int_m^n P_{2q+1}(t) \, g(t) \, dt < \frac{2}{2q+2} \left(1 - \frac{1}{2^{2q+2}}\right) B_{2q+2} \, g(t)|_m^n,$$

and then equality (2.35) follows.

(ii) If $B_{2q+2} > 0$, in the same way, we have $\sum_{k=1}^{n-m} \alpha_k > 0$. The corresponding result is

$$\frac{2}{2q+2} \left(1 - \frac{1}{2^{2q+2}}\right) B_{2q+2} \, g(t)|_m^n < \int_m^n P_{2q+1}(t) \, g(t) \, dt < 0,$$

and we still have (2.37).

For $n = \infty$, by the same way, we have equality (2.38).   □

Example 2.3. If $m, n \in \mathbb{N}$, $q \in \mathbb{N}_0$, $2m < n$,

$$g(t) = \frac{1}{\max\{2m, t\}} = \begin{cases} \frac{1}{2m}, & t \leq 2m, \\ \frac{1}{t}, & t > 2m, \end{cases}$$

$g(t)|_m^n = \frac{1}{n} - \frac{1}{2m}$, then, by (2.37) and (2.38), we find

$$\tilde{\delta}_q(m, n) = \frac{1}{(2q+1)!} \int_m^n P_{2q+1}(t) \, \frac{1}{\max\{2m, t\}} \, dt$$

$$= \frac{2B_{2q+2} \, \tilde{\varepsilon}_q}{(2q+2)!} \left(1 - \frac{1}{2^{2q+2}}\right) \left(\frac{1}{n} - \frac{1}{2m}\right), \quad 0 < \tilde{\varepsilon}_q < 1, \quad (2.41)$$

$$\tilde{\delta}_q(m, \infty) = \frac{1}{(2q+1)!} \int_m^\infty P_{2q+1}(t) \, \frac{1}{\max\{2m, t\}} \, dt$$

$$= \frac{-B_{2q+2} \, \tilde{\varepsilon}_q}{(2q+2)!} \left(1 - \frac{1}{2^{2q+2}}\right) \frac{1}{m}, \quad 0 < \tilde{\varepsilon}_q < 1. \quad (2.42)$$

Corollary 2.5. *Assuming that* m, $n \in \mathbf{Z}$, $q \in \mathbf{N}$, $m < n$, $g'(t)$ *is a monotone piecewise smooth continuous function in* $[m, n]$, *then, we have the following estimation:*

$$\int_m^n P_{2q}(t)\, g(t)\, dt = \frac{-2B_{2q+2}\,\widetilde{\varepsilon}_q}{(2q+1)(2q+2)} \left(1 - \frac{1}{2^{2q+2}}\right) g'(t)|_m^n, \quad 0 < \widetilde{\varepsilon}_q < 1.$$
(2.43)

Setting $n = \infty$, *and* $g'(\infty) = 0$, *then, we have*

$$\int_m^\infty P_{2q}(t)\, g(t)\, dt = \frac{2B_{2q+2}\,\widetilde{\varepsilon}_q}{(2q+1)(2q+2)} \left(1 - \frac{1}{2^{2q+2}}\right) g'(m), \quad 0 < \widetilde{\varepsilon}_q < 1.$$
(2.44)

Corollary 2.6. *Assuming that* $q \in \mathbf{N}_0$, x_1, $x_2 \notin \mathbf{Z}$, $x_1 < x_2$, $g(t)$ *is a monotone piecewise smooth continuous function in* $[x_1, x_2]$, *then, we have the following estimation:*

$$I_{2q+1} = \int_{x_1}^{x_2} P_{2q+1}(t)\, g(t)\, dt$$

$$= \frac{B_{2q+2}}{q+1} \left(1 - \frac{1}{2^{2q+2}}\right) \left[\varepsilon_0 g(t)|_{x_1}^{x_2} + \varepsilon_1 g(x_1) - \varepsilon_2 g(x_2)\right], \quad (2.45)$$

where $\varepsilon_0 \in (0, 1)$, $\varepsilon_i \in (0, 1]$ $(i = 1, 2)$. *In particular, when* $q = 0$,

$$I_1 = \int_{x_1}^{x_2} P_1(t)\, g(t)\, dt = \frac{1}{8} \left[\varepsilon_0\, g(t)|_{x_1}^{x_2} + \varepsilon_1 g(x_1) - \varepsilon_2 g(x_2)\right]. \quad (2.46)$$

Proof. We define a monotone piecewise smooth continuous function $\widetilde{g}(t)$ by

$$\widetilde{g}(t) = \begin{cases} g(x_1), & t \in [[x_1], x_1), \\ g(t), & t \in [x_1, x_2], \\ g(x_2), & t \in (x_2, [x_2] + 1]. \end{cases}$$

Then, we find

$$I_{2q+1} = \int_{[x_1]}^{[x_2]+1} P_{2q+1}(t)\, \widetilde{g}(t)\, dt$$

$$- g(x_1) \int_{[x_1]}^{x_1} P_{2q+1}(t)\, dt - g(x_2) \int_{x_2}^{[x_2]+1} P_{2q+1}(t)\, dt. \quad (2.47)$$

By (2.37), it follows that

$$\int_{[x_1]}^{[x_2]+1} P_{2q+1}(t)\widetilde{g}(t)dt = \frac{2B_{2q+2}}{2q+2} \left(1 - \frac{1}{2^{2q+2}}\right) \varepsilon_0\, \widetilde{g}(t)|_{[x_1]}^{[x_2]+1}$$

$$= \frac{2B_{2q+2}}{2q+2} \left(1 - \frac{1}{2^{2q+2}}\right) \varepsilon_0 g(t)|_{x_1}^{x_2}, \quad \varepsilon_0 \in (0, 1).$$

Since, for $x_1 \notin \mathbf{Z}$, we have $0 < x_1 - [x_1] < 1$, and

$$0 < \frac{\int_0^{x_1 - [x_1]} P_{2q+1}(t)dt}{\int_0^{\frac{1}{2}} P_{2q+1}(t)dt} \leq 1,$$

there exists $\varepsilon_1 \in (0, 1]$ such that

$$\int_{[x_1]}^{x_1} P_{2q+1}(t)dt$$

$$= \int_0^{x_1 - [x_1]} P_{2q+1}(t)\, dt = \varepsilon_1 \int_0^{\frac{1}{2}} P_{2q+1}(t)\, dt$$

$$= \frac{\varepsilon_1}{2q+2} P_{2q+2}(t)|_0^{\frac{1}{2}} = \frac{\varepsilon_1}{2q+2}\left(P_{2q+2}\left(\frac{1}{2}\right) - B_{2q+2}\right)$$

$$= \frac{\varepsilon_1}{2q+2}\left(2^{-1-2q} - 2\right)B_{2q+2} = -\frac{B_{2q+2}}{q+1}\left(1 - \frac{1}{2^{2q+2}}\right)\varepsilon_1. \quad (2.48)$$

Similarly, there exists $\varepsilon_2 \in (0, 1]$ such that

$$\int_{x_2}^{[x_2]+1} P_{2q+1}(t)\, dt$$

$$= \int_{x_2 - [x_2]}^1 P_{2q+1}(t)\, dt = \varepsilon_2 \int_{\frac{1}{2}}^1 P_{2q+1}(t)\, dt$$

$$= \frac{\varepsilon_2}{2q+2} P_{2q+2}(t)|_{\frac{1}{2}}^1 = \frac{-\varepsilon_2}{2q+2}\left(P_{2q+2}\left(\frac{1}{2}\right) - B_{2q+2}\right)$$

$$= \frac{B_{2q+2}}{q+1}\left(1 - \frac{1}{2^{2q+2}}\right)\varepsilon_2. \quad (2.49)$$

Hence, by (2.47) and the above results, we have (2.45). $\qquad\square$

Note. If x_1, $x_2 \in \mathbf{Z}$, then we can find $\varepsilon_1 = \varepsilon_2 = 0$ in (2.48) and (2.49). In this case, (2.45) is an extension of (2.37).

Corollary 2.7. *Assuming that $q \in \mathbf{N}_0$, m, $n \in \mathbf{Z}$, $x \notin \mathbf{Z}$, $m < x < n$, both $g_1(t)$ $(t \in [m, x])$ and $g_2(t)$ $(t \in [x, n])$ are decreasing (increasing) piecewise smooth continuous functions, and*

$$g(t) = \begin{cases} g_1(t), & t \in [m, x), \\ g_2(t), & t \in [x, n], \end{cases}$$

then, we have

$$J_{2q+1} = \int_m^n P_{2q+1}(t)\, g(t)\, dt$$

$$= \frac{B_{2q+2}}{q+1}\left(1 - \frac{1}{2^{2q+2}}\right)\{\varepsilon_0(g_2(n) - g_1(m))$$

$$+ [\varepsilon_0 - \varepsilon_1](g_1(x) - g_2(x))\}, \; \varepsilon_0 \in (0,1), \; and\; \varepsilon_1 \in (0,1].$$

$$(2.50)$$

Setting $n = \infty$, and $g_2(\infty) = 0$, then, we have

$$\widetilde{J}_{2q+1} = \int_m^\infty P_{2q+1}(t)\, g(t)\, dt$$

$$= \frac{B_{2q+2}}{q+1}\left(1 - \frac{1}{2^{2q+2}}\right)\{-\varepsilon_0 g_1(m)$$

$$+ [\varepsilon_0 - \varepsilon_1](g_1(x) - g_2(x))\}, \varepsilon_0 \in (0,1), \varepsilon_1 \in (0,1].$$

$$(2.51)$$

In particular, when $q = 0$, it follows that

$$J_1 = \int_m^n P_1(t)\, g(t)\, dt = \frac{1}{8}[\varepsilon_0(g_2(n) - g_1(m))$$

$$+ (\varepsilon_0 - \varepsilon_1)(g_1(x) - g_2(x))], \quad (2.52)$$

$$\widetilde{J}_1 = \int_m^\infty P_1(t)\, g(t)\, dt$$

$$= \frac{1}{8}[-\varepsilon_0\, g_1(m) + (\varepsilon_0 - \varepsilon_1)(g_1(x) - g_2(x))], \varepsilon_0 \in (0,1), \varepsilon_1 \in (0,1].$$

$$(2.53)$$

Proof.   Define a decreasing (increasing) piecewise smooth continuous function $\widetilde{g}(t)$ as follows:

$$\widetilde{g}(t) = \begin{cases} g_1(t) - g_1(x) + g_2(x), & t \in [m,x), \\ g_2(t), & t \in [x,n]. \end{cases}$$

Then, by (2.37) and (2.48), it follows that

$$J_{2q+1} = \int_m^n P_{2q+1}(t)\, \widetilde{g}(t)\, dt + (g_1(x) - g_2(x))\int_m^x P_{2q+1}(t)\, dt$$

$$= \frac{2B_{2q+2}}{2q+2}\left(1 - \frac{1}{2^{2q+2}}\right)\varepsilon_0\, \widetilde{g}(t)|_m^n$$

$$- \frac{B_{2q+2}}{q+1}\left(1 - \frac{1}{2^{2q+2}}\right)\varepsilon_1(g_1(x) - g_2(x))$$

$$= \frac{2B_{2q+2}}{2q+2}\left(1 - \frac{1}{2^{2q+2}}\right)\varepsilon_0\left(g_2(n) - g_1(m) + g_1(x) - g_2(x)\right)$$

$$- \frac{B_{2q+2}}{q+1}\left(1 - \frac{1}{2^{2q+2}}\right)\varepsilon_1(g_1(x) - g_2(x))$$

$$= \frac{B_{2q+2}}{q+1}\left(1 - \frac{1}{2^{2q+2}}\right)\{\varepsilon_0(g_2(n) - g_1(m))$$

$$+ [\varepsilon_0 - \varepsilon_1](g_1(x) - g_2(x))\},$$

where

$$\varepsilon_0 \in (0,1), \quad \varepsilon_1 \in (0,1].$$

Hence, we have (2.50). In the same way, we have (2.51).   □

Note. If $x \in \mathbf{Z}$, then, in (2.50) and (2.51), we find $\varepsilon_1 = 0$. Hence, for $g_2(t) = 0$ in (2.52), we have

$$\int_m^x P_1(t)\, g_1(t)\, dt = \frac{\varepsilon_0}{8}(-g_1(m) + g_1(x)) - \frac{\varepsilon_1}{8}g_1(x), \tag{2.54}$$

where

$$x > m; \quad \varepsilon_0 \in (0,1) \quad \text{and} \quad \varepsilon_1 \in [0,1];$$

for $g_1(t) = 0$ in (2.52) and (2.53), we have

$$\int_x^n P_1(t)\, g_2(t)\, dt = \frac{\varepsilon_0}{8}(g_2(n) - g_2(x)) + \frac{\varepsilon_1}{8}\, g_2(x), \tag{2.55}$$

where

$$x > m; \quad \varepsilon_0 \in (0,1) \quad \text{and} \quad \varepsilon_1 \in [0,1],$$

$$\int_x^\infty P_1(t)\, g_2(t)\, dt = \frac{\varepsilon_1 - \varepsilon_2}{8}\, g_2(x), \; x \in \mathbf{R}; \; \varepsilon_0 \in (0,1) \text{ and } \varepsilon_1 \in [0,1]. \tag{2.56}$$

2.3.3   *Estimations of $\delta_q(m,n)$ and Some Applications*

Setting $g(t) = f^{(2q+1)}(t)$ in Theorem 2.4 and Theorem 2.9, we have the following corollaries:

Corollary 2.8. *Assuming that $m, n, q \in \mathbf{N}_0$, $m < n$, $f(t) \in C^{2q+4}[m,n]$ with*

$$(-1)^k f^{(2q+1+k)}(t) \geq 0 \;(\leq 0), \quad t \in [m,n] \quad (k = 1,3),$$

and there exist two intervals $I_k \subset [m, n]$ such that

$$(-1)^k f^{(2q+1+k)}(t) > 0 \quad (< 0), \quad t \in I_k \quad (k = 1, 3),$$

then, we have

$$\delta_q(m, n) = \frac{1}{(2q+1)!} \int_m^n P_{2q+1}(t) \, f^{(2q+1)}(t) dt$$

$$= \varepsilon_q \frac{B_{2q+2}}{(2q+2)!} \, f^{(2q+1)}(t)|_m^n, \quad 0 < \varepsilon_q < 1. \tag{2.57}$$

Setting $n = \infty$, and in addition, $f^{(2q+1+k)}(\infty) = 0 \ (k = 0, 2)$, then, we have

$$\delta_q(m, \infty) = \frac{1}{(2q+1)!} \int_m^\infty P_{2q+1}(t) f^{(2q+1)}(t) \, dt$$

$$= \varepsilon_q \frac{-B_{2q+2}}{(2q+2)!} \, f^{(2q+1)}(m), \quad 0 < \varepsilon_q < 1. \tag{2.58}$$

In particular, when $q = 0$, it follows that

$$\delta_0(m, \infty) = \int_m^\infty P_1(t) f'(t) \, dt = -\frac{\varepsilon_0}{12} \, f'(m), \quad 0 < \varepsilon_0 < 1. \tag{2.59}$$

Corollary 2.9. *Assuming that m, n, $q \in \mathbf{N}_0$, $m < n$, $f^{(2q+1)}(t)$ is a monotone piecewise smooth continuous function in $[m, n]$, then, we have*

$$\delta_q(m, n) = \frac{1}{(2q+1)!} \int_m^n P_{2q+1}(t) \, f^{(2q+1)}(t) \, dt$$

$$= \widetilde{\varepsilon}_q \frac{2B_{2q+2}}{(2q+2)!} \left(1 - \frac{1}{2^{2q+2}}\right) f^{(2q+1)}(t)|_m^n, \quad 0 < \widetilde{\varepsilon}_q < 1. \tag{2.60}$$

Setting $n = \infty$ and $f^{(2q+1)}(\infty) = 0$, then, we have

$$\delta_q(m, \infty) = \frac{1}{(2q+1)!} \int_m^\infty P_{2q+1}(t) \, f^{(2q+1)}(t) \, dt$$

$$= \widetilde{\varepsilon}_q \frac{-2B_{2q+2}}{(2q+2)!} \left(1 - \frac{1}{2^{2q+2}}\right) f^{(2q+1)}(m), \quad 0 < \widetilde{\varepsilon}_q < 1. \tag{2.61}$$

In particular, when $q = 0$, it follows that

$$\delta_0(m, \infty) = \int_m^\infty P_1(t) f'(t) \, dt = -\frac{\widetilde{\varepsilon}_0}{8} f'(m), \quad 0 < \widetilde{\varepsilon}_0 < 1. \tag{2.62}$$

For $q = 0$ in Corollary 2.8, in view of (2.13), we have the following corollary:

Corollary 2.10. *Assuming that m, $n \in \mathbf{N}_0$, $m < n$, $f(t) \in C^4[m, n]$ with*

$$(-1)^k f^{(1+k)}(t) \geq 0 \ (\leq 0), \quad t \in [m, n] \ (k = 1, 3),$$

and there exist two intervals $I_k \subset [m, n]$ such that

$$(-1)^k f^{(1+k)}(t) > 0 \ (< 0), \quad t \in I_k \quad (k = 1, 3),$$

then, we have

$$\sum_{k=m}^{n} f(k) = \int_{m}^{n} f(t)dt + \frac{1}{2}(f(n) + f(m)) + \frac{\varepsilon_0}{12}f'(t)|_m^n, \quad 0 < \varepsilon_0 < 1. \quad (2.63)$$

Setting $n = \infty$ and $f^{(k)}(\infty) = 0$ $(k = 0, 2)$, and both of $\sum_{k=m}^{\infty} f(k)$ and $\int_m^\infty f(t)dt$ are convergent, then, we have

$$\sum_{k=m}^{\infty} f(k) = \int_{m}^{\infty} f(t)dt + \frac{1}{2}f(m) - \frac{\varepsilon_0}{12}f'(m), \quad 0 < \varepsilon_0 < 1. \quad (2.64)$$

In particular, when $f'(m) < 0$, the following inequalities follow:

$$\int_{m}^{\infty} f(t)dt + \frac{1}{2}f(m) < \sum_{k=m}^{\infty} f(k) < \int_{m}^{\infty} f(t)dt + \frac{1}{2}f(m) - \frac{1}{12}f'(m). \quad (2.65)$$

For $q = 0$ in Corollary 2.4, in view of (2.13), we have the following corollary:

Corollary 2.11. *Assuming that m, $n \in \mathbf{N}_0$, $m < n$, $f'(t)$ is a monotone piecewise smooth continuous function in $[m, n]$, then, we have*

$$\sum_{k=m}^{n} f(k) = \int_{m}^{n} f(t)dt + \frac{1}{2}(f(n) + f(m)) + \frac{\widetilde{\varepsilon_0}}{8}f'(t)|_m^n, \quad 0 < \widetilde{\varepsilon}_0 < 1. \quad (2.66)$$

Setting $n = \infty$ and $f'(\infty) = 0$, and both $\sum_{k=m}^{\infty} f(k)$ and $\int_m^\infty f(t)dt$ are convergent, then, we have

$$\sum_{k=m}^{\infty} f(k) = \int_{m}^{\infty} f(t) \, dt + \frac{1}{2}f(m) - \frac{\widetilde{\varepsilon_0}}{8}f'(m), \quad 0 < \widetilde{\varepsilon}_0 < 1. \quad (2.67)$$

In particular, when $f'(m) < 0$, the following inequalities follow:

$$\int_{m}^{\infty} f(t)dt + \frac{1}{2}f(m) < \sum_{k=m}^{\infty} f(k) < \int_{m}^{\infty} f(t)dt + \frac{1}{2}f(m) - \frac{1}{8}f'(m). \quad (2.68)$$

Example 2.4. If $f(t) = \frac{1}{(1+t)t^{1/2}}$ $(t > 0)$, $m \in \mathbf{N}$, then, we obtain $f(t) > 0$,

$$f'(t) = \frac{-1}{(1+t)^2 t^{1/2}} - \frac{1}{2(1+t)t^{3/2}} = -\frac{3t+1}{2(1+t)^2 t^{3/2}} < 0,$$

$f''(t) > 0$, $f'''(t) < 0$, and

$$\int_{m}^{\infty} \frac{1}{(1+t)t^{1/2}} dt = 2 \arctan\left(\frac{1}{\sqrt{m}}\right).$$

By (2.65), when $m \in \mathbf{N}$, we have

$$2\arctan\left(\frac{1}{\sqrt{m}}\right) + \frac{1}{2(1+m)\sqrt{m}} < \sum_{k=m}^{\infty} \frac{1}{(1+k)\sqrt{k}}$$

$$< 2\arctan\left(\frac{1}{\sqrt{m}}\right) + \frac{1}{2(1+m)\sqrt{m}} + \frac{3m+1}{24(1+m)^2\sqrt{m^3}}. (2.69)$$

Example 2.5. If $f(t) = \frac{1}{t^2+1}$ $(t > 0)$, $m \in \mathbf{N}$, then, we obtain $f(t) > 0$, $f'(t) = \frac{-2t}{(t^2+1))^2} < 0$,

$$\int_m^{\infty} \frac{1}{t^2+1} \, dt = \frac{\pi}{2} - \arctan m.$$

By (2.68), when $m \in \mathbf{N}$, we have

$$\frac{\pi}{2} - \arctan m + \frac{1}{2(m^2+1)} < \sum_{k=m}^{\infty} \frac{1}{k^2+1}$$

$$< \frac{\pi}{2} - \arctan m + \frac{1}{2(m^2+1)} + \frac{m}{4(m^2+1)^2}.$$

$$(2.70)$$

In particular, when $m = 1$, we find

$$\frac{\pi}{4} + \frac{5}{4} < \sum_{k=0}^{\infty} \frac{1}{k^2+1} < \frac{\pi}{4} + \frac{21}{16}. \tag{2.71}$$

2.4 Two Classes of Series Estimations

2.4.1 *One Class of Convergent Series Estimation*

For $n = \infty$ in (2.16), by Corollary 2.8 and Corollary 2.9, we have the following theorem:

Theorem 2.4. *Assuming that* $m, n, q \in \mathbf{N}_0$, $m < n$, $f(t) \in C^{2q+1}[m, n]$, $f(\infty) = f^{(2k-1)}(\infty) = 0$ $(k = 1, 2, \cdots, q+1)$, $\delta_q(m, \infty)$ *is convergent, if both* $\sum_{k=m}^{\infty} f(k)$ *and* $\int_m^{\infty} f(t)dt$ *are convergent, then, we have the following estimation:*

$$\sum_{k=m}^{\infty} f(k) = \int_m^{\infty} f(t)dt + \frac{1}{2}f(m) - \sum_{k=1}^{q} \frac{1}{(2k)!} B_{2k} f^{(2k-1)}(m) + \delta_q(m, \infty),$$

$$(2.72)$$

where

$$\delta_q(m, \infty) = \frac{1}{(2q+1)!} \int_m^{\infty} P_{2q+1}(t) \, f^{(2q+1)}(t) \, dt, \tag{2.73}$$

and $\delta_q(m, \infty)$ satisfies the following recursion formulas

$$\delta_0(m, \infty) = \int_m^\infty P_1(t) f'(t) dt,$$

$$\delta_q(m, \infty) = \frac{1}{(2q)!} B_{2q} f^{(2q-1)}(m) + \delta_{q-1}(m, \infty), \quad (q \in \mathbf{N}). \quad (2.74)$$

(i) If $(-1)^k f^{(2q+1+k)}(t) \geq 0 (\leq 0)$, $t \in [m, n]$ $(k = 1, 3)$, and there exist two intervals $I_k \subset [m, n]$, such that

$$(-1)^k f^{(2q+1+k)}(t) > 0 (< 0), \quad t \in I_k \ (k = 1, 3),$$

then, we have

$$\delta_q(m, \infty) = \varepsilon_q \frac{-B_{2q+2}}{(2q+2)!} f^{(2q+1)}(m), \quad 0 < \varepsilon_q < 1 \ (q \in \mathbf{N}_0); \quad (2.75)$$

(ii) if $f^{(2q+1)}(t)$ is a monotone piecewise smooth continuous function in $[m, n]$, then, we have

$$\delta_q(m, \infty) = \frac{-\widetilde{\varepsilon}_q B_{2q+2}}{(2q+2)!} \frac{2^{2q+2} - 1}{2^{2q+1}} f^{(2q+1)}(m), \quad 0 < \widetilde{\varepsilon}_q < 1 \ (q \in \mathbf{N}_0). \quad (2.76)$$

Example 2.6. If $0 < a \leq 1$, $p > 1$, $f(t) = \frac{1}{(t+a)^p}$ $(t > 0)$, then, we find

$$f^{(k)}(t) = \binom{-p}{k} \frac{k!}{(t+a)^{p+k}}.$$

For $m, a \in \mathbf{N}_0$, by (2.22) and (2.75), we have

$$\sum_{k=m}^\infty \frac{1}{(k+a)^p} = \frac{1}{(p-1)(m+a)^{p-1}} + \frac{1}{2(m+a)^p}$$

$$- \sum_{k=1}^q \frac{B_{2k}}{2k} \frac{\binom{-p}{2k-1}}{(m+a)^{p+2k-1}} - \frac{\varepsilon_q B_{2q+2}}{2q+2} \frac{\binom{-p}{2q+1}}{(m+a)^{p+2q+1}}, (2.77)$$

where $0 < \varepsilon_q < 1$. In particular, if $a = 1$, replacing $m + 1$ by m, we find (see Yang [153])

$$\sum_{k=m}^\infty \frac{1}{k^p} = \frac{1}{(p-1) m^{p-1}} + \frac{1}{2 m^p}$$

$$- \sum_{k=1}^q \frac{B_{2k}}{2k} \frac{\binom{-p}{2k-1}}{m^{p+2k-1}} - \frac{\varepsilon_q B_{2q+2}}{2q+2} \frac{\binom{-p}{2q+1}}{m^{p+2q+1}}, \quad 0 < \varepsilon_q < 1 \ (m \in \mathbf{N}).$$

$$(2.78)$$

2.4.2   One Class of Finite Sum Estimation on Divergence Series

For $m \in \mathbf{N}_0$, setting $\delta_q(m) = \delta_q(m, \infty)$. If $F'(t) = f(t)$, then, we define the constant β_m by

$$\beta_m = -F(m) + \frac{1}{2}f(m) - \sum_{k=1}^{q} \frac{B_{2k}}{(2k)!} f^{(2k-1)}(m) + \delta_q(m). \qquad (2.79)$$

By (2.16), when $m < n$, we find

$$\sum_{k=m}^{n} f(k) = F(n) + \frac{1}{2}f(n) + \sum_{k=1}^{q} \frac{B_{2k}}{(2k)!} f^{(2k-1)}(n) + \beta_m - \delta_q(n), \qquad (2.80)$$

where

$$\delta_q(n) = \frac{1}{(2q+1)!} \int_{n}^{\infty} P_{2q+1}(t) \, f^{(2q+1)}(t) \, dt, \qquad (2.81)$$

since $\delta_q(\infty) = 0$, it follows that

$$\beta_m = \lim_{n \to \infty} \left[\sum_{k=m}^{n} f(k) - F(n) - \frac{1}{2} f(n) - \sum_{k=1}^{q} \frac{B_{2k}}{(2k)!} f^{(2k-1)}(n) \right]$$

$$= \sum_{k=m}^{n} f(k) - F(n) - \frac{1}{2}f(n) - \sum_{k=1}^{q} \frac{B_{2k}}{(2k)!} f^{(2k-1)}(n) + \delta_q(n). \qquad (2.82)$$

Note. If both $\sum_{k=m}^{\infty} f(k)$ and $\int_{m}^{\infty} f(x)dx$ are convergent and
$$f^{(2k-1)}(\infty) = 0 \quad (k = 1, 2, \cdots, q),$$
then, by (2.79), $\sum_{k=m}^{\infty} f(k) = \beta_m + F(\infty)$. Hence, in view of (2.78), we still have (2.72).

Example 2.7. If $f(t) = \frac{\ln t}{t-1}$ $(t > 0)$, since, for $n > 1$, we find

$$\int_{1}^{n} f(t) \, dt = \int_{1}^{n} \frac{\ln t}{t - 1} \, dt = \int_{\frac{1}{n}}^{1} \frac{\ln u}{(u - 1)u} \, du$$

$$= -\int_{\frac{1}{n}}^{1} + \frac{\ln u}{u} \, du + \int_{\frac{1}{n}}^{1} \frac{\ln u}{u - 1} \, du$$

$$= \frac{1}{2}(\ln n)^2 + \sum_{k=1}^{\infty} \int_{\frac{1}{n}}^{1} \frac{(1 - u)^{k-1}}{k} \, du$$

$$= \frac{1}{2}(\ln n)^2 + \sum_{k=1}^{\infty} \frac{1}{k^2} \left(1 - \frac{1}{n} \right)^k,$$

$$f'(n) = \frac{1}{n(n - 1)} - \frac{\ln n}{(n - 1)^2},$$

$$f'''(n) = \frac{2}{n^3(n - 1)} + \frac{3}{n^2(n - 1)^2} + \frac{6}{n(n - 1)^3} - \frac{6 \ln n}{(n - 1)^4},$$

then, by (2.79) and (2.75) (for $m = q = 1$), we have

$$\sum_{k=1}^{n} \frac{\ln k}{k-1} = \frac{1}{2}(\ln n)^2 + \sum_{k=1}^{\infty} \frac{1}{k^2}\left(1 - \frac{1}{n}\right)^k + \frac{1}{2}\frac{\ln n}{n-1}$$

$$+ \frac{1}{12}\frac{1}{n(n-1)} - \frac{\ln n}{12(n-1)^2} + \beta_1$$

$$- \frac{\varepsilon}{720}\left[\frac{2}{n^3(n-1)} + \frac{3}{n^2(n-1)^2} + \frac{6}{n(n-1)^3} - \frac{6\ln n}{(n-1)^4}\right],$$

$$\beta_1 = \lim_{n \to \infty}\left[\sum_{k=1}^{n} \frac{\ln k}{k-1} - \frac{1}{2}(\ln n)^2 - \sum_{k=1}^{\infty} \frac{1}{k^2}\left(1 - \frac{1}{n}\right)^k\right]$$

$$= \sum_{k=1}^{n} \frac{\ln k}{k-1} - \frac{1}{2}(\ln n)^2 - \sum_{k=1}^{\infty} \frac{1}{k^2}\left(1 - \frac{1}{n}\right)^k - \frac{1}{2}\frac{\ln n}{n-1}$$

$$- \frac{1}{12}\frac{1}{n(n-1)} + \frac{\ln n}{12(n-1)^2}$$

$$+ \frac{\varepsilon}{720}\left[\frac{2}{n^3(n-1)} + \frac{3}{n^2(n-1)^2} + \frac{6}{n(n-1)^3} - \frac{6\ln n}{(n-1)^4}\right],$$

$$(2.83)$$

where $0 < \varepsilon < 1$. If $n = 3$ in (2.83), we find

$$-\sum_{k=1}^{\infty} \frac{1}{k^2}\left(\frac{2}{3}\right)^k < -\sum_{k=1}^{20} \frac{1}{k^2}\left(\frac{2}{3}\right)^k,$$

$$-\sum_{k=1}^{\infty} \frac{1}{k^2}\left(\frac{2}{3}\right)^k = -\sum_{k=1}^{20} \frac{1}{k^2}\left(\frac{2}{3}\right)^k - \sum_{k=21}^{\infty} \frac{1}{k^2}\left(1 - \frac{1}{3}\right)^k$$

$$> -\sum_{k=1}^{20} \frac{1}{k^2}\left(\frac{2}{3}\right)^k - \int_{20}^{\infty} \frac{1}{x^2}\left(\frac{2}{3}\right)^x dx$$

$$> -\sum_{k=1}^{20} \frac{1}{k^2}\left(\frac{2}{3}\right)^k - \left(\frac{2}{3}\right)^{20} \int_{20}^{\infty} \frac{1}{x^2} dx$$

$$= -\sum_{k=1}^{20} \frac{1}{k^2}\left(\frac{2}{3}\right)^k - \frac{1}{20}\left(\frac{2}{3}\right)^{20}. \qquad (2.84)$$

In view of (2.83) and (2.84), we can obtain $0.539902 < \beta_1 < 0.539976$, and then, it follows that $\beta_1 = 0.5399^+$.

Example 2.8. If $0 < a \leq 1$, $f(t) = \frac{1}{t+a}$ $(t \geq 0)$, $F(t) = \ln(t + a)$, then

$$f^{(k)}(t) = \frac{(-1)^k}{(t+a)^{k+1}} k!.$$

When $m = 0$ in (2.76), we find (see [154] and [155]) that

$$\beta_0 = \lim_{n \to \infty} \left[\sum_{k=0}^{n} f(k) - F(n) \right]$$

$$= \lim_{n \to \infty} \left[\sum_{k=0}^{n} \frac{1}{k+a} - \ln(n+a) \right] = \gamma_0(a). \tag{2.85}$$

We call $\gamma_0(a)$ the Stieltjes constant (see Pan and Pan [62]).

By (2.80), (2.81) and (2.75), for $n \in \mathbf{N}$, we find (see Yang and Li [154]) that

$$\sum_{k=0}^{n} \frac{1}{k+a} = \gamma_0(a) + \ln(n+a) + \frac{1}{2(n+a)}$$

$$- \sum_{k=1}^{q} \frac{B_{2k}}{2k(n+a)^{2k}} - \frac{\varepsilon_q B_{2q+2}}{2(q+1)(n+a)^{2q+2}}, \quad 0 < \varepsilon_q < 1. \tag{2.86}$$

In particular, for $a = 1$, replacing $n+1$ by n, we have the following estimation of the harmonic series (see Yang and Wang [155]):

$$\sum_{k=1}^{n} \frac{1}{k} = \gamma + \ln n + \frac{1}{2n} - \sum_{k=1}^{q} \frac{B_{2k}}{2 k\, n^{2k}} - \frac{\varepsilon_q B_{2q+2}}{2(q+1)\, n^{2q+2}}, \quad 0 < \varepsilon_q < 1, \tag{2.87}$$

where $\gamma\ (= \gamma_0(1) = 0.5772156649^{+})$ is called the *Euler constant*.

Example 2.9. If $s \in \mathbf{R}\backslash\{1\}$, $0 < a \le 1$, $f(t) = \frac{1}{(t+a)^s}$ $(t \ge 0)$, setting $\zeta(s,a) = \beta_0$, then, by (2.80), (2.81) and (2.75), for $n \in \mathbf{N}$, $q \ge \frac{1-s}{2}$, we have an estimation of the Hurwitz ζ–function $\zeta(s,a)$ in the real axis as follows (see Titchmarsh [71]):

$$\zeta(s,a) = \sum_{k=0}^{n} \frac{1}{(k+a)^s} - \frac{1}{1-s}(n+a)^{1-s} - \frac{1}{2(n+a)^s}$$

$$- \sum_{k=1}^{q} \frac{\binom{-s}{2k-1} B_{2k}}{2k(n+a)^{s+2k-1}} - \frac{\varepsilon_q \binom{-s}{2q+1} B_{2q+2}}{2(q+1)(n+a)^{s+2q+1}}, \quad 0 < \varepsilon_q < 1. \tag{2.88}$$

In particular, for $a = 1$, $q \ge \frac{1-s}{2}$, we have an estimation of the Riemann ζ–function, $\zeta(s) = \zeta(s,1)$ in the real axis as follows (see Zhu and Yang [180]):

$$\zeta(s) = \sum_{k=1}^{n} \frac{1}{k^s} - \frac{1}{1-s} n^{1-s} - \frac{1}{2n^s}$$

$$- \sum_{k=1}^{q} \frac{\binom{-s}{2k-1} B_{2k}}{2k\, n^{s+2k-1}} - \frac{\varepsilon_q \binom{-s}{2q+1} B_{2q+2}}{2(q+1)\, n^{s+2q+1}}, \quad 0 < \varepsilon_q < 1. \tag{2.89}$$

Example 2.10. If $0 < a \leq 1$, $f(t) = \ln(t+a)$ $(t \geq 0)$, and $F(t) = (t+a)\ln(t+a) - t$, then,

$$f^{(k)}(t) = \frac{(-1)^{k-1}(k-1)!}{(t+a)^k}.$$

Using (2.80), (2.81) and (2.75), we obtain

$$\beta_0(a) = \lim_{n\to\infty} \left[\sum_{k=0}^{n} \ln(k+a) - \left(n+a+\frac{1}{2}\right)\ln(n+a) + n \right],$$

$$ln \prod_{k=0}^{n}(k+a) = \sum_{k=0}^{n} \ln(k+a) = \beta_0(a) + \left(n+a+\frac{1}{2}\right)\ln(n+a) - n$$

$$+ \sum_{k=1}^{q} \frac{B_{2k}}{2k(2k-1)(n+a)^{2k-1}} + \frac{\varepsilon_q B_{2q+2}}{2(q+1)(2q+1)(n+a)^{2q+1}},$$

$$(2.90)$$

where $0 < \varepsilon_q < 1$. Equivalently, since $e^{\beta_0(a)} = \frac{\sqrt{\pi}}{\Gamma(a)}$ (see Xie [93]), we find

$$\prod_{k=0}^{n}(k+a) = \frac{\sqrt{\pi}}{\Gamma(a)}\sqrt{n+a}\left(\frac{n+a}{e}\right)^{n+a}$$

$$\times \exp\left\{ \sum_{k=1}^{q} \frac{B_{2k}}{2k(2k-1)(n+a)^{2k-1}} + \frac{\varepsilon_q B_{2q+2}}{2(q+1)(2q+1)(n+a)^{2q+1}} \right\},$$

$$(2.91)$$

where $0 < \varepsilon_q < 1$. In particular, for $a = 1$, replacing $n+1$ by n, we have the following extended Stirling formula (see Knopp [51] and Yang [156]):

$$n! = \sqrt{2\pi n}\left(\frac{n}{e}\right)^n \exp\left\{ \sum_{k=1}^{q} \frac{B_{2k}}{2k(2k-1)\,n^{2k-1}} + \frac{\varepsilon_q\, B_{2q+2}}{2(q+1)(2q+1)\,n^{2q+1}} \right\}.$$

$$(2.92)$$

When $q = 1$, we have the following inequality:

$$\sqrt{2\pi n}\left(\frac{n}{e}\right)^n \exp\left(\frac{1}{12n}\right) < n! < \sqrt{2\pi n}\left(\frac{n}{e}\right)^n \exp\left\{ \frac{1}{12n}\left(1 - \frac{1}{30n^2}\right) \right\}.$$

$$(2.93)$$

Chapter 3

A Half-Discrete Hilbert-Type Inequality with a General Homogeneous Kernel

"One of the properties inherent in mathematics is that any real progress is accompanied by the discovery and development of new methods and simplifications of previous procedures.... The unified character of mathematics lies in its very nature; indeed, mathematics is the foundation of all exact natural sciences."

David Hilbert

"In great mathematics there is a very high degree of unexpectedness, combined with inevitability and economy."

G. H. Hardy

3.1 Introduction

This chapter is devoted to half-discrete Hilbert-type inequalities with general homogeneous kernels and their many extensions based on the way of weight functions and techniques of real analysis. Special attention is given to the proofs of the best possible constant factors of half-discrete Hilbert-type inequalities. Included are several equivalent inequalities, operator expressions, and the reverses of the Hilbert-type inequalities with many extensions and particular examples.

3.2   Some Preliminary Lemmas

3.2.1   *Definition of Weight Functions and Related Lemmas*

Lemma 3.1. *Suppose that* λ, λ_1, $\lambda_2 \in \mathbf{R}$, $\lambda = \lambda_1 + \lambda_2$, $k_\lambda(x,y)$ *is a non-negative finite measurable homogeneous function of degree* $-\lambda$ *in* $\mathbf{R}^2_+$, *satisfying*

$$k_\lambda(ux, uy) = u^{-\lambda} \, k_\lambda(x,y), \quad u, x, y > 0,$$

and $v(y)$ *is a strict increasing differentiable function in* $[n_0, \infty)(n_0 \in \mathbf{N})$ *with* $v(n_0) > 0$ *and* $v(\infty) = \infty$. *Define two weight functions* $\omega_{\lambda_2}(n)$ *and* $\varpi_{\lambda_1}(x)$ *as follows:*

$$\omega_{\lambda_2}(n) = [v(n)]^{\lambda_2} \int_0^\infty k_\lambda(x, v(n)) x^{\lambda_1-1} dx, \quad n \geq n_0 \quad (n \in \mathbf{N}), \quad (3.1)$$

$$\varpi_{\lambda_1}(x) = x^{\lambda_1} \sum_{n=n_0}^\infty k_\lambda(x, v(n))[v(n)]^{\lambda_2-1} v'(n), \quad x \in (0, \infty). \quad (3.2)$$

Then we have

$$\omega_{\lambda_2}(n) = k(\lambda_1) = \int_0^\infty k_\lambda(t,1) \, t^{\lambda_1-1} dt = \int_0^\infty k_\lambda(1,t) \, t^{\lambda_2-1} dt. \quad (3.3)$$

Moreover, we set

$$f(x,y) = x^{\lambda_1} k_\lambda(x, v(y))[v(y)]^{\lambda_2-1} \, v'(y)$$

and the following conditions:

Condition (i). $v(y)$ is strictly increasing in $[n_0 - 1, \infty)$ with $v(n_0 - 1) \geq 0$, and for any fixed $x > 0$, $f(x,y)$ is decreasing with respect to $y \in [n_0 - 1, \infty)$ and strictly decreasing in an interval $I \subset (n_0 - 1, \infty)$.

Condition (ii). $v(y)$ is strictly increasing in $[n_0 - \frac{1}{2}, \infty)$ with $v(n_0 - \frac{1}{2}) \geq 0$, and for any fixed $x > 0$, $f(x,y)$ is convex with respect to $y \in [n_0 - \frac{1}{2}, \infty)$ and strictly convex in an interval $\widetilde{I} \subset (n_0 - \frac{1}{2}, \infty)$.

Condition (iii). There exists a constant $\beta > 0$, such that $v(y)$ is strictly increasing in $[n_0 - \beta, \infty)$ with $v(n_0 - \beta) \geq 0$, and for fixed $x > 0$, $f(x,y)$ is a piecewise smooth continuous function with respect to $y \in [n_0 - \beta, \infty)$, satisfying

$$R(x) = \int_{n_0-\beta}^{n_0} f(x,y) dy - \frac{1}{2} f(x, n_0) - \int_{n_0}^\infty P_1(y) f'_y(x,y) dy > 0, \quad (3.4)$$

where $P_1(y)(= y - [y] - \frac{1}{2})$ is the Bernoulli function of order one (see equation (2.8) in Chapter 2).

If $k(\lambda_1) \in \mathbf{R}_+$ and one of the above three conditions is satisfied, then we have

$$\varpi_{\lambda_1}(x) < k(\lambda_1), \quad x \in (0, \infty). \tag{3.5}$$

Proof. Setting $t = \frac{x}{v(n)}$ in (3.1), since $\lambda = \lambda_1 + \lambda_2$, we find

$$\omega_{\lambda_2}(n) = [v(n)]^{\lambda_2} \int_0^\infty k_\lambda \left(tv(n), v(n)\right) (tv(n))^{\lambda_1 - 1} v(n) dt$$

$$= \int_0^\infty k_\lambda(t, 1) \, t^{\lambda_1 - 1} dt = k(\lambda_1).$$

Setting $t = \frac{v(n)}{x}$ in (3.1), we also obtain

$$\omega_{\lambda_2}(n) = [v(n)]^{\lambda_2} \int_0^\infty k_\lambda \left(\frac{v(n)}{t}, v(n) \right) \left(\frac{v(n)}{t} \right)^{\lambda_1 - 1} \frac{v(n)}{t^2} \, dt$$

$$= \int_0^\infty k_\lambda(1, t) \, t^{\lambda_2 - 1} dt,$$

and then (3.3) follows.

(i) If Condition (i) is satisfied, then for any $n \geq n_0$, we have

$$f(x, n) \leq \int_{n-1}^n f(x, y) dy,$$

and there exists an integer $n_1 \geq n_0$ such that

$$f(x, n_1) < \int_{n_1 - 1}^{n_1} f(x, y) dy.$$

Since $v(n_0 - 1) \geq 0$ and $k(\lambda_1) \in R_+$, we find

$$\varpi_{\lambda_1}(x) = \sum_{n=n_0}^\infty f(x, n) < \sum_{n=n_0}^\infty \int_{n-1}^n f(x, y) dy$$

$$= \int_{n_0 - 1}^\infty f(x, y) dy = x^{\lambda_1} \int_{n_0 - 1}^\infty k_\lambda(x, v(y))[v(y)]^{\lambda_2 - 1} v'(y) dy$$

$$= \int_{\frac{v(n_0 - 1)}{x}}^\infty k_\lambda(1, t) \, t^{\lambda_2 - 1} dt \quad (t = v(y)/x)$$

$$\leq \int_0^\infty k_\lambda(1, t) \, t^{\lambda_2 - 1} dt = k(\lambda_1),$$

and then (3.5) follows.

(ii) If Condition (ii) is satisfied, then, by Hermite-Hadamard's inequality (see Kuang [47]), we have

$$f(x, n) \leq \int_{n - \frac{1}{2}}^{n + \frac{1}{2}} f(x, y) dy, n \geq n_0,$$

and there exists a positive integer $n_2 \geq n_0$ such that

$$f(x, n_2) < \int_{n_2-\frac{1}{2}}^{n_2+\frac{1}{2}} f(x, y)dy.$$

Since $v(n_0 - \frac{1}{2}) \geq 0$ and $k(\lambda_1) \in R_+$, we find

$$\begin{aligned}
\varpi_{\lambda_1}(x) &= \sum_{n=n_0}^{\infty} f(x, n) < \sum_{n=n_0}^{\infty} \int_{n-\frac{1}{2}}^{n+\frac{1}{2}} f(x, y)dy \\
&= \int_{n_0-\frac{1}{2}}^{\infty} f(x, y)dy \\
&= x^{\lambda_1} \int_{n_0-\frac{1}{2}}^{\infty} k_{\lambda}(x, v(y))[v(y)]^{\lambda_2-1}v'(y)dy \\
&= \int_{\frac{v(n_0-\frac{1}{2})}{x}}^{\infty} k_{\lambda}(1, t)t^{\lambda_2-1}dt \leq \int_0^{\infty} k_{\lambda}(1, t)t^{\lambda_2-1}dt = k(\lambda_1),
\end{aligned}$$

and then (3.5) follows.

(iii) If Condition (iii) is satisfied and $k(\lambda_1) \in \mathbf{R}_+$, then by the Euler-Maclaurin summation formula (see equation (2.14) in Chapter 2) for $q = 0$, $m = n_0$, $n \to \infty$, we have

$$\begin{aligned}
\varpi_{\lambda_1}(x) &= \sum_{n=n_0}^{\infty} f(x, n) = \int_{n_0}^{\infty} f(x, y)dy + \frac{1}{2}f(x, n_0) + \int_{n_0}^{\infty} P_1(y)f'_y(x, y)dy \\
&= \int_{n_0-\beta}^{\infty} f(x, y)dy - R(x) = \int_{\frac{v(n_0-\beta)}{x}}^{\infty} k_{\lambda}(1, t)\, t^{\lambda_2-1}dt - R(x) \\
&\leq k(\lambda_1) - R(x) < k(\lambda_1),
\end{aligned}$$

and then (3.5) follows. $\square$

Lemma 3.2. *Let the assumptions of Lemma 3.1 be fulfilled and additionally, let $p \in \mathbf{R}\backslash\{0, 1\}$, $\frac{1}{p} + \frac{1}{q} = 1$, $a_n \geq 0$, $n \geq n_0(n \in \mathbf{N})$, $f(x)$ be a non-negative measurable function in $\mathbf{R}_+$. Then*

(i) for $p > 1$, we have the following inequalities:

$$J = \left\{ \sum_{n=n_0}^{\infty} \frac{v'(n)}{[v(n)]^{1-p\lambda_2}} \left[\int_0^{\infty} k_{\lambda}(x, v(n))f(x)dx \right]^p \right\}^{\frac{1}{p}}$$

$$\leq [k(\lambda_1)]^{\frac{1}{q}} \left\{ \int_0^{\infty} \varpi_{\lambda_1}(x)x^{p(1-\lambda_1)-1} f^p(x)dx \right\}^{\frac{1}{p}}, \tag{3.6}$$

and

$$\widetilde{L} = \left\{ \int_0^\infty \frac{[\varpi_{\lambda_1}(x)]^{1-q}}{x^{1-q\lambda_1}} \left[\sum_{n=n_0}^\infty k_\lambda(x, v(n)) a_n \right]^q dx \right\}^{\frac{1}{q}}$$

$$\leq \left\{ k(\lambda_1) \sum_{n=n_0}^\infty \frac{[v(n)]^{q(1-\lambda_2)-1}}{[v'(n)]^{q-1}} a_n^q \right\}^{\frac{1}{q}} ; \tag{3.7}$$

(ii) for $p < 0$ or $0 < p < 1$, we have the reverses of (3.6) and (3.7)
Note. *If $a_n = 0$, then $a_n^q = 0$, for any $q \in \mathbf{R}\backslash\{0, 1\}$.*

Proof. (i) For $p > 1$, by Hölder's inequality with weight (see Kuang [47]) and (3.3), it follows that

$$\left[\int_0^\infty k_\lambda(x, v(n)) f(x) \, dx \right]^p$$

$$= \left\{ \int_0^\infty k_\lambda(x, v(n)) \left[\frac{x^{(1-\lambda_1)/q} [v'(n)]^{1/p}}{[v(n)]^{(1-\lambda_2)/p}} f(x) \right] \left[\frac{[v(n)]^{(1-\lambda_2)/p}}{x^{(1-\lambda_1)/q} [v'(n)]^{1/p}} \right] dx \right\}^p$$

$$\leq \int_0^\infty k_\lambda(x, v(n)) \frac{x^{(1-\lambda_1)(p-1)} v'(n)}{[v(n)]^{1-\lambda_2}} f^p(x) \, dx$$

$$\times \left\{ \int_0^\infty k_\lambda(x, v(n)) \frac{[v(n)]^{(1-\lambda_2)(q-1)}}{x^{1-\lambda_1} [v'(n)]^{q-1}} dx \right\}^{p-1}$$

$$= \left\{ \frac{\omega_{\lambda_2}(n)[v(n)]^{q(1-\lambda_2)-1}}{[v'(n)]^{q-1}} \right\}^{p-1} \int_0^\infty k_\lambda(x, v(n)) \frac{x^{(1-\lambda_1)(p-1)} v'(n)}{[v(n)]^{1-\lambda_2}} f^p(x) \, dx$$

$$= \frac{[k(\lambda_1)]^{p-1}}{[v(n)]^{p\lambda_2-1} v'(n)} \int_0^\infty k_\lambda(x, v(n)) \frac{x^{(1-\lambda_1)(p-1)} v'(n)}{[v(n)]^{1-\lambda_2}} f^p(x) \, dx.$$

Then, by the Lebesgue term by term integration theorem (see Kuang [49]), we have

$$J \leq [k(\lambda_1)]^{\frac{1}{q}} \left\{ \sum_{n=n_0}^\infty \int_0^\infty k_\lambda(x, v(n)) \frac{x^{(1-\lambda_1)(p-1)} v'(n)}{[v(n)]^{1-\lambda_2}} f^p(x) \, dx \right\}^{\frac{1}{p}}$$

$$= [k(\lambda_1)]^{\frac{1}{q}} \left\{ \int_0^\infty \sum_{n=n_0}^\infty k_\lambda(x, v(n)) \frac{x^{(1-\lambda_1)(p-1)} v'(n)}{[v(n)]^{1-\lambda_2}} f^p(x) \, dx \right\}^{\frac{1}{p}}$$

$$= [k(\lambda_1)]^{\frac{1}{q}} \left\{ \int_0^\infty \varpi_{\lambda_1}(x) \, x^{p(1-\lambda_1)-1} f^p(x) dx \right\}^{\frac{1}{p}},$$

and then (3.6) follows.

Hence, by Hölder's inequality with weight, it follows that

$$
\left[\sum_{n=n_0}^{\infty} k_\lambda(x, v(n)) a_n \right]^q
$$

$$
= \left\{ \sum_{n=n_0}^{\infty} k_\lambda(x, v(n)) \left[\frac{x^{(1-\lambda_1)/q}[v'(n)]^{1/p}}{[v(n)]^{(1-\lambda_2)/p}} \right] \left[\frac{[v(n)]^{(1-\lambda_2)/p} a_n}{x^{(1-\lambda_1)/q}[v'(n)]^{1/p}} \right] \right\}^q
$$

$$
\leq \left\{ \sum_{n=n_0}^{\infty} k_\lambda(x, v(n)) \frac{x^{(1-\lambda_1)(p-1)} v'(n)}{[v(n)]^{1-\lambda_2}} \right\}^{q-1}
$$

$$
\times \sum_{n=n_0}^{\infty} k_\lambda(x, v(n)) \frac{[v(n)]^{(1-\lambda_2)(q-1)}}{x^{1-\lambda_1}[v'(n)]^{q-1}} a_n^q
$$

$$
= \frac{x^{1-q\lambda_1}}{[\varpi_{\lambda_1}(x)]^{1-q}} \sum_{n=n_0}^{\infty} k_\lambda(x, v(n)) \frac{[v(n)]^{(1-\lambda_2)(q-1)}}{x^{1-\lambda_1}[v'(n)]^{q-1}} a_n^q.
$$

Then, by the Lebesgue term by term integration theorem, we have

$$
\widetilde{L} \leq \left\{ \int_0^\infty \sum_{n=n_0}^{\infty} k_\lambda(x, v(n)) \frac{[v(n)]^{(1-\lambda_2)(q-1)}}{x^{1-\lambda_1}[v'(n)]^{q-1}} a_n^q \, dx \right\}^{\frac{1}{q}}
$$

$$
= \left\{ \sum_{n=n_0}^{\infty} \int_0^\infty k_\lambda(x, v(n)) \frac{[v(n)]^{(1-\lambda_2)(q-1)}}{x^{1-\lambda_1}[v'(n)]^{q-1}} a_n^q \, dx \right\}^{\frac{1}{q}}
$$

$$
= \left\{ \sum_{n=n_0}^{\infty} \omega_{\lambda_2}(n) \frac{[v(n)]^{q(1-\lambda_2)-1}}{[v'(n)]^{q-1}} a_n^q \right\}^{\frac{1}{q}},
$$

and then, in view of (3.3), we have (3.7).

(ii) For $p < 0$ or $0 < p < 1$, by the reverse Hölder's inequality with weight (see Kuang [47]) and in the same way, we obtain the reverses of (3.6) and (3.7). $\qquad\square$

3.2.2   *Estimations about Two Series*

Lemma 3.3. *Let $v(y)$ be a strictly increasing differentiable functions in $[n_0, \infty)$ $(n_0 \in \mathbf{N})$ with $v(n_0) > 0$ and $v(\infty) = \infty$, and $\frac{v'(y)}{v(y)}(> 0)$ be decreasing in $[n_0, \infty)$. Then, for $\varepsilon > 0$, we have*

$$
A(\varepsilon) = \sum_{n=n_0}^{\infty} \frac{v'(n)}{[v(n)]^{1+\varepsilon}} = \frac{1}{\varepsilon}(1 + o(1)) \quad (\varepsilon \to 0^+). \tag{3.8}
$$

Proof.   In view of the assumptions, $\frac{v'(y)}{[v(y)]^{1+\varepsilon}}(=\frac{v'(y)}{[v(y)]}\cdot\frac{1}{[v(y)]^{\varepsilon}})$ is still decreasing in $[n_0, \infty)$, we find

$$\frac{1}{\varepsilon}(1+o_1(1)) = \frac{1}{\varepsilon[v(n_0)]^{\varepsilon}} = \int_{n_0}^{\infty} \frac{v'(y)}{[v(y)]^{1+\varepsilon}}dy$$

$$\leq A(\varepsilon) = \frac{v'(n_0)}{[v(n_0)]^{1+\varepsilon}} + \sum_{n=n_0+1}^{\infty} \frac{v'(n)}{[v(n)]^{1+\varepsilon}}$$

$$\leq \frac{v'(n_0)}{[v(n_0)]^{1+\varepsilon}} + \int_{n_0}^{\infty} \frac{v'(y)}{[v(y)]^{1+\varepsilon}}dy$$

$$= \frac{1}{\varepsilon}\left\{\frac{1}{[v(n_0)]^{\varepsilon}} + \frac{\varepsilon v'(n_0)}{[v(n_0)]^{1+\varepsilon}}\right\}$$

$$= \frac{1}{\varepsilon}(1+o_2(1)) \quad (\varepsilon \to 0^+),$$

and then, we have equality (3.8). □

Lemma 3.4. *Let the assumptions of Lemma 3.1 be fulfilled and additionally, let $p \in \mathbf{R}\backslash\{0,1\}$, $\frac{1}{p} + \frac{1}{q} = 1$, $\frac{v'(y)}{v(y)}(> 0)$ be decreasing in $[n_0, \infty)$. If there exist constants $\delta < \lambda_1$ and $L > 0$, such that*

$$k_\lambda(t,1) \leq L\left(\frac{1}{t^\delta}\right), \quad t \in \left(0, \frac{1}{v(n_0)}\right] \tag{3.9}$$

then for $0 < \varepsilon < \min\{|p|, |q|\}(\lambda_1 - \delta)$, we have

$$B(\varepsilon) = \sum_{n=n_0}^{\infty} \frac{v'(n)}{[v(n)]^{1+\varepsilon}} \int_0^{\frac{1}{v(n)}} k_\lambda(t,1)t^{\lambda_1-\frac{\varepsilon}{p}-1}dt = O(1)(\varepsilon \to 0^+). \tag{3.10}$$

Proof.   In view of (3.9), and since

$$\frac{v'(y)}{[v(y)]^{\lambda_1-\delta+\frac{\varepsilon}{q}+1}} = \frac{v'(y)}{v(y)} \cdot \frac{1}{[v(y)]^{\lambda_1-\delta+\frac{\varepsilon}{q}}}$$

is still decreasing in $[n_0, \infty)$, we find

$$0 < B(\varepsilon) \leq L \sum_{n=n_0}^{\infty} \frac{v'(n)}{[v(n)]^{1+\varepsilon}} \int_0^{\frac{1}{v(n)}} t^{\lambda_1-\delta-\frac{\varepsilon}{p}-1}dt$$

$$= \frac{L}{\lambda_1 - \delta - \frac{\varepsilon}{p}} \sum_{n=n_0}^{\infty} \frac{v'(n)}{[v(n)]^{\lambda_1-\delta+\frac{\varepsilon}{q}+1}}$$

$$= \frac{L}{\lambda_1 - \delta - \frac{\varepsilon}{p}} \left[\frac{v'(n_0)}{[v(n_0)]^{\lambda_1-\delta+\frac{\varepsilon}{q}+1}} + \sum_{n=n_0+1}^{\infty} \frac{v'(n)}{[v(n)]^{\lambda_1-\delta+\frac{\varepsilon}{q}+1}}\right]$$

$$\leq \frac{L}{\lambda_1 - \delta - \frac{\varepsilon}{p}} \left[\frac{v'(n_0)}{[v(n_0)]^{\lambda_1-\delta+\frac{\varepsilon}{q}+1}} + \int_{n_0}^{\infty} \frac{v'(y)}{[v(y)]^{\lambda_1-\delta+\frac{\varepsilon}{q}+1}}dy\right]$$

$$= \frac{L}{\lambda_1 - \delta - \frac{\varepsilon}{p}} \left[\frac{v'(n_0)}{[v(n_0)]^{\lambda_1-\delta+\frac{\varepsilon}{q}+1}} + \frac{[v(n_0)]^{-\lambda_1+\delta-\frac{\varepsilon}{q}}}{\lambda_1 - \delta + \frac{\varepsilon}{q}}\right] < \infty,$$

and then (3.10) follows.　　　　　　　　　　　　　　　　　$\square$

Lemma 3.5. *If C is the set of complex numbers and $C_\infty = C \cup \{\infty\}$, $z_k \in C\{z | \text{Re} z \geq 0, \text{Im} z = 0\}$ $(k = 1, 2, \cdots, n)$ are distinct points, the function $f(z)$ is analytic in C_∞ except for $z_i (i = 1, 2, \cdots, n)$, and $z = \infty$ is a zero point of $f(z)$ whose order is not less than 1, then for $\alpha \in \mathbf{R}$, we have*

$$\int_0^\infty f(x)\, x^{\alpha-1}\, dx = \frac{2\pi i}{1 - e^{2\pi\alpha i}} \sum_{k=1}^n \text{Res}[f(z) z^{\alpha-1}, z_k], \qquad (3.11)$$

where $0 < \text{Im} \ln z = \arg z < 2\pi$. In particular, if $z_k (k = 1, \cdots, n)$ are all poles of order 1, setting $\varphi_k(z) = (z - z_k)f(z)$, $(\varphi_k(z_k) \neq 0)$, then

$$\int_0^\infty f(x)\, x^{\alpha-1}\, dx = \frac{\pi}{\sin \pi\alpha} \sum_{k=1}^n (-z_k)^{\alpha-1} \varphi_k(z_k). \qquad (3.12)$$

Proof.　　Using results of Pang *et al.* [63] on page 118, we have (3.11). We find

$$1 - e^{2\pi\alpha i} = 1 - \cos 2\pi\alpha - i \sin 2\pi\alpha$$
$$= -2i \sin \pi\alpha(\cos \pi\alpha + i \sin \pi\alpha) = -2i\, e^{i\pi\alpha}\, \sin \pi\alpha.$$

In particular, since $f(z)\, z^{\alpha-1} = \frac{1}{z - z_k}(\varphi_k(z)\, z^{\alpha-1})$, it is obvious that

$$\text{Res}[f(z)\, z^{\alpha-1}, -a_k] = z_k{}^{\alpha-1}\varphi_k(z_k) = -e^{i\pi\alpha}\, (-z_k)^{\alpha-1}\, \varphi_k(z_k).$$

Then, by (3.11), we obtain (3.12).　　　　　　　　　　　　　　　　$\square$

Example 3.1. (i) For $k_\lambda(t, 1) = \frac{1}{t^\lambda + 1}$ $(\lambda, \lambda_1, \lambda_2 > 0, \lambda_1 + \lambda_2 = \lambda)$, we find

$$k(\lambda_1) = \int_0^\infty \frac{1}{t^\lambda + 1}\, t^{\lambda_1-1}\, dt = \frac{1}{\lambda} \int_0^\infty \frac{1}{u + 1}\, u^{\frac{\lambda_1}{\lambda}-1}\, du \qquad (u = t^\lambda)$$
$$= \frac{\pi}{\lambda\, \sin(\frac{\pi\lambda_1}{\lambda})} \in \mathbf{R}_+.$$

Setting $\delta = \frac{\lambda_1}{2}$ $(< \lambda_1 < \lambda)$, it follows that

$$k_\lambda(t, 1) = \frac{1}{t^\lambda + 1} \leq \frac{1}{t^\delta}, \quad \left(t \in \left(0, \frac{1}{v(n_0)}\right]\right).$$

(ii) For $k_\lambda(t, 1) = \frac{1}{(t+1)^\lambda}$ $(\lambda, \lambda_1, \lambda_2 > 0, \lambda_1 + \lambda_2 = \lambda)$, we find

$$k(\lambda_1) = \int_0^\infty \frac{1}{(t + 1)^\lambda}\, t^{\lambda_1-1}\, dt = B(\lambda_1, \lambda_2) \in \mathbf{R}_+.$$

Setting $\delta = \frac{\lambda_1}{2}$ $(< \lambda_1 < \lambda)$, it follows that

$$k_\lambda(t,1) = \frac{1}{(t+1)^\lambda} \le \frac{1}{t^\delta} \quad \left(t \in \left(0, \frac{1}{v(n_0)}\right] \right).$$

(iii) For $k_\lambda(t,1) = \frac{\ln t}{t^\lambda - 1}$ $(\lambda, \lambda_1, \lambda_2 > 0, \lambda_1 + \lambda_2 = \lambda)$, we find

$$k(\lambda_1) = \int_0^\infty \frac{\ln t}{t^\lambda - 1} t^{\lambda_1 - 1} \, dt = \frac{1}{\lambda^2} \int_0^\infty \frac{\ln u}{u - 1} u^{\frac{\lambda_1}{\lambda} - 1} du, \quad (u = t^\lambda)$$

$$= \left[\frac{\pi}{\lambda \, \sin(\frac{\pi \lambda_1}{\lambda})} \right]^2 \in \mathbf{R}_+.$$

Setting $\delta = \frac{\lambda_1}{2}$ $(< \lambda_1 < \lambda)$, in view of $\frac{t^\delta \ln t}{t^\lambda - 1} \to 0$ as $t \to 0^+$, there exists a constant $L > 0$ such that

$$k_\lambda(t,1) = \frac{\ln t}{t^\lambda - 1} \le \frac{L}{t^\delta}, \quad \left(t \in \left(0, \frac{1}{v(n_0)}\right] \right).$$

(iv) For $s \in \mathbf{N}$, $k_{\lambda s}(t,1) = \prod_{k=1}^s \frac{1}{a_k t^\lambda + 1}$ $(0 < a_1 < \cdots < a_s, \lambda, \lambda_1, \lambda_2 > 0, \lambda_1 + \lambda_2 = \lambda s)$, by (3.12), we find

$$k(\lambda_1) = \int_0^\infty \prod_{k=1}^s \frac{1}{a_k t^\lambda + 1} t^{\lambda_1 - 1} \, dt$$

$$= \frac{1}{\lambda} \int_0^\infty \prod_{k=1}^s \frac{1}{a_k + u} u^{\frac{\lambda_2}{\lambda} - 1} \, du$$

$$= \frac{\pi}{\lambda \, \sin(\frac{\pi \lambda_2}{\lambda})} \sum_{k=1}^s a_k^{\frac{\lambda_2}{\lambda} - 1} \prod_{j=1(j \ne k)}^s \frac{1}{a_j - a_k} \in \mathbf{R}_+.$$

Setting $\delta = \frac{\lambda_1}{2}$ $(< \lambda_1 < \lambda s)$, there exists a constant $L = \frac{1}{(v(n_0))^\delta} > 0$

$$k_{\lambda s}(t,1) = \prod_{k=1}^s \frac{1}{a_k t^\lambda + 1} \le 1 \le \frac{L}{t^\delta}, \quad \left(t \in \left(0, \frac{1}{v(n_0)}\right] \right).$$

(v) For $k_\lambda(t,1) = \frac{1}{t^\lambda + 2t^{\lambda/2} \cos \beta + 1}$ $(\lambda > 0, 0 < \beta \le \frac{\pi}{2}, \lambda_1, \lambda_2 > 0, \lambda_1 + \lambda_2 = \lambda)$, setting $z_1 = -e^{i\beta}$, $z_2 = -e^{-i\beta}$, by (3.12), we find

$$k(\lambda_1) = \int_0^\infty \frac{t^{\lambda_1 - 1}}{t^\lambda + 2t^{\lambda/2} \cos \beta + 1} \, dt$$

$$= \frac{2}{\lambda} \int_0^\infty \frac{u^{\frac{2\lambda_1}{\lambda} - 1}}{u^2 + 2u \cos \beta + 1} \, du, \quad \left(u = t^{\lambda/2} \right)$$

$$= \frac{\pi}{\sin \left(\frac{2\pi \lambda_1}{\lambda} \right)} \left[(e^{i\beta})^{\frac{2\lambda_1}{\lambda} - 1} \frac{1}{e^{-i\beta} - e^{i\beta}} + (e^{-i\beta})^{\frac{2\lambda_1}{\lambda} - 1} \frac{1}{e^{i\beta} - e^{-i\beta}} \right]$$

$$= \frac{2\pi \sin \beta (1 - \frac{2\lambda_1}{\lambda})}{\lambda \sin \beta \sin \left(\frac{2\pi \lambda_1}{\lambda} \right)} \in \mathbf{R}_+.$$

Setting $\delta = \frac{\lambda_1}{2}$ ($< \lambda_1 < \lambda$), in view of $\frac{t^\delta}{t^\lambda + 2t^{\lambda/2}\cos\beta + 1} \to 0$ as $t \to 0^+$, there exists a constant $L > 0$ such that

$$k_\lambda(t,1) = \frac{1}{t^\lambda + 2t^{\lambda/2}\cos\beta + 1} \le L\left(\frac{1}{t^\delta}\right) \ \left(t \in \left(0, \frac{1}{v(n_0)}\right]\right).$$

(vi) For $k_\lambda(t,1) = \frac{1}{t^\lambda + bt^{\lambda/2} + c}$ ($c > 0$, $0 \le b \le 2\sqrt{c}$, $\lambda > 0$, $\lambda_1 = \lambda_2 = \frac{\lambda}{2}$), we find

$$k(\lambda_1) = \int_0^\infty \frac{t^{\frac{\lambda}{2}-1}dt}{t^\lambda + bt^{\lambda/2} + c} = \frac{2}{\lambda}\int_0^\infty \frac{du}{u^2 + bu + c}, \qquad (u = t^{\lambda/2})$$

$$= \frac{2}{\lambda}\begin{cases} \frac{\pi}{\sqrt{c}}, & b = 0, \\ \frac{4}{\sqrt{4c-b^2}}\arctan\frac{\sqrt{4c-b^2}}{b}, & 0 < b < 2\sqrt{c}, \\ \frac{2}{\sqrt{c}}, & b = 2\sqrt{c}. \end{cases}$$

Setting $\delta = \frac{\lambda}{4}$ ($< \lambda_1 < \lambda$), in view of $\frac{t^\delta}{t^\lambda + bt^{\lambda/2} + c} \to 0$ as $t \to 0^+$, there exists a constant $L > 0$ such that

$$k_\lambda(t,1) = \frac{1}{t^\lambda + bt^{\lambda/2} + c} \le L\left(\frac{1}{t^\delta}\right), \ \left(t \in \left(0, \frac{1}{v(n_0)}\right]\right).$$

(vii) For $k_0(t,1) = \frac{1}{e^{\gamma/t^\lambda} - Ae^{-\gamma/t^\lambda}}$, ($-1 \le A < 1$, $\beta > 0$ ($A = 1$, $\beta > 1$), $\gamma > 0$, $\lambda_1 = -\lambda_2 = -\beta\lambda < 0 < \lambda$), in view of the following formula (see Wang and Guo [73])

$$\int_0^\infty e^{-\alpha u}\, u^{\beta-1}\, du = \frac{1}{\alpha^\beta}\,\Gamma(\beta) \quad (\alpha,\ \beta > 0), \tag{3.13}$$

we find

$$k(\lambda_1) = \int_0^\infty \frac{t^{-\beta\lambda-1}\, dt}{e^{\gamma/t^\lambda} - Ae^{-\gamma/t^\lambda}} = \frac{1}{\lambda}\int_0^\infty \frac{u^{\beta-1}\, du}{e^{\gamma u}(1 - Ae^{-2\gamma u})}$$

$$= \frac{1}{\lambda}\int_0^\infty \sum_{k=0}^\infty A^k e^{-\gamma(2k+1)u}\, u^{\beta-1}\, du$$

$$= \frac{1}{\lambda}\sum_{k=0}^\infty A^k \int_0^\infty e^{-\gamma(2k+1)u}\, u^{\beta-1}\, du$$

$$= \frac{\Gamma(\beta)}{\lambda\gamma^\beta}\sum_{k=0}^\infty \frac{A^k}{(2k+1)^\beta} \in \mathbf{R}_+.$$

Setting $\delta = -2\beta\lambda < -\beta\lambda = \lambda_1$, in view of $\frac{t^\delta}{e^{\gamma/t^\lambda} - Ae^{-\gamma/t^\lambda}} \to 0$ ($t \to 0^+$), there exists a constant $L > 0$ such that

$$\frac{t^\delta}{e^{\gamma/t^\lambda} - Ae^{-\gamma/t^\lambda}} \le L, \ \left(t \in \left(0, \frac{1}{v(n_0)}\right]\right).$$

Then, it follows that

$$k_0(t,1) = \frac{1}{e^{\gamma/t^\lambda} - Ae^{-\gamma/t^\lambda}} \leq L, \quad \left(\frac{1}{t^\delta}\right)\left(t \in \left(0, \frac{1}{v(n_0)}\right]\right).$$

(viii) For $k_0(t,1) = \ln\left(\frac{bt^\lambda+1}{at^\lambda+1}\right)$ $(0 \leq a < b,\ \lambda > 0,\ \lambda_1 = -\lambda_2 = -\beta\lambda,\ 0 < \beta < 1)$, we find

$$k(\lambda_1) = \int_0^\infty \ln\left(\frac{bt^\lambda+1}{at^\lambda+1}\right) t^{-\beta\lambda-1}\, dt = \frac{-1}{\beta\lambda}\int_0^\infty \ln\left(\frac{bt^\lambda+1}{at^\lambda+1}\right) dt^{-\beta\lambda}$$

$$= \frac{-1}{\beta\lambda}\left[t^{-\beta\lambda}\ln\left(\frac{bt^\lambda+1}{at^\lambda+1}\right)\Big|_0^\infty - \int_0^\infty \left(\frac{b\lambda}{bt^\lambda+1} - \frac{a\lambda}{at^\lambda+1}\right) t^{(1-\beta)\lambda-1} dt\right]$$

$$= \left(\frac{b^\beta}{\beta\lambda} - \frac{a^\beta}{\beta\lambda}\right)\int_0^\infty \frac{u^{(1-\beta)-1}}{1+u}\, du = \frac{(b^\beta - a^\beta)\pi}{\beta\lambda\sin(\beta\pi)}.$$

Setting $\delta = -\beta_0\lambda < \lambda_1$ $(\beta < \beta_0 < 1)$, in view of $t^\delta \ln\left(\frac{bt^\lambda+1}{at^\lambda+1}\right) \to 0$ as $t \to 0^+$, there exists a constant $L > 0$ such that

$$t^\delta \ln\left(\frac{bt^\lambda+1}{at^\lambda+1}\right) \leq L, \quad \left(t \in \left(0, \frac{1}{v(n_0)}\right]\right).$$

Then it follows that

$$k_0(t,1) = \ln\left(\frac{bt^\lambda+1}{at^\lambda+1}\right) \leq \frac{L}{t^\delta}, \quad \left(t \in \left(0, \frac{1}{v(n_0)}\right]\right).$$

(ix) For $k_0(t,1) = \arctan t^\lambda$ $(\lambda_1 = -\lambda_2 = -\beta,\ 0 < \beta < \lambda)$, we find

$$k(\lambda_1) = \int_0^\infty t^{-\beta-1}\arctan t^\lambda\, dt = -\frac{1}{\beta}\int_0^\infty \arctan t^\lambda\, dt^{-\beta}$$

$$= -\frac{1}{\beta}\left[t^{-\beta}\arctan t^\lambda\big|_0^\infty - \int_0^\infty \frac{\lambda}{1+t^{2\lambda}} t^{(\lambda-\beta)-1}\, dt\right]$$

$$= \frac{1}{2\beta}\int_0^\infty \frac{u^{\frac{\lambda-\beta}{2\lambda}-1}}{1+u}\, du = \frac{\pi}{2\beta\cos\left(\frac{\pi\beta}{2\lambda}\right)} \in \mathbf{R}_+.$$

Setting $\delta = -\beta_0 < -\beta = \lambda_1(\beta_0 < \lambda)$, since $t^\delta \arctan t^\lambda \to 0$ as $t \to 0^+$, there exists a constant $L > 0$ such that

$$k_0(t,1) = \arctan t^\lambda \leq L\left(\frac{1}{t^\delta}\right), \quad \left(t \in \left(0, \frac{1}{v(n_0)}\right]\right).$$

(x) For $k_\lambda(t,1) = \frac{1}{(\max\{t,1\})^\lambda}$ $(\lambda, \lambda_1, \lambda_2 > 0,\ \lambda_1 + \lambda_2 = \lambda)$, we find

$$k(\lambda_1) = \int_0^\infty \frac{1}{(\max\{t,1\})^\lambda} t^{\lambda_1-1} dt$$

$$= \int_0^1 t^{\lambda_1-1} dt + \int_1^\infty \frac{1}{t^\lambda} t^{\lambda_1-1} dt = \frac{\lambda}{\lambda_1\lambda_2} \in \mathbf{R}_+.$$

Setting $\delta = \frac{\lambda_1}{2} (< \lambda_1 < \lambda)$, it follows that

$$k_\lambda(t,1) = \frac{1}{(\max\{t,1\})^\lambda} \le \frac{1}{t^\delta}, \quad \left(t \in \left(0, \frac{1}{v(n_0)} \right] \right).$$

(xi) For $k_{-\lambda}(t,1) = (\min\{t,1\})^\lambda$ $(\lambda_1, \lambda_2 < 0, \ \lambda_1 + \lambda_2 = -\lambda)$, we find

$$k(\lambda_1) = \int_0^\infty (\min\{t,1\})^\lambda t^{\lambda_1 - 1} dt$$

$$= \int_0^1 t^\lambda \, t^{\lambda_1 - 1} \, dt + \int_1^\infty t^{\lambda_1 - 1} \, dt = \frac{\lambda}{\lambda_1 \lambda_2} \in \mathbf{R}_+.$$

Setting $\delta = -\lambda \ (< \lambda_1)$, it follows that

$$k_{-\lambda}(t,1) = (\min\{t,1\})^\lambda \le t^\lambda = \frac{1}{t^\delta} \ \left(t \in \left(0, \frac{1}{v(n_0)} \right] \right).$$

(xii) For $k_0(t,1) = \left(\frac{\min\{t,1\}}{\max\{t,1\}} \right)^\lambda$ $(\lambda > 0, \ \lambda_1 = -\lambda_2 = \alpha, \ |\alpha| < \lambda)$, we find

$$k(\lambda_1) = \int_0^\infty \left(\frac{\min\{t,1\}}{\max\{t,1\}} \right)^\lambda t^{\alpha - 1} dt$$

$$= \int_0^1 t^\lambda t^{\alpha - 1} dt + \int_1^\infty t^{-\lambda} t^{\alpha - 1} dt = \frac{2\lambda}{\lambda^2 - \alpha^2} \in \mathbf{R}_+.$$

Setting $\delta = \frac{-\lambda + \lambda_1}{2} < \lambda_1 \ (|\delta| < \lambda)$, and in view of

$$t^\delta \left(\frac{\min\{t,1\}}{\max\{t,1\}} \right)^\lambda \to 0 \quad (t \to 0^+),$$

there exists a constant $L > 0$ such that $t^\delta \left(\frac{\min\{t,1\}}{\max\{t,1\}} \right)^\lambda \le L$, and then,

$$k_\lambda(t,1) = \left(\frac{\min\{t,1\}}{\max\{t,1\}} \right)^\lambda \le \frac{L}{t^\delta} \ \left(t \in \left(0, \frac{1}{v(n_0)} \right] \right).$$

3.2.3 *Some Inequalities Relating the Constant $k(\lambda_1)$*

Lemma 3.6. *Suppose that $\lambda, \lambda_1, \lambda_2 \in \mathbf{R}, \ \lambda = \lambda_1 + \lambda_2, \ k_\lambda(x,y)$ is a non-negative finite measurable homogeneous function of degree $-\lambda$ in $\mathbf{R}_+^2$ with*

$$k(\lambda_1) = \int_0^\infty k_\lambda \, (t,1) \, t^{\lambda_1 - 1} \, dt \in \mathbf{R}_+.$$

Then,

(i) for $p > 1$, $\varepsilon > 0$, we have

$$k\left(\lambda_1 - \frac{\varepsilon}{p}\right) \geq k(\lambda_1) + o(1) \quad (\varepsilon \to 0^+); \tag{3.14}$$

(ii) for $p < 0$, if there exists a constant $\delta_1 > 0$, such that $k(\lambda_1 + \delta_1) \in \mathbf{R}_+$, then, for $0 < \varepsilon < (-p)\delta_1$, we have

$$k\left(\lambda_1 - \frac{\varepsilon}{p}\right) \leq k(\lambda_1) + o(1) \quad (\varepsilon \to 0^+); \tag{3.15}$$

(iii) for $0 < p < 1$, if there exists a constant $\delta_2 > 0$, such that $k(\lambda_1 - \delta_2) \in \mathbf{R}_+$, then, for $0 < \varepsilon < p\delta_2$, we still have (3.15).

Proof. (i) For $p > 1$, by the Fatou lemma (see Kuang [49]), it follows that

$$k(\lambda_1) = \int_0^\infty \lim_{\varepsilon \to 0^+} k_\lambda(t,1) \, t^{(\lambda_1 - \frac{\varepsilon}{p})-1} \, dt \leq \lim_{\varepsilon \to 0^+} k\left(\lambda_1 - \frac{\varepsilon}{p}\right),$$

and then we have (3.14).

(ii) For $p < 0$, since $t^{-\frac{\varepsilon}{p}} \leq 1 \ (0 < t \leq 1)$, we find

$$k\left(\lambda_1 - \frac{\varepsilon}{p}\right) = \int_0^1 k_\lambda(t,1) \, t^{(\lambda_1 - \frac{\varepsilon}{p})-1} dt + \int_1^\infty k_\lambda(t,1) \, t^{(\lambda_1 - \frac{\varepsilon}{p})-1} dt$$

$$\leq \int_0^1 k_\lambda(t,1) \, t^{\lambda_1 - 1} dt + \int_1^\infty k_\lambda(t,1) \, t^{(\lambda_1 - \frac{\varepsilon}{p})-1} dt. \tag{3.16}$$

For $0 < \varepsilon < (-p)\delta_1$, we have

$$\int_1^\infty k_\lambda(t,1) \, t^{(\lambda_1 - \frac{\varepsilon}{p})-1} dt \leq \int_1^\infty k_\lambda(t,1) \, t^{(\lambda_1 + \delta_1)-1} dt \leq k(\lambda_1 + \delta_1) < \infty,$$

then, by the Lebesgue control convergence theorem, it follows that

$$\int_1^\infty k_\lambda(t,1) \, t^{(\lambda_1 - \frac{\varepsilon}{p})-1} \, dt = \int_1^\infty k_\lambda(t,1) \, t^{\lambda_1 - 1} \, dt + o(1) \quad (\varepsilon \to 0^+).$$

Hence, by (3.16), we have (3.15).

(iii) For $0 < p < 1$, since $t^{-\frac{\varepsilon}{p}} \leq 1 \ (t \geq 1)$, we find

$$k\left(\lambda_1 - \frac{\varepsilon}{p}\right) = \int_0^1 k_\lambda(t,1) \, t^{(\lambda_1 - \frac{\varepsilon}{p})-1} dt + \int_1^\infty k_\lambda(t,1) \, t^{(\lambda_1 - \frac{\varepsilon}{p})-1} dt$$

$$\leq \int_0^1 k_\lambda(t,1) \, t^{(\lambda_1 - \frac{\varepsilon}{p})-1} dt + \int_1^\infty k_\lambda(t,1) \, t^{\lambda_1 - 1} dt. \tag{3.17}$$

For $0 < \varepsilon < p\delta_2$, we have

$$\int_0^1 k_\lambda(t,1) \, t^{(\lambda_1 - \frac{\varepsilon}{p})-1} dt \leq \int_0^1 k_\lambda(t,1) \, t^{(\lambda_1 - \delta_2)-1} dt \leq k(\lambda_1 - \delta_2) < \infty,$$

then by the Lebesgue control convergence theorem (see Kuang [49]), it follows

$$\int_0^1 k_\lambda(t,1) t^{(\lambda_1 - \frac{\varepsilon}{p})-1} dt = \int_0^1 k_\lambda(t,1) t^{\lambda_1 - 1} dt + o(1) \quad (\varepsilon \to 0^+).$$

Hence, by (3.17), we have (3.15) for $0 < p < 1$. $\qquad\square$

3.3   Some Theorems and Corollaries

3.3.1   *Equivalent Inequalities and their Operator Expressions*

For $p \in \mathbf{R} \backslash \{0, 1\}$, $\frac{1}{p} + \frac{1}{q} = 1$, we set two functions $\varphi(x) = x^{p(1-\lambda_1)-1}$ $(x \in (0, \infty))$ and

$$\Psi(n) = \frac{[v(n)]^{q(1-\lambda_2)-1}}{[v'(n)]^{q-1}}, \quad n \geq n_0 \quad (n \in \mathbf{N}),$$

wherefrom, $[\varphi(x)]^{1-q} = x^{q\lambda_1-1}$ and $[\Psi(n)]^{1-p} = \frac{v'(n)}{[v(n)]^{1-p\lambda_2}}$, and then, we define two sets as follows:

$$L_{p,\varphi}(\mathbf{R}_+) = \left\{ f \big| \|f\|_{p,\varphi} = \{ \int_0^\infty x^{p(1-\lambda_1)-1} |f(x)|^p dx \}^{\frac{1}{p}} < \infty \right\},$$

$$l_{q,\Psi} = \left\{ a = \{a_n\}_{n=n_0}^\infty \big| \|a\|_{q,\Psi} = \{ \sum_{n=n_0}^\infty \frac{[v(n)]^{q(1-\lambda_2)-1}}{[v'(n)]^{q-1}} |a_n|^q \}^{\frac{1}{q}} < \infty \right\}.$$

Note. It is obvious that for $p > 1$, both of the above two sets are the normed spaces. Since for $p < 0$ or $0 < p < 1$, neither of the above two sets is normed space, then we agree on that the above sets with $\|f\|_{p,\varphi}$ and $\|a\|_{q,\Psi}$ are the formal symbols.

Theorem 3.1. *Suppose that* $\lambda, \lambda_1, \lambda_2 \in \mathbf{R}$, $\lambda = \lambda_1 + \lambda_2$, $k_\lambda(x, y)$ *is a non-negative finite measurable homogeneous function of degree* $-\lambda$ *in* $\mathbf{R}_+^2, v(y)$ *is a strictly increasing differentiable function in* $[n_0, \infty)$ $(n_0 \in \mathbf{N})$ *with* $v(n_0) > 0$ *and* $v(\infty) = \infty$

$$k(\lambda_1) = \int_0^\infty k_\lambda(t, 1) t^{\lambda_1-1} dt \in \mathbf{R}_+$$

and $\varpi_{\lambda_1}(x) < k(\lambda_1)$ $(x \in (0, \infty))$. *If* $p > 1, \frac{1}{p} + \frac{1}{q} = 1$, $f(x)$, $a_n \geq 0$, $f \in L_{p,\varphi}(\mathbf{R}_+)$, $a = \{a_n\}_{n=n_0}^\infty \in l_{q,\Psi}$, $\|f\|_{p,\varphi} > 0$, $\|a\|_{q,\Psi} > 0$, *then we have the following equivalent inequalities:*

$$I = \sum_{n=n_0}^\infty a_n \int_0^\infty k_\lambda(x, v(n)) f(x) dx = \int_0^\infty f(x) \sum_{n=n_0}^\infty a_n k_\lambda(x, v(n)) dx$$

$$< k(\lambda_1) \|f\|_{p,\varphi} \|a\|_{q,\Psi}, \tag{3.18}$$

$$J = \left\{ \sum_{n=n_0}^\infty [\Psi(n)]^{1-p} \left[\int_0^\infty k_\lambda(x, v(n)) f(x) dx \right]^p \right\}^{\frac{1}{p}} < k(\lambda_1) \|f\|_{p,\varphi}, \tag{3.19}$$

and

$$L = \left\{ \int_0^\infty [\varphi(x)]^{1-q} \left[\sum_{n=n_0}^\infty k_\lambda(x, v(n)) a_n \right]^q dx \right\}^{\frac{1}{q}} < k(\lambda_1) ||a||_{q,\Psi}. \quad (3.20)$$

Moreover, if $\frac{v'(y)}{v(y)}(> 0)$ is decreasing in $[n_0, \infty)$ and there exist constants $\delta < \lambda_1$ and $L > 0$ such that the inequality (3.9) is satisfied, then the constant factor $k(\lambda_1)$ in the above inequalities is the best possible.

Proof.   By the Lebesgue term by term integration theorem, there are two expressions for I in (3.18). In view of (3.6) and $\varpi_{\lambda_1}(x) < k(\lambda_1)$, we have (3.19).

By Hölder's inequality, we find

$$I = \sum_{n=n_0}^\infty \left[[\Psi(n)]^{-\frac{1}{q}} \int_0^\infty k_\lambda(x, v(n)) f(x) dx \right] [[\Psi(n)]^{\frac{1}{q}} a_n] \leq J ||a||_{q,\Psi}.$$

$$(3.21)$$

Then by (3.19), we have (3.18).

On the other hand, suppose that (3.18) is valid. We set

$$a_n = [\Psi(n)]^{1-p} \left[\int_0^\infty k_\lambda(x, v(n)) f(x) dx \right]^{p-1}, \quad n \geq n_0 \ (n \in \mathbf{N}).$$

Then it follows that $J^{p-1} = ||a||_{q,\Psi}$. By (3.6), we find $J < \infty$. If $J = 0$, then (3.19) is trivially valid; if $J > 0$, then by (3.18), we have

$$||a||_{q,\Psi}^q = J^p = I < k(\lambda_1) ||f||_{p,\varphi} ||a||_{q,\Psi}, \quad \text{that is,}$$

$$||a||_{q,\Psi}^{q-1} = J < k(\lambda_1) ||f||_{p,\varphi},$$

and then (3.18) is equivalent to (3.19).

In view of (3.7), since $[\varpi_{\lambda_1}(x)]^{1-q} > [k(\lambda_1)]^{1-q}$, we have (3.20).

By Hölder's inequality, we find

$$I = \int_0^\infty [[\varphi(x)]^{\frac{1}{p}} f(x)] \left[[\varphi(x)]^{-\frac{1}{p}} \sum_{n=n_0}^\infty a_n k_\lambda(x, v(n)) \right] dx \leq ||f||_{p,\varphi} L.$$

$$(3.22)$$

Then by (3.20), we have (3.18).

On the other hand, suppose that (3.18) is valid. We set

$$f(x) = [\varphi(x)]^{1-q} \left[\sum_{n=n_0}^\infty k_\lambda(x, v(n)) a_n \right]^{q-1}, \quad x \in (0, \infty).$$

Then it follows that $L^{q-1} = ||f||_{p,\varphi}$. By (3.7), we find $L < \infty$. If $L = 0$, then (3.20) is trivially valid; if $L > 0$, then, by (3.18), we have

$$||f||_{p,\varphi}^p = L^q = I < k(\lambda_1)||f||_{p,\varphi}||a||_{q,\Psi}, \quad \text{that is}$$
$$||f||_{p,\varphi}^{p-1} = L < k(\lambda_1)||a||_{q,\Psi},$$

and then, inequality (3.18) is equivalent to (3.20).

Hence, inequalities (3.18), (3.19) and (3.20) are equivalent.

For $0 < \varepsilon < p(\lambda_1 - \delta)$, setting

$$\widetilde{f}(x) = \begin{cases} 0, & 0 < x < 1, \\ x^{\lambda_1 - \frac{\varepsilon}{p} - 1}, & x \geq 1 \end{cases}$$

and

$$\widetilde{a}_n = [v(n)]^{\lambda_2 - \frac{\varepsilon}{q} - 1} v'(n), \quad n \geq n_0 \ (n \in \mathbf{N}),$$

if there exists a positive constant $k(\leq k(\lambda_1))$ such that (3.18) is still valid as we replace $k(\lambda_1)$ by k, then substitution of $\widetilde{f}(x)$ and $\widetilde{a} = \{\widetilde{a}_n\}_{n=n_0}^{\infty}$, by (3.8), it follows that

$$\widetilde{I} = \sum_{n=n_0}^{\infty} \widetilde{a}_n \int_0^{\infty} k_\lambda(x, v(n))\widetilde{f}(x)dx \ < \ k||\widetilde{f}||_{p,\varphi}||\widetilde{a}||_{q,\Psi}$$
$$= k \left(\int_1^{\infty} x^{-1-\varepsilon} dx \right)^{\frac{1}{p}} (A(\varepsilon))^{\frac{1}{q}} = \frac{k}{\varepsilon}(1 + o(1))^{\frac{1}{q}}.$$

By (3.8), (3.10) and (3.14), we find

$$\widetilde{I} = \sum_{n=n_0}^{\infty} [v(n)]^{\lambda_2 - \frac{\varepsilon}{q} - 1} v'(n) \int_1^{\infty} k_\lambda(x, v(n)) x^{\lambda_1 - \frac{\varepsilon}{p} - 1} dx$$
$$= \sum_{n=n_0}^{\infty} \frac{v'(n)}{[v(n)]^{1+\varepsilon}} \int_{\frac{1}{v(n)}}^{\infty} k_\lambda(t, 1) t^{\lambda_1 - \frac{\varepsilon}{p} - 1} dt, \quad t = x/v(n)$$
$$= \sum_{n=n_0}^{\infty} \frac{v'(n)}{[v(n)]^{1+\varepsilon}} \left[\int_0^{\infty} k_\lambda(t, 1) \, t^{(\lambda_1 - \frac{\varepsilon}{p}) - 1} \, dt - \int_0^{\frac{1}{v(n)}} k_\lambda(t, 1) \, t^{\lambda_1 - \frac{\varepsilon}{p} - 1} \, dt \right]$$
$$= A(\varepsilon)k \left(\lambda_1 - \frac{\varepsilon}{p} \right) - B(\varepsilon)$$
$$\geq \frac{1}{\varepsilon}[(1 + o(1))(k(\lambda_1) + o(1)) - \varepsilon \, O(1)].$$

Hence, in view of the above results for $\widetilde{I}$, it follows that

$$(1 + o(1))(k(\lambda_1) + o(1)) - \varepsilon \, O(1) \leq \varepsilon \widetilde{I} < k(1 + o(1))^{\frac{1}{q}},$$

and then, $k(\lambda_1) \leq k$ ($\varepsilon \to 0^+$). Therefore, the constant factor $k(\lambda_1)$ in (3.18) is the best possible.

By the equivalency, the constant factor $k(\lambda_1)$ in (3.19) and (3.20) is also the best possible. Otherwise, it leads to a contradiction by (3.21) and (3.22) that the constant factor in (3.18) is not the best possible. □

Note. If we replace the lower limit 0 of the integral to $c > 0$ in (3.18)-(3.20), then Theorem 3.1 is still valid for

$$\omega_{\lambda_2}(n) = [v(n)]^{\lambda_2} \int_c^\infty k_\lambda(x, v(n))\, x^{\lambda_1 - 1} dx \; < \; k(\lambda_1).$$

Remark 3.1. If we replace the conditions that $k(\lambda_1) \in \mathbf{R}_+$, and there exist constants $\delta < \lambda_1$ and $L > 0$ such that inequality (3.9) is satisfied to the condition that there exist $\delta_0, \eta > 0$, such that for any $\widetilde{\lambda}_1 \in [\lambda_1, \quad \lambda_1 + \delta_0)$,

$$0 < k(\widetilde{\lambda}_1)(1 - \widetilde{\theta}_{\widetilde{\lambda}_1}(x)) < \varpi_{\widetilde{\lambda}_1}(x) < k(\widetilde{\lambda}_1) < \infty \quad (x \in (0, \infty)), \qquad (3.23)$$

where, $\widetilde{\theta}_{\widetilde{\lambda}_1}(x) > 0$ ($x > 0$), and $\widetilde{\theta}_{\widetilde{\lambda}_1}(x) = O\left(\frac{1}{x^\eta}\right)$ ($1 \leq x < \infty$), then the constant factor $k(\lambda_1)$ in (3.18) is still the best possible. In fact, for $0 < \varepsilon < q\delta_0$, $\widetilde{\lambda}_1 = \lambda_1 + \frac{\varepsilon}{q}$, we find

$$
\begin{aligned}
\widetilde{I} &= \int_1^\infty x^{-\varepsilon - 1} \left[x^{\lambda_1 + \frac{\varepsilon}{q}} \sum_{n=n_0}^\infty k_\lambda(x, v(n)) \frac{v'(n)}{[v(n)]^{1 - (\lambda_2 - \frac{\varepsilon}{q})}} \right] dx \\
&= \int_1^\infty x^{-\varepsilon - 1} \varpi_{\widetilde{\lambda}_1}(x) dx \; > \; k(\widetilde{\lambda}_1) \int_1^\infty x^{-\varepsilon - 1} \left(1 - O\left(\frac{1}{x^\eta}\right) \right) dx \\
&\geq \frac{1}{\varepsilon} (k(\lambda_1) + o(1))(1 - \varepsilon\, O(1)),
\end{aligned}
$$

and then by the same way, we can still show that $k(\lambda_1)$ is the best possible constant in (3.18).

Remark 3.2. By virtue of Theorem 3.1, (i) define a half-discrete Hilbert-type operator $T : L_{p,\varphi}(\mathbf{R}_+) \to l_{p,\Psi^{1-p}}$ as follows: For $f \in L_{p,\varphi}(\mathbf{R}_+)$, we define $Tf \in l_{p,\Psi^{1-p}}$, satisfying

$$Tf(n) = \int_0^\infty k_\lambda(x, v(n))\, f(x)\, dx, \quad n \geq n_0 \ (n \in \mathbf{N}). \qquad (3.24)$$

Then, by (3.19), it follows that

$$\|Tf\|_{p.\Psi^{1-p}} \leq k(\lambda_1) \|f\|_{p,\varphi},$$

and then, T is a bounded operator with $\|T\| \leq k(\lambda_1)$. Since by Theorem 3.1, the constant factor in (3.19) is the best possible, we have $\|T\| = k(\lambda_1)$.

(ii) Define a half-discrete Hilbert-type operator $\widetilde{T} : l_{q,\Psi} \to L_{q,\varphi^{1-q}}(\mathbf{R}_+)$ as follows:

For $a \in l_{q,\Psi}$, we define $\widetilde{T}a \in L_{q,\varphi^{1-q}}(\mathbf{R}_+)$, satisfying

$$\widetilde{T}a(x) = \sum_{n=n_0}^{\infty} k_\lambda(x, v(n))a_n, \quad x \in (0, \infty). \tag{3.25}$$

Then, by (3.20), it follows that

$$\|\widetilde{T}a\|_{q,\varphi^{1-q}} \le k(\lambda_1)\|a\|_{q,\Psi},$$

and then, $\widetilde{T}$ is a bounded operator with $\|\widetilde{T}\| \le k(\lambda_1)$. Since by Theorem 3.1, the constant factor in (3.20) is the best possible, we have $\|\widetilde{T}\| = k(\lambda_1)$.

Example 3.2. In view of Remark 3.2, in particular, setting $v(n) = n$ and $\psi(n) = n^{q(1-\lambda_2)-1}$ ($n \in \mathbf{N}$), define $T : L_{p,\varphi}(\mathbf{R}_+) \to l_{p,\psi^{1-p}}$ as follows:

For $f \in L_{p,\varphi}(\mathbf{R}_+)$, we define $Tf \in l_{p,\psi^{1-p}}$, satisfying

$$Tf(n) = \int_0^\infty k_\lambda(x, n)f(x)dx, \quad n \in \mathbf{N}.$$

Also we define $\widetilde{T} : l_{q,\psi} \to L_{q,\varphi^{1-q}}(\mathbf{R}_+)$ as follows: For $a \in l_{q,\psi}$, there exists $\widetilde{T}a \in L_{q,\varphi^{1-q}}(\mathbf{R}_+)$, satisfying

$$\widetilde{T}a(x) = \sum_{n=1}^{\infty} k_\lambda(x, n)a_n, \quad x \in (0, \infty).$$

Then, by Remark 3.2, we still have

$$\|T\| = \|\widetilde{T}\| = k(\lambda_1).$$

(i) If $k_\lambda(x, n) = \frac{1}{x^\lambda + n^\lambda}$ ($\lambda_1 > 0$, $0 < \lambda_2 \le 1$, $\lambda_1 + \lambda_2 = \lambda$), since for $x > 0$,

$$f(x, y) = \frac{x^{\lambda_1}}{x^\lambda + y^\lambda} y^{\lambda_2 - 1}$$

is decreasing with respect to $y \in (0, \infty)$, then, Condition (i) is satisfied. In view of Example 3.1(i), the constant factor in (3.19) and (3.20) (for $v(n) = n, n_0 = 1$) is the best possible, we have

$$\|T\| = \|\widetilde{T}\| = k(\lambda_1) = \frac{\pi}{\lambda \sin\left(\frac{\pi\lambda_1}{\lambda}\right)}.$$

(ii) If $k_\lambda(x, n) = \frac{1}{(x+n)^\lambda}$ ($\lambda_1 > 0, 0 < \lambda_2 \le 1, \lambda_1 + \lambda_2 = \lambda$), since for $x > 0$,

$$f(x, y) = \frac{x^{\lambda_1}}{(x + y)^\lambda} y^{\lambda_2 - 1}$$

is decreasing with respect to $y \in (0, \infty)$, then Condition (i) is satisfied. In view of Example 3.1(ii), the constant factor in (3.19) and (3.20) (for $v(n) = n, n_0 = 1$) is the best possible, we have (see Yang [135])

$$||T|| = ||\widetilde{T}|| = k(\lambda_1) = B(\lambda_1, \lambda_2).$$

(iii) If $k_\lambda(x, n) = \frac{\ln(x/n)}{x^\lambda - n^\lambda}$ ($\lambda_1 > 0$, $0 < \lambda_2 \le 1$, $\lambda_1 + \lambda_2 = \lambda$), then by the decreasing property of $f(x, y)$ and Example 3.1(iii), we have

$$||T|| = ||\widetilde{T}|| = k(\lambda_1) = \left[\frac{\pi}{\lambda \sin(\frac{\pi \lambda_1}{\lambda})} \right]^2.$$

(iv) If $s \in \mathbf{N}, k_{\lambda s}(x, n) = \prod_{k=1}^{s} \frac{1}{a_k x^\lambda + n^\lambda} (0 < a_1 < \cdots < a_s, \lambda, \lambda_1 > 0, 0 < \lambda_2 \le 1, \lambda_1 + \lambda_2 = \lambda s)$, then by the decreasing property of $f(x, y)$ and Example 3.1(iv), we have

$$||T|| = ||\widetilde{T}|| = k(\lambda_1) = \frac{\pi}{\lambda \sin(\frac{\pi \lambda_2}{\lambda})} \sum_{k=1}^{s} a_k^{\frac{\lambda_2}{\lambda} - 1} \prod_{j=1(j \ne k)}^{s} \frac{1}{a_j - a_k}.$$

(v) For $k_\lambda(x, n) = \frac{1}{x^\lambda + 2(xn)^{\lambda/2} \cos \beta + n^\lambda}$ ($\lambda > 0$, $0 < \beta \le \frac{\pi}{2}$, $\lambda_1 > 0$, $0 < \lambda_2 \le 1$, $\lambda_1 + \lambda_2 = \lambda$), then by the decreasing property of $f(x, y)$ and Example 3.1(v), we have

$$||T|| = ||\widetilde{T}|| = k(\lambda_1) = \frac{2\pi \sin \beta(1 - \frac{2\lambda_1}{\lambda})}{\lambda \sin \beta \sin(\frac{2\pi \lambda_1}{\lambda})}.$$

(vi) If $k_\lambda(x, n) = \frac{1}{x^\lambda + b(xn)^{\lambda/2} + cn^\lambda}$ ($c > 0$, $0 \le b \le 2\sqrt{c}$, $0 < \lambda \le 2$, $\lambda_1 = \lambda_2 = \frac{\lambda}{2}$), then by the decreasing property of $f(x, y)$ and Example 3.1(vi), we have

$$||T|| = ||\widetilde{T}|| = \frac{2}{\lambda} \begin{cases} \frac{\pi}{\sqrt{c}}, & b = 0, \\ \frac{4}{\sqrt{4c-b^2}} \arctan \frac{\sqrt{4c-b^2}}{b}, & 0 < b < 2\sqrt{c}, \\ \frac{2}{\sqrt{c}}, & b = 2\sqrt{c}. \end{cases}$$

(vii) If $k_0(x, n) = \frac{1}{e^{\gamma(n/x)^\lambda} - Ae^{-\gamma(n/x)^\lambda}}$ ($-1 \le A < 1, \beta > 0(A = 1, \beta > 1), \gamma > 0, -1 \le \lambda_1 = -\lambda_2 = -\beta\lambda < 0 < \lambda$), then by the decreasing property of $f(x, y)$ and Example 3.1(vii), we have

$$||T|| = ||\widetilde{T}|| = k(-\beta\lambda) = \frac{\Gamma(\beta)}{\lambda \gamma^\beta} \sum_{k=0}^{\infty} \frac{A^k}{(2k + 1)^\beta}.$$

(viii) If $k_0(x, n) = \ln(\frac{bx^\lambda + n^\lambda}{ax^\lambda + n^\lambda})$ ($0 \le a < b, \lambda > 0, -1 \le \lambda_1 = -\lambda_2 = -\beta\lambda, 0 < \beta < 1$), then by the decreasing property of $f(x, y)$ and Example 3.1(viii), we have

$$||T|| = ||\widetilde{T}|| = k(-\beta\lambda) = \frac{(b^\beta - a^\beta)\pi}{\beta\lambda \sin(\beta\pi)}.$$

(ix) If $k_0(x,n) = \arctan\left(\frac{x}{n}\right)^\lambda$ ($\lambda_1 = -\lambda_2 = -\beta$, $0 < \beta < \lambda$, $\beta \le 1$), then by the decreasing property of $f(x,y)$ and Example 3.1(ix), we have

$$||T|| = ||\widetilde{T}|| = k(-\beta) = \frac{\pi}{2\beta \cos(\frac{\pi\beta}{2\lambda})}.$$

(x) If $k_\lambda(x,n) = \frac{1}{(\max\{x,n\})^\lambda}$ ($\lambda_1 > 0$, $0 < \lambda_2 \le 1$, $\lambda_1 + \lambda_2 = \lambda$), then by the decreasing property of $f(x,y)$ and Example 3.1(x), we have (see Yang [159])

$$||T|| = ||\widetilde{T}|| = k(\lambda_1) = \frac{\lambda}{\lambda_1 \lambda_2}.$$

(xi) If $k_{-\lambda}(x,n) = (\min\{x,n\})^\lambda$ ($\lambda_1 < 0$, $-1 \le \lambda_2 < 0$, $\lambda_1 + \lambda_2 = -\lambda$), then by the decreasing property of $f(x,y)$ and Example 3.1(xi), we have

$$||T|| = ||\widetilde{T}|| = k(\lambda_1) = \frac{\lambda}{\lambda_1 \lambda_2}.$$

(xii) If $k_0(x,n) = \left(\frac{\min\{x,n\}}{\max\{x,n\}}\right)^\lambda$ ($\lambda_1 = -\lambda_2 = \alpha$, $|\alpha| < \lambda \le 1 + \alpha$, $\alpha > -\frac{1}{2}$), since for $x > 0$,

$$f(x,y) = x^\alpha \left(\frac{\min\{x,y\}}{\max\{x,y\}}\right)^\lambda y^{-\alpha-1}$$
$$= \begin{cases} x^{\alpha-\lambda} y^{\lambda-\alpha-1}, & 0 < y \le x, \\ x^{\alpha+\lambda} y^{-\lambda-\alpha-1}, & y > x \end{cases}$$

is decreasing with respect to $y \in (0,\infty)$, then Condition (i) is satisfied. In view of Example 3.1(xii), the constant factor in (3.19) and (3.20) (for $v(n) = n$, $n_0 = 1$) is the best possible, we obtain

$$||T|| = ||\widetilde{T}|| = k(\alpha) = \frac{2\lambda}{\lambda^2 - \alpha^2}.$$

3.3.2   *Two Classes of Equivalent Reverse Inequalities*

Theorem 3.2. *Suppose that λ, λ_1, $\lambda_2 \in \mathbf{R}$, $\lambda = \lambda_1 + \lambda_2$, $k_\lambda(x,y)$ is a non-negative finite measurable homogeneous function of degree $-\lambda$ in $\mathbf{R}_+^2$, $v(y)$ is a strictly increasing differentiable function in $[n_0, \infty)$ ($n_0 \in \mathbf{N}$) with $v(n_0) > 0$ and $v(\infty) = \infty$*

$$k(\lambda_1) = \int_0^\infty k_\lambda(t,1)\, t^{\lambda_1 - 1}\, dt \in \mathbf{R}_+$$

and $\varpi_{\lambda_1}(x) < k(\lambda_1)$ $(x \in (0, \infty))$. *If $p < 0$, $\frac{1}{p} + \frac{1}{q} = 1$, $f(x)$, $a_n \geq 0$, $f \in L_{p,\varphi}(\mathbf{R}_+)$, $a = \{a_n\}_{n=n_0}^\infty \in l_{q,\Psi}$, $\|f\|_{p,\varphi} > 0$, $\|a\|_{q,\Psi} > 0$, then we have the following equivalent inequalities:*

$$I = \sum_{n=n_0}^\infty a_n \int_0^\infty k_\lambda(x, v(n))f(x)\,dx = \int_0^\infty f(x) \sum_{n=n_0}^\infty a_n k_\lambda(x, v(n))\,dx$$
$$> k(\lambda_1)\|f\|_{p,\varphi}\|a\|_{q,\Psi}, \tag{3.26}$$

$$J = \left\{ \sum_{n=n_0}^\infty [\Psi(n)]^{1-p} \left[\int_0^\infty k_\lambda(x, v(n))f(x)\,dx \right]^p \right\}^{\frac{1}{p}} > k(\lambda_1)\|f\|_{p,\varphi}, \tag{3.27}$$

and

$$L = \left\{ \int_0^\infty [\varphi(x)]^{1-q} \left[\sum_{n=n_0}^\infty k_\lambda(x, v(n))a_n \right]^q dx \right\}^{\frac{1}{q}} > k(\lambda_1)\|a\|_{q,\Psi}. \tag{3.28}$$

Moreover, if $\frac{v'(y)}{v(y)}(> 0)$ is decreasing in $[n_0, \infty)$ and there exists a constant $\delta_1 > 0$, such that $k(\lambda_1 + \delta_1) \in \mathbf{R}_+$, then the constant factor $k(\lambda_1)$ in the above inequalities is the best possible.

Proof. In view of the reverse of (3.6) and $\varpi_{\lambda_1}(x) < k(\lambda_1)$, for $p < 0$, we have (3.27). By the reverse Hölder's inequality (see Kuang [47]), we find

$$I = \sum_{n=n_0}^\infty \left[[\Psi(n)]^{-\frac{1}{q}} \int_0^\infty k_\lambda(x, v(n))f(x)dx \right] [[\Psi(n)]^{\frac{1}{q}} a_n] \geq J\|a\|_{q,\Psi}. \tag{3.29}$$

Then, by (3.27), we have (3.26).

On the other hand, suppose that (3.26) is valid. We set

$$a_n = [\Psi(n)]^{1-p} \left[\int_0^\infty k_\lambda(x, v(n))f(x)dx \right]^{p-1}, \quad n \geq n_0 \ (n \in \mathbf{N}).$$

Then it follows that $J^{p-1} = \|a\|_{q,\Psi}$. By the reverse of (3.6), we find $J > 0$. If $J = \infty$, then (3.27) is trivially valid; if $J < \infty$, then, by (3.26), we have

$$\|a\|_{q,\Psi}^q = J^p = I > k(\lambda_1)\|f\|_{p,\varphi}\|a\|_{q,\Psi}, \text{that is}$$
$$\|a\|_{q,\Psi}^{q-1} = J > k(\lambda_1)\|f\|_{p,\varphi},$$

and then, inequality (3.26) is equivalent to (3.27).

In view of the reverse of (3.7) and

$$[\varpi_{\lambda_1}(x)]^{1-q} < [k(\lambda_1)]^{1-q} \quad (0 < q < 1),$$

we have inequality (3.28). By the reverse Hölder's inequality, we find

$$I = \int_0^\infty [[\varphi(x)]^{\frac{1}{p}} f(x)] \left[[\varphi(x)]^{-\frac{1}{p}} \sum_{n=n_0}^\infty k_\lambda(x, v(n)) a_n \right] dx \geq \|f\|_{p,\varphi} L. \tag{3.30}$$

Then, by (3.28), we have (3.26).

On the other hand, suppose that (3.26) is valid. We set

$$f(x) = [\varphi(x)]^{1-q} \left[\sum_{n=n_0}^\infty k_\lambda(x, v(n)) a_n \right]^{q-1}, \quad x \in (0, \infty).$$

Then, it follows that $L^{q-1} = \|f\|_{p,\varphi}$. By the reverse of (3.7), we find $L > 0$. If $L = \infty$, then (3.28) is trivially valid; if $L < \infty$, then by (3.26), we have

$$\|f\|_{p,\varphi}^p = L^q = I > k(\lambda_1) \|f\|_{p,\varphi} \|a\|_{q,\Psi}, \text{that is}$$
$$\|f\|_{p,\varphi}^{p-1} = L > k(\lambda_1) \|a\|_{q,\Psi},$$

and then, (3.26) is equivalent to (3.28).

Hence, inequalities (3.26), (3.27) and (3.28) are equivalent.

For $0 < \varepsilon < (-p)\delta_1$, setting $\widetilde{f}(x)$ and $\widetilde{a}_n$ as Theorem 3.1, if there exists a positive constant $k(\geq k(\lambda_1))$, such that (3.26) is valid as we replace $k(\lambda_1)$ by k, then substitution of $\widetilde{f}(x)$ and $\widetilde{a} = \{\widetilde{a}_n\}_{n=n_0}^\infty$, by (3.8), it follows that

$$\widetilde{I} = \sum_{n=n_0}^\infty \widetilde{a}_n \int_0^\infty k_\lambda(x, v(n)) \widetilde{f}(x) dx > k \|\widetilde{f}\|_{p,\varphi} \|\widetilde{a}\|_{q,\Psi}$$
$$= k \left(\int_1^\infty x^{-1-\varepsilon} dx \right)^{\frac{1}{p}} (A(\varepsilon))^{\frac{1}{q}} = \frac{k}{\varepsilon}(1 + o(1))^{\frac{1}{q}}. \tag{3.31}$$

In view of (3.8) and (3.15), we find

$$\widetilde{I} = \sum_{n=n_0}^\infty [v(n)]^{\lambda_2 - \frac{\varepsilon}{q} - 1} v'(n) \int_1^\infty k_\lambda(x, v(n)) x^{\lambda_1 - \frac{\varepsilon}{p} - 1} dx$$
$$= \sum_{n=n_0}^\infty \frac{v'(n)}{[v(n)]^{1+\varepsilon}} \int_{\frac{1}{v(n)}}^\infty k_\lambda(t, 1) t^{\lambda_1 - \frac{\varepsilon}{p} - 1} dt, \qquad (t = x/v(n))$$
$$\leq \sum_{n=n_0}^\infty \frac{v'(n)}{[v(n)]^{1+\varepsilon}} \int_0^\infty k_\lambda(t, 1) t^{(\lambda_1 - \frac{\varepsilon}{p}) - 1} dt$$
$$= A(\varepsilon) k \left(\lambda_1 - \frac{\varepsilon}{p} \right) \leq \frac{1}{\varepsilon}(1 + o(1))(k(\lambda_1) + o(1)). \tag{3.32}$$

Hence, by (3.31) and (3.32), it follows that

$$(1 + o(1))(k(\lambda_1) + o(1)) > k(1 + o(1))^{\frac{1}{q}},$$

and then $k(\lambda_1) \geq k \ (\varepsilon \to 0^+)$. Therefore, the constant factor $k(\lambda_1)$ in (3.26) is the best possible.

By the equivalency, the constant factor $k(\lambda_1)$ in (3.27) and (3.28) is the best possible. Otherwise, it leads to a contradiction by (3.29) and (3.30) that the constant factor in (3.26) is not the best possible. $\square$

If we replace the lower limit 0 of the integral to $c \ (> 0)$ in Theorem 3.2 then it follows that

$$\omega_{\lambda_2}(n) = [v(n)]^{\lambda_2} \int_c^\infty k_\lambda(x, v(n)) x^{\lambda_1 - 1} dx < k(\lambda_1).$$

Setting

$$\omega_{\lambda_2}(n) = k(\lambda_1)(1 - \theta_{\lambda_1}(v(n))),$$

$$\theta_{\lambda_1}(v(n)) = \frac{1}{k(\lambda_1)} [v(n)]^{\lambda_2} \int_0^c k_\lambda(x, v(n)) \, x^{\lambda_1 - 1} dx > 0,$$

and

$$\widetilde{\Psi}(n) = \Psi(n)(1 - \theta_{\lambda_1}(v(n))) = \frac{[v(n)]^{q(1-\lambda_2)-1}}{[v'(n)]^{q-1}} (1 - \theta_{\lambda_1}(v(n))),$$

we have

$$\theta_{\lambda_1}(v(n)) = \frac{1}{k(\lambda_1)} \int_0^{\frac{c}{v(n)}} k_\lambda(t, 1) \, t^{\lambda_1 - 1} \, dt,$$

and the following corollary:

Corollary 3.1. *Suppose that $c > 0$, λ, λ_1, $\lambda_2 \in \mathbf{R}$, $\lambda = \lambda_1 + \lambda_2$, $k_\lambda(x,y)$ is a non-negative finite measurable homogeneous function of degree $-\lambda$ in $\mathbf{R}_+^2$, $v(y)$ is a strictly increasing differential functions in $[n_0, \infty)$ $(n_0 \in \mathbf{N})$ with $v(n_0) > 0$ and $v(\infty) = \infty$*

$$k(\lambda_1) = \int_0^\infty k_\lambda \, (t, 1) \, t^{\lambda_1 - 1} \, dt \in \mathbf{R}_+$$

and $\varpi_{\lambda_1}(x) < k(\lambda_1) \ (x \in (c, \infty))$.

If $p < 0$, $\frac{1}{p} + \frac{1}{q} = 1$, $f(x)$, $a_n \geq 0$, $f \in L_{p,\varphi}(c, \infty)$, $a = \{a_n\}_{n=n_0}^\infty \in l_{q,\widetilde{\Psi}}$, $||f||_{p,\varphi} > 0$, $||a||_{q,\widetilde{\Psi}} > 0$, then, we have the following equivalent inequalities:

$$\sum_{n=n_0}^\infty a_n \int_c^\infty k_\lambda(x, v(n)) f(x) dx$$

$$= \int_c^\infty f(x) \sum_{n=n_0}^\infty a_n k_\lambda(x, v(n)) dx > k(\lambda_1) ||f||_{p,\varphi} ||a||_{q,\widetilde{\Psi}}, \ (3.33)$$

$$\left\{ \sum_{n=n_0}^\infty [\widetilde{\Psi}(n)]^{1-p} \left[\int_c^\infty k_\lambda(x, v(n)) f(x) dx \right]^p \right\}^{\frac{1}{p}} > k(\lambda_1) ||f||_{p,\varphi},$$

$$(3.34)$$

and

$$\left\{ \int_c^\infty [\varphi(x)]^{1-q} \left[\sum_{n=n_0}^\infty k_\lambda(x, v(n)) a_n \right]^q dx \right\}^{\frac{1}{q}} > k(\lambda_1) ||a||_{q, \widetilde{\Psi}}. \qquad (3.35)$$

Moreover, if $\frac{v'(y)}{v(y)} (> 0)$ is decreasing in $[n_0, \infty)$ and there exist constants $L > 0$, $0 < \delta < \lambda_1$, such that $k(\lambda_1 + \delta) \in \mathbf{R}_+$, and

$$k_\lambda(t, 1) \le L \left(\frac{1}{t^\delta} \right), \ \left(t \in \left(0, \frac{c}{v(n_0)} \right] \right),$$

then, the constant factor $k(\lambda_1)$ in the above inequalities is the best possible.

Theorem 3.3. *Suppose that λ, λ_1, $\lambda_2 \in \mathbf{R}$, $\lambda = \lambda_1 + \lambda_2$, $k_\lambda(x, y)$ is a non-negative finite measurable homogeneous function of degree $-\lambda$ in $\mathbf{R}_+^2$, $v(y)$ is a strictly increasing differentiable function in $[n_0, \infty)$ $(n_0 \in \mathbf{N})$ with $v(n_0) > 0$, $v(\infty) = \infty$, and*

$$0 < k(\lambda_1)(1 - \theta_{\lambda_1}(x)) < \varpi_{\lambda_1}(x) < k(\lambda_1) < \infty \quad (x \in (0, \infty)),$$

where, $\theta_{\lambda_1}(x) > 0$. Setting $\widetilde{\varphi}(x) = (1 - \theta_{\lambda_1}(x))\varphi(x)$, if $0 < p < 1$, $\frac{1}{p} + \frac{1}{q} = 1$, $f(x), a_n \ge 0$, $f \in L_{p,\widetilde{\varphi}}(\mathbf{R}_+)$, $a = \{a_n\}_{n=n_0}^\infty \in l_{q,\Psi}$, $||f||_{p,\widetilde{\varphi}} > 0$, $||a||_{q,\Psi} > 0$, then, we have the following equivalent inequalities:

$$I = \sum_{n=n_0}^\infty a_n \int_0^\infty k_\lambda(x, v(n)) f(x) dx = \int_0^\infty f(x) \sum_{n=n_0}^\infty a_n k_\lambda(x, v(n)) dx$$
$$> k(\lambda_1) ||f||_{p,\widetilde{\varphi}} ||a||_{q,\Psi}, \qquad (3.36)$$

$$J = \left\{ \sum_{n=n_0}^\infty [\Psi(n)]^{1-p} \left[\int_0^\infty k_\lambda(x, v(n)) f(x) dx \right]^p \right\}^{\frac{1}{p}} > k(\lambda_1) ||f||_{p,\widetilde{\varphi}},$$
$$(3.37)$$

and

$$L_1 = \left\{ \int_0^\infty [\widetilde{\varphi}(x)]^{1-q} \left[\sum_{n=n_0}^\infty k_\lambda(x, v(n)) a_n \right]^q dx \right\}^{\frac{1}{q}} > k(\lambda_1) ||a||_{q,\Psi}.$$
$$(3.38)$$

Moreover, if $\frac{v'(y)}{v(y)} (> 0)$ is decreasing in $[n_0, \infty)$ and there exist constants $\delta_2, \eta > 0$, such that $k(\lambda_1 - \delta_2) \in \mathbf{R}_+$ and for any $\widetilde{\lambda}_1 \in (\lambda_1 - \delta_2, \lambda_1)$, $\theta_{\widetilde{\lambda}_1}(x) = O\left(\frac{1}{x^\eta} \right)$ $(x \in [1, \infty))$, then, the constant factor $k(\lambda_1)$ in the above inequalities is the best possible.

Proof. In view of the reverse of (3.6) and $\varpi_{\lambda_2}(x) < k(\lambda_1)$, for $0 < p < 1$, we have (3.37). By the reverse Hölder's inequality, we find

$$I = \sum_{n=n_0}^{\infty} \left[[\Psi(n)]^{-\frac{1}{q}} \int_0^{\infty} k_\lambda(x, v(n)) f(x) dx \right] [[\Psi(n)]^{\frac{1}{q}} a_n] \geq J \|a\|_{q,\Psi}.$$

(3.39)

Then by (3.37), we have (3.36). On the other hand, suppose that (3.36) is valid. We set

$$a_n = [\Psi(n)]^{1-p} \left[\int_0^{\infty} k_\lambda(x, v(n)) f(x) dx \right]^{p-1}, \quad n \geq n_0 \ (n \in \mathbf{N}).$$

Then it follows that $J^{p-1} = \|a\|_{q,\Psi}$. By the reverse of (3.6), we find $J > 0$. If $J = \infty$, then (3.37) is trivially valid; if $J < \infty$, then, by (3.36), we have

$$\|a\|_{q,\Psi}^q = J^p = I > k(\lambda_1) \|f\|_{p,\widetilde{\varphi}} \|a\|_{q,\Psi}, \text{that is,}$$
$$\|a\|_{q,\Psi}^{q-1} = J > k(\lambda_1) \|f\|_{p,\widetilde{\varphi}},$$

and then (3.36) is equivalent to (3.37).

In view of the reverse of (3.7) and

$$[\varpi_{\lambda_2}(x)]^{1-q} > [k(\lambda_1)(1 - \theta_{\lambda_1}(x))]^{1-q}, \ (q < 0),$$

we have (3.38). By the reverse Hölder's inequality, we find

$$I = \int_0^{\infty} [[\widetilde{\varphi}(x)]^{\frac{1}{p}} f(x)] \left[[\widetilde{\varphi}(x)]^{-\frac{1}{p}} \sum_{n=n_0}^{\infty} a_n k_\lambda(x, v(n)) \right] dx \geq \|f\|_{p,\widetilde{\varphi}} L.$$

(3.40)

Then, by (3.38), we have (3.36). On the other hand, suppose that (3.36) is valid. We set

$$f(x) = [\widetilde{\varphi}(x)]^{1-q} \left[\sum_{n=n_0}^{\infty} k_\lambda(x, v(n)) a_n \right]^{q-1}, \quad x \in (0, \infty).$$

Then, it follows that $L_1^{q-1} = \|f\|_{p,\widetilde{\varphi}}$. By the reverse of (3.7), we find $L_1 > 0$. If $L_1 = \infty$, then (3.38) is trivially valid; if $L_1 < \infty$, then by (3.36), we have

$$\|f\|_{p,\widetilde{\varphi}}^p = L_1^q = I > k(\lambda_1) \|f\|_{p,\widetilde{\varphi}} \|a\|_{q,\Psi}, \text{that is}$$
$$\|f\|_{p,\widetilde{\varphi}}^{p-1} = L_1 > k(\lambda_1) \|a\|_{q,\Psi},$$

and then (3.36) is equivalent to (3.38).

Hence, inequalities (3.36), (3.37) and (3.38) are equivalent.

For $0 < \varepsilon < p\delta_2$, setting $\widetilde{f}(x)$ and $\widetilde{a}_n$ as Theorem 3.1, if there exists a positive constant $k(\geq k(\lambda_1))$ such that (3.36) is valid as we replace $k(\lambda_1)$ by k, then substitution of $\widetilde{f}(x)$ and $\widetilde{a} = \{\widetilde{a}_n\}_{n=n_0}^{\infty}$, by (3.8), it follows that

$$\widetilde{I} = \sum_{n=n_0}^{\infty} \widetilde{a}_n \int_0^{\infty} k_\lambda(x, v(n))\widetilde{f}(x)\, dx > k\|\widetilde{f}\|_{p,\widetilde{\varphi}}\|\widetilde{a}\|_{q,\Psi}$$

$$= k\left[\int_1^{\infty}\left(1 - O\left(\frac{1}{x^\eta}\right)\right) x^{-1-\varepsilon}\, dx\right]^{\frac{1}{p}} (A(\varepsilon))^{\frac{1}{q}}$$

$$= \frac{k}{\varepsilon}(1 - \varepsilon\, O(1))^{\frac{1}{p}}(1 + o(1))^{\frac{1}{q}}.$$

By (3.8) and (3.15), we find

$$\widetilde{I} = \sum_{n=n_0}^{\infty}[v(n)]^{\lambda_2-\frac{\varepsilon}{q}-1}v'(n)\int_1^{\infty}k_\lambda(x, v(n))x^{\lambda_1-\frac{\varepsilon}{p}-1}dx$$

$$= \sum_{n=n_0}^{\infty}\frac{v'(n)}{[v(n)]^{1+\varepsilon}}\int_{\frac{1}{v(n)}}^{\infty}k_\lambda(t, 1)\, t^{\lambda_1-\frac{\varepsilon}{p}-1}\, dt, \qquad (t = x/v(n))$$

$$\leq \sum_{n=n_0}^{\infty}\frac{v'(n)}{[v(n)]^{1+\varepsilon}}\int_0^{\infty}k_\lambda(t, 1)\, t^{(\lambda_1-\frac{\varepsilon}{p})-1}dt$$

$$= A(\varepsilon)k\left(\lambda_1 - \frac{\varepsilon}{p}\right) \leq \frac{1}{\varepsilon}(1 + o(1))(k(\lambda_1) + o(1)).$$

In view of the above results, it follows that

$$(1 + o(1))(k(\lambda_1) + o(1)) > k(1 - \varepsilon\, O(1))^{\frac{1}{p}}(1 + o(1))^{\frac{1}{q}},$$

and then $k(\lambda_1) \geq k$ ($\varepsilon \to 0^+$). Therefore, the constant factor $k(\lambda_1)$ in (3.36) is the best possible.

By the equivalency, the constant factor $k(\lambda_1)$ in (3.37) and (3.38) is the best possible. Otherwise, we can imply a contradiction by (3.39) and (3.40) that the constant factor in (3.35) is not the best possible. $\qquad\square$

Note. If we replace the lower limit 0 of the integral to $c > 0$, then Theorem 3.3 is still valid. In this case, we use

$$\omega_{\lambda_2}(n) = [v(n)]^{\lambda_2}\int_c^{\infty}k_\lambda(x, v(n))x^{\lambda_1-1}dx < k(\lambda_1).$$

3.3.3   *Some Corollaries*

Corollary 3.2. *(see Yang and Chen [138]) Let the assumptions of Lemma 3.1 be fulfilled and additionally let*

$$\varpi_{\lambda_1}(x) < k(\lambda_1) = \int_0^{\infty}k_\lambda(t, 1)t^{\lambda_1-1}dt \in \mathbf{R}_+ \ (x \in (0, \infty)),$$

and for $-\infty \le b < c \le \infty$, $u(x)$ be a strictly increasing differentiable function in (b,c), with $u(b^+) = 0$ and $u(c^-) = \infty$. For $p \in \mathbf{R}\backslash\{0,1\}$, $\frac{1}{p} + \frac{1}{q} = 1$, we define the function $\Phi(x)$ as follows:

$$\Phi(x) = \frac{[u(x)]^{p(1-\lambda_1)-1}}{[u'(x)]^{p-1}} \qquad (x \in (b,c)).$$

If $f(x)$, $a_n \ge 0$, $f \in L_{p,\Phi}(b,c)$, $a = \{a_n\}_{n=n_0}^\infty \in l_{q,\Psi}$, $||f||_{p,\Phi} > 0$, $||a||_{q,\Psi} > 0$, then

(i) for $p > 1$, we have the following equivalent inequalities:

$$\sum_{n=n_0}^\infty a_n \int_b^c k_\lambda(u(x),v(n))f(x)\,dx$$

$$= \int_b^c f(x) \sum_{n=n_0}^\infty a_n k_\lambda(u(x),v(n))\,dx < k(\lambda_1)||f||_{p,\Phi}||a||_{q,\Psi}, \qquad (3.41)$$

$$\left\{\sum_{n=n_0}^\infty [\Psi(n)]^{1-p}\left[\int_b^c k_\lambda(u(x),v(n))f(x)\,dx\right]^p\right\}^{\frac{1}{p}} < k(\lambda_1)||f||_{p,\Phi},$$

$$\qquad (3.42)$$

and

$$\left\{\int_b^c [\Phi(x)]^{1-q}\left[\sum_{n=n_0}^\infty k_\lambda(u(x),v(n))a_n\right]^q dx\right\}^{\frac{1}{q}} < k(\lambda_1)||a||_{q,\Psi}; \qquad (3.43)$$

(ii) for $p < 0$, we have the equivalent reverses of (3.41), (3.42) and (3.43).

Moreover, if $\frac{v'(y)}{v(y)}(> 0)$ is decreasing in $[n_0, \infty)$ and there exist constants $\delta < \lambda_1$, $L > 0$, and $\delta_1 > 0$ such that inequality (3.9) is fulfilled, and $k(\lambda_1 + \delta_1) \in \mathbf{R}_+$, then the constant factor $k(\lambda_1)$ in the inequalities of (i) and the reverses in (ii) is the best possible.

Proof. (i) For $p > 1$, setting $x = u(t)$ on two sides of (3.18), we obtain

$$I = \sum_{n=n_0}^\infty a_n \int_b^c k_\lambda(u(t),v(n))f(u(t))u'(t)\,dt$$

$$= \int_b^c f(u(t))u'(t)\sum_{n=n_0}^\infty a_n k_\lambda(u(t),v(n))\,dt$$

$$< k(\lambda_1)\left\{\int_b^c [u(t)]^{p(1-\lambda_1)-1}f^p(u(t))u'(t)\,dt\right\}^{\frac{1}{p}}||a||_{q,\Psi}. \qquad (3.44)$$

Replacing t and $f(u(t))u'(t)$ by x and $f(x)$ in (3.44), by simplification, we obtain (3.41). On the other hand, setting $u(x) = x$ ($x \in (0, \infty)$) in (3.41), we have (3.18). It follows that (3.41) and (3.18) are equivalent and then both of them with the same best constant factor $k(\lambda_1)$. Similarly, we can prove that (3.42) and (3.19), (3.43) and (3.20) are equivalent and with the same best constant factor $k(\lambda_1)$. Since (3.18), (3.19) and (3.20) are equivalent, then, it follows that (3.41), (3.42) and (3.43) are equivalent.

(ii) For $p < 0$, in the same way, we have the equivalent reverses of (3.41), (3.42) and (3.43) with the same best constant factor $k(\lambda_1)$.   □

Note. Some particular cases about the reverse inequalities were published by Yang [160]–[162].

Remark 3.3. Define the operator $T : L_{p,\Phi}(b, c) \to l_{q,\Psi}$ as follows: For $f \in L_{p,\Phi}(b, c)$, we define $Tf \in l_{p,\Psi^{1-p}}$, satisfying

$$Tf(n) = \int_0^\infty k_\lambda(u(x), v(n))f(x)dx, n \geq n_0, \qquad n \in \mathbf{N}.$$

Also, we define the operator $\widetilde{T} : l_{q,\Psi} \to L_{p,\Phi}(b, c)$ as follows: For $a \in l_{q,\Psi}$, we define $\widetilde{T}a \in L_{p,\Phi}(b, c)$, satisfying

$$\widetilde{T}a(x) = \sum_{n=n_0}^\infty k_\lambda(u(x), v(n))a_n, \qquad x \in (b, c).$$

Then, by Corollary 3.2, we still have

$$||T|| = ||\widetilde{T}|| = k(\lambda_1).$$

Example 3.3. Setting $u(x) = \ln x$ ($x \in (1, \infty)$), $v(n) = \ln n$, $n \geq n_0 = 2$, then $\frac{v'(y)}{v(y)} = \frac{1}{y \ln y} (> 0)$ is decreasing for $y > 1$.

(i) If $k_\lambda(\ln x, \ln n) = \frac{1}{(\ln x)^\lambda + (\ln n)^\lambda}$ ($\lambda_1 > 0$, $0 < \lambda_2 \leq 1$, $\lambda_1 + \lambda_2 = \lambda$), since for $x > 1$,

$$f(x, y) = \frac{x^{\lambda_1}}{[x^\lambda + (\ln y)^\lambda]y} (\ln y)^{\lambda_2 - 1}$$

is decreasing with respect to $y \in (1, \infty)$, then Condition (i) is satisfied. In view of Example 3.1(i), the constant factor in (3.42) and (3.43) (for $u(x) = \ln x$, $v(n) = \ln n$) is the best possible, we have

$$||T|| = ||\widetilde{T}|| = k(\lambda_1) = \frac{\pi}{\lambda \sin(\frac{\pi \lambda_1}{\lambda})}.$$

(ii) If $k_\lambda(\ln x, \ln n) = \frac{1}{(\ln xn)^\lambda}$ ($\lambda_1 > 0$, $0 < \lambda_2 \leq 1$, $\lambda_1 + \lambda_2 = \lambda$), since for $x > 1$,

$$f(x, y) = \frac{x^{\lambda_1}}{(x + \ln y)^\lambda y} (\ln y)^{\lambda_2 - 1}$$

is decreasing with respect to $y \in (1, \infty)$, then Condition (i) is satisfied. In view of Example 3.1(ii), the constant factor in (3.42) and (3.43) (for $u(x) = x, v(n) = \ln n$) is the best possible, we have (see Yang [163])

$$||T|| = ||\widetilde{T}|| = k(\lambda_1) = B(\lambda_1, \lambda_2).$$

(iii) If $k_\lambda (\ln x, \ln n) = \frac{\ln[(\ln x)/(\ln n)]}{(\ln x)^\lambda - (\ln n)^\lambda}$ $(\lambda_1 > 0, 0 < \lambda_2 \leq 1, \lambda_1 + \lambda_2 = \lambda)$, then by the same way and Example 3.1(iii), we have

$$||T|| = ||\widetilde{T}|| = k(\lambda_1) = \left[\frac{\pi}{\lambda \sin(\frac{\pi \lambda_1}{\lambda})} \right]^2.$$

(iv) If $s \in \mathbf{N}$, $k_{\lambda s}(\ln x, \ln n) = \prod_{k=1}^s \frac{1}{a_k (\ln x)^\lambda + (\ln n)^\lambda}$ $(0 < a_1 < \cdots < a_s, \lambda, \lambda_1 > 0, 0 < \lambda_2 \leq 1, \lambda_1 + \lambda_2 = \lambda s)$, then by the same way and Example 3.1(iv), we have

$$||T|| = ||\widetilde{T}|| = k(\lambda_1) = \frac{\pi}{\lambda \sin(\frac{\pi \lambda_2}{\lambda})} \sum_{k=1}^s a_k^{\frac{\lambda_2}{\lambda} - 1} \prod_{j=1(j \neq k)}^s \frac{1}{a_j - a_k}.$$

(v) If $k_\lambda(\ln x, \ln n) = \frac{1}{(\ln x)^\lambda + 2(\ln x \ln n)^{\lambda/2} \cos \beta + c(\ln n)^\lambda}$ $(\lambda > 0, 0 < \beta \leq \frac{\pi}{2}, \lambda_1 > 0, 0 < \lambda_2 \leq 1, \lambda_1 + \lambda_2 = \lambda)$, then by the same way and Example 3.1(v), we have

$$||T|| = ||\widetilde{T}|| = k(\lambda_1) = \frac{2\pi \sin \beta (1 - \frac{2\lambda_1}{\lambda})}{\lambda \sin \beta \sin(\frac{2\pi \lambda_1}{\lambda})}.$$

(vi) If $k_\lambda(\ln x, \ln n) = \frac{1}{(\ln x)^\lambda + b(\ln x \ln n)^{\lambda/2} + c(\ln n)^\lambda}$ $(c > 0, 0 \leq b \leq 2\sqrt{c}, 0 < \lambda \leq 2, \lambda_1 = \lambda_2 = \frac{\lambda}{2})$, then by the same way and Example 3.1(vi), we have

$$||T|| = ||\widetilde{T}|| = \frac{2}{\lambda} \begin{cases} \frac{\pi}{\sqrt{c}}, & b = 0, \\ \frac{4}{\sqrt{4c-b^2}} \arctan \frac{\sqrt{4c-b^2}}{b}, & 0 < b < 2\sqrt{c}, \\ \frac{2}{\sqrt{c}}, & b = 2\sqrt{c}. \end{cases}$$

(vii) If $k_0(\ln x, \ln n) = \frac{1}{e^{\gamma(\ln n/\ln x)^\lambda} - Ae^{-\gamma(\ln n/\ln x)^\lambda}}$ $(-1 \leq A < 1, \beta > 0(A = 1, \beta > 1), \gamma > 0, -1 \leq \lambda_1 = -\lambda_2 = -\beta\lambda < 0 < \lambda)$, then by the same way and Example 3.1(vii), we have

$$||T|| = ||\widetilde{T}|| = k(-\beta\lambda) = \frac{\Gamma(\beta)}{\lambda \gamma^\beta} \sum_{k=0}^\infty \frac{A^k}{(2k+1)^\beta}.$$

(viii) If $k_0(\ln x, \ln n) = \ln \left[\frac{b(\ln x)^\lambda + (\ln n)^\lambda}{a(\ln x)^\lambda + (\ln n)^\lambda} \right]$ $(0 \leq a < b, \lambda > 0, -1 \leq \lambda_1 = -\lambda_2 = -\beta\lambda, 0 < \beta < 1)$, then by the same way and Example 3.1(viii), we have

$$||T|| = ||\widetilde{T}|| = k(-\beta\lambda) = \frac{(b^\beta - a^\beta)\pi}{\beta\lambda \sin(\beta\pi)}.$$

(ix) If $k_0(\ln x,\ \ln n) = \arctan\left(\frac{\ln x}{\ln n}\right)^\lambda$ ($\lambda_1 = -\lambda_2 = -\beta$, $0 < \beta < \lambda$, $\beta \leq 1$), then by the same way and Example 3.1(ix), we have

$$||T|| = ||\widetilde{T}|| = k(-\beta) = \frac{\pi}{2\beta\cos(\frac{\pi\beta}{2\lambda})}.$$

(x) If $k_\lambda(\ln x,\ \ln n) = \frac{1}{(\max\{\ln x,\ln n\})^\lambda}$ ($\lambda_1 > 0$, $0 < \lambda_2 \leq 1$, $\lambda_1 + \lambda_2 = \lambda$), then by the same way and Example 3.1(x), we have

$$||T|| = ||\widetilde{T}|| = k(\lambda_1) = \frac{\lambda}{\lambda_1\lambda_2}.$$

(xi) If $k_{-\lambda}(\ln x,\ \ln n) = (\min\{\ln x,\ln n\})^\lambda$ ($\lambda_1 < 0$, $-1 \leq \lambda_2 < 0$, $\lambda_1 + \lambda_2 = -\lambda$), then by the same way and Example 3.1(xi), we have

$$||T|| = ||\widetilde{T}|| = k(\lambda_1) = \frac{\lambda}{\lambda_1\lambda_2}.$$

(xii) If $k_0(\ln x,\ \ln n) = \left(\frac{\min\{\ln x,\ln n\}}{\max\{\ln x,\ln n\}}\right)^\lambda$ ($\lambda_2 = -\lambda_1, |\lambda_1| < \lambda \leq 1 + \lambda_1, \lambda_1 > -\frac{1}{2}$), then, by the same way and Example 3.1(xii), we have

$$||T|| = ||\widetilde{T}|| = k(\lambda_1) = \frac{2\lambda}{\lambda^2 - \lambda_1^2}.$$

Example 3.4. For $0 \leq \gamma_1,\ \gamma_2 \leq \frac{1}{2}$, setting $u(x) = x - \gamma_1 (x \in (\gamma_1, \infty))$, $v(n) = n - \gamma_2$, $n \geq n_0 = 1$, then $\frac{v'(y)}{v(y)} = \frac{1}{y - \gamma_2}(> 0)$ is decreasing for $y > \frac{1}{2}$.

(i) If $k_\lambda(x - \gamma_1, n - \gamma_2) = \frac{1}{(x-\gamma_1)^\lambda + (n-\gamma_2)^\lambda}$ ($\lambda_1 > 0$, $0 < \lambda_2 \leq \min\{1, 2 - \lambda\}$, $\lambda_1 + \lambda_2 = \lambda$), for $x > 0$,

$$f(x,y) = \frac{x^{\lambda_1}}{x^\lambda + (y - \gamma_2)^\lambda}(y - \gamma_2)^{\lambda_2 - 1}$$

is strictly convex satisfying $f''_{y^2}(x,y) > 0$ with respect to $y \in \left(\frac{1}{2}, \infty\right)$, then Condition (ii) is satisfied. In view of Example 3.1(i), the constant factor in (3.42) and (3.43) (for $u(x) = x - \gamma_1$, $v(n) = n - \gamma_2$) is the best possible, we have

$$||T|| = ||\widetilde{T}|| = \frac{\pi}{\lambda\sin(\frac{\pi\lambda_1}{\lambda})}.$$

(ii) If $k_\lambda(x - \gamma_1,\ n - \gamma_2) = \frac{1}{(x - \gamma_1 + n - \gamma_2)^\lambda}$ ($\lambda_1 > 0$, $0 < \lambda_2 \leq 1$, $\lambda_1 + \lambda_2 = \lambda$), by the same way and Example 3.1(ii), we have (see Huang and Yang [38])

$$||T|| = ||\widetilde{T}|| = B(\lambda_1, \lambda_2).$$

(iii) If $k_\lambda(x - \gamma_1, \, n - \gamma_2) = \frac{\ln[(x-\gamma_1)/(n-\gamma_2)]}{(x-\gamma_1)^\lambda - (n-\gamma_2)^\lambda}$ ($\lambda_1 > 0$, $0 < \lambda_2 \leq 1$, $\lambda_1 + \lambda_2 = \lambda$), then, by the same way and Example 3.1(iii), we have

$$||T|| = ||\widetilde{T}|| = \left[\frac{\pi}{\lambda \sin(\frac{\pi \lambda_1}{\lambda})} \right]^2.$$

(iv) If $s \in \mathbf{N}$, $k_{\lambda s}(x - \gamma_1, n - \gamma_2) = \prod_{k=1}^{s} \frac{1}{a_k(x-\gamma_1)^\lambda + (n-\gamma_2)^\lambda}$ ($0 < a_1 < \cdots < a_s$, $\lambda_1 > 0$, $0 < \lambda_2 \leq \min\{1, 2-\lambda\}$, $\lambda_1 + \lambda_2 = \lambda s$), then by the same way and Example 3.1(iv), we have

$$||T|| = ||\widetilde{T}|| = k(\lambda_1) = \frac{\pi}{\lambda \sin(\frac{\pi \lambda_2}{\lambda})} \sum_{k=1}^{s} a_k^{\frac{\lambda_2}{\lambda}-1} \prod_{j=1(j \neq k)}^{s} \frac{1}{a_j - a_k}.$$

(v) If $k_\lambda(x - \gamma_1, \, n - \gamma_2) = \frac{1}{(x-\gamma_1)^\lambda + 2[(x-\gamma_1)(n-\gamma_2)]^{\lambda/2} \cos \beta + (n-\gamma_2)^\lambda}$ ($\lambda > 0$, $0 < \beta \leq \frac{\pi}{2}$, $\lambda_1 > 0$, $0 < \lambda_2 \leq \min\{1, 2-\lambda\}$, $\lambda_1 + \lambda_2 = \lambda$), then by the same way and Example 3.1(v), we have

$$||T|| = ||\widetilde{T}|| = k(\lambda_1) = \frac{2\pi \sin \beta(1 - \frac{2\lambda_1}{\lambda})}{\lambda \sin \beta \sin(\frac{2\pi \lambda_1}{\lambda})}.$$

(vi) If $k_\lambda(x - \gamma_1, \, n - \gamma_2) = \frac{1}{(x-\gamma_1)^\lambda + b[(x-\gamma_1)(n-\gamma_2)]^{\lambda/2} + c(n-\gamma_2)^\lambda}$ ($c > 0$, $0 \leq b \leq 2\sqrt{c}$, $0 < \lambda \leq \frac{4}{3}$, $\lambda_1 = \lambda_2 = \frac{\lambda}{2}$), then by the same way and Example 3.1(vi), we have

$$||T|| = ||\widetilde{T}|| = \frac{2}{\lambda} \begin{cases} \frac{\pi}{\sqrt{c}}, & b = 0, \\ \frac{4}{\sqrt{4c-b^2}} \arctan \frac{\sqrt{4c-b^2}}{b}, & 0 < b < 2\sqrt{c}, \\ \frac{2}{\sqrt{c}}, & b = 2\sqrt{c}. \end{cases}$$

(vii) If $k_0(x - \gamma_1, \, n - \gamma_2) = \frac{1}{e^{\gamma[n-\gamma_2)/(x-\gamma_1)]^\lambda} - Ae^{-\gamma[n-\gamma_2)/(x-\gamma_1)]^\lambda}}$ ($0 \leq A < 1$, $\beta > 0(A = 1, \beta > 1)$, $\gamma > 0$, $-1 \leq \lambda_1 = -\lambda_2 = -\beta\lambda < 0 < \lambda \leq 1$), then, for $x > 0$,

$$f(x, y) = \frac{x^{-\beta\lambda}}{e^{\frac{\gamma}{x^\lambda}(y-\gamma_2)^\lambda} - Ae^{\frac{-\gamma}{x^\lambda}(y-\gamma_2)^\lambda}} (y - \gamma_2)^{\beta\lambda-1}$$

is strictly convex. In fact, setting $g(x, u) = \dfrac{1}{e^{\frac{\gamma}{x^\lambda}u} - A e^{\frac{-\gamma}{x^\lambda}u}}$, we find

$$g'_u(x, u) = -\frac{\frac{\lambda\gamma}{x^\lambda}(e^{\frac{\gamma}{x^\lambda}u} + A e^{\frac{-\gamma}{x^\lambda}u})}{(e^{\frac{\gamma}{x^\lambda}u} - A e^{\frac{-\gamma}{x^\lambda}u})^2} < 0,$$

$$g''_{u^2}(x, u) = \frac{2(\frac{\lambda\gamma}{x^\lambda})^2(e^{\frac{\gamma}{x^\lambda}u} + A e^{\frac{-\gamma}{x^\lambda}u})^2}{(e^{\frac{\gamma}{x^\lambda}u} - A e^{\frac{-\gamma}{x^\lambda}u})^3} - \frac{(\frac{\lambda\gamma}{x^\lambda})^2(e^{\frac{\gamma}{x^\lambda}u} - A e^{\frac{-\gamma}{x^\lambda}u})}{(e^{\frac{\gamma}{x^\lambda}u} - A e^{\frac{-\gamma}{x^\lambda}u})^2}$$

$$= \frac{(\frac{\lambda\gamma}{x^\lambda})^2[2(e^{\frac{\gamma}{x^\lambda}u} + A e^{\frac{-\gamma}{x^\lambda}u})^2 - (e^{\frac{\gamma}{x^\lambda}u} - A e^{\frac{-\gamma}{x^\lambda}u})^2]}{(e^{\frac{\gamma}{x^\lambda}u} - A e^{\frac{-\gamma}{x^\lambda}u})^3}$$

$$= \frac{(\frac{\lambda\gamma}{x^\lambda})^2[(e^{\frac{\gamma}{x^\lambda}u} + A e^{\frac{-\gamma}{x^\lambda}u})^2 + 4A]}{(e^{\frac{\gamma}{x^\lambda}u} - A e^{\frac{-\gamma}{x^\lambda}u})^3} > 0 \quad (0 \le A \le 1),$$

and then, $g''_{y^2}(x, (y - \gamma_2)^\lambda) > 0 \ (0 < \lambda \le 1)$. Hence,

$$f''_{y^2}(x, y) = \left[x^{-\beta\lambda} g\left(x, (y - \gamma_2)^\lambda\right)(y - \gamma_2)^{\beta\lambda - 1}\right]''_{y^2} > 0, \quad \left(y \in \left(\frac{1}{2}, \infty\right)\right).$$

Then Condition (ii) is satisfied and in view of Example 3.1(vii), we have

$$||T|| = ||\widetilde{T}|| = k(-\beta\lambda) = \frac{\Gamma(\beta)}{\lambda\gamma^\beta} \sum_{k=0}^{\infty} \frac{A^k}{(2k+1)^\beta}.$$

(viii) If $k_0(x - \gamma_1, \ n - \gamma_2) = \ln[\frac{b(x - \gamma_1)^\lambda + (n - \gamma_2)^\lambda}{a(x - \gamma_1)^\lambda + (n - \gamma_2)^\lambda}] \ (0 \le a < b, \ 0 < \lambda \le 1, \ -1 \le \lambda_1 = -\lambda_2 = -\beta\lambda, \ 0 < \beta < 1)$, then, by the same way and Example 3.1(viii), we have

$$||T|| = ||\widetilde{T}|| = \frac{(b^\beta - a^\beta)\pi}{\beta\lambda\sin(\beta\pi)}.$$

(ix) If $k_0(x - \gamma_1, \ n - \gamma_2) = \arctan\left(\frac{x - \gamma_1}{n - \gamma_2}\right)^\lambda \ (\lambda_1 = -\lambda_2 = -\beta, \ 0 < \beta < \lambda, \ \beta \le \min\{1, 2 - \lambda\})$, then by the same way and Example 3.1(ix), we have

$$||T|| = ||\widetilde{T}|| = \frac{\pi}{2\beta\cos(\frac{\pi\beta}{2\lambda})}.$$

Example 3.5. For $a \ge \frac{2}{3}$, setting $u(x) = \ln ax \ (x \in (\frac{1}{a}, \infty))$, $v(n) = \ln an, \ n \ge n_0 = 2$, then $\frac{v'(y)}{v(y)} = \frac{1}{y \ln ay}(> 0)$ is decreasing for $y > \frac{3}{2}$.

(i) If $k_\lambda(\ln ax, \ \ln an) = \frac{1}{(\ln ax)^\lambda + (\ln an)^\lambda} \ (\lambda_1 > 0, \ 0 < \lambda_2 \le \min\{1, \ 2 - \lambda\}, \ \lambda_1 + \lambda_2 = \lambda)$, since for $x > 0$,

$$f(x, y) = \frac{x^{\lambda_1}}{[x^\lambda + (\ln ay)^\lambda]y}(\ln ay)^{\lambda_2 - 1}$$

is strictly convex satisfying $f''_{y^2}(x,y) > 0$ with respect to $y \in \left(\frac{3}{2}, \infty\right)$, then Condition (ii) is satisfied. In view of Example 3.1(i), the constant factor in (3.42) and (3.43) (for $u(x) = \ln ax, v(n) = \ln an$) is the best possible, we have

$$||T|| = ||\widetilde{T}|| = \frac{\pi}{\lambda \sin(\frac{\pi \lambda_1}{\lambda})}.$$

(ii) If $k_\lambda(\ln ax, \ln an) = \frac{1}{(\ln a^2 xn)^\lambda}$ $(\lambda_1 > 0,\ 0 < \lambda_2 \leq 1,\ \lambda_1 + \lambda_2 = \lambda)$, by the same way and Example 3.1(ii), we have (see Chen and Yang [7])

$$||T|| = ||\widetilde{T}|| = B(\lambda_1, \lambda_2).$$

(iii) If $k_\lambda(\ln ax,\ \ln an) = \frac{\ln[(\ln ax)/(\ln an)]}{(\ln ax)^\lambda - (\ln an)^\lambda}$ $(\lambda_1 > 0,\ 0 < \lambda_2 \leq 1,\ \lambda_1 + \lambda_2 = \lambda)$, then by the same way and Example 3.1(iii), we have

$$||T|| = ||\widetilde{T}|| = \left[\frac{\pi}{\lambda \sin(\frac{\pi \lambda_1}{\lambda})}\right]^2.$$

(iv) If $s \in \mathbf{N}$, $k_{\lambda s}(\ln ax,\ \ln an) = \prod_{k=1}^s \frac{1}{a_k(\ln ax)^\lambda + (\ln an)^\lambda}$ $(0 < a_1 < \cdots < a_s,\ \lambda,\ \lambda_1 > 0,\ 0 < \lambda_2 \leq \min\{1,\ 2 - \lambda\},\ \lambda_1 + \lambda_2 = \lambda s)$, then, by the same way and Example 3.1(iv), we have

$$||T|| = ||\widetilde{T}|| = k(\lambda_1) = \frac{\pi}{\lambda \sin(\frac{\pi \lambda_2}{\lambda})} \sum_{k=1}^s a_k^{\frac{\lambda_2}{\lambda} - 1} \prod_{j=1(j \neq k)}^s \frac{1}{a_j - a_k}.$$

(v) If $k_\lambda(\ln ax, \ln an) = \frac{1}{(\ln ax)^\lambda + 2(\ln ax \ln an)^{\lambda/2} \cos \beta + (\ln an)^\lambda}$ $(\lambda > 0,\ 0 < \beta \leq \frac{\pi}{2},\ \lambda_1 > 0,\ 0 < \lambda_2 \leq \min\{1, 2 - \lambda\},\ \lambda_1 + \lambda_2 = \lambda)$, then by the same way and Example 3.1(v), we have

$$||T|| = ||\widetilde{T}|| = k(\lambda_1) = \frac{2\pi \sin \beta (1 - \frac{2\lambda_1}{\lambda})}{\lambda \sin \beta \sin(\frac{2\pi \lambda_1}{\lambda})}.$$

(vi) If $k_\lambda(\ln ax,\ \ln an) = \frac{1}{(\ln ax)^\lambda + b(\ln ax \ln an)^{\lambda/2} + c(\ln an)^\lambda}$ $(c > 0,\ 0 \leq b \leq 2\sqrt{c},\ 0 < \lambda \leq \frac{4}{3},\ \lambda_1 = \lambda_2 = \frac{\lambda}{2})$, then by the same way and Example 3.1(vi), we have

$$||T|| = ||\widetilde{T}|| = \frac{2}{\lambda} \begin{cases} \frac{\pi}{\sqrt{c}}, & b = 0, \\ \frac{4}{\sqrt{4c - b^2}} \arctan \frac{\sqrt{4c - b^2}}{b}, & 0 < b < 2\sqrt{c}, \\ \frac{2}{\sqrt{c}}, & b = 2\sqrt{c}. \end{cases}$$

(vii) If $k_0(\ln ax,\ \ln an) = \frac{1}{e^{\gamma[\ln an]/\ln ax]^\lambda} - A e^{-\gamma[\ln an]/\ln ax]^\lambda}}$ $(0 \leq A < 1, \beta > 0(A = 1, \beta > 1), \gamma > 0,\ -1 \leq \lambda_1 = -\lambda_2 = -\beta\lambda < 0 < \lambda \leq 1)$, then by the same way and Example 3.1(vii), we have

$$||T|| = ||\widetilde{T}|| = k(-\beta\lambda) = \frac{\Gamma(\beta)}{\lambda \gamma^\beta} \sum_{k=0}^\infty \frac{A^k}{(2k + 1)^\beta}.$$

(viii) If $k_0(\ln ax, \ln an) = \ln[\frac{b(\ln ax)^\lambda + (\ln an)^\lambda}{a(\ln ax)^\lambda + (\ln an)^\lambda}]$ $(0 \le a < b, 0 < \lambda \le 1,$ $-1 \le \lambda_1 = -\lambda_2 = -\beta\lambda, 0 < \beta < 1)$, then by the same way and Example 3.1(viii), we have

$$||T|| = ||\widetilde{T}|| = \frac{(b^\beta - a^\beta)\pi}{\beta\lambda \sin(\beta\pi)}.$$

(ix) If $k_0(\ln ax, \ln an) = \arctan(\frac{\ln ax}{\ln an})^\lambda$ $(\lambda_1 = -\lambda_2 = -\beta, 0 < \beta < \lambda, \beta \le \min\{1, 2-\lambda\})$, then by the same way and Example 3.1(ix), we have

$$||T|| = ||\widetilde{T}|| = \frac{\pi}{2\beta \cos(\frac{\pi\beta}{2\lambda})}.$$

Note. (1) For $\gamma_1 = \gamma_2 = 0$ in Example 3.4(i)-(ix), we obtain the corresponding results in Example 3.11; for $a = 1$ in Example 3.5(i)-(ix), we obtain the corresponding results in Example 3.3. Since For $u(x) = x - \gamma_1$, $v(n) = n - \gamma_2$ or $u(x) = \ln ax$, $v(n) = \ln an$ in Example 3.1(x)-(xii), the corresponding $f(x, y)$ is not convex, so we cannot use Condition (ii) to derive some similar results.

(2) For $a \le \frac{1}{2}$, setting $u(x) = \ln(x-a)$ $(x \in (a+1, \infty))$, $v(n) = \ln(n-a)$, $n \ge n_0 = 2$, then $\frac{v'(y)}{v(y)} = \frac{1}{(y-a)\ln(y-a)}(> 0)$ is decreasing for $y > \frac{3}{2}$. In this case, we can still obtain some similar results of Example 3.5(i)-(ix).

By the same way, we still have

Corollary 3.3. *Let the assumptions of Theorem 3.2 be fulfilled and additionally, $u(x)$ is a strictly increasing differentiable function in (b, c), with $u(b^+) = 0$ and $u(c^-) = \infty$. For $0 < p < 1$, $\frac{1}{p} + \frac{1}{q} = 1$, setting $\widetilde{\Phi}(x) = \Phi(x)(1 - \theta_{\lambda_1}(u(x)))$ $(x \in (b, c))$, if $f(x)$, $a_n \ge 0$, $f \in L_{p,\widetilde{\Phi}}(b, c)$, $a = \{a_n\}_{n=n_0}^\infty \in l_{q,\Psi}$, $||f||_{p,\widetilde{\Phi}} > 0$, $||a||_{q,\Psi} > 0$, then, we have the following equivalent inequalities:*

$$\sum_{n=n_0}^\infty a_n \int_b^c k_\lambda(u(x), v(n))f(x)dx$$

$$= \int_b^c f(x) \sum_{n=n_0}^\infty a_n k_\lambda(u(x), v(n)) \, dx > k(\lambda_1)||f||_{p,\widetilde{\Phi}}||a||_{q,\Psi}, \quad (3.45)$$

$$\left\{\sum_{n=n_0}^\infty [\Psi(n)]^{1-p} \left[\int_b^c k_\lambda(u(x), v(n))f(x) \, dx\right]^p\right\}^{\frac{1}{p}} > k(\lambda_1)||f||_{p,\widetilde{\Phi}},$$

$$(3.46)$$

and

$$\left\{\int_b^c [\widetilde{\Phi}(x)]^{1-q} \left[\sum_{n=n_0}^\infty k_\lambda(u(x), v(n))a_n\right]^q \, dx\right\}^{\frac{1}{q}} > k(\lambda_1)||a||_{q,\Psi}. \quad (3.47)$$

Moreover, if $\frac{v'(y)}{v(y)}(>0)$ is decreasing in $[n_0, \infty)$ and there exist constants η, $\delta_2 > 0$, such that $k(\lambda_1 - \delta_2) \in \mathbf{R}_+$ and for any $\widetilde{\lambda}_1 \in (\lambda_1 - \delta_2, \lambda_1)$, $\theta_{\widetilde{\lambda}_1}(x) = O\left(\frac{1}{x^\eta}\right)$ $(x \geq 1)$, then the constant factor $k(\lambda_1)$ in the above inequalities is the best possible.

In view of Remark 3.1, Corollary 3.1 and Theorem 3.3, we still have

Corollary 3.4. *Suppose that λ, λ_1, $\lambda_2 \in \mathbf{R}$, $\lambda = \lambda_1 + \lambda_2$, $k_\lambda(x, y)$ is a non-negative finite measurable homogeneous function of degree $-\lambda$ in $\mathbf{R}_+^2$, $v(y)$ is a strictly increasing differentiable function in $[n_0, \infty)$ $(n_0 \in \mathbf{N})$ with $v(n_0) > 0$ and $v(\infty) = \infty$, $u(x)$ is a strict increasing differentiable function in (b, c), with $u(b^+) = 0$ and $u(c^-) = \infty$, $\frac{v'(y)}{v(y)}(>0)$ is decreasing in $[n_0, \infty)$, and there exist constants δ_0, $\eta > 0$, such that for any $\widetilde{\lambda}_1 \in (\lambda_1 - \delta_0, \lambda_1 + \delta_0)(\widetilde{\lambda}_2 = \lambda - \widetilde{\lambda}_1)$,*

$$0 < k(\widetilde{\lambda}_1)(1 - \theta_{\widetilde{\lambda}_1}(x)) < \varpi_{\widetilde{\lambda}_1}(x) < k(\widetilde{\lambda}_1) < \infty \ (x \in (0, \infty)), \qquad (3.48)$$

where, $\theta_{\widetilde{\lambda}_1}(x) > 0$ $(x \in (0, \infty))$, and $\theta_{\widetilde{\lambda}_1}(x) = O\left(\frac{1}{x^\eta}\right)$ $(x \in [1, \infty))$.

(i) For $p > 1(p < 0)$, $\frac{1}{p} + \frac{1}{q} = 1$, $f(x)$, $a_n \geq 0$, $f \in L_{p,\Phi}(b, c)$, $a = \{a_n\}_{n=n_0}^\infty \in l_{q,\Psi}$, $||f||_{p,\Phi} > 0$, $||a||_{q,\Psi} > 0$, if $p > 1$, then we have the equivalent inequalities (3.41), (3.42) and (3.43), with the best constant factor $k(\lambda_1)$; if $p < 0$, then we have the equivalent reverses of (3.41), (3.42) and (3.43), with the same best constant factor $k(\lambda_1)$.

(ii) For $0 < p < 1$, $\frac{1}{p} + \frac{1}{q} = 1$, if $f(x)$, $a_n \geq 0$, $f \in L_{p,\widetilde{\Phi}}(b, c)$, $a = \{a_n\}_{n=n_0}^\infty \in l_{q,\Psi}$, $||f||_{p,\widetilde{\Phi}} > 0$, $||a||_{q,\Psi} > 0$, then we have the equivalent inequalities (3.45), (3.46) and (3.47) with the same best constant factor $k(\lambda_1)$.

In particular, for $\lambda = 0$, $\lambda_1 = \alpha$, $\lambda_2 = -\alpha$ in Corollary 3.1, Theorem 3.3 and Corollary 3.2, setting

$$\Phi_0(x) = \frac{[u(x)]^{p(1-\alpha)-1}}{[u'(x)]^{p-1}} \ (x \in (b, c)),$$

$$\Psi_0(n) = \frac{[v(n)]^{q(1+\alpha)-1}}{[v'(n)]^{q-1}} \ (n \geq n_0, n \in \mathbf{N}),$$

and $\widetilde{\Phi}_0(x) = (1 - \theta_\alpha(u(x)))\Phi_0(x)$, we have the following corollary:

Corollary 3.5. *(see Yang and Krnic [157]) Suppose that $k_0(x, y)$ is a non-negative finite measurable homogeneous function of degree 0 in $\mathbf{R}_+^2$,*

$$k(\alpha) = \int_0^\infty k_0(t, 1)t^{\alpha-1}dt \in \mathbf{R}_+ \ (\alpha \in \mathbf{R}),$$

$v(y)$ *is a strict increasing differentiable function in* $[n_0, \infty)(n_0 \in \mathbf{N})$ *with* $v(n_0) > 0$ *and* $v(\infty) = \infty$, $u(x)$ *is a strict increasing differentiable function in* (b, c), *with* $u(b^+) = 0$ *and* $u(c^-) = \infty$. *If* $p > 1$, *or* $p < 0$, $\frac{1}{p} + \frac{1}{q} = 1$, $f(x)$, $a_n \geq 0$, $f \in L_{p,\Phi_0}(b, c)$, $a = \{a_n\}_{n=n_0}^\infty \in l_{q,\Psi_0}$, $||f||_{p,\Phi_0} > 0$, $||a||_{q,\Psi_0} > 0$, *then*

  (i) for $p > 1$, *we have the following equivalent inequalities:*

$$\sum_{n=n_0}^\infty a_n \int_b^c k_0(u(x), v(n)) f(x)\, dx$$

$$= \int_b^c f(x) \sum_{n=n_0}^\infty a_n k_0(u(x), v(n))\, dx < k(\alpha) ||f||_{p,\Phi_0} ||a||_{q,\Psi_0}, \quad (3.49)$$

$$\left\{ \sum_{n=n_0}^\infty [\Psi_0(n)]^{1-p} \left[\int_b^c k_0(u(x), v(n)) f(x) dx \right]^p \right\}^{\frac{1}{p}} < k(\alpha) ||f||_{p,\Phi_0},$$

$$(3.50)$$

and

$$\left\{ \int_0^\infty [\Phi_0(x)]^{1-q} \left[\sum_{n=n_0}^\infty k_0(u(x), v(n)) a_n \right]^q dx \right\}^{\frac{1}{q}} < k(\alpha) ||a||_{q,\Psi_0}; \quad (3.51)$$

  (ii) for $p < 0$, *we have the equivalent reverses of (3.49), (3.50) and (3.51).*

  Moreover, if $\frac{v'(y)}{v(y)}(> 0)$ is decreasing in $[n_0, \infty)$ and there exist constants $\eta < \alpha$, $L > 0$, and $\delta_1 > 0$, such that

$$k_0(t, 1) \leq L\left(\frac{1}{t^\eta}\right) \left(t \in \left(0, \frac{1}{v(n_0)}\right]\right), \quad (3.52)$$

and $k(\alpha + \delta_1) \in \mathbf{R}_+$, then the constant factor $k(\alpha)$ in the inequalities of (i) and the reverses in (ii) is the best possible.

Corollary 3.6. *Let the assumptions of Corollary 3.5 be fulfilled. If* $0 < p < 1$, $\frac{1}{p} + \frac{1}{q} = 1$, $f(x)$, $a_n \geq 0$, $f \in L_{p,\tilde{\Phi}_0}(b, c)$, $a = \{a_n\}_{n=n_0}^\infty \in l_{q,\Psi_0}$, $||f||_{p,\tilde{\Phi}_0} > 0$, $||a||_{q,\Psi_0} > 0$, *then we have the following equivalent* *inequalities:*

$$\sum_{n=n_0}^{\infty} a_n \int_b^c k_0(u(x), v(n)) f(x)\, dx$$

$$= \int_b^c f(x) \sum_{n=n_0}^{\infty} a_n k_0(u(x), v(n)) dx > k(\alpha) \|f\|_{p,\widetilde{\Phi}_0} \|a\|_{q,\Psi_0}, \quad (3.53)$$

$$\left\{ \sum_{n=n_0}^{\infty} [\Psi_0(n)]^{1-p} \left[\int_b^c k_0(u(x), v(n)) f(x) dx \right]^p \right\}^{\frac{1}{p}} > k(\alpha) \|f\|_{p,\widetilde{\Phi}_0}, $$

$$(3.54)$$

and

$$\left\{ \int_0^{\infty} [\widetilde{\Phi}_0(x)]^{1-q} \left[\sum_{n=n_0}^{\infty} k_0(u(x), v(n)) a_n \right]^q dx \right\}^{\frac{1}{q}} > k(\alpha) \|a\|_{q,\Psi_0}. \quad (3.55)$$

Moreover, if $\frac{v'(y)}{v(y)} (> 0)$ is decreasing in $[n_0, \infty)$ and there exists a constant $\delta_2 > 0$, such that $k(\alpha - \delta_2) \in \mathbf{R}_+$, then the constant factor $k(\alpha)$ in the above inequalities is the best possible.

Corollary 3.7. *Suppose that $k_0(x, y)$ is a non-negative finite measurable homogeneous function of degree 0 in $\mathbf{R}_+^2$,*

$$k(\alpha) = \int_0^{\infty} k_0(t, 1)\, t^{\alpha-1} dt \qquad (\alpha \in \mathbf{R}),$$

$v(y)$ is a strictly increasing differentiable function in $[n_0, \infty)(n_0 \in \mathbf{N})$ with $v(n_0) > 0$ and $v(\infty) = \infty, u(x)$ is a strictly increasing differentiable function in (b, c), with $u(b^+) = 0$ and $u(c^-) = \infty$, $\frac{v'(y)}{v(y)}$ is decreasing in $[n_0, \infty)$, and there exist constants $\delta < \alpha$ and $\delta_0, \eta > 0$, such that for any $\widetilde{\alpha} \in (\alpha - \delta_0, \alpha + \delta_0)$,

$$0 < k(\widetilde{\alpha})(1 - \theta_{\widetilde{\alpha}}(x)) < \varpi_{\widetilde{\alpha}}(x) < k(\widetilde{\alpha}) < \infty, \qquad (x \in (0, \infty)), \quad (3.56)$$

where, $\theta_{\widetilde{\alpha}}(x) > 0 \ (x \in (0, \infty)), \theta_{\widetilde{\alpha}}(x) = O\left(\frac{1}{x^{\eta}}\right) \ (x \in [1, \infty))$.

(i) If $p > 1, \frac{1}{p} + \frac{1}{q} = 1, f(x), a_n \geq 0, f \in L_{p,\Phi_0}(b, c), a = \{a_n\}_{n=n_0}^{\infty} \in l_{q,\Psi_0}, \|f\|_{p,\Phi_0} > 0, \|a\|_{q,\Psi_0} > 0$, then we have the equivalent inequalities (3.49), (3.50) and (3.51), with the best constant factor $k(\alpha)$.

(ii) If $p < 0$, then we have the equivalent reverses of (3.49), (3.50) and (3.51) with the same best constant factor $k(\alpha)$.

(iii) If $0 < p < 1, \frac{1}{p} + \frac{1}{q} = 1, f(x), a_n \geq 0, f \in L_{p,\widetilde{\Phi}_0}(b, c), a = \{a_n\}_{n=n_0}^{\infty} \in l_{q,\Psi_0}, \|f\|_{p,\widetilde{\Phi}_0} > 0, \|a\|_{q,\Psi_0} > 0$, then, we have the equivalent

inequalities (3.53), (3.54) and (3.55) with the same best constant factor $k(\alpha)$.

For $v(x) = u(x) = x$ $(x \in (0, \infty))$, $n_0 = 1$, in Corollary 3.15, setting $\varphi(x) = x^{p(1-\lambda_1)-1}$ $(x \in (0, \infty))$, $\psi(n) = n^{q(1-\lambda_2)-1}$ $(n \in \mathbf{N})$, we have

$$\omega(\lambda_2, n) = n^{\lambda_2} \int_0^\infty k_\lambda(x, n) x^{\lambda_1-1} \, dx = k(\lambda_1), \qquad n \in \mathbf{N}, \quad (3.57)$$

$$\varpi(\lambda_1, x) = x^{\lambda_1} \sum_{n=1}^\infty k_\lambda(x, n) n^{\lambda_2-1}, \qquad x \in (0, \infty), \quad (3.58)$$

and the following corollary:

Corollary 3.8. *Suppose that* $\lambda, \lambda_1, \lambda_2 \in \mathbf{R}$, $\lambda = \lambda_1 + \lambda_2$, $k_\lambda(x, y)$ *is a non-negative finite measurable homogeneous function of degree* $-\lambda$ *in* $\mathbf{R}_+^2$, *there exist constants* $\delta_0, \eta > 0$, *such that for any* $\widetilde{\lambda}_1 \in (\lambda_1 - \delta_0, \lambda_1 + \delta_0)(\widetilde{\lambda}_2 = \lambda - \widetilde{\lambda}_1)$,

$$0 < k(\widetilde{\lambda}_1)(1 - \theta_{\widetilde{\lambda}_1}(x)) < \varpi(\widetilde{\lambda}_1, x) < k(\widetilde{\lambda}_1) < \infty \ (x \in (0, \infty)), \quad (3.59)$$

where, $\theta_{\widetilde{\lambda}_1}(x) > 0$ $(x \in (0, \infty))$, *and* $\theta_{\widetilde{\lambda}_1}(x) = O\left(\frac{1}{x^\eta}\right)$ $(x \in [1, \infty))$.

(i) For $p > 1$ or $p < 0$, $\frac{1}{p} + \frac{1}{q} = 1$, $f(x)$, $a_n \geq 0$, $f \in L_{p,\varphi}(\mathbf{R}_+)$, $a = \{a_n\}_{n=1}^\infty \in l_{q,\psi}$, $\|f\|_{p,\varphi} > 0$, $\|a\|_{q,\psi} > 0$, if $p > 1$, then we have the following equivalent inequalities:

$$\sum_{n=1}^\infty a_n \int_0^\infty k_\lambda(x, n) f(x) dx$$

$$= \int_0^\infty f(x) \sum_{n=1}^\infty a_n k_\lambda(x, n) \, dx < k(\lambda_1) \|f\|_{p,\varphi} \|a\|_{q,\psi}, \quad (3.60)$$

$$\left\{ \sum_{n=1}^\infty [\psi(n)]^{1-p} \left[\int_0^\infty k_\lambda(x, n) f(x) \, dx \right]^p \right\}^{\frac{1}{p}} < k(\lambda_1) \|f\|_{p,\varphi},$$

$$(3.61)$$

and

$$\left\{ \int_0^\infty [\varphi(x)]^{1-q} \left[\sum_{n=1}^\infty k_\lambda(x, n) a_n \right]^q \, dx \right\}^{\frac{1}{q}} < k(\lambda_1) \|a\|_{q,\psi}, \quad (3.62)$$

with the best constant factor $k(\lambda_1)$; if $p < 0$, then we have the equivalent reverses of (3.60), (3.61) and (3.62), with the same best constant factor $k(\lambda_1)$.

(ii) For $0 < p < 1$, $\frac{1}{p} + \frac{1}{q} = 1$, setting $\widetilde{\varphi}(x) = (1 - \theta(x))\varphi(x)$, if $f(x)$, $a_n \geq 0$, $f \in L_{p,\widetilde{\varphi}}(\mathbf{R}_+)$, $a = \{a_n\}_{n=1}^{\infty} \in l_{q,\psi}$, $\|f\|_{p,\widetilde{\varphi}} > 0$, $\|a\|_{q,\psi} > 0$, then, we have the following equivalent inequalities:

$$\sum_{n=1}^{\infty} a_n \int_0^{\infty} k_\lambda(x,n) f(x) dx$$

$$= \int_0^{\infty} f(x) \sum_{n=1}^{\infty} a_n k_\lambda(x,n) \, dx > k(\lambda_1)\|f\|_{p,\widetilde{\varphi}}\|a\|_{q,\psi}, \quad (3.63)$$

$$\left\{ \sum_{n=1}^{\infty} [\psi(n)]^{1-p} \left[\int_0^{\infty} k_\lambda(x,n) f(x) \, dx \right]^p \right\}^{\frac{1}{p}} > k(\lambda_1)\|f\|_{p,\widetilde{\varphi}},$$

$$(3.64)$$

and

$$\left\{ \int_0^{\infty} [\widetilde{\varphi}(x)]^{1-q} \left[\sum_{n=1}^{\infty} k_\lambda(x,n) a_n \right]^q \, dx \right\}^{\frac{1}{q}} > k(\lambda_1)\|a\|_{q,\psi}, \quad (3.65)$$

where the constant factor $k(\lambda_1)$ is the best possible.

Note. As a particular case of (3.60), we may refer to a half-discrete Hilbert-type inequality with the homogeneous kernel of degree -4μ as

$$k_{4\mu}(x,n) = \frac{1}{(ax^\mu + bn^\mu)^2(cx^\mu + dn^\mu)^2}$$

where

$$a,b,c,d > 0, \quad bc \neq ae, \quad \text{and} \quad 0 < \mu < \frac{1}{2}$$

and a best constant factor as

$$\frac{bc + ae}{\mu(bc - ae)^2} \left[\frac{\ln(ae) - \ln(bc)}{ae - bc} - \frac{2}{ae + bc} \right],$$

which was published by Xie [83].

If we replace the lower limit 0 of integral to $c > 0$, then, the results of Corollary 3.8 is still value for $p > 1$ and $0 < p < 1$. For $p < 0$, we may refer to Corollary 3.13 for $v(n) = n$, $n_0 = 1$.

In particular, for $\lambda = 0$, $\lambda_1 = \alpha$, $\lambda_2 = -\alpha$, in Corollary 3.21, setting $\varphi_0(x) = x^{p(1-\alpha)-1}$ $(x \in (0,\infty))$, $\psi_0(n) = n^{q(1+\alpha)-1}$ $(n \in \mathbf{N})$, we have

$$\omega(-\alpha,n) = n^{-\alpha} \int_0^{\infty} k_0(x,n) x^{\alpha-1} dx = k(\alpha), \quad n \in \mathbf{N}, \quad (3.66)$$

$$\varpi(\alpha,x) = x^{\alpha} \sum_{n=1}^{\infty} k_0(x,n) n^{-\alpha-1}, \quad x \in (0,\infty), \quad (3.67)$$

and the following corollary:

Corollary 3.9. *Suppose that $k_0(x, y)$ is a non-negative finite measurable homogeneous function of degree 0 in $\mathbf{R}_+^2$,*

$$k(\alpha) = \int_0^\infty k_0(t, 1)t^{\alpha-1}dt \ (\alpha \in \mathbf{R}),$$

there exist constants δ_0, $\eta > 0$ such that for any $\widetilde{\alpha} \in (\alpha - \delta_0, \ \alpha + \delta_0)$,

$$0 < k(\widetilde{\alpha})(1 - \theta_{\widetilde{\alpha}}(x)) < \varpi(\widetilde{\alpha}, x) < k(\widetilde{\alpha}) < \infty(x \in (0, \infty)), \qquad (3.68)$$

where, $\theta_{\widetilde{\alpha}}(x) > 0$ $(x \in (0, \infty))$, $\theta_{\widetilde{\alpha}}(x) = O\left(\frac{1}{x^\eta}\right)$ $(x \in [1, \infty))$.

(i) *For $p > 1$ or $p < 0$, $\frac{1}{p} + \frac{1}{q} = 1$, $f(x)$, $a_n \geq 0$, $f \in L_{p,\varphi_0}(\mathbf{R}_+)$, $a = \{a_n\}_{n=1}^\infty \in l_{q,\psi_0}$, $||f||_{p,\varphi_0} > 0$, $||a||_{q,\psi_0} > 0$, if $p > 1$, then we have the following equivalent inequalities:*

$$\sum_{n=1}^\infty a_n \int_0^\infty k_0(x, n)f(x) \ dx$$

$$= \int_0^\infty f(x) \sum_{n=1}^\infty a_n k_0(x, n) \ dx < k(\alpha)||f||_{p,\varphi_0}||a||_{q,\psi_0}, \qquad (3.69)$$

$$\left\{\sum_{n=1}^\infty [\psi_0(n)]^{1-p} \left[\int_0^\infty k_0(x, n)f(x)dx\right]^p\right\}^{\frac{1}{p}} < k(\alpha)||f||_{p,\varphi_0}, \tag{3.70}$$

and

$$\left\{\int_0^\infty [\varphi_0(x)]^{1-q} \left[\sum_{n=1}^\infty k_0(x, n)a_n\right]^q dx\right\}^{\frac{1}{q}} < k(\alpha)||a||_{q,\psi_0}, \qquad (3.71)$$

with the best constant factor $k(\alpha)$; if $p < 0$, then we have the equivalent reverses of (3.69), (3.70) and (3.71), with the same best constant factor $k(\alpha)$.

(ii) *For $0 < p < 1$, $\frac{1}{p} + \frac{1}{q} = 1$, setting $\widetilde{\varphi}_0(x) = (1 - \theta_\alpha(x))\varphi_0(x)$, if $f(x)$, $a_n \geq 0$, $f \in L_{p,\widetilde{\varphi}_0}(\mathbf{R}_+)$, $a = \{a_n\}_{n=1}^\infty \in l_{q,\psi_0}$, $||f||_{p,\widetilde{\varphi}_0} > 0$, $||a||_{q,\psi_0} > 0$, then, we have the following equivalent inequalities:*

$$\sum_{n=1}^\infty a_n \int_0^\infty k_0(x, n)f(x) \ dx$$

$$= \int_0^\infty f(x) \sum_{n=1}^\infty a_n k_0(x, n) \ dx > k(\alpha)||f||_{p,\widetilde{\varphi}_0}||a||_{q,\psi_0}, \qquad (3.72)$$

$$\left\{\sum_{n=1}^\infty [\psi_0(n)]^{1-p} \left[\int_0^\infty k_0(x, n)f(x) \ dx\right]^p\right\}^{\frac{1}{p}} > k(\alpha)||f||_{p,\widetilde{\varphi}_0}, \tag{3.73}$$

and

$$\left\{ \int_0^\infty [\widetilde{\varphi}_0(x)]^{1-q} \left[\sum_{n=1}^\infty k_0(x,n)a_n \right]^q dx \right\}^{\frac{1}{q}} > k(\alpha)||a||_{q,\psi_0}, \qquad (3.74)$$

where the constant factor $k(\alpha)$ is the best possible.

3.3.4 *Some Particular Examples*

Applying Condition (i) and Corollaries 3.8-3.9, we have

Example 3.6. We set $k_\lambda(x,y) = \frac{1}{A(\max\{x,y\})^\lambda + B(\min\{x,y\})^\lambda}$ $((x,y) \in \mathbf{R}_+^2; \ 0 < B \le A, \ \lambda_1 > 0, \ 0 < \lambda_2 < 1, \ \lambda_1 + \lambda_2 = \lambda)$. There exists a constant $0 < \delta_0 < \min\{\lambda_1, \lambda_2, 1 - \lambda_2\}$ such that for any $\widetilde{\lambda}_1 \in (\lambda_1 - \delta_0, \ \lambda_1 + \delta_0) \subset (0, \lambda), \ \widetilde{\lambda}_2 = \lambda - \widetilde{\lambda}_1 \in (\lambda_2 - \delta_0, \ \lambda_2 + \delta_0) \subset (0, 1)$. For $x > 0$,

$$\begin{aligned}
f(x,y) &= \frac{x^{\widetilde{\lambda}_1} y^{\widetilde{\lambda}_2 - 1}}{A(\max\{x,y\})^\lambda + B(\min\{x,y\})^\lambda} \\
&= \begin{cases} \frac{x^{\widetilde{\lambda}_1} y^{\widetilde{\lambda}_2 - 1}}{Ax^\lambda + By^\lambda}, & 0 < y \le x, \\ \frac{x^{\widetilde{\lambda}_1} y^{\widetilde{\lambda}_2 - 1}}{Ay^\lambda + Bx^\lambda}, & y > x \end{cases}
\end{aligned}$$

is strictly decreasing with respect to $y > 0$. We find

$$\begin{aligned}
k(\widetilde{\lambda}_1) &= \int_0^\infty \frac{1}{A(\max\{t,1\})^\lambda + B(\min\{t,1\})^\lambda} t^{\widetilde{\lambda}_1 - 1} dt \\
&= \int_0^1 \frac{1}{A + Bt^\lambda} t^{\widetilde{\lambda}_1 - 1} dt + \int_1^\infty \frac{1}{At^\lambda + B} t^{\widetilde{\lambda}_1 - 1} dt \\
&= \frac{1}{A} \int_0^1 \frac{1}{1 + \frac{B}{A} t^\lambda} \left(t^{\widetilde{\lambda}_1 - 1} + t^{\widetilde{\lambda}_2 - 1} \right) dt \\
&= \frac{1}{A} \int_0^1 \sum_{k=0}^\infty \left(-\frac{B}{A} \right)^k t^{\lambda k} \left(t^{\widetilde{\lambda}_1 - 1} + t^{\widetilde{\lambda}_2 - 1} \right) dt \\
&= \frac{1}{A} \sum_{k=0}^\infty \left(-\frac{B}{A} \right)^k \int_0^1 \left(t^{\lambda k + \widetilde{\lambda}_1 - 1} + t^{\lambda k + \widetilde{\lambda}_2 - 1} \right) dt \\
&= \frac{1}{A} \sum_{k=0}^\infty \left(-\frac{B}{A} \right)^k \left(\frac{1}{\lambda k + \widetilde{\lambda}_1} + \frac{1}{\lambda k + \widetilde{\lambda}_2} \right) \in \mathbf{R}_+,
\end{aligned}$$

and obtain

$$\varpi(\widetilde{\lambda}_1, x) = x^{\widetilde{\lambda}_1} \sum_{n=1}^{\infty} \frac{n^{\widetilde{\lambda}_2 - 1}}{A(\max\{x, n\})^{\lambda} + B(\min\{x, n\})^{\lambda}}$$

$$> x^{\widetilde{\lambda}_1} \int_1^{\infty} \frac{y^{\widetilde{\lambda}_2 - 1}}{A(\max\{x, y\})^{\lambda} + B(\min\{x, y\})^{\lambda}} \, dy$$

$$= k(\widetilde{\lambda}_1)(1 - \theta_{\widetilde{\lambda}_1}(x)) > 0,$$

$$\theta_{\widetilde{\lambda}_1}(x) = \frac{x^{\widetilde{\lambda}_1}}{k(\widetilde{\lambda}_1)} \int_0^1 \frac{y^{\widetilde{\lambda}_2 - 1} dy}{A(\max\{x, y\})^{\lambda} + B(\min\{x, y\})^{\lambda}} > 0.$$

For $x \geq 1$, it follows that

$$0 < \theta_{\widetilde{\lambda}_1}(x) = \frac{x^{\widetilde{\lambda}_1}}{k(\widetilde{\lambda}_1)} \int_0^1 \frac{y^{\widetilde{\lambda}_2 - 1}}{Ax^{\lambda} + By^{\lambda}} \, dy$$

$$\leq \frac{x^{\widetilde{\lambda}_1}}{k(\widetilde{\lambda}_1)} \int_0^1 \frac{y^{\widetilde{\lambda}_2 - 1}}{Ax^{\lambda}} \, dy$$

$$= \frac{1}{Ak(\widetilde{\lambda}_1)\widetilde{\lambda}_2 x^{\widetilde{\lambda}_2}} \leq \frac{1}{Ak(\widetilde{\lambda}_1)\widetilde{\lambda}_2 x^{\eta}},$$

that is $\theta_{\widetilde{\lambda}_1}(x) = O\left(\frac{1}{x^{\eta}}\right)$ $(\eta = \lambda_2 - \delta_0 > 0)$.

By Condition (i), it follows that

$$\varpi(\widetilde{\lambda}_1, x) = \sum_{n=1}^{\infty} \frac{x^{\widetilde{\lambda}_1} n^{\widetilde{\lambda}_2 - 1}}{A(\max\{x, n\})^{\lambda} + B(\min\{x, n\})^{\lambda}} < k(\widetilde{\lambda}_1).$$

Hence, we obtain (3.59).

By Corollary 3.8, for $k_{\lambda}(x, n) = \frac{1}{A(\max\{x,n\})^{\lambda} + B(\min\{x,n\})^{\lambda}}$ $(0 < B \leq A$, $\lambda_1 > 0$, $0 < \lambda_2 < 1$, $\lambda_1 + \lambda_2 = \lambda)$, if $p > 1$, we have equivalent inequalities (3.60), (3.61) and (3.62); if $p < 0$, we have equivalent reverses of (3.60), (3.61) and (3.62); if $0 < p < 1$, we have equivalent inequalities (3.63), (3.64) and (3.65). All the inequalities are with the same best constant factor

$$k(\lambda_1) = \frac{\lambda}{A} \sum_{k=0}^{\infty} \left(-\frac{B}{A}\right)^k \frac{2k+1}{(\lambda k + \lambda_1)(\lambda k + \lambda_2)}. \tag{3.75}$$

Example 3.7. We set $k_0(x, y) = \frac{(\min\{x,y\})^{\lambda}}{A(\max\{x,y\})^{\lambda} + B(\min\{x,y\})^{\lambda}}$ $((x, y) \in \mathbf{R}_+^2$; $0 < B \leq A, \lambda > 0$, $\max\{-\lambda, \lambda - 1\} < \alpha < \lambda)$. There exists a constant $0 < \delta_0 < \min\{\lambda - \alpha, \lambda + \alpha, \alpha - \lambda + 1\}$, such that for any $\widetilde{\alpha} \in (\alpha - \delta_0, \alpha + \delta_0)$, $\max\{-\lambda, \lambda - 1\} < \widetilde{\alpha} < \lambda$. For $x > 0$,

$$f(x, y) = \frac{(\min\{x, y\})^{\lambda} x^{\widetilde{\alpha}} y^{-\widetilde{\alpha} - 1}}{A(\max\{x, y\})^{\lambda} + B(\min\{x, y\})^{\lambda}}$$

$$= \begin{cases} \frac{x^{\widetilde{\alpha}} y^{\lambda - \widetilde{\alpha} - 1}}{Ax^{\lambda} + By^{\lambda}}, & 0 < y \leq x, \\ \frac{x^{\lambda + \widetilde{\alpha}} y^{-\widetilde{\alpha} - 1}}{Ay^{\lambda} + Bx^{\lambda}} & y > x \end{cases}$$

is strictly decreasing with respect to $y > 0$. We find

$$
k(\widetilde{\lambda}_1) = \int_0^\infty \frac{(\min\{t,1\})^\lambda}{A(\max\{t,1\})^\lambda + B(\min\{t,1\})^\lambda} t^{\widetilde{\alpha}-1}\, dt
$$

$$
= \int_0^1 \frac{1}{A + Bt^\lambda} t^{\lambda+\widetilde{\alpha}-1}\, dt + \int_1^\infty \frac{1}{At^\lambda + B} t^{\widetilde{\alpha}-1}\, dt
$$

$$
= \frac{1}{A} \int_0^1 \frac{1}{1 + \frac{B}{A}t^\lambda} (t^{\lambda+\widetilde{\alpha}-1} + t^{\lambda-\widetilde{\alpha}-1})\, dt
$$

$$
= \frac{1}{A} \int_0^1 \sum_{k=0}^\infty \left(-\frac{B}{A}\right)^k t^{\lambda k} (t^{\lambda+\widetilde{\alpha}-1} + t^{\lambda-\widetilde{\alpha}-1})\, dt
$$

$$
= \frac{1}{A} \sum_{k=0}^\infty \left(-\frac{B}{A}\right)^k \int_0^1 (t^{\lambda k+\lambda+\widetilde{\alpha}-1} + t^{\lambda k+\lambda-\widetilde{\alpha}-1})\, dt
$$

$$
= \frac{1}{A} \sum_{k=1}^\infty \left(-\frac{B}{A}\right)^{k-1} \left(\frac{1}{\lambda k + \widetilde{\alpha}} + \frac{1}{\lambda k - \widetilde{\alpha}}\right) \in \mathbf{R}_+,
$$

and then, we obtain

$$
\varpi(\widetilde{\alpha}, x) = x^{\widetilde{\alpha}} \sum_{n=1}^\infty \frac{(\min\{x,n\})^\lambda n^{-\widetilde{\alpha}-1}}{A(\max\{x,n\})^\lambda + B(\min\{x,n\})^\lambda}
$$

$$
> x^{\widetilde{\alpha}} \int_1^\infty \frac{(\min\{x,y\})^\lambda y^{-\widetilde{\alpha}-1}}{A(\max\{x,y\})^\lambda + B(\min\{x,y\})^\lambda}\, dy
$$

$$
= k(\widetilde{\alpha})(1 - \theta_{\widetilde{\alpha}}(x)) > 0,
$$

$$
\theta_{\widetilde{\alpha}}(x) = \frac{x^{\widetilde{\alpha}}}{k(\widetilde{\alpha})} \int_0^1 \frac{(\min\{x,y\})^\lambda y^{-\widetilde{\alpha}-1}\, dy}{A(\max\{x,y\})^\lambda + B(\min\{x,y\})^\lambda} > 0.
$$

For $x \geq 1$, it follows that

$$
0 < \theta_{\widetilde{\alpha}}(x) = \frac{x^{\widetilde{\alpha}}}{k(\widetilde{\alpha})} \int_0^1 \frac{y^{\lambda-\widetilde{\alpha}-1}}{Ax^\lambda + By^\lambda}\, dy
$$

$$
\leq \frac{x^{\widetilde{\alpha}}}{k(\widetilde{\alpha})} \int_0^1 \frac{y^{\lambda-\widetilde{\alpha}-1}}{Ax^\lambda}\, dy
$$

$$
= \frac{1}{Ak(\widetilde{\alpha})(\lambda - \widetilde{\alpha})x^{\lambda-\widetilde{\alpha}}} \leq \frac{1}{Ak(\widetilde{\alpha})(\lambda - \widetilde{\alpha})x^\eta},
$$

that is $\theta_{\widetilde{\alpha}}(x) = O\left(\frac{1}{x^\eta}\right)$ ($\eta = \lambda - \alpha - \delta_0 > 0$). By Condition (i), it follows that

$$
\varpi(\widetilde{\alpha}, x) = x^{\widetilde{\alpha}} \sum_{n=1}^\infty \frac{(\min\{x,n\})^\lambda n^{-\widetilde{\alpha}-1}}{A(\max\{x,n\})^\lambda + B(\min\{x,n\})^\lambda} < k(\widetilde{\alpha}).
$$

Hence, we obtain (3.68).

By Corollary 3.9, for $k_0(x, n) = \frac{(\min\{x,n\})^\lambda}{A(\max\{x,n\})^\lambda + B(\min\{x,n\})^\lambda}$ $(0 < B \leq A, \ \lambda > 0, \ \max\{-\lambda, \lambda - 1\} < \alpha < \lambda)$, if $p > 1$, we have equivalent inequalities (3.69), (3.70) and (3.71); if $p < 0$, we have equivalent reverses of (3.69), (3.70) and (3.71); if $0 < p < 1$, we have equivalent inequalities (3.72), (3.73) and (3.74). All the inequalities are with the same best constant factor

$$k(\lambda_1) = \frac{2\lambda}{A} \sum_{k=1}^\infty \left(-\frac{B}{A}\right)^{k-1} \frac{k}{(\lambda^2 k^2 - \alpha^2)}. \tag{3.76}$$

Example 3.8. (See He and Li [22]).

We set $k_\lambda(x, y) = \frac{(\min\{x,y\})^\beta}{(\max\{x,y\})^{\lambda+\beta}}$ $((x, y) \in \mathbf{R}_+^2; \ \lambda + 2\beta > 0, \ \lambda_1 > -\beta, \ -\beta < \lambda_2 < 1 - \beta, \ \lambda_1 + \lambda_2 = \lambda)$. There exists a constant $0 < \delta_0 < \min\{\lambda_1 + \beta, \ \lambda_2 + \beta, \ 1 - \beta - \lambda_2\}$, such that for any $\widetilde{\lambda}_1 \in (\lambda_1 - \delta_0, \lambda_1 + \delta_0) \subset (-\beta, \lambda + \beta), \ \widetilde{\lambda}_2 = \lambda - \widetilde{\lambda}_1 \in (\lambda_2 - \delta_0, \lambda_2 + \delta_0) \subset (-\beta, 1 - \beta)$. For $x > 0$,

$$f(x, y) = \frac{(\min\{x, y\})^\beta x^{\widetilde{\lambda}_1} y^{\widetilde{\lambda}_2 - 1}}{(\max\{x, y\})^{\lambda + \beta}}$$

$$= \begin{cases} \frac{y^{\widetilde{\lambda}_2 + \beta - 1}}{x^{\lambda + \beta - \widetilde{\lambda}_1}}, & 0 < y \leq x, \\ \frac{x^{\widetilde{\lambda}_1 + \beta}}{y^{\lambda + \beta - \widetilde{\lambda}_2 + 1}}, & y > x \end{cases}$$

is strictly decreasing with respect to $y > 0$. We find

$$k(\widetilde{\lambda}_1) = \int_0^\infty \frac{(\min\{t, 1\})^\beta}{(\max\{t, 1\})^{\lambda+\beta}} t^{\widetilde{\lambda}_1 - 1} \, dt$$

$$= \int_0^1 t^{\widetilde{\lambda}_1 + \beta - 1} dt + \int_1^\infty \frac{1}{t^{\lambda+\beta}} t^{\widetilde{\lambda}_1 - 1} \, dt$$

$$= \frac{1}{\widetilde{\lambda}_1 + \beta} + \frac{1}{\widetilde{\lambda}_2 + \beta} \in \mathbf{R}_+,$$

and then, we have

$$\varpi(\widetilde{\lambda}_1, x) = x^{\widetilde{\lambda}_1} \sum_{n=1}^\infty \frac{(\min\{x, n\})^\beta n^{\widetilde{\lambda}_2 - 1}}{(\max\{x, n\})^{\lambda + \beta}}$$

$$> x^{\widetilde{\lambda}_1} \int_1^\infty \frac{(\min\{x, y\})^\beta y^{\widetilde{\lambda}_2 - 1}}{(\max\{x, y\})^{\lambda + \beta}} \, dy$$

$$= k(\widetilde{\lambda}_1)(1 - \theta_{\widetilde{\lambda}_1}(x)) > 0,$$

$$\theta_{\widetilde{\lambda}_1}(x) = \frac{x^{\widetilde{\lambda}_1}}{k(\widetilde{\lambda}_1)} \int_0^1 \frac{(\min\{x, y\})^\beta y^{\widetilde{\lambda}_2 - 1}}{(\max\{x, y\})^{\lambda + \beta}} dy > 0.$$

For $x \geq 1$, it follows that

$$0 < \theta_{\widetilde{\lambda}_1}(x) = \frac{x^{\widetilde{\lambda}_1}}{k(\widetilde{\lambda}_1)} \int_0^1 \frac{y^{\widetilde{\lambda}_2+\beta-1}}{x^{\lambda+\beta}} \, dy$$

$$= \frac{1}{k(\widetilde{\lambda}_1)(\widetilde{\lambda}_2+\beta)x^{\widetilde{\lambda}_2+\beta}} \leq \frac{1}{k(\widetilde{\lambda}_1)(\widetilde{\lambda}_2+\beta)x^{\eta}},$$

that is, $\theta_{\widetilde{\lambda}_1}(x) = O\left(\frac{1}{x^\eta}\right) (\eta = \lambda_2 + \beta - \delta_0 > 0)$. By Condition (i), it follows that

$$\varpi(\widetilde{\lambda}_1, x) = x^{\widetilde{\lambda}_1} \sum_{n=1}^{\infty} \frac{(\min\{x,n\})^\beta n^{\widetilde{\lambda}_2-1}}{(\max\{x,n\})^{\lambda+\beta}} < k(\widetilde{\lambda}_1).$$

Hence, we obtain (3.60).

By Corollary 3.8, for $k_\lambda(x,n) = \frac{(\min\{x,n\})^\beta}{(\max\{x,n\})^{\lambda+\beta}}$ $(\lambda + 2\beta > 0, \ \lambda_1 > -\beta, \ -\beta < \lambda_2 < 1-\beta, \ \lambda_1+\lambda_2 = \lambda)$, if $p > 1$, we have equivalent inequalities (3.60), (3.61) and (3.62); if $p < 0$, we have equivalent reverses of (3.60), (3.61) and (3.62); if $0 < p < 1$, we have equivalent inequalities (3.63), (3.64) and (3.65). All the inequalities are with the same best constant factor

$$k(\widetilde{\lambda}_1) = \frac{2\beta+\lambda}{(\beta+\lambda_1)(\beta+\lambda_2)}. \tag{3.77}$$

Note. In particular of this example, for (i) $\beta = 0$, $k_\lambda(x,n) = \frac{1}{(\max\{x,n\})^\lambda}(\lambda > 0, \lambda_1 > 0, 0 < \lambda_2 < 1, \lambda_1 + \lambda_2 = \lambda)$; (ii) $\lambda = -\beta$, $k_{-\beta}(x,n) = (\min\{x,n\})^\beta(\beta > 0, \lambda_1 > -\beta, -\beta < \lambda_2 < 1-\beta, \lambda_1 + \lambda_2 = -\beta)$; (iii) $\lambda = 0$, $k_0(x,n) = (\frac{\min\{x,n\}}{\max\{x,n\}})^\beta(\beta > 0, \lambda_1 > -\beta, -\beta < \lambda_2 < 1 - \beta, \lambda_1 + \lambda_2 = 0)$, we obtain respectively some results of Example 3.2 (viii)-(x).

3.3.5 *Applying Condition (iii) and Corollary 3.8*

Theorem 3.4. *We set $k_\lambda(x,y) = \frac{1}{(x+y)^\lambda}((x,y) \in \mathbf{R}_+^2; 0 < \lambda < 5, 0 < \lambda_1 < 3, 0 < \lambda_2 < 2, \lambda_1 + \lambda_2 = \lambda)$. There exists a constant $0 < \delta_0 < \min\{\lambda_1, \lambda_2, 3 - \lambda_1, 2 - \lambda_2\}$, such that for any $\widetilde{\lambda}_1 \in (\lambda_1 - \delta_0, \lambda_1 + \delta_0) \subset (0,3), \widetilde{\lambda}_2 = \lambda - \widetilde{\lambda}_1 \in (\lambda_2 - \delta_0, \lambda_2 + \delta_0) \subset (0,2)$. We find*

$$k(\widetilde{\lambda}_1) = \int_0^\infty \frac{1}{(t+1)^\lambda} t^{\widetilde{\lambda}_1-1} dt = B(\widetilde{\lambda}_1, \widetilde{\lambda}_2) \in \mathbf{R}_+.$$

In the following, we show that

$$0 < k(\widetilde{\lambda}_1)(1 - \theta_{\widetilde{\lambda}_1}(x)) < \varpi(\widetilde{\lambda}_1, x) = \sum_{n=1}^\infty \frac{x^{\widetilde{\lambda}_1} n^{\widetilde{\lambda}_2-1}}{(x+n)^\lambda} < k(\widetilde{\lambda}_1), \tag{3.78}$$

$$\theta_{\widetilde{\lambda}_1}(x) > 0(x > 0), \quad \theta_{\widetilde{\lambda}_1}(x) = O\left(\frac{1}{x^\eta}\right) (x \geq 1; \eta = \lambda_2 - \delta_0 > 0). \tag{3.79}$$

Setting $\widetilde{\lambda}_1 = \frac{\lambda}{R}$, $\widetilde{\lambda}_2 = \frac{\lambda}{S}$ $\left(R > 1, \frac{1}{R} + \frac{1}{S} = 1\right)$, it follows that $\widetilde{\lambda}_2 = \frac{\lambda}{S} < 2$ and $0 < \lambda < \min\{2S, 5\}$.

(1) For $g(x,y) = \frac{1}{(x+y)^\lambda} y^{\frac{\lambda}{S}-1}$, we show that

$$\widetilde{r}(x) = \int_0^1 g(x,y)dy - \frac{1}{2}g(x,1) - \int_1^\infty P_1(y)g_y'(x,y)\,dy > 0. \qquad (3.80)$$

In fact, since $0 < \lambda < 2S$, it follows that $\frac{1}{(S+\lambda)(2S+\lambda)} \geq \frac{1}{12S^2}$ and

$$\int_0^1 g(x,y)dy = \int_0^1 \frac{1}{(x+y)^\lambda} y^{\frac{\lambda}{S}-1}\,dy = \frac{S}{\lambda}\int_0^1 \frac{1}{(x+y)^\lambda}\,dy^{\frac{\lambda}{S}}$$

$$= \frac{S}{\lambda(x+1)^\lambda} + S\int_0^1 \frac{1}{(x+y)^{\lambda+1}} y^{\frac{\lambda}{S}}\,dy$$

$$= \frac{S}{\lambda(x+1)^\lambda} + \frac{S^2}{S+\lambda}\int_0^1 \frac{1}{(x+y)^{\lambda+1}}\,dy^{\frac{\lambda}{S}+1}$$

$$= \frac{S}{\lambda(x+1)^\lambda} + \frac{S^2}{(S+\lambda)(x+1)^{\lambda+1}}$$

$$\quad + \frac{S^2(\lambda+1)}{S+\lambda}\int_0^1 \frac{y^{\frac{\lambda}{S}+1}dy}{(x+y)^{\lambda+2}}$$

$$= \frac{S}{\lambda(x+1)^\lambda} + \frac{S^2}{(S+\lambda)(x+1)^{\lambda+1}}$$

$$\quad + \frac{S^3(\lambda+1)}{(S+\lambda)(2S+\lambda)}\int_0^1 \frac{dy^{\frac{\lambda}{S}+2}}{(x+y)^{\lambda+2}}$$

$$> \frac{S}{\lambda(x+1)^\lambda} + \frac{S^2}{(S+\lambda)(x+1)^{\lambda+1}}$$

$$\quad + \frac{S^3(\lambda+1)}{(S+\lambda)(2S+\lambda)(x+1)^{\lambda+2}}$$

$$> \frac{S}{\lambda(x+1)^\lambda} + \frac{S^2}{(S+\lambda)(x+1)^{\lambda+1}} + \frac{S(\lambda+1)}{12(x+1)^{\lambda+2}}.$$

For $x > 0$, we find

$$g_y'(x,y) = \frac{-\lambda}{(x+y)^{\lambda+1} y^{1-\frac{\lambda}{S}}} - \frac{1-(\lambda/S)}{(x+y)^\lambda y^{2-\frac{\lambda}{S}}}$$

$$= \frac{-[1+(\lambda/R)]}{(x+y)^\lambda y^{2-\frac{\lambda}{S}}} + \frac{\lambda x}{(x+y)^{\lambda+1} y^{2-\frac{\lambda}{S}}}.$$

Since $0 < \lambda < 5$, we find $\lambda x < 5(x+1)$ $(x > 0)$ and using equations (2.74) and (2.75) in Chapter 2 (for $m = q = 1$), it follows that

$$
-\int_1^\infty P_1(y) g_y'(x,y)\, dy
$$

$$
= \int_1^\infty \frac{[1 + (\lambda/R)]P_1(y)}{(x+y)^\lambda\, y^{2-\frac{\lambda}{S}}}\, dy - \int_1^\infty \frac{\lambda x P_1(y)\, dy}{(x+y)^{\lambda+1}\, y^{2-\frac{\lambda}{S}}} \quad (3.81)
$$

$$
= \int_1^\infty \frac{[1 + (\lambda/R)]P_1(y)}{(x+y)^\lambda y^{2-\frac{\lambda}{S}}}\, dy
$$

$$
+ \left\{ \frac{\lambda x}{12(x+1)^{\lambda+1}} - \frac{\lambda x}{6}\int_1^\infty P_3(y)\left[\frac{1}{(x+y)^{\lambda+1} y^{2-\frac{\lambda}{S}}} \right]_y'' dy \right\}
$$

$$
> -\frac{1 + (\lambda/R)}{12(x+1)^\lambda} + \frac{\lambda(x+1) - \lambda}{12(x+1)^{\lambda+1}} - \frac{5(x+1)}{720}
$$

$$
\times \left[\frac{(\lambda+1)(\lambda+2)}{(x+1)^{\lambda+3}} + \frac{2(\lambda+1)(2S-\lambda)}{S(x+1)^{\lambda+2}} + \frac{(2S-\lambda)(3S-\lambda)}{S^2(x+1)^{\lambda+1}} \right]
$$

$$
\geq \left(\frac{-1}{12} + \frac{\lambda}{12S} \right)\frac{1}{(x+1)^\lambda} - \frac{\lambda}{12(x+1)^{\lambda+1}}
$$

$$
- \frac{1}{144}\left[(\lambda+1)\frac{7}{(x+1)^{\lambda+2}} + \frac{24}{(x+1)^{\lambda+1}} + \frac{6}{(x+1)^\lambda} \right].
$$

In view of $-\frac{1}{2}g(x,1) = \frac{-1}{2(x+1)^\lambda}$, by (3.80) and the above results, we obtain

$$
\widetilde{r}(x) > \frac{S}{\lambda(x+1)^\lambda} + \frac{S^2}{(S+\lambda)(x+1)^{\lambda+1}} + \frac{S(\lambda+1)}{12(x+1)^{\lambda+2}}
$$

$$
- \left(\frac{1}{2} + \frac{1}{12} - \frac{\lambda}{12S} \right)\frac{1}{(x+1)^\lambda} - \frac{\lambda}{12(x+1)^{\lambda+1}}
$$

$$
- \frac{1}{144}\left[\frac{7(\lambda+1)}{(x+1)^{\lambda+2}} + \frac{24}{(x+1)^{\lambda+1}} + \frac{6}{(x+1)^\lambda} \right]
$$

$$
= \frac{h(\lambda)}{24\lambda S(x+1)^\lambda} + \left(\frac{S^2}{S+\lambda} - \frac{\lambda}{12} - \frac{1}{6} \right)\frac{1}{(x+1)^{\lambda+1}}
$$

$$
+ \left(\frac{S}{12} - \frac{7}{144} \right)\frac{\lambda+1}{(x+1)^{\lambda+2}}, \quad (3.82)
$$

where, $h(\lambda) = 2\lambda^2 - 15S\lambda + 24S^2 (0 < \lambda < 2S)$.

We find $h'(\lambda) = 4\lambda - 15S < 8S - 15S < 0$,

$$
h(\lambda) > h(2S) = 2(2S)^2 - 15S(2S) + 24S^2 = 2S^2 > 0,
$$

$$
\frac{S^2}{S+\lambda} - \frac{\lambda}{12} - \frac{1}{6} > \frac{S}{3} - \frac{S}{6} - \frac{1}{6} = \frac{S-1}{6} > 0,
$$

and $\frac{S}{12} - \frac{7}{144} > \frac{1}{12} - \frac{7}{144} > 0$. Then, by inequality (3.82), it follows $\widetilde{r}(x) > 0$.

(2) In view of

$$f(x,y) = \frac{x^{\widetilde{\lambda}_1}}{(x+y)^\lambda} \, y^{\widetilde{\lambda}_2 - 1} = x^{\widetilde{\lambda}_1} \, g(x,y),$$

by (3.4) and (3.80), we have $\widetilde{R}(x) = x^{\widetilde{\lambda}_1} \widetilde{r}(x) > 0$.

By Condition (iii), it follows that

$$\varpi(\widetilde{\lambda}_1, x) = x^{\widetilde{\lambda}_1} \sum_{n=1}^{\infty} \frac{n^{\widetilde{\lambda}_2 - 1}}{(x+n)^\lambda} = k(\widetilde{\lambda}_1) - \widetilde{R}(x) < k(\widetilde{\lambda}_1).$$

We find

$$\int_0^1 \frac{1}{(x+y)^\lambda} \, y^{\frac{\lambda}{S} - 1} \, dy < \int_0^1 \frac{1}{x^\lambda} \, y^{\frac{\lambda}{S} - 1} \, dy = \frac{S}{\lambda x^\lambda},$$

and by (3.81) and (2.22) in Chapter 2, it follows that

$$-\int_1^\infty P_1(y) g_y'(x,y) dy$$

$$= \int_1^\infty \frac{[1 + (\lambda/R)] P_1(y)}{(x+y)^\lambda \, y^{2-\frac{\lambda}{S}}} \, dy - \int_1^\infty \frac{\lambda x P_1(y)}{(x+y)^{\lambda+1} \, y^{2-\frac{\lambda}{S}}} \, dy$$

$$< 0 + \frac{\lambda x}{12(x+1)^{\lambda+1}}.$$

Then, by (3.80), we have

$$\widetilde{r}(x) < \frac{S}{\lambda x^\lambda} - \frac{1}{2(x+1)^\lambda} + \frac{\lambda x}{12(x+1)^{\lambda+1}}. \tag{3.83}$$

Setting

$$\theta_{\widetilde{\lambda}_1}(x) = \frac{1}{k(\widetilde{\lambda}_1)} \left[\widetilde{R}(x) + x^{\frac{\lambda}{R}} \frac{1}{(x+1)^\lambda} \right] > 0,$$

we find

$$\varpi(\widetilde{\lambda}_1, x) > \varpi(\widetilde{\lambda}_1, x) - \frac{x^{\frac{\lambda}{R}}}{(x+1)^\lambda} = k(\widetilde{\lambda}_1)(1 - \theta_{\widetilde{\lambda}_1}(x)) > 0.$$

By (3.83), for $x \geq 1$, we obtain

$$0 < x^{\lambda_2 - \delta_0} \theta_{\widetilde{\lambda}_1}(x) \leq x^{\frac{\lambda}{S}} \theta_{\widetilde{\lambda}_1}(x) = \frac{1}{k(\widetilde{\lambda}_1)} \left[x^\lambda \, \widetilde{r}(x) + \frac{x^\lambda}{(x+1)^\lambda} \right]$$

$$< \frac{1}{k(\widetilde{\lambda}_1)} \left[\frac{S}{\lambda} + \frac{x^\lambda}{2(x+1)^\lambda} + \frac{\lambda x^{\lambda+1}}{12(x+1)^{\lambda+1}} \right]$$

$$\rightarrow \frac{1}{k(\widetilde{\lambda}_1)} \left[\frac{S}{\lambda} + \frac{1}{2} + \frac{\lambda}{12} \right], \qquad \text{as} \quad x \rightarrow \infty,$$

that is, $\theta_{\widetilde{\lambda}_1}(x) = O\left(\frac{1}{x^\eta}\right)$ $(x \geq 1;\ \eta = \lambda_2 - \delta_0 > 0)$. Hence we show that (3.78) and (3.79) are valid.

By Corollary 3.8, for $k_\lambda(x,n) = \frac{1}{(x+n)^\lambda}$ $(0 < \lambda < 5,\ 0 < \lambda_1 < 3,\ 0 < \lambda_2 < 2,\ \lambda_1 + \lambda_2 = \lambda)$, if $p > 1$, we have equivalent inequalities (3.60), (3.61) and (3.62); if $p < 0$, we have equivalent reverses of (3.60), (3.61) and (3.62); if $0 < p < 1$, we have equivalent inequalities (3.63), (3.64) and (3.65). All the inequalities are with the same best constant factor

$$k(\lambda_1) = B(\lambda_1, \lambda_2). \tag{3.84}$$

Example 3.9. We set $k_1(x,y) = \frac{|\ln(x/y)|}{x+y}$ $((x,y) \in \mathbf{R}_+^2;\ 0 < \lambda_1 < 1,\ \lambda_2 = 1 - \lambda_1)$. There exists a constant $0 < \delta_0 < \min\{\lambda_1,\ \lambda_2,\ 1-\lambda_1\}$, such that for any $\widetilde{\lambda}_1 \in (\lambda_1 - \delta_0,\ \lambda_1 + \delta_0) \subset (0,1)$, $\widetilde{\lambda}_2 = \lambda - \widetilde{\lambda}_1 \in (\lambda_2 - \delta_0,\ \lambda_2 + \delta_0) \subset (0,1)$. There exists a constant $\delta_1 > 0$, such that $\eta = \lambda_2 - \delta_0 - \delta_1 > 0$. By the Lebesgue term by term integration theorem, it follows that

$$
\begin{aligned}
k(\widetilde{\lambda}_1) &= \int_0^\infty \frac{|\ln t|\, t^{\widetilde{\lambda}_1 - 1}}{t+1}\, dt \\
&= \int_0^1 \frac{(-\ln t)\, t^{\widetilde{\lambda}_1 - 1}}{t+1}\, dt + \int_1^\infty \frac{(\ln t)\, t^{\widetilde{\lambda}_1 - 1}}{t+1}\, dt \\
&= \int_0^1 (-\ln t) \sum_{k=0}^\infty (-1)^k \left(t^{k+\widetilde{\lambda}_1 - 1} + t^{k+\widetilde{\lambda}_2 - 1}\right)\, dt \\
&= \sum_{k=0}^\infty (-1)^k \int_0^1 (-\ln t) \left(t^{k+\widetilde{\lambda}_1 - 1} + t^{k+\widetilde{\lambda}_2 - 1}\right)\, dt \\
&= \sum_{k=0}^\infty (-1)^k \int_0^1 (-\ln t) \left(t^{k+\widetilde{\lambda}_1 - 1} + t^{k+\widetilde{\lambda}_2 - 1}\right)\, dt \\
&= \sum_{k=0}^\infty (-1)^k \left[\frac{1}{(k+\widetilde{\lambda}_1)^2} + \frac{1}{(k+\widetilde{\lambda}_2)^2}\right] \in \mathbf{R}_+.
\end{aligned}
$$

Since $t^{\delta_0} k_1(t,1) = \frac{t^{\delta_0}|\ln t|}{t+1} \to 0$ $(t \to 0^+)$, we still find

$$k_1(t,1) = \frac{-\ln t}{t+1} \leq L\left(\frac{1}{t^{\delta_0}}\right) \qquad (t \in (0,1];\ L > 0).$$

In the following, we show that

$$\omega(\widetilde{\lambda}_2, n) = n^{\widetilde{\lambda}_2} \int_1^\infty \frac{|\ln \frac{x}{n}| n^{\widetilde{\lambda}_1 - 1}}{x+n}\, x^{\widetilde{\lambda}_1 - 1}\, dx < k(\widetilde{\lambda}_1) \quad (n \in \mathbf{N}), \tag{3.85}$$

$$0 < k(\widetilde{\lambda}_1)(1 - \theta_{\widetilde{\lambda}_1}(x))$$

$$< \varpi(\widetilde{\lambda}_1, x) = x^{\widetilde{\lambda}_1} \sum_{n=1}^\infty \frac{|\ln \frac{x}{n}| n^{\widetilde{\lambda}_2 - 1}}{x+n} < k(\widetilde{\lambda}_1) \quad (x \geq 1), \tag{3.86}$$

and

$$0 < \theta_{\widetilde{\lambda}_1}(x) = O\left(\frac{1}{x^\eta}\right) \quad (x \geq 1; \ \eta = \lambda_2 - \delta_0 - \delta_1 > 0). \tag{3.87}$$

(1) For $n \in \mathbf{N}$, putting $t = \frac{x}{n}$ in the integral of (3.85), we find

$$\omega(\lambda_2, n) = \int_{\frac{1}{n}}^{\infty} \frac{|\ln t|}{t+1} t^{\widetilde{\lambda}_1-1} \, dt < \int_0^\infty \frac{|\ln t|}{t+1} t^{\widetilde{\lambda}_1-1} \, dt = k(\widetilde{\lambda}_1),$$

and then (3.85) follows.

(2) For $g(x, y) = \frac{|\ln(x/y)|}{x+y} y^{\widetilde{\lambda}_2-1}$ $(x, y \geq 1)$, in the following, we prove that

$$\widetilde{r}(x) = \int_0^1 g(x, y) dy - \frac{1}{2} g(x, 1) - \int_1^\infty P_1(y) g_y'(x, y) dy > 0. \tag{3.88}$$

In fact, putting $t = \frac{y}{x}$, we obtain

$$\int_0^1 g(x, y) \, dy = \int_0^1 \frac{|\ln(x/y)|}{x+y} y^{\widetilde{\lambda}_2-1} \, dy$$

$$= \frac{1}{x^{\widetilde{\lambda}_1}} \int_0^{\frac{1}{x}} \frac{-\ln t}{t+1} t^{\widetilde{\lambda}_2-1} \, dt$$

$$> \frac{1}{x^{\widetilde{\lambda}_1}} \int_0^{\frac{1}{x}} \frac{-\ln t}{\frac{1}{x}+1} t^{\widetilde{\lambda}_2-1} \, dt$$

$$= \frac{1}{1+x} \left(\frac{\ln x}{\widetilde{\lambda}_2} + \frac{1}{\widetilde{\lambda}_2^2}\right). \tag{3.89}$$

For $x, y \geq 1$, since $\frac{\ln(y/x)}{y-x}$ is decreasing with respect to y, we set two decreasing functions for $y \in [1, \infty)$ as follows:

$$h_1(x, y) = \frac{1}{(y+x)y^{\widetilde{\lambda}_1+1}} + \frac{2x \ln(y/x)}{(y-x)(y+x)^2 y^{\widetilde{\lambda}_1}} + \frac{2\widetilde{\lambda}_1 x \ln(y/x)}{(y-x)(y+x)y^{\widetilde{\lambda}_1+1}},$$

$$h_2(x, y) = \frac{\ln(y/x)}{(y-x)(y+x)y^{\widetilde{\lambda}_1}} + \frac{\widetilde{\lambda}_1 \ln(y/x)}{(y-x)y^{\widetilde{\lambda}_1+1}}.$$

For $1 \leq y < x$, $g(x, y) = -\frac{\ln(y/x)}{y+x} y^{\widetilde{\lambda}_2-1}$, it follows that

$$g_y'(x, y) = \frac{-1}{(y+x)y^{\widetilde{\lambda}_1+1}} + \frac{\ln(y/x)}{(y+x)^2 y^{\widetilde{\lambda}_1}} + \frac{\widetilde{\lambda}_1 \ln(y/x)}{(y+x)y^{\widetilde{\lambda}_1+1}}$$

$$= h_2(x, y) - h_1(x, y);$$

for $y \geq x$, $g(x, y) = \frac{\ln(y/x)}{y+x} y^{\widetilde{\lambda}_2-1}$, we find $g_y'(x, y) = h_1(x, y) - h_2(x, y)$.

We define two functions $h(x, y)$ and $\widetilde{h}(x, y)$ as follows:

$$h(x, y) = \begin{cases} h_2(x, y), \ 1 \le y < x, \\ h_1(x, y), \quad y \ge x, \end{cases}$$

$$\widetilde{h}(x, y) = \begin{cases} h_1(x, y), \ 1 \le y < x, \\ h_2(x, y), \quad y \ge x. \end{cases}$$

Since $-g_y'(x, y) = \widetilde{h}(x, y) - h(x, y)$, (2.39) in Chapter 2, it follows that

$$\int_1^\infty P_1(y)\widetilde{h}(x, y) \, dy = \frac{\varepsilon_0}{8}(-h_1(x, 1) + h_1(x, x) - h_2(x, x))$$

$$+ (h_1(x, x) - h_2(x, x)) \int_1^x P_1(t) dt$$

$$= \frac{\varepsilon_0}{8}\left(-h_1(x, 1) + \frac{1}{2 \, x^{\widetilde{\lambda}_1 + 1}}\right) + \frac{1}{2 \, x^{\widetilde{\lambda}_1 + 1}} \int_1^x P_1(t) dt \quad (\varepsilon_0 \in (0, 1)),$$

$$-\int_1^\infty P_1(y)h(x, y) dy = -\frac{\varepsilon_0'}{8}(-h_2(x, 1) + h_2(x, x) - h_1(x, x))$$

$$- (h_2(x, x) - h_1(x, x)) \int_1^x P_1(t) dt$$

$$= \frac{\varepsilon_0'}{8}\left(h_2(x, 1) + \frac{1}{2x^{\widetilde{\lambda}_1 + 1}}\right) + \frac{1}{2x^{\widetilde{\lambda}_1 + 1}} \int_1^x P_1(t) dt \quad (\varepsilon_0' \in (0, 1)).$$

Then, by equation (2.48) in Chapter 2 (for $q = 0$), we have

$$\int_1^x P_1(t) dt = -\frac{\varepsilon_1}{8} \quad (\varepsilon_1 \in [0, 1]),$$

and

$$-\int_1^\infty P_1(y)g_y'(x, y) dy$$

$$= \int_1^\infty P_1(y)\widetilde{h}(x, y) dy - \int_1^\infty P_1(y)h(x, y) dy$$

$$= \frac{\varepsilon_0}{8}\left(-h_1(x, 1) + \frac{1}{2x^{\widetilde{\lambda}_1 + 1}}\right) + \frac{\varepsilon_0'}{8}\left(h_2(x, 1) + \frac{1}{2x^{\widetilde{\lambda}_1 + 1}}\right)$$

$$= -\frac{\varepsilon_1}{8x^{\widetilde{\lambda}_1 + 1}} (\varepsilon_0 \in (0, 1), \varepsilon_0' \in (0, 1), \qquad \varepsilon_1 \in [0, 1]). \qquad (3.90)$$

Since, we have

$$-h_1(x, 1) + \frac{1}{2x^{\widetilde{\lambda}_1 + 1}} \le -h_1(x, x) + \frac{1}{2x^{\widetilde{\lambda}_1 + 1}} < 0,$$

then, by (3.90), we obtain

$$-\int_1^\infty P_1(y)g_y'(x,y)dy > \frac{1}{8}\left(-h_1(x,1) + \frac{1}{2x^{\widetilde{\lambda}_1+1}}\right) - \frac{1}{8x^{\widetilde{\lambda}_1+1}}$$

$$= \frac{-1}{8(1+x)} + \frac{x\ln x}{4(1-x)(1+x)^2}$$

$$+ \frac{\widetilde{\lambda}_1 x\ln x}{4(1-x)(1+x)} - \frac{1}{16x^{\widetilde{\lambda}_1+1}}. \quad (3.91)$$

Since $-\frac{1}{2}g(x,1) = -\frac{\ln x}{2(1+x)}$, in view of (3.88), (3.89) and (3.91), (i) for $1 \le x < 2$, $\frac{\ln x}{x-1} \le 1$, we find

$$\widetilde{r}(x) > \frac{\ln x}{\widetilde{\lambda}_2(1+x)} + \frac{1}{\widetilde{\lambda}_2^2(1+x)} - \frac{\ln x}{2(1+x)} - \frac{1}{8(1+x)}$$

$$- \frac{x\ln x}{4(x-1)(1+x)^2} - \frac{\widetilde{\lambda}_1 x\ln x}{4(x-1)(1+x)} - \frac{1}{16x^{\widetilde{\lambda}_1+1}} \quad (3.92)$$

$$\ge \frac{\ln x}{1+x}\left[\frac{1}{\widetilde{\lambda}_2} - \frac{1}{2(1+x)}\right] + \frac{1}{\widetilde{\lambda}_2^2(1+x)} - \frac{1}{8(1+x)}$$

$$- \frac{x}{4(1+x)^2} - \frac{\widetilde{\lambda}_1 x}{4(1+x)} - \frac{1}{16x^{\widetilde{\lambda}_1+1}}$$

$$\ge \frac{1}{1+x} - \frac{1}{8(1+x)} - \frac{x}{4(1+x)^2} - \frac{x}{4(1+x)} - \frac{1}{16}$$

$$= \frac{7}{8(1+x)} + \frac{1}{4(1+x)^2} - \frac{5}{16}$$

$$> \frac{7}{8(1+2)} + \frac{1}{4(1+2)^2} - \frac{5}{16} = \frac{1}{114} > 0;$$

(ii) for $x \ge 2$, $-\frac{x}{x-1} \ge -2$, by (3.92), we have

$$\widetilde{r}(x) > \frac{\ln x}{\widetilde{\lambda}_2(1+x)} + \frac{1}{1+x} - \frac{\ln x}{2(1+x)} - \frac{1}{8(1+x)}$$

$$- \frac{\ln x}{2(1+x)^2} - \frac{\widetilde{\lambda}_1 \ln x}{2(1+x)} - \frac{1}{32}$$

$$\ge \frac{\ln x}{x+1}\left[\frac{1}{\widetilde{\lambda}_2} + \frac{\widetilde{\lambda}_2}{2} - \frac{7}{6}\right] + \frac{7}{8(1+x)} - \frac{1}{32}$$

$$\ge \frac{\ln x}{x+1}\left[\sqrt{2} - \frac{7}{6}\right] + \frac{7}{8(1+2)} - \frac{1}{32} > 0.$$

Hence, (3.88) follows.

(3) In view of

$$f(x,y) = \frac{x^{\tilde{\lambda}_1}|\ln(x/y)|}{x+y} y^{\tilde{\lambda}_2-1} = x^{\tilde{\lambda}_1} g(x,y),$$

by (3.4) and (3.88), we have $\tilde{R}(x) = x^{\tilde{\lambda}_1}\tilde{r}(x) > 0$.

By Condition (iii), it follows that

$$\varpi(\tilde{\lambda}_1, x) = x^{\tilde{\lambda}_1} \sum_{n=1}^{\infty} \frac{|\ln(x/n)|n^{\tilde{\lambda}_2-1}}{x+n}$$

$$= k(\tilde{\lambda}_1) - \tilde{R}(x) < k(\tilde{\lambda}_1) \qquad (x \geq 1).$$

We find

$$\int_0^1 g(x,y)dy = \frac{1}{x^{\tilde{\lambda}_1}} \int_0^{\frac{1}{x}} \frac{-\ln t}{t+1} t^{\tilde{\lambda}_2-1} dt$$

$$< \int_0^{\frac{1}{x}} (-\ln t) t^{\tilde{\lambda}_2-1} dt = \left(\frac{\ln x}{\tilde{\lambda}_2} + \frac{1}{\tilde{\lambda}_2^2} \right) \frac{1}{x},$$

and by (3.90), we obtain

$$-\int_1^{\infty} P_1(y)g_y'(x,y) \, dy < \frac{1}{8} \left(h_2(x,1) + \frac{1}{2x^{\tilde{\lambda}_1+1}} \right)$$

$$= \frac{\ln x}{8(x-1)(1+x)} + \frac{\tilde{\lambda}_1 \ln x}{8(x-1)} + \frac{1}{16x^{\tilde{\lambda}_1+1}}.$$

Hence by (3.88), it follows that

$$\tilde{R}(x) = x^{\tilde{\lambda}_1}\tilde{r}(x) < x^{\tilde{\lambda}_1} \left[\left(\frac{\ln x}{\tilde{\lambda}_2} + \frac{1}{\tilde{\lambda}_2^2} \right) \frac{1}{x} - \frac{\ln x}{2(1+x)} \right.$$

$$\left. + \frac{\ln x}{8(x-1)(1+x)} + \frac{\tilde{\lambda}_1 \ln x}{8(x-1)} + \frac{1}{16x^{\tilde{\lambda}_1+1}} \right]. \quad (3.93)$$

Setting $\theta_{\tilde{\lambda}_1}(x) = \frac{1}{k(\tilde{\lambda}_1)}[\tilde{R}(x) + x^{\tilde{\lambda}_1}\frac{\ln x}{x+1}] > 0$, we obtain

$$\varpi(\tilde{\lambda}_1, x) > \varpi(\tilde{\lambda}_1, x) - \frac{x^{\tilde{\lambda}_1} \ln x}{x+1} = k(\tilde{\lambda}_1)(1 - \theta_{\tilde{\lambda}_1}(x)) > 0.$$

For $\eta = \lambda_2 - \delta_0 - \delta_1 > 0$, by (3.93), we find

$$0 < x^{\eta} \theta_{\tilde{\lambda}_1}(x) \leq x^{\tilde{\lambda}_2-\delta_1} \theta_{\tilde{\lambda}_1}(x) = \frac{1}{k(\tilde{\lambda}_1)} \left[x^{1-\delta_1} \tilde{r}(x) + \frac{x^{1-\delta_1} \ln x}{x+1} \right]$$

$$< \frac{1}{k(\tilde{\lambda}_1)} \left[\left(\frac{\ln x}{\tilde{\lambda}_2} + \frac{1}{\tilde{\lambda}_2^2} \right) \frac{1}{x^{\delta_1}} - \frac{x^{1-\delta_1} \ln x}{2(1+x)} + \frac{x^{1-\delta_1} \ln x}{8(x-1)(1+x)} \right.$$

$$\left. + \frac{\tilde{\lambda}_1 x^{1-\delta_1} \ln x}{8(x-1)} + \frac{x^{1-\delta_1}}{16x^{\tilde{\lambda}_1+1}} + \frac{x^{1-\delta_1} \ln x}{x+1} \right] \to 0 \quad (x \to \infty),$$

that is, $\theta_{\widetilde{\lambda}_1}(x) = O\left(\frac{1}{x^\eta}\right)$ $(x \geq 1; \eta = \lambda_2 - \delta_0 - \delta_1 > 0)$. Hence, we show that (3.86) and (3.87) are valid.

By the Note of Corollary 3.8 $(c = 1)$, for $k_1(x, n) = \frac{|\ln(x/n)|}{x+n}$ $(x \geq 1,\ n \in \mathbf{N}; 0 < \lambda_1 < 1,\ \lambda_2 = 1 - \lambda_1)$, if $p > 1$, we have equivalent inequalities (3.60), (3.61) and (3.62); if $p < 0$, we have equivalent inequalities (3.33), (3.34) and (3.35)(for $v(n) = n$, $n_0 = 1$); if $0 < p < 1$, we have equivalent inequalities (3.63), (3.64) and (3.65). All the inequalities are with the same best constant factor

$$k(\lambda_1) = \sum_{k=0}^{\infty}(-1)^k\left[\frac{1}{(k+\lambda_1)^2} + \frac{1}{(k+\lambda_2)^2}\right]. \tag{3.94}$$

Example 3.10. (See Yang and Chen [126]).

We set $k_\lambda(x, y) = \frac{|\ln(x/y)|}{(\max\{x,y\})^\lambda}$ $((x, y) \in \mathbf{R}_+^2;\ \lambda_1 > 0,\ 0 < \lambda_2 < 1,$ $\lambda_2 + \lambda_1 = \lambda \leq 4)$. There exists a constant $0 < \delta_0 < \min\{\lambda_1, \lambda_2, 1 - \lambda_2\}$, such that for any $\widetilde{\lambda}_1 \in (\lambda_1 - \delta_0,\ \lambda_1 + \delta_0) \subset (0, \lambda)$, $\widetilde{\lambda}_2 = \lambda - \widetilde{\lambda}_1 \in (\lambda_2 - \delta_0,\ \lambda_2 + \delta_0) \subset (0, 1)$. Since $\lambda_2 - \delta_0 > 0$, there exists $\delta_1 > 0$, such that $\eta = \lambda_2 - \delta_0 - \delta_1 > 0$. It follows that

$$k(\widetilde{\lambda}_1) = \int_0^\infty \frac{|\ln t|t^{\widetilde{\lambda}_1 - 1}}{(\max\{t, 1\})^\lambda}dt$$

$$= \int_0^1 (-\ln t)t^{\widetilde{\lambda}_1 - 1}dt + \int_1^\infty \frac{(\ln t)t^{\widetilde{\lambda}_1 - 1}}{t^\lambda}dt$$

$$= \int_0^1 (-\ln t)(t^{\widetilde{\lambda}_1 - 1} + t^{\widetilde{\lambda}_2 - 1})dt = \int_0^1 (-\ln t)d\left(\frac{1}{\widetilde{\lambda}_1}t^{\widetilde{\lambda}_1} + \frac{1}{\widetilde{\lambda}_2}t^{\widetilde{\lambda}_2}\right)$$

$$= (-\ln t)\left(\frac{1}{\widetilde{\lambda}_1}t^{\widetilde{\lambda}_1} + \frac{1}{\widetilde{\lambda}_2}t^{\widetilde{\lambda}_2}\right)\Big|_0^1 + \int_0^1 \left(\frac{1}{\widetilde{\lambda}_1}t^{\widetilde{\lambda}_1 - 1} + \frac{1}{\widetilde{\lambda}_2}t^{\widetilde{\lambda}_2 - 1}\right)dt$$

$$= \frac{1}{\widetilde{\lambda}_1^2} + \frac{1}{\widetilde{\lambda}_2^2} \in \mathbf{R}_+.$$

Since $t^{\delta_0}k_1(t, 1) = \frac{t^{\delta_0}|\ln t|}{(\max\{t,1\})^\lambda} \to 0$ $(t \to 0^+)$, we still find that

$$k_1(t, 1) = -\ln t \leq L\left(\frac{1}{t^{\delta_0}}\right) \quad (t \in (0, 1]; L > 0).$$

In the following, we show that

$$\omega(\widetilde{\lambda}_2, n) = n^{\widetilde{\lambda}_2} \int_1^\infty \frac{|\ln \frac{x}{n}| n^{\widetilde{\lambda}_1 - 1}}{(\max\{x, n\})^\lambda} x^{\widetilde{\lambda}_1 - 1} dx < k(\widetilde{\lambda}_1) \quad (n \in \mathbf{N}), \quad (3.95)$$

$$0 < k(\widetilde{\lambda}_1)(1 - \theta_{\widetilde{\lambda}_1}(x))$$

$$< \varpi(\widetilde{\lambda}_1, x) = x^{\widetilde{\lambda}_1} \sum_{n=1}^\infty \frac{|\ln \frac{x}{n}| n^{\widetilde{\lambda}_2 - 1}}{(\max\{x, n\})^\lambda} < k(\widetilde{\lambda}_1) \quad (x \geq 1), \quad (3.96)$$

$$0 < \theta_{\widetilde{\lambda}_1}(x) = O\left(\frac{1}{x^\eta}\right) \quad (x \geq 1; \ \eta = \lambda_2 - \delta_0 - \delta_1 > 0). \quad (3.97)$$

(1) For $n \in \mathbf{N}$, putting $t = \frac{x}{n}$ in the integral of (3.95), we find

$$\omega(\lambda_2, n) = \int_{\frac{1}{n}}^\infty \frac{|\ln t| \, t^{\widetilde{\lambda}_1 - 1}}{(\max\{t, 1\})^\lambda} dt$$

$$< \int_0^\infty \frac{|\ln t| \, t^{\widetilde{\lambda}_1 - 1}}{(\max\{t, 1\})^\lambda} dt = k(\widetilde{\lambda}_1),$$

and then, inequality (3.95) follows.

(2) Setting

$$g(x, y) = \frac{|\ln(x/y)|}{(\max\{x, y\})^\lambda} y^{\widetilde{\lambda}_2 - 1} \quad (x \geq 1, y > 0),$$

in the following, we prove that

$$\widetilde{r}(x) = \int_0^1 g(x, y) dy - \frac{1}{2} g(x, 1) - \int_1^\infty P_1(y) g_y'(x, y) dy > 0. \quad (3.98)$$

In fact, putting $t = \frac{y}{x}$, we obtain

$$\int_0^1 g(x, y) dy = \int_0^1 \frac{\ln(x/y) \, y^{\widetilde{\lambda}_2 - 1}}{x^\lambda} dy = \frac{1}{x^{\widetilde{\lambda}_1}} \int_0^{\frac{1}{x}} (-\ln t) \, t^{\widetilde{\lambda}_2 - 1} dt$$

$$= \frac{1}{\widetilde{\lambda}_2 x^\lambda} \left(\ln x + \frac{1}{\widetilde{\lambda}_2}\right). \quad (3.99)$$

For $x, y \geq 1$, we set four positive decreasing functions with respect to $y \in [1, \infty)$ as follows:

$$h_1(x, y) = \frac{1}{y^{2 + \widetilde{\lambda}_1}} + (1 + \widetilde{\lambda}_1) \frac{x \ln(y/x)}{(y - x) y^{2 + \widetilde{\lambda}_1}},$$

$$h_2(x, y) = (1 - \widetilde{\lambda}_2) \frac{\ln(y/x)}{x^\lambda (y - x) y^{1 - \widetilde{\lambda}_2}},$$

$$\widetilde{h}_1(x, y) = (1 + \widetilde{\lambda}_1) \frac{\ln(y/x)}{(y - x) y^{1 + \widetilde{\lambda}_1}},$$

and

$$\widetilde{h}_2(x,y) = \frac{1}{x^\lambda y^{2-\widetilde{\lambda}_2}} + (1-\widetilde{\lambda}_2)\frac{\ln(y/x)}{x^{\lambda-1}(y-x)y^{2-\widetilde{\lambda}_2}}.$$

Then, for $1 \le y < x$, $g(x,y) = -\frac{\ln(y/x)}{x^\lambda}y^{\widetilde{\lambda}_2-1}$, it follows that

$$g'_y(x,y) = \frac{-1}{x^\lambda y^{2-\widetilde{\lambda}_2}} + (1-\widetilde{\lambda}_2)\frac{(y-x)\ln(y/x)}{x^\lambda(y-x)y^{2-\widetilde{\lambda}_2}}$$

$$= h_2(x,y) - \widetilde{h}_2(x,y);$$

for $y \ge x$, $g(x,y) = \ln(y/x)y^{-\widetilde{\lambda}_1-1}$, we find

$$g'_y(x,y) = \frac{1}{y^{2+\widetilde{\lambda}_1}} - (1+\widetilde{\lambda}_1)\frac{(y-x)\ln(y/x)}{(y-x)y^{2+\widetilde{\lambda}_1}}$$

$$= h_1(x,y) - \widetilde{h}_1(x,y).$$

We define two functions $h(x,y)$ and $\widetilde{h}(x,y)$ as follows:

$$h(x,y) = \begin{cases} h_2(x,y), & 1 \le y < x, \\ h_1(x,y), & y \ge x, \end{cases}$$

$$\widetilde{h}(x,y) = \begin{cases} \widetilde{h}_2(x,y), & 1 \le y < x, \\ \widetilde{h}_1(x,y), & y \ge x. \end{cases}$$

By equation (2.39) in Chapter 2, it follows that

$$\int_1^\infty P_1(y)\widetilde{h}(x,y)dy = \frac{\varepsilon_0}{8}(-\widetilde{h}_2(x,1) + \widetilde{h}_2(x,x) - \widetilde{h}_1(x,x))$$

$$+ (\widetilde{h}_2(x,x) - \widetilde{h}_1(x,x))\int_1^x P_1(t)dt$$

$$= \frac{\varepsilon_0}{8}\left(-\widetilde{h}_2(x,1) + \frac{1-\lambda}{x^{\widetilde{\lambda}_1+2}}\right) + \frac{1-\lambda}{x^{\widetilde{\lambda}_1+2}}\int_1^x P_1(t)dt \quad (\varepsilon_0 \in (0,1)),$$

$$-\int_1^\infty P_1(y)h(x,y)dy = -\frac{\varepsilon_0'}{8}(-h_2(x,1) + h_2(x,x) - h_1(x,x))$$

$$-(h_2(x,x) - h_1(x,x))\int_1^x P_1(t)dt$$

$$= \frac{\varepsilon_0'}{8}\left(h_2(x,1) + \frac{1+\lambda}{x^{\widetilde{\lambda}_1+2}}\right) + \frac{1+\lambda}{x^{\widetilde{\lambda}_1+2}}\int_1^x P_1(t)dt \quad (\varepsilon_0' \in (0,1)).$$

Since $-g'_y(x,y) = \widetilde{h}(x,y) - h(x,y)$, and by equation (2.48) in Chapter 2 (for $q = 0$),

$$\int_1^x P_1(t)dt = -\frac{\varepsilon_1}{8} \quad (\varepsilon_1 \in [0,1]),$$

we find

$$-\int_1^\infty P_1(y)g_y'(x,y)\,dy = \int_1^\infty P_1(y)\widetilde{h}(x,y)dy - \int_1^\infty P_1(y)h(x,y)dy$$

$$= \frac{\varepsilon_0}{8}\left(-\widetilde{h}_2(x,1) + \frac{1-\lambda}{x^{\widetilde{\lambda}_1+2}}\right)$$

$$+\frac{\varepsilon_0'}{8}\left(h_2(x,1) + \frac{1+\lambda}{x^{\widetilde{\lambda}_1+2}}\right) - \frac{\varepsilon_1}{4x^{\widetilde{\lambda}_1+2}}.$$

$$(3.100)$$

Since we have

$$-\widetilde{h}_2(x,1) + \frac{1-\lambda}{x^{\widetilde{\lambda}_1+2}} \le -\widetilde{h}_2(x,x) + \frac{1-\lambda}{x^{\widetilde{\lambda}_1+2}} < 0,$$

then, by equality (3.100), we find

$$-\int_1^\infty P_1(y)g_y'(x,y)dy > \frac{1}{8}\left(-\widetilde{h}_2(x,1) + \frac{1-\lambda}{x^{\widetilde{\lambda}_1+2}}\right) - \frac{1}{4x^{\widetilde{\lambda}_1+2}}$$

$$= -\frac{1}{8x^\lambda} - \frac{(1-\widetilde{\lambda}_2)\ln x}{8x^{\lambda-1}(x-1)} + \frac{1-\lambda}{8x^{\widetilde{\lambda}_1+2}} - \frac{1}{4x^{\widetilde{\lambda}_1+2}}.$$

$$(3.101)$$

For $-\frac{1}{2}g(x,1) = -\frac{\ln x}{2x^\lambda}$, $0 < \lambda \le 4$ and $0 < \widetilde{\lambda}_2 < 1$, in view of (3.98), (3.99) and (3.101), we obtain

$$\widetilde{r}(x) > \frac{\ln x}{\widetilde{\lambda}_2 x^\lambda} + \frac{1}{\widetilde{\lambda}_2^2 x^\lambda} - \frac{\ln x}{2x^\lambda} - \frac{1}{8x^\lambda} - \frac{(1-\widetilde{\lambda}_2)\ln x}{8x^{\lambda-1}(x-1)} - \frac{1+\lambda}{8x^{\widetilde{\lambda}_1+2}}$$

$$\ge \frac{1}{x^\lambda}\left[\left(1-\frac{1}{2}\right)\ln x + \frac{7}{8} - \frac{(1-\widetilde{\lambda}_2)x\ln x}{8(x-1)} - \frac{5}{8}\right]$$

$$\ge \frac{1}{x^\lambda}\left[\frac{\ln x}{2} + \frac{1}{4} - \frac{x\ln x}{8(x-1)}\right].$$

$$(3.102)$$

(i) If $1 \le x < \frac{5}{2}$, then, $\frac{x\ln x}{x-1} \le 2$, by (3.102), we have

$$\widetilde{r}(x) > \frac{1}{x^\lambda}\left(\frac{\ln x}{2} + \frac{1}{4} - \frac{1}{4}\right) > 0;$$

(ii) if $x \ge \frac{5}{2}$, then, $\frac{x}{x-1} \le \frac{5}{3}$, by (3.102), we have

$$\widetilde{r}(x) > \frac{1}{x^\lambda}\left(\frac{\ln x}{2} + \frac{1}{4} - \frac{5\ln x}{24}\right) > 0.$$

Hence, (3.97) follows for $x \ge 1$.

(3) In view of

$$f(x,y) = \frac{x^{\tilde{\lambda}_1}|\ln(x/y)|}{(\max\{x,y\})^\lambda} y^{\tilde{\lambda}_2 - 1} = x^{\tilde{\lambda}_1} g(x,y),$$

by (3.4) and (3.98), we have $\tilde{R}(x) = x^{\tilde{\lambda}_1} \tilde{r}(x) > 0$.

By Condition (iii), it follows that

$$\varpi(\tilde{\lambda}_1, x) = \sum_{n=1}^{\infty} \frac{x^{\tilde{\lambda}_1}|\ln(x/n)|n^{\tilde{\lambda}_2 - 1}}{(\max\{x,n\})^\lambda}$$

$$= k(\tilde{\lambda}_1) - \tilde{R}(x) < k(\tilde{\lambda}_1) \quad (x \geq 1).$$

By equality (3.100), we obtain

$$-\int_1^\infty P_1(y)g'_y(x,y)dy < \frac{1}{8}\left(h_2(x,1) + \frac{1+\lambda}{x^{\tilde{\lambda}_1+2}}\right)$$

$$= \frac{(1-\tilde{\lambda}_2)\ln x}{8x^\lambda(x-1)} + \frac{1+\lambda}{8x^{\tilde{\lambda}_1+2}}.$$

Hence, inequality by (3.98), it follows that

$$\tilde{R}(x) = x^{\tilde{\lambda}_1}\tilde{r}(x) < x^{\tilde{\lambda}_1}\left[\frac{1}{\tilde{\lambda}_2 x^\lambda}\left(\ln x + \frac{1}{\tilde{\lambda}_2}\right) - \frac{\ln x}{2x^\lambda}\right.$$

$$\left. + \frac{(1-\tilde{\lambda}_2)\ln x}{8x^\lambda(x-1)} + \frac{1+\lambda}{8x^{\tilde{\lambda}_1+2}}\right]. \quad (3.103)$$

Setting $\theta_{\tilde{\lambda}_1}(x) = \frac{1}{k(\tilde{\lambda}_1)}(\tilde{R}(x) + \frac{\ln x}{x^{\tilde{\lambda}_2}}) > 0$, we obtain

$$\varpi(\tilde{\lambda}_1, x) > \varpi(\tilde{\lambda}_1, x) - \frac{\ln x}{x^{\tilde{\lambda}_2}} = k(\tilde{\lambda}_1)(1 - \theta_{\tilde{\lambda}_1}(x)) > 0.$$

For $\eta = \lambda_2 - \delta_0 - \delta_1 > 0$ ($\delta_1 > 0$), by (3.103), since $x \geq 1$, we obtain

$$0 < x^\eta \theta_{\tilde{\lambda}_1}(x) \leq x^{\tilde{\lambda}_2 - \delta_1}\theta_{\tilde{\lambda}_1}(x) = \frac{1}{k(\tilde{\lambda}_1)}\left(x^{\lambda - \delta_1}\tilde{r}(x) + \frac{\ln x}{x^{\delta_1}}\right)$$

$$< \frac{1}{k(\tilde{\lambda}_1)}\left[\left(\frac{\ln x}{\tilde{\lambda}_2} + \frac{1}{\tilde{\lambda}_2^2}\right)\frac{1}{x^{\delta_1}} + \frac{\ln x}{2x^{\delta_1}}\right.$$

$$\left. + \frac{(1-\tilde{\lambda}_2)\ln x}{8x^{\delta_1}(x-1)} + \frac{1+\lambda}{8x^{2-\tilde{\lambda}_2+\delta_1}}\right] \to 0 \quad (x \to \infty),$$

that is, $\theta_{\tilde{\lambda}_1}(x) = O\left(\frac{1}{x^\eta}\right)$ ($x \geq 1$; $\eta = \lambda_2 - \delta_0 - \delta_1 > 0$). Hence, we show that (3.96) and (3.97) are valid.

By the Note of Corollary 3.8 ($c = 1$), for $k_\lambda(x,n) = \frac{|\ln(x/n)|}{(\max\{x,n\})^\lambda}(x \geq 1, n \in \mathbf{N}; \lambda_1 > 0, 0 < \lambda_2 < 1, \lambda_2 + \lambda_1 = \lambda \leq 4)$, if $p > 1$, we have equivalent inequalities (3.60), (3.61) and (3.62); if $p < 0$, we have equivalent inequalities (3.33), (3.34) and (3.35)(for $v(n) = n$, $n_0 = 1$); if $0 < p < 1$, we have equivalent inequalities (3.63), (3.64) and (3.65). All the inequalities are with the same best constant factor

$$k(\lambda_1) = \frac{1}{\lambda_1^2} + \frac{1}{\lambda_2^2}. \tag{3.104}$$

3.3.6 Applying Condition (iii) and Corollary 3.4

Example 3.11. (See Yang [158])

We set $k_\lambda(x, y) = \frac{1}{x^\lambda + y^\lambda}$ $((x, y) \in \mathbf{R}_+^2;$ $\lambda_1 > 0$, $0 < \lambda_2 < 2$, $\lambda_1 + \lambda_2 = \lambda)$. There exists a constant $0 < \delta_0 < \min\{\lambda_1, \lambda_2, 2 - \lambda_2\}$, such that for any $\widetilde{\lambda}_1 \in (\lambda_1 - \delta_0, \lambda_1 + \delta_0) \subset (0, \lambda)$, $\widetilde{\lambda}_2 = \lambda - \widetilde{\lambda}_1 \in (\lambda_2 - \delta_0, \lambda_2 + \delta_0) \subset (0, 2)$. It follows that

$$k(\widetilde{\lambda}_1) = \int_0^\infty \frac{t^{\widetilde{\lambda}_1 - 1}}{t^\lambda + 1} dt = \int_0^\infty \frac{u^{(\widetilde{\lambda}_1/\lambda) - 1}}{\lambda(u + 1)} du$$

$$= \frac{\pi}{\lambda \sin(\pi \widetilde{\lambda}_1/\lambda)} \in \mathbf{R}_+.$$

For $\gamma < 1$, it is obvious that

$$\omega(\widetilde{\lambda}_2, n) = (n - \gamma)^{\widetilde{\lambda}_2} \int_0^\infty \frac{x^{\widetilde{\lambda}_1 - 1} dx}{x^\lambda + (n - \gamma)^\lambda} = k(\widetilde{\lambda}_1) \quad (n \in \mathbf{N}).$$

In the following, we show that, for

$$\gamma \le \gamma(\widetilde{\lambda}_1) = 1 - \frac{1}{4}\left(\widetilde{\lambda}_1 + \sqrt{3\widetilde{\lambda}_1^2 + 2\widetilde{\lambda}_1}\right) < 1,$$

it follows that

$$0 < k(\widetilde{\lambda}_1)(1 - \theta_{\widetilde{\lambda}_1}(x))$$

$$< \varpi(\widetilde{\lambda}_1, x) = \sum_{n=1}^\infty \frac{x^{\widetilde{\lambda}_1}(n - \gamma)^{\widetilde{\lambda}_2 - 1}}{x^\lambda + (n - \gamma)^\lambda} < k(\widetilde{\lambda}_1)(x > 0), \tag{3.105}$$

$$0 < \theta_{\widetilde{\lambda}_1}(x) = O\left(\frac{1}{x^\eta}\right) \quad (x \ge 1; \eta = \lambda_2 - \delta_0 > 0). \tag{3.106}$$

(1) Setting

$$g(x, y) = \frac{(y - \gamma)^{\widetilde{\lambda}_2 - 1}}{x^\lambda + (y - \gamma)^\lambda} \quad (x > 0, y > \gamma),$$

in the following, we prove that

$$\widetilde{R}(x) = x^{\widetilde{\lambda}_1}\left[\int_\gamma^1 g(x, y) dy - \frac{1}{2}g(x, 1) - \int_1^\infty P_1(y)g_y'(x, y) dy\right] > 0. \tag{3.107}$$

In fact, we obtain

$$\int_\gamma^1 g(x,y)dy = \int_\gamma^1 \frac{(y-\gamma)^{\widetilde{\lambda}_2-1}dy}{x^\lambda+(y-\gamma)^\lambda} = \frac{1}{x^{\widetilde{\lambda}_1}}\int_0^{\frac{1-\gamma}{x}}\frac{t^{\widetilde{\lambda}_2-1}dt}{1+t^\lambda}, \qquad (3.108)$$

$-\frac{1}{2}g(x,1) = -\frac{1}{2}\frac{(1-\gamma)^{\widetilde{\lambda}_2-1}}{x^\lambda+(1-\gamma)^\lambda}$, and

$$\begin{aligned}
g_y'(x,y) &= -\frac{\lambda(y-\gamma)^{\widetilde{\lambda}_2+\lambda-2}}{[x^\lambda+(y-\gamma)^\lambda]^2} + \frac{(\widetilde{\lambda}_2-1)(y-\gamma)^{\widetilde{\lambda}_2-2}}{x^\lambda+(y-\gamma)^\lambda} \\
&= -\frac{\lambda[(y-\gamma)^\lambda+x^\lambda-x^\lambda](y-\gamma)^{\widetilde{\lambda}_2-2}}{[x^\lambda+(y-\gamma)^\lambda]^2} + \frac{(\widetilde{\lambda}_2-1)(y-\gamma)^{\widetilde{\lambda}_2-2}}{x^\lambda+(y-\gamma)^\lambda} \\
&= -\frac{(\lambda-\widetilde{\lambda}_2+1)(y-\gamma)^{\widetilde{\lambda}_2-2}}{x^\lambda+(y-\gamma)^\lambda} + \frac{\lambda x^\lambda(y-\gamma)^{\widetilde{\lambda}_2-2}}{[x^\lambda+(y-\gamma)^\lambda]^2}.
\end{aligned}$$

Then by equation (2.39) in Chapter 2, for $\lambda > 0, \widetilde{\lambda}_2 < 2$, we have

$$\begin{aligned}
-\int_1^\infty P_1(y)g_y'(x,y)dy = &\int_1^\infty P_1(y)\frac{(\lambda-\widetilde{\lambda}_2+1)(y-\gamma)^{\widetilde{\lambda}_2-2}}{x^\lambda+(y-\gamma)^\lambda}dy \\
&- \int_1^\infty P_1(y)\frac{\lambda x^\lambda(y-\gamma)^{\widetilde{\lambda}_2-2}}{[x^\lambda+(y-\gamma)^\lambda]^2}dy \\
&> -\frac{1}{8}\frac{(\lambda-\widetilde{\lambda}_2+1)(1-\gamma)^{\widetilde{\lambda}_2-2}}{x^\lambda+(1-\gamma)^\lambda} + 0. \qquad (3.109)
\end{aligned}$$

Then, by (3.107), (3.108) and (3.109), setting

$$A = \left[\frac{1-\gamma}{2} + \frac{1}{8}(\widetilde{\lambda}_1+1)\right](1-\gamma)^{\widetilde{\lambda}_2-2},$$

we have

$$\widetilde{R}(x) > h(x) = \int_0^{\frac{1-\gamma}{x}}\frac{t^{\widetilde{\lambda}_2-1}dt}{1+t^\lambda} - \frac{Ax^{\widetilde{\lambda}_1}}{x^\lambda+(1-\gamma)^\lambda}. \qquad (3.110)$$

Since, for $\gamma \le \gamma(\widetilde{\lambda}_1)$,

$$1-\gamma \ge \frac{1}{4}\left(\widetilde{\lambda}_1+\sqrt{3\widetilde{\lambda}_1^2+2\widetilde{\lambda}_1}\right),$$

$$B(\gamma) = (1-\gamma)^2 - \frac{\widetilde{\lambda}_1}{2}(1-\gamma) - \frac{1}{8}(\widetilde{\lambda}_1^2+\widetilde{\lambda}_1) \ge 0,$$

then, we find

$$\begin{aligned}
h'(x) &= -\frac{(1-\gamma)(\frac{1-\gamma}{x})^{\widetilde{\lambda}_2-1}}{x^2[1+(\frac{1-\gamma}{x})^\lambda]} + \frac{\lambda Ax^{\widetilde{\lambda}_1+\lambda-1}}{[x^\lambda+(1-\gamma)^\lambda]^2} - \frac{A\widetilde{\lambda}_1 x^{\widetilde{\lambda}_1-1}}{x^\lambda+(1-\gamma)^\lambda} \\
&= -[(1-\gamma)^{\widetilde{\lambda}_2}-A\widetilde{\lambda}_1]\frac{x^{\widetilde{\lambda}_1-1}}{x^\lambda+(1-\gamma)^\lambda} - \frac{\lambda A(1-\gamma)^\lambda x^{\widetilde{\lambda}_1-1}}{[x^\lambda+(1-\gamma)^\lambda]^2} \\
&< -B(\gamma)\frac{(1-\gamma)^{\widetilde{\lambda}_2-2}x^{\widetilde{\lambda}_1-1}}{x^\lambda+(1-\gamma)^\lambda} < 0 \qquad (A,\ \lambda > 0).
\end{aligned}$$

By (3.110), it follows that $\widetilde{R}(x) > h(\infty) = 0 (x > 0)$. Hence, (3.107) holds.

(2) In view of $f(x,y) = x^{\widetilde{\lambda}_1} g(x,y)$, by (3.4) and Condition (iii), it follows that

$$\varpi(\widetilde{\lambda}_1, x) = \sum_{n=1}^{\infty} \frac{x^{\widetilde{\lambda}_1}(n-\gamma)^{\widetilde{\lambda}_2-1}}{x^\lambda + (n-\gamma)^\lambda} = k(\widetilde{\lambda}_1) - \widetilde{R}(x) < k(\widetilde{\lambda}_1) \quad (x > 0).$$

By (3.109), we still obtain

$$-\int_1^\infty P_1(y)g_y'(x,y)dy < \frac{1}{8} \frac{\lambda x^\lambda (1-\gamma)^{\widetilde{\lambda}_2-2}}{[x^\lambda + (1-\gamma)^\lambda]^2},$$

and by (3.107), it follows that

$$\widetilde{R}(x) < \frac{1}{\widetilde{\lambda}_2}\left(\frac{1-\gamma}{x}\right)^{\widetilde{\lambda}_2} - \frac{(1-\gamma)^{\widetilde{\lambda}_2-1}x^{\widetilde{\lambda}_1}}{2[x^\lambda+(1-\gamma)^\lambda]} + \frac{1}{8}\frac{\lambda x^{\lambda+\widetilde{\lambda}_1}(1-\gamma)^{\widetilde{\lambda}_2-2}}{[x^\lambda+(1-\gamma)^\lambda]^2}. \quad (3.111)$$

Setting

$$\theta_{\widetilde{\lambda}_1}(x) = \frac{1}{k(\widetilde{\lambda}_1)}\left[\widetilde{R}(x) + \frac{x^{\widetilde{\lambda}_1}(1-\gamma)^{\widetilde{\lambda}_2-1}}{x^\lambda+(1-\gamma)^\lambda}\right] > 0,$$

we find

$$\varpi(\widetilde{\lambda}_1, x) > \varpi(\widetilde{\lambda}_1, x) - \frac{x^{\widetilde{\lambda}_1}(1-\gamma)^{\widetilde{\lambda}_2-1}}{x^\lambda+(1-\gamma)^\lambda} = k(\widetilde{\lambda}_1)(1-\theta_{\widetilde{\lambda}_1}(x)) > 0.$$

For $\eta = \lambda_2 - \delta_0 > 0$, $x \geq 1$, by (3.111), we obtain

$$0 < x^\eta\,\theta_{\widetilde{\lambda}_1}(x) \leq x^{\widetilde{\lambda}_2}\,\theta_{\widetilde{\lambda}_1}(x) = \frac{1}{k(\widetilde{\lambda}_1)}\left[x^{\widetilde{\lambda}_2}\,\widetilde{R}(x) + \frac{x^\lambda(1-\gamma)^{\widetilde{\lambda}_2-1}}{x^\lambda+(1-\gamma)^\lambda}\right]$$

$$< \frac{1}{k(\widetilde{\lambda}_1)}\left\{\frac{x^{\widetilde{\lambda}_2}}{\widetilde{\lambda}_2}\left(\frac{1-\gamma}{x}\right)^{\widetilde{\lambda}_2} - \frac{(1-\gamma)^{\widetilde{\lambda}_2-1}x^\lambda}{2[x^\lambda+(1-\gamma)^\lambda]}\right.$$

$$\left. + \frac{1}{8}\frac{\lambda x^{2\lambda}(1-\gamma)^{\widetilde{\lambda}_2-2}}{[x^\lambda+(1-\gamma)^\lambda]^2} + \frac{x^\lambda(1-\gamma)^{\widetilde{\lambda}_2-1}}{x^\lambda+(1-\gamma)^\lambda}\right\} \to \text{constant} \quad (x \to \infty),$$

that is, $\theta_{\widetilde{\lambda}_1}(x) = O\left(\frac{1}{x^\eta}\right)$ $(x \geq 1; \eta = \lambda_2 - \delta_0 > 0)$. Hence, we show that (3.105) and (3.106) are valid.

Setting $u(x) = x - \gamma$ $(x > \gamma)$, by Corollary 3.4, for

$$\gamma \leq 1 - \frac{1}{4}\left(\lambda_1 + \sqrt{3\lambda_1^2 + 2\lambda_1}\right),$$

$k_\lambda(u(x), v(n)) = \frac{1}{(x-\gamma)^\lambda+(n-\gamma)^\lambda}$ $(x > \gamma, n \in \mathbf{N}; \lambda_1 > 0, 0 < \lambda_2 < 2, \lambda_2 + \lambda_1 = \lambda)$, if $p > 1$, we have equivalent inequalities (3.41), (3.42) and (3.43); if $p < 0$, we have equivalent reverses of (3.41), (3.42) and (3.43); if $0 < p < 1$, we have equivalent inequalities (3.45), (3.46) and (3.47). All the inequalities are with the same best constant factor

$$k(\lambda_1) = \frac{\pi}{\lambda\sin(\pi\lambda_1/\lambda)}. \quad (3.112)$$

Note. If $0 = \gamma \le \gamma(\lambda_1) = 1 - \frac{1}{4}(\lambda_1 + \sqrt{3\lambda_1^2 + 2\lambda_1})$, then we find that $0 < \lambda_1 \le \frac{1}{2}(\sqrt{57} - 5) = 1.27^+$ and $0 < \lambda = \lambda_1 + \lambda_2 < 1.27^+ + 2 = 3.27^+$, and we get some results similar to Example 3.2(i). But the conditions are different.

Example 3.12. We set $k_\lambda(x,y) = \frac{(\min\{x,y\})^\beta}{(\max\{x,y\})^{\lambda+\beta}}((x,y) \in \mathbf{R}_+^2; 0 < 2\beta + \lambda \le 4, \lambda_1 > -\beta, -\beta < \lambda_2 < 1 - \beta, \lambda_1 + \lambda_2 = \lambda)$. There exists a constant $0 < \delta_0 < \min\{\beta + \lambda_1, \beta + \lambda_2, 1 - \beta + \lambda_2\}$, such that for any $\widetilde{\lambda}_1 \in (\lambda_1 - \delta_0, \lambda_1 + \delta_0) \subset (-\beta, \lambda + \beta)$, $\widetilde{\lambda}_2 = \lambda - \widetilde{\lambda}_1 \in (\lambda_2 - \delta_0, \lambda_2 + \delta_0) \subset (-\beta, 1 - \beta)$. It follows that

$$k(\widetilde{\lambda}_1) = \int_0^\infty \frac{(\min\{t,1\})^\beta}{(\max\{t,1\})^{\lambda+\beta}} t^{\widetilde{\lambda}_1-1} dt = \frac{2\beta + \lambda}{(\beta + \widetilde{\lambda}_1)(\beta + \widetilde{\lambda}_2)} \in \mathbf{R}_+.$$

For $0 \le \gamma < 1$, it is obvious that

$$\omega(\widetilde{\lambda}_2, n) = (n - \gamma)^{\widetilde{\lambda}_2} \int_0^\infty \frac{(\min\{x, n - \gamma\})^\beta x^{\widetilde{\lambda}_1-1}}{(\max\{x, n - \gamma\})^{\lambda+\beta}} dx = k(\widetilde{\lambda}_1) \ (n \in \mathbf{N}).$$

In the following, we show that

$$\gamma(\widetilde{\lambda}_2) = 1 - \frac{\widetilde{\lambda}_2 + \beta}{4} \left[1 + \sqrt{\frac{2(\lambda + 2\beta + 1)}{\widetilde{\lambda}_2 + \beta} - 1} - 1 \right] \in (0, 1),$$

and for $0 \le \gamma \le \gamma(\widetilde{\lambda}_2)$,

$$0 < k(\widetilde{\lambda}_2)(1 - \theta_{\widetilde{\lambda}_1}(x)) < \varpi(\widetilde{\lambda}_1, x)$$

$$= \sum_{n=1}^{\infty} \frac{x^{\widetilde{\lambda}_1}(\min\{x, n - \gamma\})^\beta}{(\max\{x, n - \gamma\})^{\lambda+\beta}} (n - \gamma)^{\widetilde{\lambda}_2-1} < k(\widetilde{\lambda}_1) \quad (x > 0), (3.113)$$

$$0 < \theta_{\widetilde{\lambda}_1}(x) = O\left(\frac{1}{x^\eta}\right) \ (x \ge 1; \ \eta = \lambda_2 + \beta - \delta_0 > 0). \tag{3.114}$$

(1) It is obvious that $\gamma(\widetilde{\lambda}_2) < 1$. We find that $\gamma(\widetilde{\lambda}_2) > 0$ is equivalent to

$$\frac{2(\lambda + 2\beta + 1)}{\widetilde{\lambda}_2 + \beta} - 1 < \left(\frac{4}{\widetilde{\lambda}_2 + \beta} - 1\right)^2,$$

or

$$\frac{8}{(\widetilde{\lambda}_2 + \beta)^2} - \frac{\lambda + 2\beta + 5}{\widetilde{\lambda}_2 + \beta} + 1 > 0.$$

Setting $y = \frac{1}{\widetilde{\lambda}_2 + \beta} \ (> 1)$,

$$f(y) = 8y^2 - (\lambda + 2\beta + 5)y + 1 \quad (y \ge 1),$$

since, $f(1) = 4 - (\lambda + 2\beta) \geq 0(\lambda + 2\beta \leq 4)$, we find

$$f'(y) = 16y^2 - (\lambda + 2\beta + 5) \geq 16 - (\lambda + 2\beta + 5) > 0(y \geq 1),$$

then, it follows that $f(y) > f(1) = 0(y > 1)$, that is, $\gamma(\widetilde{\lambda}_2) > 0$.

(2) Setting

$$g(x, y) = \frac{(\min\{x, y - \gamma\})^\beta}{(\max\{x, y - \gamma\})^{\lambda + \beta}}(y - \gamma)^{\widetilde{\lambda}_2 - 1} \quad (x > 0, \ y > \gamma),$$

in the following, we prove that, for $0 \leq \gamma \leq \gamma(\widetilde{\lambda}_2)$,

$$\widetilde{R}(x) = x^{\widetilde{\lambda}_1}\left[\int_\gamma^1 g(x, y)dy - \frac{1}{2}g(x, 1) - \int_1^\infty P_1(y)g_y'(x, y)dy\right] > 0.$$

$$(3.115)$$

In fact, we obtain

$$g(x, y) = \begin{cases} x^{-\lambda - \beta}(y - \gamma)^{\widetilde{\lambda}_2 + \beta - 1}, & \gamma < y < x + \gamma, \\ x^\beta(y - \gamma)^{-\widetilde{\lambda}_1 - \beta - 1}, & y \geq x + \gamma, \end{cases}$$

$$-g_y'(x, y) = \begin{cases} h_1(x, y), & \gamma < y < x + \gamma, \\ h_2(x, y), & y \geq x + \gamma, \end{cases}$$

$$h_1(x, y) = (1 - \widetilde{\lambda}_2 - \beta)x^{-\lambda - \beta}(y - \gamma)^{\widetilde{\lambda}_2 + \beta - 2},$$

$$h_2(x, y) = (\widetilde{\lambda}_1 + \beta + 1)x^\beta(y - \gamma)^{-\widetilde{\lambda}_1 - \beta - 2}.$$

For $0 < x < 1 - \gamma$, $y \geq 1 \geq x + \gamma$, by (2.22) in Chapter 2, we have

$$-\int_1^\infty P_1(y)g_y'(x, y)dy = \int_1^\infty P_1(y)h_2(x, y)dy$$

$$> -\frac{(\widetilde{\lambda}_1 + \beta + 1)x^\beta}{12(1 - \gamma)^{\widetilde{\lambda}_1 + \beta + 2}} > -\frac{(\widetilde{\lambda}_1 + \beta + 1)x^\beta}{8(1 - \gamma)^{\widetilde{\lambda}_1 + \beta + 2}}; \qquad (3.116)$$

for $x \geq 1 - \gamma$, by equation (2.38) in Chapter 2 (for $\varepsilon_0 \in (0, 1), \varepsilon_1 \in [0, 1]$), we find

$$-\int_1^\infty P_1(y)g_y'(x, y)dy$$

$$= \frac{\varepsilon_0}{8}(-h_1(x, 1) + h_1(x, x + \gamma) - h_2(x, x + \gamma))$$

$$\quad - \frac{\varepsilon_1}{8}(h_1(x, x + \gamma) - h_2(x, x + \gamma))$$

$$= -\frac{\varepsilon_0}{8}\left[(1 - \widetilde{\lambda}_2 - \beta)x^{-\lambda - \beta}(1 - \gamma)^{\widetilde{\lambda}_2 + \beta - 2} + \frac{2\beta + \lambda}{x^{\widetilde{\lambda}_1 + 2}}\right] + \frac{2\beta + \lambda}{8x^{\widetilde{\lambda}_1 + 2}}\varepsilon_1$$

$$> -\frac{1}{8}(1 - \widetilde{\lambda}_2 - \beta)\frac{(1 - \gamma)^{\widetilde{\lambda}_2 + \beta - 2}}{x^{\widetilde{\lambda}_2 + \beta}} - \frac{2\beta + \lambda}{8x^{\widetilde{\lambda}_1 + 2}}. \qquad (3.117)$$

We still find

$$-\frac{1}{2}g(x,1) = \begin{cases} -\frac{1}{2}\frac{x^\beta}{(1-\gamma)^{\widetilde{\lambda}_1+\beta+1}}, & 0 < x < 1-\gamma, \\ -\frac{1}{2}(1-\gamma)^{\widetilde{\lambda}_2+\beta-1}\frac{1}{x^{\lambda+\beta}}, & x \geq 1-\gamma. \end{cases} \tag{3.118}$$

For $0 < x < 1-\gamma$,

$$\int_\gamma^1 g(x,y)dy = \int_\gamma^{x+\gamma} \frac{1}{x^{\lambda+\beta}}(y-\gamma)^{\widetilde{\lambda}_2+\beta-1}dy + \int_{\gamma+x}^1 x^\beta(y-\gamma)^{-\widetilde{\lambda}_1-\beta-1}dy$$

$$= \frac{1}{(\widetilde{\lambda}_2+\beta)x^{\widetilde{\lambda}_1}} + \frac{1}{(\widetilde{\lambda}_1+\beta)x^{\widetilde{\lambda}_1}} - \frac{x^\beta}{(\widetilde{\lambda}_1+\beta)(1-\gamma)^{\widetilde{\lambda}_1+\beta}}; \tag{3.119}$$

for $x \geq 1-\gamma$,

$$\int_\gamma^1 g(x,y)dy = \int_\gamma^1 \frac{1}{x^{\lambda+\beta}}(y-\gamma)^{\widetilde{\lambda}_2+\beta-1}dy = \frac{(1-\gamma)^{\widetilde{\lambda}_2+\beta}}{(\widetilde{\lambda}_2+\beta)x^{\lambda+\beta}}. \tag{3.120}$$

(i) For $0 < x < 1-\gamma$, $0 \leq \gamma \leq \gamma(\widetilde{\lambda}_2)$,

$$1-\gamma \geq \frac{\widetilde{\lambda}_2+\beta}{4}\left[1+\sqrt{\frac{2(\lambda+2\beta+1)}{\widetilde{\lambda}_2+\beta}}-1\right],$$

it is obvious that

$$A(\gamma) = (1-\gamma)^2 - \frac{\widetilde{\lambda}_2+\beta}{2}(1-\gamma) - \frac{1}{8}(\widetilde{\lambda}_2+\beta)(\widetilde{\lambda}_1+\beta+1) \geq 0,$$

and then, by (3.115), (3.116), (3.118) and (3.120), in view of $0 < x < 1-\gamma$, $\beta + \widetilde{\lambda}_1 > 0$ and $\beta + \widetilde{\lambda}_2 > 0$, it follows that

$$\widetilde{R}(x) > \frac{1}{\widetilde{\lambda}_2+\beta} + \frac{1}{\widetilde{\lambda}_1+\beta} - \frac{x^{\beta+\widetilde{\lambda}_1}}{(\widetilde{\lambda}_1+\beta)(1-\gamma)^{\widetilde{\lambda}_1+\beta}}$$

$$- \frac{x^{\beta+\widetilde{\lambda}_1}}{2(1-\gamma)^{\widetilde{\lambda}_1+\beta+1}} - \frac{(\widetilde{\lambda}_1+\beta+1)x^{\beta+\widetilde{\lambda}_1}}{8(1-\gamma)^{\widetilde{\lambda}_1+\beta+2}}$$

$$> \frac{1}{\widetilde{\lambda}_2+\beta} + \frac{1}{\widetilde{\lambda}_1+\beta} - \frac{(1-\gamma)^{\beta+\widetilde{\lambda}_1}}{(\widetilde{\lambda}_1+\beta)(1-\gamma)^{\widetilde{\lambda}_1+\beta}}$$

$$- \frac{(1-\gamma)^{\beta+\widetilde{\lambda}_1}}{2(1-\gamma)^{\widetilde{\lambda}_1+\beta+1}} - \frac{(\widetilde{\lambda}_1+\beta+1)(1-\gamma)^{\beta+\widetilde{\lambda}_1}}{8(1-\gamma)^{\widetilde{\lambda}_1+\beta+2}}$$

$$= \frac{A(\gamma)}{(\widetilde{\lambda}_2+\beta)(1-\gamma)^2} \geq 0;$$

(ii) for $x \geq 1 - \gamma$, $0 \leq \gamma \leq \gamma(\widetilde{\lambda}_2)$, in view of $\frac{1}{x^2} \leq \frac{(1-\gamma)^{\widetilde{\lambda}_2+\beta-2}}{x^{\widetilde{\lambda}_2+\beta}}$, we still find

$$\widetilde{R}(x) > \frac{(1-\gamma)^{\widetilde{\lambda}_2+\beta}}{(\widetilde{\lambda}_2+\beta)x^{\widetilde{\lambda}_2+\beta}} - \frac{1}{2}(1-\gamma)^{\widetilde{\lambda}_2+\beta-1}\frac{1}{x^{\widetilde{\lambda}_2+\beta}}$$

$$- \frac{1}{8}(1-\widetilde{\lambda}_2-\beta)\frac{(1-\gamma)^{\widetilde{\lambda}_2+\beta-2}}{x^{\widetilde{\lambda}_2+\beta}} - \frac{2\beta+\lambda}{8}\frac{(1-\gamma)^{\widetilde{\lambda}_2+\beta-2}}{x^{\widetilde{\lambda}_2+\beta}}$$

$$= A(\gamma)\frac{(1-\gamma)^{\widetilde{\lambda}_2+\beta-2}}{(\widetilde{\lambda}_2+\beta)x^{\widetilde{\lambda}_2+\beta}} \geq 0.$$

Hence, (3.115) follows for $x > 0$.

(3) By (3.4) and Condition (iii), it follows that

$$\varpi(\widetilde{\lambda}_1, x) = k(\widetilde{\lambda}_1) - \widetilde{R}(x) < k(\widetilde{\lambda}_1) \ (x > 0).$$

Setting

$$\theta_{\widetilde{\lambda}_1}(x) = \frac{1}{k(\widetilde{\lambda}_1)}\left[\widetilde{R}(x) + \frac{x^{\widetilde{\lambda}_1}(\min\{x, 1-\gamma\})^{\beta}}{(\max\{x, 1-\gamma\})^{\lambda+\beta}}(1-\gamma)^{\widetilde{\lambda}_2-1}\right] > 0,$$

we obtain

$$\varpi(\widetilde{\lambda}_1, x) > \varpi(\widetilde{\lambda}_1, x) - \frac{x^{\widetilde{\lambda}_1}(\min\{x, 1-\gamma\})^{\beta}}{(\max\{x, 1-\gamma\})^{\lambda+\beta}}(1-\gamma)^{\widetilde{\lambda}_2-1}$$

$$= k(\widetilde{\lambda}_1)(1-\theta_{\widetilde{\lambda}_1}(x)) > 0.$$

For $x \geq 1 \geq 1 - \gamma \ (\gamma \geq 0)$, by (3.117), we still obtain

$$-\int_1^\infty P_1(y)g_y'(x,y)dy < \frac{2\beta+\lambda}{8x^{\widetilde{\lambda}_1+2}},$$

and by (3.115), (3.118) and (3.120), it follows that

$$\widetilde{R}(x) < \frac{(1-\gamma)^{\widetilde{\lambda}_2+\beta}}{(\widetilde{\lambda}_2+\beta)x^{\widetilde{\lambda}_2+\beta}} - \frac{1}{2}(1-\gamma)^{\widetilde{\lambda}_2+\beta-1}\frac{1}{x^{\widetilde{\lambda}_2+\beta}} + \frac{2\beta+\lambda}{8x^2}. \qquad (3.121)$$

For $\eta = \lambda_2 + \beta - \delta_0 > 0$, $x \geq 1 \geq 1 - \gamma$, it follows from (3.121) that

$$0 < x^\eta \, \theta_{\widetilde{\lambda}_1}(x) \leq x^{\widetilde{\lambda}_2+\beta} \, \theta_{\widetilde{\lambda}_1}(x)$$

$$= \frac{1}{k(\widetilde{\lambda}_1)}\left[x^{\widetilde{\lambda}_2+\beta}\widetilde{R}(x) + (1-\gamma)^{\widetilde{\lambda}_2+\beta-1}\right]$$

$$< \frac{1}{k(\widetilde{\lambda}_1)}\left[\frac{(1-\gamma)^{\widetilde{\lambda}_2+\beta}}{\widetilde{\lambda}_2+\beta} - \frac{1}{2}(1-\gamma)^{\widetilde{\lambda}_2+\beta-1}\right.$$

$$\left. + \frac{2\beta+\lambda}{8x^{2-\widetilde{\lambda}_2-\beta}} + (1-\gamma)^{\widetilde{\lambda}_2+\beta-1}\right] \to \text{constant} \quad (x \to \infty),$$

namely, $\theta_{\tilde{\lambda}_1}(x) = O\left(\frac{1}{x^\eta}\right) (x \geq 1; \eta = \lambda_2 + \beta - \delta_0 > 0)$. Hence we show that (3.113) and (3.114) are valid.

Setting $u(x) = x - \mu(x > \mu)$, by Corollary 3.4, for

$$0 \leq \gamma \leq \gamma(\lambda_2) = 1 - \frac{\lambda_2 + \beta}{4}\left[1 + \sqrt{\frac{2(\lambda + 2\beta + 1)}{\lambda_2 + \beta}} - 1\right],$$

and

$$k_\lambda(u(x), v(n)) = \frac{(\min\{x - \mu, n - \gamma\})^\beta}{(\max\{x - \mu, n - \gamma\})^{\lambda + \beta}},$$

$(x > \mu,\ n \in \mathbf{N};\ 0 < 2\beta + \lambda \leq 4,\ \lambda_1 > -\beta,\ -\beta < \lambda_2 < 1 - \beta,\ \lambda_1 + \lambda_2 = \lambda)$, if $p > 1$, we have equivalent inequalities (3.41), (3.42) and (3.43); if $p < 0$, we have equivalent reverses of (3.41), (3.42) and (3.43); if $0 < p < 1$, we have equivalent inequalities (3.45), (3.46) and (3.47). All the inequalities are with the same best constant factor

$$k(\lambda_1) = \frac{2\beta + \lambda}{(\beta + \lambda_1)(\beta + \lambda_2)}. \tag{3.122}$$

Chapter 4

A Half-Discrete Hilbert-Type Inequality with a Non-Homogeneous Kernel

"... we have always found, even with the most famous inequalities, that we have a little new to add."

G. H. Hardy

"There is no branch of mathematics, however abstract, which may not some day be applied to phenomena of the real world."

Nikolai Lobatchevsky

4.1 Introduction

The main objective of this chapter is to derive a half-discrete Hilbert-type inequality with a general non-homogeneous kernel and its extensions using the way of weight functions and methods of real analysis. The best possible factors involved in inequalities are also determined. Included are equivalent inequalities and their operator expressions, two classes of reverse inequality, many extensions and particular examples.

4.2 Some Preliminary Lemmas

4.2.1 *Definition of Weight Functions and Some Related Lemmas*

Lemma 4.1. *Suppose that $\alpha \in \mathbf{R}, h(t)$ is a non-negative finite measurable function in $\mathbf{R}_+$, $v(y)$ is a strictly increasing differentiable function* in $[n_0, \infty)(n_0 \in \mathbf{N})$ *with* $v(n_0) > 0$ *and* $v(\infty) = \infty$. *Define two weight functions* $\omega_\alpha(n)$ *and* $\varpi_\alpha(x)$ *as follows:*

$$\omega_\alpha(n) = [v(n)]^\alpha \int_0^\infty h(xv(n))x^{\alpha-1}dx, n \geq n_0 \quad (n \in \mathbf{N}), \qquad (4.1)$$

$$\varpi_\alpha(x) = x^\alpha \sum_{n=n_0}^\infty h(xv(n))[v(n)]^{\alpha-1}v'(n), \quad x \in (0, \infty). \qquad (4.2)$$

Then, we have

$$\omega_\alpha(n) = k(\alpha) = \int_0^\infty h(t)\, t^{\alpha-1}dt. \qquad (4.3)$$

Moreover, we set

$$f(x, y) = x^\alpha h(xv(y))[v(y)]^{\alpha-1}v'(y)$$

and use the following conditions:

Condition (i) $v(y)$ is strictly increasing in $[n_0-1, \infty)$ with $v(n_0-1) \geq 0$, and for any fixed $x > 0$, $f(x, y)$ is decreasing with respect to $y \in [n_0-1, \infty)$ and strictly decreasing in an interval $I \subset (n_0-1, \infty)$.

Condition (ii) $v(y)$ is strictly increasing in $[n_0-\frac{1}{2}, \infty)$ with $v(n_0-\frac{1}{2}) \geq 0$, and for any fixed $x > 0$, $f(x, y)$ is convex with respect to $y \in [n_0-\frac{1}{2}, \infty)$ and strictly convex in an interval $\widetilde{I} \subset (n_0-\frac{1}{2}, \infty)$.

Condition (iii) There exists a constant $\beta > 0$, such that $v(y)$ is strictly increasing in $[n_0-\beta, \infty)$ with $v(n_0-\beta) \geq 0$, and for any fixed $x > 0$, $f(x, y)$ is a piecewise smooth continuous function with respect to $y \in [n_0-\beta, \infty)$, satisfying

$$R(x) = \int_{n_0-\beta}^{n_0} f(x, y)dy - \frac{1}{2}f(x, n_0) - \int_{n_0}^\infty P_1(y)f'_y(x, y)dy > 0, \qquad (4.4)$$

where, $P_1(y)(= y - [y] - \frac{1}{2})$ is the Bernoulli function of 1-order equation (2.8) in Chapter 2.

If $k(\alpha) \in \mathbf{R}_+$ and one of the above three conditions is satisfied, then, we have

$$\varpi_\alpha(x) < k(\alpha) \quad (x \in (0, \infty)). \qquad (4.5)$$

Proof. Setting $t = xv(n)$ in (4.1), we find

$$\omega_\alpha(n) = [v(n)]^\alpha \int_0^\infty h(t)\left(\frac{t}{v(n)}\right)^{\alpha-1}\frac{1}{v(n)}dt$$

$$= \int_0^\infty h(t)\, t^{\alpha-1}dt = k(\alpha),$$

then, (4.3) follows.

(i) If Condition (i) is satisfied then for any $n \geq n_0$ ($n \in \mathbf{N}$), it follows that

$$f(x,n) \leq \int_{n-1}^{n} f(x,y)dy,$$

and there exists an integer $n_1 \geq n_0$, such that

$$f(x,n_1) < \int_{n_1-1}^{n_1} f(x,y)dy.$$

Since $v(n_0 - 1) \geq 0$ and $k(\alpha) \in \mathbf{R}_+$, we find

$$\varpi_{\lambda_1}(x) = \sum_{n=n_0}^{\infty} f(x,n) < \sum_{n=n_0}^{\infty} \int_{n-1}^{n} f(x,y)dy$$

$$= \int_{n_0-1}^{\infty} f(x,y)dy = x^{\alpha} \int_{n_0-1}^{\infty} h(xv(y))[v(y)]^{\alpha-1}v'(y)dy, \quad (t=xv(y))$$

$$= \int_{xv(n_0-1)}^{\infty} h(t)t^{\alpha-1}dt \leq \int_{0}^{\infty} h(t)t^{\alpha-1}dt = k(\alpha),$$

then, (4.5) follows.

(ii) If Condition (ii) is satisfied, then by the Hermite-Hadamard's inequality (see Kuang [47]), it follows that

$$f(x,n) \leq \int_{n-\frac{1}{2}}^{n+\frac{1}{2}} f(x,y)\, dy, \quad n \geq n_0,$$

and there exists a positive integer $n_2 \geq n_0$, such that

$$f(x,n_2) < \int_{n_2-\frac{1}{2}}^{n_2+\frac{1}{2}} f(x,y)dy.$$

Since $v(n_0 - \frac{1}{2}) \geq 0$ and $k(\alpha) \in \mathbf{R}_+$, we find

$$\varpi_{\lambda_1}(x) = \sum_{n=n_0}^{\infty} f(x,n) < \sum_{n=n_0}^{\infty} \int_{n-\frac{1}{2}}^{n+\frac{1}{2}} f(x,y)dy$$

$$= \int_{n_0-\frac{1}{2}}^{\infty} f(x,y)dy = x^{\alpha} \int_{n_0-\frac{1}{2}}^{\infty} h(xv(y))[v(y)]^{\alpha-1}v'(y)dy$$

$$= \int_{xv(n_0-\frac{1}{2})}^{\infty} h(t)t^{\alpha-1}dt, \quad (t=xv(y))$$

$$\leq \int_{0}^{\infty} h(t)t^{\alpha-1}dt = k(\alpha),$$

then, (4.5) follows.

(iii) If Condition (iii) is satisfied and $k(\alpha) \in \mathbf{R}_+$, then by the Euler-Maclaurin summation formula (see equation (2.16) in Chapter 2) (for $q = 0$, $m = n_0$, $n \to \infty$), we have

$$\varpi_{\lambda_1}(x) = \sum_{n=n_0}^{\infty} f(x, n)$$

$$= \int_{n_0}^{\infty} f(x, y)dy + \frac{1}{2}f(x, n_0) + \int_{n_0}^{\infty} P_1(y)f_y'(x, y)dy$$

$$= \int_{n_0-\beta}^{\infty} f(x, y)dy - R(x) = \int_{xv(n_0-\beta)}^{\infty} h(t)t^{\alpha-1}dt - R(x), \quad (t = xv(y))$$

$$\leq k(\alpha) - R(x) < k(\alpha),$$

then, (4.5) follows.   □

Lemma 4.2. *Let the assumptions of Lemma 4.1 be fulfilled and additionally, $p \in \mathbf{R}\backslash\{0, 1\}$, $\frac{1}{p} + \frac{1}{q} = 1$, $a_n \geq 0$, $n \geq n_0(n \in \mathbf{N})$, $f(x)$ is a nonnegative measurable function in $\mathbf{R}_+$. Then*
(i) for $p > 1$, we have the following inequalities:

$$J = \left\{ \sum_{n=n_0}^{\infty} \frac{v'(n)}{[v(n)]^{1-p\alpha}} \left[\int_0^{\infty} h(xv(n))f(x)dx \right]^p \right\}^{\frac{1}{p}}$$

$$\leq [k(\alpha)]^{\frac{1}{q}} \left\{ \int_0^{\infty} \varpi_\alpha(x)x^{p(1-\alpha)-1}f^p(x)dx \right\}^{\frac{1}{p}}, \tag{4.6}$$

and

$$\widetilde{L} = \left\{ \int_0^{\infty} \frac{[\varpi_\alpha(x)]^{1-q}}{x^{1-q\alpha}} \left[\sum_{n=n_0}^{\infty} h(xv(n))a_n \right]^q dx \right\}^{\frac{1}{q}}$$

$$\leq \left\{ k(\alpha) \sum_{n=n_0}^{\infty} \frac{[v(n)]^{q(1-\alpha)-1}}{[v'(n)]^{q-1}}a_n^q \right\}^{\frac{1}{q}}; \tag{4.7}$$

(ii) for $p < 0$ or $0 < p < 1$, we have the reverses of (4.6) and (4.7).

Proof. (i) For $p > 1$, by Hölder's inequality with weight (see Kuang [47]) and (4.3), it follows that

$$\left[\int_0^{\infty} h(xv(n))f(x)dx \right]^p = \left\{ \int_0^{\infty} h(xv(n)) \left[\frac{x^{(1-\alpha)/q}[v'(n)]^{1/p}}{[v(n)]^{(1-\alpha)/p}}f(x) \right] \right.$$

$$\times \left. \left[\frac{[v(n)]^{(1-\alpha)/p}}{x^{(1-\alpha)/q}[v'(n)]^{1/p}} \right] dx \right\}^p$$

$$\leq \int_0^\infty h(xv(n)) \frac{x^{(1-\alpha)(p-1)} v'(n)}{[v(n)]^{1-\alpha}} f^p(x) dx$$

$$\times \left\{ \int_0^\infty h(xv(n)) \frac{[v(n)]^{(1-\alpha)(q-1)}}{x^{1-\alpha}[v'(n)]^{q-1}} dx \right\}^{p-1}$$

$$= \left\{ \frac{\omega_\alpha(n)[v(n)]^{q(1-\alpha)-1}}{[v'(n)]^{q-1}} \right\}^{p-1}$$

$$\times \int_0^\infty h(xv(n)) \frac{x^{(1-\alpha)(p-1)} v'(n)}{[v(n)]^{1-\alpha}} f^p(x) dx$$

$$= \frac{[k(\alpha)]^{p-1}}{[v(n)]^{p\alpha-1} v'(n)} \int_0^\infty h(xv(n)) \frac{x^{(1-\alpha)(p-1)} v'(n)}{[v(n)]^{1-\alpha}} f^p(x) dx.$$

Then by Lebesgue term by term integration theorem (see Kuang [47]), we have

$$J \leq [k(\alpha)]^{\frac{1}{q}} \left\{ \sum_{n=n_0}^\infty \int_0^\infty h(xv(n)) \frac{x^{(1-\alpha)(p-1)} v'(n)}{[v(n)]^{1-\alpha}} f^p(x) dx \right\}^{\frac{1}{p}}$$

$$= [k(\alpha)]^{\frac{1}{q}} \left\{ \int_0^\infty \sum_{n=n_0}^\infty h(xv(n)) \frac{x^{(1-\alpha)(p-1)} v'(n)}{[v(n)]^{1-\alpha}} f^p(x) dx \right\}^{\frac{1}{p}}$$

$$= [k(\alpha)]^{\frac{1}{q}} \left\{ \int_0^\infty \varpi_\alpha(x) x^{p(1-\alpha)-1} f^p(x) dx \right\}^{\frac{1}{p}},$$

then (4.6) follows.

Still by Hölder's inequality with weight, it follows that

$$\left[\sum_{n=n_0}^\infty h(xv(n)) a_n \right]^q = \left\{ \sum_{n=n_0}^\infty h(xv(n)) \left[\frac{x^{(1-\alpha)/q}[v'(n)]^{1/p}}{[v(n)]^{(1-\alpha)/p}} \right] \right.$$

$$\left. \times \frac{[v(n)]^{(1-\alpha)/p}}{x^{(1-\alpha)/q}[v'(n)]^{1/p}} a_n \right\}^q$$

$$\leq \left\{ \sum_{n=n_0}^\infty h(xv(n)) \frac{x^{(1-\alpha)(p-1)} v'(n)}{[v(n)]^{1-\alpha}} \right\}^{q-1}$$

$$\times \sum_{n=n_0}^\infty h(xv(n)) \frac{[v(n)]^{(1-\alpha)(q-1)}}{x^{1-\alpha}[v'(n)]^{q-1}} a_n^q$$

$$= \frac{x^{1-q\alpha}}{[\varpi_\alpha(x)]^{1-q}} \sum_{n=n_0}^\infty h(xv(n)) \frac{[v(n)]^{(1-\alpha)(q-1)}}{x^{1-\alpha}[v'(n)]^{q-1}} a_n^q.$$

Then by the Lebesgue term by term integration theorem, we have

$$\widetilde{L} \leq \left\{ \int_0^\infty \sum_{n=n_0}^\infty h(xv(n)) \frac{[v(n)]^{(1-\alpha)(q-1)}}{x^{1-\alpha}[v'(n)]^{q-1}} a_n^q dx \right\}^{\frac{1}{q}}$$

$$= \left\{ \sum_{n=n_0}^\infty \int_0^\infty h(xv(n)) \frac{[v(n)]^{(1-\alpha)(q-1)}}{x^{1-\alpha}[v'(n)]^{q-1}} a_n^q dx \right\}^{\frac{1}{q}}$$

$$= \left\{ \sum_{n=n_0}^\infty \omega_\alpha(n) \frac{[v(n)]^{q(1-\alpha)-1}}{[v'(n)]^{q-1}} a_n^q \right\}^{\frac{1}{q}},$$

and then, in view of (4.3), we have (4.7).

(ii) For $p < 0$ or $0 < p < 1$, by the reverse Hölder's inequality with weight (see Kuang [47]) and in the same way, we obtain the reverses of (4.6) and (4.7). $\qquad\square$

4.2.2   *Estimations of Two Series and Examples*

Lemma 4.3. *Assuming that $v(y)$ is a strictly increasing differentiable function in $[n_0, \infty)(n_0 \in \mathbf{N})$ with $v(n_0) > 0$ and $v(\infty) = \infty$, if $\frac{v'(y)}{v(y)}(> 0)$ is decreasing in $[n_0, \infty)$, then for $\varepsilon > 0$, we have*

$$A(\varepsilon) = \sum_{n=n_0}^\infty \frac{v'(n)}{[v(n)]^{1+\varepsilon}} = \frac{1}{\varepsilon}(1 + o(1)) \qquad (\varepsilon \to 0^+). \qquad (4.8)$$

Lemma 4.4. *Let the assumptions of Lemma 4.1 be fulfilled and additionally, let $p \in \mathbf{R}\backslash\{0,1\}$, $\frac{1}{p} + \frac{1}{q} = 1$, $\frac{v'(y)}{v(y)}(> 0)$ be decreasing in $[n_0, \infty)(n_0 \in \mathbf{N})$ and there exist constants $\delta > \alpha$ and $L > 0$, such that*

$$h(t) \leq L\left(\frac{1}{t^\delta}\right), \qquad (t \in [v(n_0), \infty)). \qquad (4.9)$$

Then, for $0 < \varepsilon < \min\{|p|, |q|\}(\delta - \alpha)$, we have

$$B(\varepsilon) = \sum_{n=n_0}^\infty \frac{v'(n)}{[v(n)]^{1+\varepsilon}} \int_{v(n)}^\infty h(t) t^{\alpha+\frac{\varepsilon}{p}-1} dt = O(1) \quad (\varepsilon \to 0^+). \qquad (4.10)$$

Proof. In view of (4.9), we find

$$0 < B(\varepsilon) \le L \sum_{n=n_0}^{\infty} \frac{v'(n)}{[v(n)]^{1+\varepsilon}} \int_{v(n)}^{\infty} t^{\alpha-\delta+\frac{\varepsilon}{p}-1} dt$$

$$= \frac{L}{\delta-\alpha-\frac{\varepsilon}{p}} \sum_{n=n_0}^{\infty} \frac{v'(n)}{[v(n)]^{\delta-\alpha+\frac{\varepsilon}{q}+1}}$$

$$= \frac{L}{\delta-\alpha-\frac{\varepsilon}{p}} \left[\frac{v'(n_0)}{[v(n_0)]^{\delta-\alpha+\frac{\varepsilon}{q}+1}} + \sum_{n=n_0+1}^{\infty} \frac{v'(n)}{[v(n)]^{\delta-\alpha+\frac{\varepsilon}{q}+1}} \right]$$

$$\le \frac{L}{\delta-\alpha-\frac{\varepsilon}{p}} \left[\frac{v'(n_0)}{[v(n_0)]^{\delta-\alpha+\frac{\varepsilon}{q}+1}} + \int_{n_0}^{\infty} \frac{v'(y)dy}{[v(y)]^{\delta-\alpha+\frac{\varepsilon}{q}+1}} \right]$$

$$= \frac{L}{\delta-\alpha-\frac{\varepsilon}{p}} \left[\frac{v'(n_0)}{[v(n_0)]^{\delta-\alpha+\frac{\varepsilon}{q}+1}} + \frac{[v(n_0)]^{\alpha-\delta-\frac{\varepsilon}{q}}}{\delta-\alpha+\frac{\varepsilon}{q}} \right] < \infty,$$

and then (4.10) follows. $\qquad\square$

Example 4.1. (i) For $h(t) = \frac{1}{1+t^\lambda}$ $(0 < \alpha < \lambda)$, since

$$k(\alpha) = \int_0^{\infty} \frac{1}{1+t^\lambda} t^{\alpha-1} dt = \frac{\pi}{\lambda \sin \pi(\frac{\alpha}{\lambda})} \in \mathbf{R}_+,$$

setting $\delta = \frac{\lambda+\alpha}{2} > \alpha$ $(\delta < \lambda)$, we find

$$h(t) = \frac{1}{t^\lambda+1} \le \frac{1}{t^\delta} \quad (t \in [v(n_0), \infty)).$$

(ii) For $h(t) = \frac{1}{(1+t)^\lambda}$ $(0 < \alpha < \lambda)$, since

$$k(\alpha) = \int_0^{\infty} \frac{1}{(1+t)^\lambda} t^{\alpha-1} dt = B(\alpha, \lambda-\alpha) \in \mathbf{R}_+,$$

setting $\delta = \frac{\lambda+\alpha}{2} > \alpha$ $(\delta < \lambda)$, we find

$$h(t) = \frac{1}{(1+t)^\lambda} \le \frac{1}{t^\delta} \quad (t \in [v(n_0), \infty)).$$

(iii) For $h(t) = \frac{\ln t}{t^\lambda-1}$ $(0 < \alpha < \lambda)$, since

$$k(\alpha) = \int_0^{\infty} \frac{\ln t}{t^\lambda-1} t^{\alpha-1} dt = \left[\frac{\pi}{\lambda \sin(\frac{\pi\alpha}{\lambda})} \right]^2 \in \mathbf{R}_+,$$

setting $\delta = \frac{\lambda+\alpha}{2} > \alpha$ $(\delta < \lambda)$, we find $\frac{t^\delta \ln t}{t^\lambda-1} \to 0$ $(t \to \infty)$, and there exists a constant $L > 0$, such that

$$h(t) = \frac{\ln t}{t^\lambda-1} \le L\left(\frac{1}{t^\delta}\right) \quad (t \in [v(n_0), \infty)).$$

(iv) For $s \in \mathbf{N}$, $k_{\lambda s}(t,1) = \prod_{k=1}^{s} \frac{1}{a_k t^\lambda + 1}(0 < a_1 < \cdots < a_s, \lambda > 0, 0 < \alpha < s)$, by (3.12), we find

$$k(\alpha\lambda) = \int_0^\infty \prod_{k=1}^{s} \frac{1}{a_k t^\lambda + 1} t^{\alpha\lambda - 1} dt$$

$$= \frac{1}{\lambda} \int_0^\infty \prod_{k=1}^{s} \frac{1}{a_k + u} u^{(s-\alpha)-1} du$$

$$= \frac{\pi}{\lambda \sin(\pi\alpha)} \sum_{k=1}^{s} a_k^{s-\alpha-1} \prod_{j=1(j\neq k)}^{s} \frac{1}{a_k - a_j} \in \mathbf{R}_+.$$

Setting $\delta = \frac{s+\alpha}{2}\lambda > \alpha\lambda \ (< s\lambda)$, since $t^\delta k_{\lambda s}(t,1) \to 0 \ (t \to \infty)$, there exists a constant $L > 0$, such that

$$k_{\lambda s}(t,1) = \prod_{k=1}^{s} \frac{1}{a_k t^\lambda + 1} \leq \frac{L}{t^\delta} \ (t \in [v(n_0), \infty)).$$

(v) For $h(t) = \frac{1}{t^\lambda + 2t^{\lambda/2} \cos\beta + 1}(0 < \beta \leq \frac{\pi}{2}, \lambda > 0, 0 < \alpha < 1)$, by Example 3.1(v), we find

$$k(\alpha\lambda) = \int_0^\infty \frac{t^{\alpha\lambda-1}}{t^\lambda + 2t^{\lambda/2} \cos\beta + 1} dt, \qquad (u = t^{\lambda/2})$$

$$= \frac{2}{\lambda} \int_0^\infty \frac{u^{2\alpha-1} du}{u^2 + 2u \cos\beta + 1}$$

$$= \frac{2\pi \sin\beta(1 - 2\alpha)}{\lambda \sin\beta \sin(2\pi\alpha)} \in \mathbf{R}_+.$$

Setting $\delta = \lambda \ (> \alpha\lambda)$, since $\frac{t^\delta}{t^\lambda + 2t^{\lambda/2} \cos\beta + 1} \to 1 \ (t \to \infty)$, there exists a constant $L > 0$, such that

$$h(t) = \frac{1}{t^\lambda + 2t^{\lambda/2} \cos\beta + 1} \leq L\left(\frac{1}{t^\delta}\right) \quad (t \in [v(n_0), \infty)).$$

(vi) For $h(t) = \frac{1}{t^\lambda + bt^{\lambda/2} + c}$ $(c > 0, \ 0 \leq b \leq 2\sqrt{c}, \ \alpha = \frac{\lambda}{2} > 0)$, we find

$$k\left(\frac{\lambda}{2}\right) = \int_0^\infty \frac{t^{\frac{\lambda}{2}-1} dt}{t^\lambda + bt^{\lambda/2} + c} = \frac{2}{\lambda} \int_0^\infty \frac{du}{u^2 + bu + c}, \quad (u = t^{\lambda/2})$$

$$= \frac{2}{\lambda} \begin{cases} \frac{\pi}{\sqrt{c}}, & b = 0, \\ \frac{4}{\sqrt{4c-b^2}} \arctan \frac{\sqrt{4c-b^2}}{b}, & 0 < b < 2\sqrt{c}, \\ \frac{2}{\sqrt{c}}, & b = 2\sqrt{c}. \end{cases}$$

Setting $\delta = \lambda > \frac{\lambda}{2}$, since $\frac{t^\delta}{t^\lambda + bt^{\lambda/2} + c} \to 1 (t \to \infty)$, there exists a constant $L > 0$, such that

$$h(t) = \frac{1}{t^\lambda + bt^{\lambda/2} + c} \leq L\left(\frac{1}{t^\delta}\right) \quad (t \in [v(n_0), \infty)).$$

(vii) For $h(t) = \frac{1}{e^{\gamma t^\lambda} - Ae^{-\gamma t^\lambda}} (-1 \leq A < 1, \beta > 0(A = 1, \beta > 1), \gamma, \lambda > 0, \alpha = \beta\lambda)$, we find

$$k(\alpha) = \int_0^\infty \frac{t^{\beta\lambda-1}dt}{e^{\gamma t^\lambda} - Ae^{-\gamma t^\lambda}} = \frac{1}{\lambda} \int_0^\infty \frac{u^{\beta-1}du}{e^{\gamma u}(1 - Ae^{-2\gamma u})}$$

$$= \frac{1}{\lambda} \int_0^\infty \sum_{k=0}^\infty A^k e^{-\gamma(2k+1)u} u^{\beta-1} du$$

$$= \frac{1}{\lambda} \sum_{k=0}^\infty A^k \int_0^\infty e^{-\gamma(2k+1)u} u^{\beta-1} du$$

$$= \frac{\Gamma(\beta)}{\lambda\gamma^\beta} \sum_{k=0}^\infty \frac{A^k}{(2k+1)^\beta} \in \mathbf{R}_+.$$

Setting $\delta = \beta_0\lambda > \beta\lambda = \alpha$ $(\beta_0 > \beta)$, since $\frac{t^\delta}{e^{\gamma t^\lambda} - Ae^{-\gamma t^\lambda}} \to 0$ $(t \to \infty)$, there exists a constant $L > 0$, such that

$$\frac{t^\delta}{e^{\gamma t^\lambda} - Ae^{-\gamma t^\lambda}} \leq L \quad (t \in [v(n_0), \infty)).$$

Then it follows that

$$h(t) = \frac{1}{e^{\gamma t^\lambda} - Ae^{-\gamma t^\lambda}} \leq L\left(\frac{1}{t^\delta}\right) \quad (t \in [v(n_0), \infty)).$$

(viii) For $h(t) = \ln\left(\frac{bt^\lambda+1}{at^\lambda+1}\right)$ $(0 \leq a < b, \lambda > 0, \alpha = -\beta\lambda, 0 < \beta < 1)$, we find

$$k(\alpha) = \int_0^\infty \ln\left(\frac{bt^\lambda + 1}{at^\lambda + 1}\right) t^{-\beta\lambda-1}dt = \frac{-1}{\beta\lambda} \int_0^\infty \ln\left(\frac{bt^\lambda + 1}{at^\lambda + 1}\right) dt^{-\beta\lambda}$$

$$= \frac{-1}{\beta\lambda}\left[t^{-\beta\lambda} \ln\left(\frac{bt^\lambda + 1}{at^\lambda + 1}\right)\Big|_0^\infty - \int_0^\infty \left(\frac{\lambda}{bt^\lambda + 1} - \frac{\lambda}{at^\lambda + 1}\right) t^{(1-\beta)\lambda-1}dt\right]$$

$$= \left(\frac{b^\beta}{\beta\lambda} - \frac{a^\beta}{\beta\lambda}\right) \int_0^\infty \frac{u^{(1-\beta)-1}}{1+u}du = \frac{(b^\beta - a^\beta)\pi}{\beta\lambda\sin(\beta\pi)}.$$

Setting $\delta = -\beta_0\lambda > -\beta\lambda = \alpha(0 < \beta_0 < \beta)$, in view of $t^\delta \ln(\frac{bt^\lambda+1}{at^\lambda+1}) \to 0$ $(t \to \infty)$, there exists a constant $L > 0$, such that

$$t^\delta \ln\left(\frac{bt^\lambda + 1}{at^\lambda + 1}\right) \leq L \quad (t \in [v(n_0), \infty)).$$

Then it follows that

$$h(t) = \ln\left(\frac{bt^\lambda + 1}{at^\lambda + 1}\right) \leq \frac{L}{t^\delta} \quad (t \in [v(n_0), \infty)).$$

(ix) For $h(t) = \arctan \frac{1}{t^\lambda}$ $(0 < \alpha < \lambda)$, we find

$$k(\alpha) = \int_0^\infty t^{\alpha-1} \arctan \frac{1}{t^\lambda} dt = \frac{\pi}{2\alpha \cos(\frac{\pi\alpha}{2\lambda})} \in \mathbf{R}_+.$$

Setting $\delta = \alpha_0 \in (\alpha, \lambda)$, since $t^\delta \arctan \frac{1}{t^\lambda} \to 0$ $(t \to \infty)$, there exists a constant $L > 0$, such that

$$h(t) = \arctan \frac{1}{t^\lambda} \le L \left(\frac{1}{t^\delta}\right) \quad (t \in [v(n_0), \infty)).$$

(x) For $h(t) = \frac{1}{(\max\{1,t\})^\lambda}$ $(0 < \alpha < \lambda)$, since

$$k(\alpha) = \int_0^\infty \frac{1}{(\max\{1,t\})^\lambda} t^{\alpha-1} dt = \frac{\lambda}{\alpha(\lambda - \alpha)} \in \mathbf{R}_+,$$

setting $\delta = \frac{\lambda+\alpha}{2} > \alpha$ $(\delta < \lambda)$, we find

$$h(t) = \frac{1}{(\max\{1,t\})^\lambda} \le \frac{1}{t^\delta} \quad (t \in [v(n_0), \infty)).$$

(xi) For $h(t) = (\min\{1,t\})^\lambda$ $(-\lambda < \alpha < 0)$, since

$$k(\alpha) = \int_0^\infty (\min\{1,t\})^\lambda t^{\alpha-1} dt = \frac{\lambda}{(-\alpha)(\alpha + \lambda)} \in \mathbf{R}_+,$$

setting $\delta = \frac{\alpha}{2} > \alpha$ $(-\lambda < \delta < 0)$, there exists a constant $L > 0$, such that

$$h(t) = (\min\{1,t\})^\lambda \le L \left(\frac{1}{t^\delta}\right) \quad (t \in [v(n_0), \infty)).$$

(xii) For $h(t) = (\frac{\min\{1,t\}}{\max\{1,t\}})^\lambda$ $(\lambda > 0, -\lambda < \alpha < \lambda)$, since

$$k(\alpha) = \int_0^\infty \left(\frac{\min\{1,t\}}{\max\{1,t\}}\right)^\lambda t^{\alpha-1} dt = \frac{2\lambda}{\lambda^2 - \alpha^2} \in \mathbf{R}_+,$$

setting $\delta = \frac{\lambda+\alpha}{2} > \alpha$ $(-\lambda < \alpha < \delta < \lambda)$, we find

$$h(t) = \left(\frac{\min\{1,t\}}{\max\{1,t\}}\right)^\lambda \le \frac{1}{t^\delta} \quad (t \in [v(n_0), \infty)).$$

4.2.3 Some Inequalities Relating the Constant $k(\alpha)$

Lemma 4.5. *Suppose that* $\alpha \in \mathbf{R}$, $h(t)$ *is a non-negative finite measurable function in* $\mathbf{R}_+$ *with*

$$k(\alpha) = \int_0^\infty h(t) t^{\alpha-1} dt \in \mathbf{R}_+.$$

Then

(i) for $p > 1$, $\varepsilon > 0$, we have

$$k\left(\alpha + \frac{\varepsilon}{p}\right) \geq k(\alpha) + o(1) \quad (\varepsilon \to 0^+); \tag{4.11}$$

(ii) for $p < 0$, if there exists a constant $\delta_1 > 0$, such that $k(\alpha - \delta_1) \in \mathbf{R}_+$, then, for $0 < \varepsilon < (-p)\delta_1$, we have

$$k\left(\alpha + \frac{\varepsilon}{p}\right) \leq k(\alpha) + o(1) \quad (\varepsilon \to 0^+); \tag{4.12}$$

(iii) for $0 < p < 1$, if there exists a constant $\delta_2 > 0$, such that $k(\alpha + \delta_2) \in \mathbf{R}_+$, then for $0 < \varepsilon < p\delta_2$, we still have (4.12).

Proof.   (i) For $p > 1$, by Fatou's lemma (see Kuang [49]), it follows that

$$k(\alpha) = \int_0^\infty \lim_{\varepsilon \to 0^+} h(t)t^{(\alpha + \frac{\varepsilon}{p})-1}dt \leq \lim_{\varepsilon \to 0^+} k\left(\alpha + \frac{\varepsilon}{p}\right),$$

and then we have (4.11).

(ii) For $p < 0$, since $t^{\frac{\varepsilon}{p}} \leq 1$ $(1 \leq t < \infty)$, we find

$$k\left(\alpha + \frac{\varepsilon}{p}\right) = \int_0^1 h(t)t^{(\alpha + \frac{\varepsilon}{p})-1}dt + \int_1^\infty h(t)t^{(\alpha + \frac{\varepsilon}{p})-1}dt$$

$$\leq \int_0^1 h(t)t^{(\alpha + \frac{\varepsilon}{p})-1}dt + \int_1^\infty h(t)t^{\alpha-1}dt. \tag{4.13}$$

For $0 < \varepsilon < (-p)\delta_1$, we have

$$\int_0^1 h(t)t^{(\alpha + \frac{\varepsilon}{p})-1}dt \leq \int_0^1 h(t)t^{(\alpha - \delta_1)-1}dt \leq k(\alpha - \delta_1) < \infty,$$

then by the Lebesgue control convergence theorem (see Kuang [49]), it follows that

$$\int_0^1 h(t)t^{(\alpha + \frac{\varepsilon}{p})-1}dt = \int_0^1 h(t)t^{\alpha-1}dt + o(1) \quad (\varepsilon \to 0^+).$$

Hence, by (4.13), we have (4.12).

(iii) For $0 < p < 1$, since $t^{\frac{\varepsilon}{p}} \leq 1$ $(0 < t \leq 1)$, we find

$$k\left(\alpha + \frac{\varepsilon}{p}\right) = \int_0^1 h(t)t^{(\alpha + \frac{\varepsilon}{p})-1}dt + \int_1^\infty h(t)t^{(\alpha + \frac{\varepsilon}{p})-1}dt$$

$$\leq \int_0^1 h(t)t^{\alpha-1}dt + \int_1^\infty h(t)t^{(\alpha + \frac{\varepsilon}{p})-1}dt. \tag{4.14}$$

For $0 < \varepsilon < p\delta_2$, we have

$$\int_1^\infty h(t)t^{(\alpha + \frac{\varepsilon}{p})-1}dt \leq \int_1^\infty h(t)t^{(\alpha + \delta_2)-1}dt \leq k(\alpha + \delta_2) < \infty,$$

then by the Lebesgue control convergence theorem, it follows that

$$\int_1^\infty h(t)t^{(\alpha + \frac{\varepsilon}{p})-1}dt = \int_1^\infty h(t)t^{\alpha-1}dt + o(1)(\varepsilon \to 0^+).$$

Hence, by (4.14), we have (4.12) for $0 < p < 1$.   □

4.3   Some Theorems and Corollaries

4.3.1   *Equivalent Inequalities and their Operator Expressions*

We set two functions $\varphi(x) = x^{p(1-\alpha)-1}$ $(x \in (0, \infty))$ and

$$\Psi(n) = \frac{[v(n)]^{q(1-\alpha)-1}}{[v'(n)]^{q-1}} \quad (n \geq n_0, n \in \mathbf{N}),$$

wherefrom, $[\varphi(x)]^{1-q} = x^{q\alpha-1}$ and $[\Psi(n)]^{1-p} = \frac{v'(n)}{[v(n)]^{1-p\alpha}}$.

For $p \in \mathbf{R} \backslash \{0, 1\}$, $\frac{1}{p} + \frac{1}{q}$, we define two sets as follows:

$$L_{p,\varphi}(\mathbf{R}_+) = \left\{ f; ||f||_{p,\varphi} = \left\{ \int_0^\infty x^{p(1-\alpha)-1} |f(x)|^p dx \right\}^{\frac{1}{p}} < \infty \right\},$$

$$l_{q,\Psi} = \left\{ a = \{a_n\}_{n=n_0}^\infty; ||a||_{q,\Psi} = \left\{ \sum_{n=n_0}^\infty \frac{[v(n)]^{q(1-\alpha)-1}}{[v'(n)]^{q-1}} |a_n|^q \right\}^{\frac{1}{q}} < \infty \right\}.$$

Note. It is obvious that for $p > 1$, both of the above two sets are the normed spaces. For $p < 0$ or $0 < p < 1$, neither of the above two sets is the normed space, and we agree on that the sets with the normed expressions of $||f||_{p,\varphi}$ and $||a||_{q,\Psi}$ are the formal symbols in these cases.

Theorem 4.1. *(see Yang [170]) Suppose that $\alpha \in \mathbf{R}$, $h(t)$ is a non-negative finite measurable function in $\mathbf{R}_+, v(y)$ is a strict increasing differentiable function in $[n_0, \infty)$ $(n_0 \in \mathbf{N})$ with $v(n_0) > 0$ and $v(\infty) = \infty$,*

$$k(\alpha) = \int_0^\infty h(t) t^{\alpha-1} dt \in \mathbf{R}_+$$

and $\varpi_\alpha(x) < k(\alpha)$ $(x \in (0, \infty))$. If $p > 1$, $\frac{1}{p} + \frac{1}{q} = 1$, $f(x)$, $a_n \geq 0$, $f \in L_{p,\varphi}(\mathbf{R}_+)$, $a = \{a_n\}_{n=n_0}^\infty \in l_{q,\Psi}$, $||f||_{p,\varphi} > 0$, $||a||_{q,\Psi} > 0$, then we have the following equivalent inequalities:

$$I = \sum_{n=n_0}^\infty a_n \int_0^\infty h(xv(n))f(x)dx = \int_0^\infty f(x) \sum_{n=n_0}^\infty a_n h(xv(n))dx$$

$$< k(\alpha)||f||_{p,\varphi}||a||_{q,\Psi}, \tag{4.15}$$

$$J = \left\{ \sum_{n=n_0}^\infty [\Psi(n)]^{1-p} \left[\int_0^\infty h(xv(n))f(x)dx \right]^p \right\}^{\frac{1}{p}} < k(\alpha)||f||_{p,\varphi}, \tag{4.16}$$

and

$$L = \left\{ \int_0^\infty [\varphi(x)]^{1-q} \left[\sum_{n=n_0}^\infty h(xv(n))a_n \right]^q dx \right\}^{\frac{1}{q}} < k(\alpha)||a||_{q,\Psi}. \tag{4.17}$$

Moreover, if $\frac{v'(y)}{v(y)}(>0)$ is decreasing in $[n_0,\infty)$ and there exists constants $\delta < \lambda_1$ and $L > 0$, such that (4.9) is satisfied, then the constant factor $k(\alpha)$ in the above inequalities is the best possible.

Proof. By the Lebesgue term by term integration theorem (see Kuang [49]), there are two expressions for I in (4.15). In view of (4.6), since $\varpi_\alpha(x) < k(\alpha)$, we have (4.16).

By Hölder's inequality, we find

$$I = \sum_{n=n_0}^{\infty} \left[[\Psi(n)]^{-\frac{1}{q}} \int_0^{\infty} h(xv(n))f(x)dx \right] [[\Psi(n)]^{\frac{1}{q}} a_n] \leq J\|a\|_{q,\Psi}. \quad (4.18)$$

Then, by (4.16), we have (4.15).

On the other hand, suppose that (4.15) is valid. We set

$$a_n = [\Psi(n)]^{1-p} \left[\int_0^{\infty} h(xv(n))f(x)dx \right]^{p-1}, \quad n \geq n_0(n \in \mathbf{N}).$$

Then it follows that $J^{p-1} = \|a\|_{q,\Psi}$. By (4.6), we find $J < \infty$. If $J = 0$, then (4.16) is trivially valid; if $J > 0$, then, by (4.15), we have

$$\|a\|_{q,\Psi}^q = J^p = I < k(\alpha)\|f\|_{p,\varphi}\|a\|_{q,\Psi}, \text{that is,}$$

$$\|a\|_{q,\Psi}^{q-1} = J < k(\alpha)\|f\|_{p,\varphi},$$

and then, (4.15) is equivalent to (4.16).

In view of (4.7), since $[\varpi_\alpha(x)]^{1-q} > [k(\alpha)]^{1-q}$, we have (4.17).

By Hölder's inequality, we find

$$I = \int_0^{\infty} [[\varphi(x)]^{\frac{1}{p}} f(x)] \left[[\varphi(x)]^{-\frac{1}{p}} \sum_{n=n_0}^{\infty} a_n h(xv(n)) \right] dx \leq \|f\|_{p,\varphi} L. \quad (4.19)$$

Then, by (4.17), we have (4.15).

On the other hand, suppose that (4.15) is valid. We set

$$f(x) = [\varphi(x)]^{1-q} \left[\sum_{n=n_0}^{\infty} h(xv(n))a_n \right]^{q-1}, x \in (0,\infty).$$

Then, it follows that $L^{q-1} = \|f\|_{p,\varphi}$. By (4.7), we find $L < \infty$. If $L = 0$, then (4.17) is trivially valid; if $L > 0$, then, by (4.15), we have

$$\|f\|_{p,\varphi}^p = L^q = I < k(\alpha)\|f\|_{p,\varphi}\|a\|_{q,\Psi}, \text{ that is,}$$

$$\|f\|_{p,\varphi}^{p-1} = L < k(\alpha)\|a\|_{q,\Psi},$$

and then, (4.15) is equivalent to (4.17).

Hence, (4.15), (4.16) and (4.17) are equivalent.

For $0 < \varepsilon < p(\delta - \alpha)$, setting $\widetilde{f}(x)$ and $\widetilde{a}_n$ as follows:

$$\widetilde{f}(x) = \begin{cases} x^{\alpha + \frac{\varepsilon}{p} - 1}, & 0 < x \leq 1, \\ 0, & x > 1, \end{cases}$$

$$\widetilde{a}_n = [v(n)]^{\alpha - \frac{\varepsilon}{q} - 1} v'(n), \quad n \geq n_0,$$

if there exists a positive constant $k(\leq k(\alpha))$, such that (4.15) is still valid as we replace $k(\alpha)$ by k, then substitution of $\widetilde{f}(x)$ and $\widetilde{a} = \{\widetilde{a}_n\}_{n=n_0}^{\infty}$, by (4.8), it follows that

$$\widetilde{I} = \sum_{n=n_0}^{\infty} \widetilde{a}_n \int_0^{\infty} h(xv(n))\widetilde{f}(x)dx < k\|\widetilde{f}\|_{p,\varphi}\|\widetilde{a}\|_{q,\Psi}$$

$$= k \left(\int_0^1 x^{-1+\varepsilon} dx \right)^{\frac{1}{p}} (A(\varepsilon))^{\frac{1}{q}} = \frac{k}{\varepsilon}(1 + o(1))^{\frac{1}{q}}. \tag{4.20}$$

By (4.8), (4.9) and (4.10), we find

$$\widetilde{I} = \sum_{n=n_0}^{\infty} [v(n)]^{\alpha - \frac{\varepsilon}{q} - 1} v'(n) \int_0^1 h(xv(n))x^{\alpha + \frac{\varepsilon}{p} - 1} dx \qquad (t = xv(n))$$

$$= \sum_{n=n_0}^{\infty} \frac{v'(n)}{[v(n)]^{1+\varepsilon}} \int_0^{v(n)} h(t)t^{\alpha + \frac{\varepsilon}{p} - 1} dt$$

$$= \sum_{n=n_0}^{\infty} \frac{v'(n)}{[v(n)]^{1+\varepsilon}} \left[\int_0^{\infty} h(t)t^{\alpha + \frac{\varepsilon}{p} - 1} dt - \int_{v(n)}^{\infty} h(t)t^{\alpha + \frac{\varepsilon}{p} - 1} dt \right]$$

$$= A(\varepsilon)k(\alpha + \frac{\varepsilon}{p}) - B(\varepsilon)$$

$$\geq \frac{1}{\varepsilon}[(1 + o(1))(k(\alpha) + o(1)) - \varepsilon O(1)]. \tag{4.21}$$

Hence, in view of (4.20) and (4.21), it follows that

$$(1 + o(1))(k(\alpha) + o(1)) - \varepsilon O(1) < k(1 + o(1))^{\frac{1}{q}},$$

and then $k(\alpha) \leq k(\varepsilon \to 0^+)$. Therefore, the constant factor $k(\alpha)$ in (4.15) is the best possible.

By the equivalency, the constant factor $k(\alpha)$ in (4.16) ((4.17) is also the best possible, otherwise, it leads to a contradiction by (4.18), (4.19) that the constant factor in (4.15) is not the best possible. $\qquad \square$

Note. If we change the upper limits, ∞ of the integral to $c > 0$, then Theorem 4.1 is still valid. In this case,

$$\omega_\alpha(n) = [v(n)]^\alpha \int_0^c h(xv(n))x^{\alpha-1}dx < k(\alpha).$$

Remark 4.1. If we replace the conditions that $k(\alpha) \in \mathbf{R}_+$, and there exist constants $\delta > \alpha$ and $L > 0$, such that (4.9) is fulfilled to the condition that there exist $\delta_0, \eta > 0$, such that for any $\widetilde{\alpha} \in (\alpha - \delta_0, \alpha]$,

$$0 < k(\widetilde{\alpha})(1 - \theta(x)) < \varpi_{\widetilde{\alpha}}(x) < k(\widetilde{\alpha}) < \infty, \tag{4.22}$$

where $\theta(x) = O(x^\eta)$ $(0 < x \leq 1)$, then the constant factor $k(\alpha)$ in (4.15) is still the best possible. In fact, for $0 < \varepsilon < q\delta_0, \widetilde{\alpha} = \alpha - \frac{\varepsilon}{q}$, we find

$$\widetilde{I} = \int_0^1 x^{\varepsilon-1} \left[x^{\alpha - \frac{\varepsilon}{q}} \sum_{n=n_0}^\infty h(xv(n)) \frac{v'(n)}{[v(n)]^{1-(\alpha-\frac{\varepsilon}{q})}} \right] dx$$

$$= \int_0^1 x^{\varepsilon-1}\varpi_{\widetilde{\alpha}}(x)dx > k(\widetilde{\alpha}) \int_0^1 x^{\varepsilon-1}(1 - O(x^\eta))dx$$

$$\geq \frac{1}{\varepsilon}(k(\alpha) + o(1))(1 - \varepsilon O(1)),$$

and then by the same way, we can still show that $k(\alpha)$ is the best possible constant factor of (4.15).

Remark 4.2. In view of Theorem 4.1, (i) define a half-discrete Hilbert-type operator $T : L_{p,\varphi}(\mathbf{R}_+) \rightarrow l_{p,\Psi^{1-p}}$ as follows:
For $f \in L_{p,\varphi}(\mathbf{R}_+)$, there exists a $Tf \in l_{p,\Psi^{1-p}}$, satisfying

$$Tf(n) = \int_0^\infty h(xv(n))f(x)dx, \quad n \geq n_0, \quad n \in \mathbf{N}. \tag{4.23}$$

Then, by (4.16), it follows that $\|Tf\|_{p,\Psi^{1-p}} \leq k(\alpha)\|f\|_{p,\varphi}$ and then T is a bounded operator with $\|T\| \leq k(\alpha)$. Since by Theorem 4.1, the constant factor in (4.16) is the best possible, we have $\|T\| = k(\alpha)$.

(ii) Define a half-discrete Hilbert-type operator $\widetilde{T} : l_{q,\Psi} \rightarrow L_{q,\varphi^{1-q}}(\mathbf{R}_+)$ as follows: For $a \in l_{q,\Psi}$, there exists a $\widetilde{T}a \in L_{q,\varphi^{1-q}}(\mathbf{R}_+)$, satisfying

$$\widetilde{T}a(x) = \sum_{n=n_0}^\infty h(xv(n))a_n, \quad x \in (0, \infty). \tag{4.24}$$

Then, by (4.17), it follows that $\|\widetilde{T}a\|_{q,\varphi^{1-q}} \leq k(\alpha)\|a\|_{q,\Psi}$ and then $\widetilde{T}$ is a bounded operator with $\|\widetilde{T}\| \leq k(\alpha)$. Since by Theorem 4.1, the constant factor in (4.17) is the best possible, we have $\|\widetilde{T}\| = k(\alpha)$.

Example 4.2. If $v(n) = \ln n, h(x \ln n) = \frac{1}{(\ln e n^x)^\lambda}$ $(0 < \alpha < \lambda,\ \alpha \le 1)$, since for $x > 0$,

$$f(x,y) = \frac{x^\alpha}{(1 + x \ln y)^\lambda} (\ln y)^{\alpha - 1}$$

is decreasing with respect to $y \in (1, \infty)$, then Condition (i) is satisfied. In view of Example 4.1(ii), the constant factor in (4.16) and (4.17) (for $v(n) = n, n_0 = 1$) is the best possible, we have

$$||T|| = ||\widetilde{T}|| = k(\alpha) = B(\alpha, \lambda - \alpha).$$

If $h(x \ln n) = \frac{1}{e^{(\ln n^x)^\lambda}}$ $(\lambda > 0, 0 < \alpha \le 1)$, then by the same way and Example 4.1(v), we have

$$||T|| = ||\widetilde{T}|| = \frac{1}{\lambda}\Gamma(\frac{\alpha}{\lambda}).$$

In particular of Remark 4.2, setting $v(n) = n$ and $\psi(n) = n^{q(1-\alpha)-1}(n \in \mathbf{N})$, we define an operator $T : L_{p,\varphi}(\mathbf{R}_+) \to l_{p,\psi^{1-p}}$ as follows:
For $f \in L_{p,\varphi}(\mathbf{R}_+)$, there exists a $Tf \in l_{p,\psi^{1-p}}$, satisfying

$$Tf(n) = \int_0^\infty h(xn)f(x)dx, \quad n \in \mathbf{N}.$$

Also we define an operator $\widetilde{T} : l_{q,\psi} \to L_{q,\varphi^{1-q}}(\mathbf{R}_+)$ as follows:
For $a \in l_{q,\psi}$, there exists a $\widetilde{T}a \in L_{q,\varphi^{1-q}}(\mathbf{R}_+)$, satisfying

$$\widetilde{T}a(x) = \sum_{n=1}^\infty h(xn)a_n, \quad x \in (0, \infty).$$

Then by Remark 4.2, we still have $||T|| = ||\widetilde{T}|| = k(\alpha)$.

We set some particular kernels as follows:
(i) If $h(xn) = \frac{1}{1+(xn)^\lambda}$ $(0 < \alpha < \lambda,\ \alpha \le 1)$, since for $x > 0$,

$$f(x,y) = \frac{x^\alpha}{1 + (xy)^\lambda} y^{\alpha - 1}$$

is decreasing with respect to $y \in (0, \infty)$, then Condition (i) is satisfied. In view of Example 4.1(i), the constant factor in (4.16) and (4.17) (for $v(n) = n, n_0 = 1$) is the best possible, we have

$$||T|| = ||\widetilde{T}|| = k(\alpha) = \frac{\pi}{\lambda \sin \pi(\frac{\alpha}{\lambda})}.$$

(ii) If $h(xn) = \frac{1}{(1+xn)^\lambda}$ $(0 < \alpha < \lambda, \alpha \le 1)$, since, for $x > 0$,

$$f(x,y) = \frac{x^\alpha}{(1 + xy)^\lambda} y^{\alpha - 1}$$

is decreasing with respect to $y \in (0, \infty)$, then Condition (i) is satisfied. In view of Example 4.1(ii), the constant factor in (4.16) and (4.17) (for $v(n) = n, n_0 = 1$) is the best possible, we have

$$||T|| = ||\widetilde{T}|| = k(\alpha) = B(\alpha, \lambda - \alpha).$$

(iii) If $h(xn) = \frac{\ln(xn)}{(xn)^\lambda - 1} (0 < \alpha < \lambda, \alpha \leq 1)$, then by the same way and Example 4.1(iii), we have (see Zhong [181])

$$||T|| = ||\widetilde{T}|| = \left[\frac{\pi}{\lambda \sin \pi(\frac{\alpha}{\lambda})} \right]^2.$$

(iv) If $h(xn) = \prod_{k=1}^s \frac{1}{a_k (xn)^\lambda + 1} (0 < a_1 < \cdots < a_s, \lambda > 0, 0 < \alpha < s, \alpha\lambda \leq 1)$, then, by the same way and Example 4.1(iv), we have

$$||T|| = ||\widetilde{T}|| = \frac{\pi}{\lambda \sin(\pi\alpha)} \sum_{k=1}^s a_k^{s-\alpha-1} \prod_{j=1(j\neq k)}^s \frac{1}{a_k - a_j}.$$

(v) If $h(xn) = \frac{1}{(xn)^\lambda + 2(xn)^{\lambda/2} \cos\beta + 1} (\lambda > 0, 0 < \beta \leq \frac{\pi}{2}, 0 < \alpha < 1)$, then by the same way and Example 4.1(v), we find

$$||T|| = ||\widetilde{T}|| = \frac{2\pi \sin\beta(1 - 2\alpha)}{\lambda \sin\beta \sin(2\pi\alpha)} \in \mathbf{R}_+.$$

(vi) If $h(xn) = \frac{1}{(xn)^\lambda + b(xn)^{\lambda/2} + c} (c > 0, 0 \leq b \leq 2\sqrt{c}, 0 < \alpha < \frac{\lambda}{2} \leq 1)$, then, by the same way and Example 4.1(vi), we have

$$||T|| = ||\widetilde{T}|| = \frac{2}{\lambda} \begin{cases} \frac{\pi}{\sqrt{c}}, & b = 0, \\ \frac{4}{\sqrt{4c-b^2}} \arctan \frac{\sqrt{4c-b^2}}{b}, & 0 < b < 2\sqrt{c}, \\ \frac{2}{\sqrt{c}}, & b = 2\sqrt{c}. \end{cases}$$

(vii) If $h(xn) = \frac{1}{e^{\gamma(xn)^\lambda} - Ae^{-\gamma(xn)^\lambda}} (-1 \leq A < 1, \beta > 0(A = 1, \beta > 1), \gamma, \lambda > 0, \alpha = \beta\lambda \leq 1)$, then by the same way and Example 4.1(vii), we have

$$||T|| = ||\widetilde{T}|| = \frac{\Gamma(\beta)}{\lambda\gamma^\beta} \sum_{k=0}^\infty \frac{A^k}{(2k + 1)^\beta}.$$

(viii) If $h(xn) = \ln \left[\frac{b(xn)^\lambda + 1}{a(xn)^\lambda + 1} \right] (0 \leq a < b, \lambda > 0), \alpha = -\beta\lambda (0 < \beta < 1)$, then, by the same way and Example 4.1(viii), we have

$$||T|| = ||\widetilde{T}|| = \frac{(b^\beta - a^\beta)\pi}{\beta\lambda \sin(\beta\pi)}.$$

(ix) If $h(xn) = \arctan \frac{1}{(xn)^\lambda} (0 < \alpha < \lambda, \alpha \leq 1)$, then, by the same way and Example 4.1(ix), we have

$$||T|| = ||\widetilde{T}|| = \frac{\pi}{2\alpha \sin \frac{\pi}{2}(1 - \frac{\alpha}{\lambda})}.$$

(x) If $h(xn) = \frac{1}{(\max\{1, xn\})^\lambda}$, $0 < \alpha < \lambda$, $\alpha \le 1$, then, by the same way and Example 4.1(x), we have (see Yang [166])

$$||T|| = ||\tilde{T}|| = \frac{\lambda}{\alpha(\lambda - \alpha)}.$$

(xi) If $h(xn) = (\min\{1, xn\})^\lambda (0 < -\alpha < \lambda \le 1 - \alpha)$, then by the same way and Example 4.1(xi), we have (see Yang [166])

$$||T|| = ||\tilde{T}|| = \frac{\lambda}{(-\alpha)(\lambda + \alpha)}.$$

(xii) If $h(xn) = \left(\frac{\min\{1, xn\}}{\max\{1, xn\}}\right)^\lambda$ $(|\alpha| < \lambda \le 1 - \alpha, \alpha < \frac{1}{2})$, since, for $x > 0$,

$$f(x, y) = x^\alpha \left(\frac{\min\{1, xy\}}{\max\{1, xy\}}\right)^\lambda y^{\alpha - 1}$$

$$= \begin{cases} x^{\alpha + \lambda} y^{\lambda + \alpha - 1}, & 0 < y \le \frac{1}{x}, \\ x^{\alpha - \lambda} y^{\alpha - \lambda - 1}, & y > \frac{1}{x} \end{cases}$$

is decreasing with respect to $y \in (0, \infty)$, then Condition (i) is satisfied. In view of Example 4.1(xii), the constant factor in (4.16) and (4.17) (for $v(n) = n, n_0 = 1$) is the best possible, we have

$$||T|| = ||\tilde{T}|| = k(\alpha) = \frac{2\lambda}{\lambda^2 - \alpha^2}.$$

4.3.2   *Two Classes of Equivalent Reverses*

Theorem 4.2. *Suppose that $\alpha \in \mathbf{R}$, $h(t)$ is a non-negative finite measurable function in $\mathbf{R}_+$, $v(y)$ is a strictly increasing differentiable functions in $[n_0, \infty)(n_0 \in \mathbf{N})$ with $v(n_0) > 0$ and $v(\infty) = \infty$,*

$$k(\alpha) = \int_0^\infty h(t) t^{\alpha - 1} dt \in \mathbf{R}_+$$

and $\varpi_\alpha(x) < k(\alpha)$ $(x \in (0, \infty))$. If $p < 0$, $\frac{1}{p} + \frac{1}{q} = 1$, $f(x)$, $a_n \ge 0$, $f \in L_{p,\varphi}(\mathbf{R}_+)$, $a = \{a_n\}_{n=n_0}^\infty \in l_{q,\Psi}$, $||f||_{p,\varphi} > 0$, $||a||_{q,\Psi} > 0$, then, we have the following equivalent inequalities:

$$I = \sum_{n=n_0}^\infty a_n \int_0^\infty h(xv(n)) f(x) dx = \int_0^\infty f(x) \sum_{n=n_0}^\infty a_n h(xv(n)) dx$$

$$> k(\alpha) ||f||_{p,\varphi} ||a||_{q,\Psi}, \tag{4.25}$$

$$J = \left\{ \sum_{n=n_0}^\infty [\Psi(n)]^{1-p} \left[\int_0^\infty h(xv(n)) f(x) dx \right]^p \right\}^{\frac{1}{p}} > k(\alpha) ||f||_{p,\varphi}, \tag{4.26}$$

and

$$L = \left\{ \int_0^\infty [\varphi(x)]^{1-q} \left[\sum_{n=n_0}^\infty h(xv(n))a_n \right]^q dx \right\}^{\frac{1}{q}} > k(\alpha)||a||_{q,\Psi}. \quad (4.27)$$

Moreover, if $\frac{v'(y)}{v(y)}(> 0)$ is decreasing in $[n_0, \infty)$ and there exists a constant $\delta_1 > 0$, such that $k(\alpha + \delta_1) \in \mathbf{R}_+$, then the constant factor $k(\alpha)$ in the above inequalities is the best possible.

Proof. In view of the reverse of (4.6) and $\varpi_\alpha(x) < k(\alpha)$, for $p < 0$, we have (4.26). By the reverse Hölder's inequality (see Kuang [47]), we find

$$I = \sum_{n=n_0}^\infty \left[[\Psi(n)]^{-\frac{1}{q}} \int_0^\infty h(xv(n))f(x)dx \right] [[\Psi(n)]^{\frac{1}{q}}a_n] \geq J||a||_{q,\Psi}. \quad (4.28)$$

Then, by (4.26), we have (4.25).

On the other hand, suppose that (4.25) is valid. We set

$$a_n = [\Psi(n)]^{1-p} \left[\int_0^\infty h(xv(n))f(x)dx \right]^{p-1}, n \geq n_0, n \in \mathbf{N}.$$

Then, it follows that $J^{p-1} = ||a||_{q,\Psi}$. By the reverse of (4.6) and the assumption, we find $J > 0$. If $J = \infty$, then (4.26) is trivially valid; if $J < \infty$, then, by (4.25), we have

$$||a||_{q,\Psi}^q = J^p = I > k(\alpha)||f||_{p,\varphi}||a||_{q,\Psi}, \text{that is}$$
$$||a||_{q,\Psi}^{q-1} = J > k(\alpha)||f||_{p,\varphi},$$

and then, (4.25) is equivalent to (4.26).

In view of the reverse of (4.7), since $[\varpi_\alpha(x)]^{1-q} < [k(\alpha)]^{1-q}$, we have (4.27).

By the reverse Hölder's inequality, we find

$$I = \int_0^\infty [[\varphi(x)]^{\frac{1}{p}}f(x)] \left[[\varphi(x)]^{-\frac{1}{p}} \sum_{n=n_0}^\infty h(xv(n))a_n \right] dx \geq ||f||_{p,\varphi} L. \quad (4.29)$$

Then, by (4.27), we have (4.25).

On the other hand, suppose that (4.25) is valid. We set

$$f(x) = [\varphi(x)]^{1-q} \left[\sum_{n=n_0}^\infty h(xv(n))a_n \right]^{q-1}, \quad x \in (0, \infty).$$

Then it follows that $L^{q-1} = ||f||_{p,\varphi}$. By the reverse of (4.7), we find $L > 0$. If $L = \infty$, then (4.27) is trivially valid; if $L < \infty$, then, by (4.25), we have

$$||f||_{p,\varphi}^p = L^q = I > k(\alpha)||f||_{p,\varphi}||a||_{q,\Psi}, \text{ s.t.}$$
$$||f||_{p,\varphi}^{p-1} = L > k(\alpha)||a||_{q,\Psi},$$

and then, (4.25) is equivalent to (4.27).

Hence, inequalities (4.25), (4.26) and (4.27) are equivalent.

For $\varepsilon > 0$, setting $\widetilde{f}(x)$ and $\widetilde{a}_n$ as Theorem 4.1, if there exists a positive constant $k(\geq k(\alpha))$, such that (4.25) is valid as we replace $k(\alpha)$ by k, then substitution of $\widetilde{f}(x)$ and $\widetilde{a} = \{\widetilde{a}_n\}_{n=n_0}^{\infty}$, by (4.8), it follows that

$$\widetilde{I} = \sum_{n=n_0}^{\infty} \widetilde{a}_n \int_0^{\infty} h(xv(n))\widetilde{f}(x)dx > k||\widetilde{f}||_{p,\varphi}||\widetilde{a}||_{q,\Psi}$$

$$= k\left(\int_0^1 x^{-1+\varepsilon}dx\right)^{\frac{1}{p}} (A(\varepsilon))^{\frac{1}{q}} = \frac{k}{\varepsilon}(1 + o(1))^{\frac{1}{q}}.$$

By (4.8) and (4.12), we find

$$\widetilde{I} = \sum_{n=n_0}^{\infty} [v(n)]^{\alpha - \frac{\varepsilon}{q} - 1} v'(n) \int_0^1 h(xv(n))x^{\alpha + \frac{\varepsilon}{p} - 1}dx, \quad (t = xv(n))$$

$$= \sum_{n=n_0}^{\infty} \frac{v'(n)}{[v(n)]^{1+\varepsilon}} \int_0^{v(n)} h(t)t^{\alpha + \frac{\varepsilon}{p} - 1}dt$$

$$\leq \sum_{n=n_0}^{\infty} \frac{v'(n)}{[v(n)]^{1+\varepsilon}} \int_0^{\infty} h(t)t^{\alpha + \frac{\varepsilon}{p} - 1}dt$$

$$= A(\varepsilon)k(\alpha + \frac{\varepsilon}{p}) \leq \frac{1}{\varepsilon}(1 + o(1))(k(\alpha) + o(1)).$$

Hence, in view of the above results, it follows that

$$(1 + o(1))(k(\alpha) + o(1)) > k(1 + o(1))^{\frac{1}{q}},$$

and then $k(\alpha) \geq k(\varepsilon \to 0^+)$. Therefore, the constant factor $k(\alpha)$ in (4.25) is the best possible.

By the equivalency, the constant factor $k(\alpha)$ in (4.26) ((4.27)) is the best possible, otherwise, we can imply a contradiction by (4.28)((4.29)) that the constant factor in (4.25) is not the best possible. □

If we change the upper limit ∞ of the integral to $c(> 0)$ in Theorem 4.2, then it follows that

$$\omega_\alpha(n) = [v(n)]^\alpha \int_0^c h(xv(n))x^{\alpha - 1}dx < k(\alpha).$$

Setting

$$\omega_\alpha(n) = k(\alpha)(1 - \widetilde{\theta}_\alpha(v(n))),$$

$$\widetilde{\theta}_\alpha(v(n)) = \frac{1}{k(\alpha)}[v(n)]^\alpha \int_c^{\infty} h(xv(n))x^{\alpha - 1}dx > 0,$$

and

$$\widetilde{\psi}(n) = \psi(n)(1 - \widetilde{\theta}_\alpha(v(n))) = \frac{[v(n)]^{q(1-\alpha)-1}}{[v'(n)]^{q-1}}\,(1 - \widetilde{\theta}_\alpha(v(n))),$$

we have

$$\widetilde{\theta}_\alpha(v(n)) = \frac{1}{k(\alpha)} \int_{cv(n)}^\infty h(t)\,t^{\alpha-1}dt,$$

and the following corollary:

Corollary 4.1. *Suppose that $c > 0$, $\alpha \in \mathbf{R}$, $h(t)$ is a non-negative finite measurable function in $\mathbf{R}_+$, $v(y)$ is a strictly increasing differentiable functions in $[n_0, \infty)(n_0 \in \mathbf{N})$ with $v(n_0) > 0$ and $v(\infty) = \infty$,*

$$k(\alpha) = \int_0^\infty h(t)\,t^{\alpha-1}dt \in \mathbf{R}_+$$

and $\varpi_\alpha(x) < k(\alpha)(x \in (0,c))$. If $p < 0, \frac{1}{p} + \frac{1}{q} = 1, f(x), a_n \geq 0, f \in L_{p,\varphi}(0,c)$, $a = \{a_n\}_{n=n_0}^\infty \in l_{q,\widetilde{\psi}}, ||f||_{p,\varphi} > 0, ||a||_{q,\widetilde{\psi}} > 0$, then, we have the following equivalent inequalities:

$$\sum_{n=n_0}^\infty a_n \int_0^c h(xv(n))f(x)dx$$

$$= \int_0^c f(x) \sum_{n=n_0}^\infty a_n h(xv(n))dx > k(\alpha)||f||_{p,\varphi}||a||_{q,\widetilde{\psi}}, \qquad (4.30)$$

$$\left\{\sum_{n=n_0}^\infty [\widetilde{\psi}(n)]^{1-p} \left[\int_0^c h(xv(n))f(x)dx\right]^p\right\}^{\frac{1}{p}} > k(\alpha)||f||_{p,\varphi}, \qquad (4.31)$$

and

$$\left\{\int_0^c [\varphi(x)]^{1-q} \left[\sum_{n=n_0}^\infty h(xv(n))a_n\right]^q dx\right\}^{\frac{1}{q}} > k(\alpha)||a||_{q,\widetilde{\psi}}. \qquad (4.32)$$

Moreover, if $\frac{v'(y)}{v(y)}(> 0)$ is decreasing in $[n_0, \infty)$ and there exist constants $\delta_1 > 0, L > 0$ and $\delta > \alpha$, such that $k(\alpha - \delta_1) \in \mathbf{R}_+$, and

$$h(t) \leq L\left(\frac{1}{t^\delta}\right) \qquad (t \in [cv(n_0), \infty)),$$

then, the constant factor $k(\alpha)$ in the above inequalities is the best possible.

Theorem 4.3. *Suppose that $\alpha \in \mathbf{R}$, $h(t)$ is a non-negative finite measurable function in $\mathbf{R}_+$, $v(y)$ is a strictly increasing differentiable function in*

$[n_0, \infty)(n_0 \in \mathbf{N})$ *with* $v(n_0) > 0$ *and* $v(\infty) = \infty$, $k(\alpha) = \int_0^\infty h(t)t^{\alpha-1}dt \in \mathbf{R}_+$ *and*

$$0 < k(\alpha)(1 - \theta(x)) < \varpi_\alpha(x) < k(\alpha) \ (x \in (0, \infty)).$$

Setting $\widetilde{\varphi}(x) = (1 - \theta(x))\varphi(x)$, *if* $0 < p < 1, \frac{1}{p} + \frac{1}{q} = 1, f(x), a_n \geq 0, f \in L_{p,\widetilde{\varphi}}(\mathbf{R}_+)$, $a = \{a_n\}_{n=n_0}^\infty \in l_{q,\Psi}$, $\|f\|_{p,\widetilde{\varphi}} > 0, \|a\|_{q,\Psi} > 0$, *then, we have the following equivalent inequalities:*

$$I = \sum_{n=n_0}^\infty a_n \int_0^\infty h(xv(n))f(x)dx = \int_0^\infty f(x) \sum_{n=n_0}^\infty a_n h(xv(n))dx$$

$$> k(\alpha)\|f\|_{p,\widetilde{\varphi}}\|a\|_{q,\Psi}, \tag{4.33}$$

$$J = \left\{ \sum_{n=n_0}^\infty [\Psi(n)]^{1-p} \left[\int_0^\infty h(xv(n))f(x)dx \right]^p \right\}^{\frac{1}{p}} > k(\alpha)\|f\|_{p,\widetilde{\varphi}}, \tag{4.34}$$

and

$$L = \left\{ \int_0^\infty [\widetilde{\varphi}(x)]^{1-q} \left[\sum_{n=n_0}^\infty h(xv(n))a_n \right]^q dx \right\}^{\frac{1}{q}} > k(\alpha)\|a\|_{q,\Psi}. \tag{4.35}$$

Moreover, if $\frac{v'(y)}{v(y)}(> 0)$ is decreasing in $[n_0, \infty)$ and there exist constants $\eta, \delta_2 > 0$, such that $\theta(x) = O\left(\frac{1}{x^\eta}\right)(0 < x \leq 1)$ and $k(\alpha + \delta_2) \in \mathbf{R}_+$, then the constant factor $k(\alpha)$ in the above inequalities is the best possible.

Proof. In view of the reverse of (4.6) and $\varpi_\alpha(x) < k(\alpha)$, for $0 < p < 1$, we have (4.34). By the reverse Hölder's inequality, we find

$$I = \sum_{n=n_0}^\infty \left[[\Psi(n)]^{-\frac{1}{q}} \int_0^\infty h(xv(n))f(x)dx \right] [[\Psi(n)]^{\frac{1}{q}} a_n] \geq J\|a\|_{q,\Psi}. \tag{4.36}$$

Then, by (4.34), we have (4.33).

On the other hand, suppose that (4.33) is valid. We set

$$a_n = [\Psi(n)]^{1-p} \left[\int_0^\infty h(xv(n))f(x)dx \right]^{p-1}, n \geq n_0(n \in \mathbf{N}).$$

Then it follows that $J^{p-1} = \|a\|_{q,\Psi}$. By the reverse of (4.6) with the assumptions, we find $J > 0$. If $J = \infty$, then (4.34) is trivially valid; if $J < \infty$, then, by (4.33), we have

$$\|a\|_{q,\Psi}^q = J^p = I > k(\alpha)\|f\|_{p,\widetilde{\varphi}}\|a\|_{q,\Psi}, \text{that is}$$

$$\|a\|_{q,\Psi}^{q-1} = J > k(\alpha)\|f\|_{p,\widetilde{\varphi}},$$

and then, inequality (4.33) is equivalent to (4.34).

In view of the reverse of (4.7), since

$$[\varpi_\alpha(x)]^{1-q} > [k(\alpha)(1 - \theta(x))]^{1-q} (q < 0),$$

we have (4.35). By the reverse Hölder's inequality, we find

$$I = \int_0^\infty [[\widetilde{\varphi}(x)]^{\frac{1}{p}} f(x)] \left[[\widetilde{\varphi}(x)]^{-\frac{1}{p}} \sum_{n=n_0}^\infty h(xv(n))a_n \right] dx \geq ||f||_{p,\widetilde{\varphi}} L. \quad (4.37)$$

Then, by (4.35), we have (4.33).

On the other hand, suppose that (4.33) is valid. We set

$$f(x) = [\widetilde{\varphi}(x)]^{1-q} \left[\sum_{n=n_0}^\infty h(xv(n))a_n \right]^{q-1}, x \in (0, \infty).$$

Then it follows that $L^{q-1} = ||f||_{p,\widetilde{\varphi}}$. By the reverse of (4.7) and the assumption, we find $L > 0$. If $L = \infty$, then (4.35) is trivially valid; if $L < \infty$, then, by (4.33), we have

$$||f||_{p,\widetilde{\varphi}}^p = L^q = I > k(\alpha)||f||_{p,\widetilde{\varphi}}||a||_{q,\Psi}, \text{ that is}$$

$$||f||_{p,\widetilde{\varphi}}^{p-1} = L > k(\alpha)||a||_{q,\Psi},$$

and then, (4.33) is equivalent to (4.35).

Hence, inequalities (4.33), (4.34) and (4.35) are equivalent.

For $\varepsilon > 0$, setting $\widetilde{f}(x)$ and $\widetilde{a}_n$ as Theorem 4.1, if there exists a positive constant $k(\geq k(\alpha))$, such that (4.33) is valid as we replace $k(\alpha)$ by k, then, substitution of $\widetilde{f}(x)$ and $\widetilde{a} = \{\widetilde{a}_n\}_{n=n_0}^\infty$, by (4.8), it follows that

$$\widetilde{I} = \sum_{n=n_0}^\infty \widetilde{a}_n \int_0^\infty h(xv(n))\widetilde{f}(x)dx > k||\widetilde{f}||_{p,\widetilde{\varphi}}||\widetilde{a}||_{q,\Psi}$$

$$= k \left[\int_0^1 \left(1 - O\left(\frac{1}{x^\eta}\right) \right) x^{-1-\varepsilon}dx \right]^{\frac{1}{p}} (A(\varepsilon))^{\frac{1}{q}}$$

$$= \frac{k}{\varepsilon}(1 - \varepsilon O(1))^{\frac{1}{p}}(1 + o(1))^{\frac{1}{q}}. \quad (4.38)$$

By (4.8) and (4.12), we find

$$\widetilde{I} = \sum_{n=n_0}^\infty [v(n)]^{\alpha-\frac{\varepsilon}{q}-1} v'(n) \int_0^1 h(xv(n))x^{\alpha+\frac{\varepsilon}{p}-1}dx \quad (t = xv(n))$$

$$= \sum_{n=n_0}^\infty \frac{v'(n)}{[v(n)]^{1+\varepsilon}} \int_0^{v(n)} h(t)t^{\alpha+\frac{\varepsilon}{p}-1}dt$$

$$\leq \sum_{n=n_0}^{\infty} \frac{v'(n)}{[v(n)]^{1+\varepsilon}} \int_0^{\infty} h(t) t^{(\alpha + \frac{\varepsilon}{p}) - 1} dt$$

$$= A(\varepsilon) k\left(\alpha + \frac{\varepsilon}{p}\right) \leq \frac{1}{\varepsilon} (1 + o(1))(k(\alpha) + o(1)). \tag{4.39}$$

Hence, in view of (4.38) and (4.39), it follows that

$$(1 + o(1))(k(\alpha) + o(1)) > k(1 - \varepsilon O(1))^{\frac{1}{p}} (1 + o(1))^{\frac{1}{q}},$$

and then, $k(\alpha) \geq k(\varepsilon \to 0^+)$. Therefore, the constant factor $k(\alpha)$ in (4.33) is the best possible.

By the equivalency, the constant factor $k(\alpha)$ in (4.34)((4.35)) is the best possible, otherwise, we can imply a contradiction by (4.36) ((4.37)) that the constant factor in (4.33) is not the best possible. $\square$

Note. If we change the upper limit ∞ of the integral to $c > 0$, then Theorem 4.3 is still value. In this case,

$$\omega_\alpha(n) = [v(n)]^\alpha \int_0^c h(xv(n)) x^{\alpha - 1} dx < k(\alpha).$$

4.3.3   *Some Corollaries*

Corollary 4.2. *Let the assumptions of Lemma 4.1 be fulfilled and additionally, let*

$$\varpi_\alpha(x) < k(\alpha) = \int_0^{\infty} h(t) t^{\alpha - 1} dt \in \mathbf{R}_+ (x \in (0, \infty)),$$

and for $-\infty \leq b < c \leq \infty$, $u(x)$ be a strictly increasing differentiable function in (b, c), with $u(b^+) = 0$ and $u(c^-) = \infty$. For $p \in \mathbf{R} \backslash \{0, 1\}, \frac{1}{p} + \frac{1}{q} = 1$, we set

$$\Phi(x) = \frac{[u(x)]^{p(1-\alpha)-1}}{[u'(x)]^{p-1}} (x \in (b, c)).$$

If $f(x), a_n \geq 0$, $f \in L_{p,\Phi}(b, c)$, $a = \{a_n\}_{n=n_0}^{\infty} \in l_{q,\Psi}, ||f||_{p,\Phi} > 0, ||a||_{q,\Psi} > 0$, then

(i) *for $p > 1$, we have the following equivalent inequalities:*

$$\sum_{n=n_0}^{\infty} a_n \int_b^c h(u(x)v(n)) f(x) dx$$

$$= \int_b^c f(x) \sum_{n=n_0}^{\infty} a_n h(u(x)v(n)) dx < k(\alpha) ||f||_{p,\Phi} ||a||_{q,\Psi}, \tag{4.40}$$

$$\left\{\sum_{n=n_0}^{\infty}[\Psi(n)]^{1-p}\left[\int_b^c h(u(x)v(n))f(x)dx\right]^p\right\}^{\frac{1}{p}} < k(\alpha)\|f\|_{p,\Phi}, \qquad (4.41)$$

and

$$\left\{\int_b^c[\Phi(x)]^{1-q}\left[\sum_{n=n_0}^{\infty}h(u(x)v(n))a_n\right]^q dx\right\}^{\frac{1}{q}} < k(\alpha)\|a\|_{q,\Psi}; \qquad (4.42)$$

(ii) for $p < 0$, we have the equivalent reverses of (4.40), (4.41) and (4.42).

Moreover, if $\frac{v'(y)}{v(y)}(> 0)$ is decreasing in $[n_0, \infty)$ and there exist constants $\delta < \lambda_1$, $L > 0$, and $\delta_1 > 0$, such that (4.9) is satisfied and $k(\alpha - \delta_1) \in \mathbf{R}_+$, then the constant factor $k(\alpha)$ in the above inequalities is the best possible.

Proof. (i) For $p > 1$, putting $x = u(t)$ in two sides of (4.15), we have

$$I = \sum_{n=n_0}^{\infty} a_n \int_b^c h(u(t)v(n))f(u(t))u'(t)dt$$

$$= \int_b^c f(u(t))u'(t) \sum_{n=n_0}^{\infty} a_n h(u(t)v(n))dt$$

$$< k(\alpha)\left\{\int_b^c[u(t)]^{p(1-\alpha)-1}f^p(u(t))u'(t)dt\right\}^{\frac{1}{p}}\|a\|_{q,\Psi}. \qquad (4.43)$$

Changing t and $f(u(t))u'(t)$ to x and $f(x)$ in (4.43), by simplification, we obtain (4.40). On the other hand, setting $u(x) = x(x \in (0, \infty))$ in (4.40), we have (4.15). It follows that (4.40) and (4.15) are equivalent and then both of them are with the same best constant factor $k(\alpha)$. By the same way, we can prove that (4.41) and (4.16) ((4.42) and (4.17)) are equivalent and with the same best constant factor $k(\alpha)$. Since (4.15), (4.16) and (4.17) are equivalent, then it follows that (4.40), (4.41) and (4.42) are equivalent.

(ii) For $p < 0$, in the same way, we have the equivalent reverses of (4.40), (4.41) and (4.42) with the same best constant factor $k(\alpha)$. □

Example 4.3. Define an operator $T : L_{p,\Psi}(b,c) \to l_{q,\Psi}$ as follows: For $f \in L_{p,\Phi}(b,c)$, there exists a $Tf \in l_{p,\Psi^{1-p}}$, satisfying

$$Tf(n) = \int_b^c h(u(x)v(n))f(x)dx, n \geq n_0, \qquad n \in \mathbf{N}.$$

Also we define an operator $\widetilde{T} : l_{q,\Psi} \to L_{p,\Phi}(b,c)$ as follows:
For $a \in l_{q,\Psi}$, there exists a $\widetilde{T}a \in L_{p,\Phi}(b,c)$, satisfying

$$\widetilde{T}a(x) = \sum_{n=n_0}^{\infty} h(u(x)v(n))a_n, x \in (b,c).$$

Then, by Corollary 4.2, we still have $\|T\| = \|\widetilde{T}\| = k(\alpha)$.

For examples, (1) setting $u(x) = \ln x (x \in (1,\infty))$, $v(n) = \ln n, n \geq n_0 = 2$, then $\frac{v'(y)}{v(y)} = \frac{1}{y \ln y} (> 0)$ is decreasing for $y > 1$.

(i) If $h(\ln x \ln n) = \frac{1}{1+(\ln x \ln n)^\lambda}$ $(0 < \alpha < \lambda, \alpha \leq 1)$, since for $x > 0$,

$$f(x,y) = \frac{x^\alpha}{1 + (x \ln y)^\lambda} (\ln y)^{\alpha-1}$$

is decreasing with respect to $y \in (1,\infty)$, then, Condition (i) is satisfied. In view of Example 4.1(i), the constant factor in (4.41) and (4.42) (for $u(x) = \ln x, v(n) = \ln n, n_0 = 2$) is the best possible, we have

$$||T|| = ||\widetilde{T}|| = k(\alpha) = \frac{\pi}{\lambda \sin \pi(\frac{\alpha}{\lambda})}.$$

(ii) If $h(\ln x \ln n) = \frac{1}{(1+\ln x \ln n)^\lambda}$ $(0 < \alpha < \lambda, \alpha \leq 1)$, since for $x > 0$,

$$f(x,y) = \frac{x^\alpha}{(1 + x \ln y)^\lambda} y^{\alpha-1}$$

is decreasing with respect to $y \in (1,\infty)$, then, Condition (i) is satisfied. In view of Example 4.1(ii), the constant factor in (4.41) and (4.42) (for $u(x) = \ln x (x > 1), v(n) = n, n_0 = 1$) is the best possible, we have (see Yang [169])

$$||T|| = ||\widetilde{T}|| = k(\alpha) = B(\alpha, \lambda - \alpha).$$

(iii) If $h(\ln x \ln n) = \frac{\ln(\ln x \ln n)}{(\ln x \ln n)^\lambda - 1}$ $(0 < \alpha < \lambda, \alpha \leq 1)$, then by the same way and Example 4.1(iii), we have

$$||T|| = ||\widetilde{T}|| = \left[\frac{\pi}{\lambda \sin \pi(\frac{\alpha}{\lambda})}\right]^2.$$

(iv) If $h(\ln x \ln n) = \prod_{k=1}^s \frac{1}{a_k (\ln x \ln n)^\lambda + 1}$ $(0 < a_1 < \cdots < a_s, \lambda > 0, \ 0 < \alpha < s, \alpha\lambda \leq 1)$, then by the same way and Example 4.1(iv), we have

$$||T|| = ||\widetilde{T}|| = \frac{\pi}{\lambda \sin(\pi\alpha)} \sum_{k=1}^s a_k^{s-\alpha-1} \prod_{j=1(j\neq k)}^s \frac{1}{a_k - a_j}.$$

(v) If $h(\ln x \ln n) = \frac{1}{(\ln x \ln n)^\lambda + 2(\ln x \ln n)^{\lambda/2} \cos \beta + 1}$ $(x > 1, n \in \mathbf{N}\backslash\{1\}$; $\lambda > 0, 0 < \beta \leq \frac{\pi}{2}, 0 < \alpha < 1, \alpha\lambda \leq 1)$, then, by the same way and Example 4.1(v), we find

$$||T|| = ||\widetilde{T}|| = \frac{2\pi \sin \beta(1 - 2\alpha)}{\lambda \sin \beta \sin(2\pi\alpha)} \in \mathbf{R}_+.$$

(vi) If $h(\ln x \ln n) = \frac{1}{(\ln x \ln n)^\lambda + b(\ln x \ln n)^{\lambda/2} + c}(c > 0, \ 0 \le b \le 2\sqrt{c}, \ 0 < \alpha = \frac{\lambda}{2} \le 1)$, then by the same way and Example 4.1(vi), we have

$$||T|| = ||\widetilde{T}|| = \frac{2}{\lambda} \begin{cases} \frac{\pi}{\sqrt{c}}, & b = 0, \\ \frac{4}{\sqrt{4c-b^2}} \arctan \frac{\sqrt{4c-b^2}}{b}, & 0 < b < 2\sqrt{c}, \\ \frac{2}{\sqrt{c}}, & b = 2\sqrt{c}. \end{cases}$$

(vii) If $h(\ln x \ln n) = \frac{1}{e^{\gamma(\ln x \ln n)^\lambda} - Ae^{-\gamma(\ln x \ln n)^\lambda}}$ $(-1 \le A < 1, \ \beta > 0(A = 1, \beta > 1), \ \gamma, \lambda > 0, \alpha = \beta\lambda \le 1)$, then, by the same way and Example 4.1(vii), we have

$$||T|| = ||\widetilde{T}|| = \frac{\Gamma(\beta)}{\lambda \gamma^\beta} \sum_{k=0}^{\infty} \frac{A^k}{(2k+1)^\beta}.$$

(viii) If $h(\ln x \ln n) = \ln \left[\frac{b(\ln x \ln n)^\lambda + 1}{a(\ln x \ln n)^\lambda + 1}\right]$ $(0 \le a < b, \lambda > 0, \alpha = -\beta\lambda, 0 < \beta < 1)$, then, by the same way and Example 4.1(vii), we have

$$||T|| = ||\widetilde{T}|| = \frac{(b^\beta - a^\beta)\pi}{\beta\lambda \sin(\beta\pi)}.$$

(ix) If $h(\ln x \ln n) = \arctan \frac{1}{(\ln x \ln n)^\lambda}(0 < \alpha < \lambda, \alpha \le 1)$, then by the same way and Example 4.1(ix), we have

$$||T|| = ||\widetilde{T}|| = \frac{\pi}{2\alpha \cos(\frac{\pi\alpha}{2\lambda})}.$$

(x) If $h(\ln x \ln n) = \frac{1}{(\max\{1, \ln x \ln n\})^\lambda}(0 < \alpha < \lambda, \alpha \le 1)$, then, by the same way and Example 4.1(x), we have

$$||T|| = ||\widetilde{T}|| = \frac{\lambda}{\alpha(\lambda - \alpha)}.$$

(xi) If $h(xn) = (\min\{1, \ln x \ln n\})^\lambda(0 < -\alpha < \lambda \le 1 - \alpha)$, then, by the same way and Example 4.1(xi), we have

$$||T|| = ||\widetilde{T}|| = \frac{\lambda}{(-\alpha)(\lambda + \alpha)}.$$

(xii) If $h(xn) = \left(\frac{\min\{1, \ln x \ln n\}}{\max\{1, \ln x \ln n\}}\right)^\lambda$ $(|\alpha| < \lambda \le 1 - \alpha(\alpha < \frac{1}{2}))$, then by the same way and Example 4.1(xii), we have

$$||T|| = ||\widetilde{T}|| = k(\alpha) = \frac{2\lambda}{\lambda^2 - \alpha^2}.$$

(2) For $0 \le \gamma_1, \gamma_2 \le \frac{1}{2}$, setting $u(x) = x - \gamma_1$ $(x \in (\gamma_1, \infty))$, $v(n) = n - \gamma_2, n \ge n_0 = 1$, then, $\frac{v'(y)}{v(y)} = \frac{1}{y - \gamma_2}(> 0)$ is decreasing for $y > \frac{1}{2}$.

(i) If $h((x-\gamma_1)(n-\gamma_2)) = \frac{1}{1+(x-\gamma_1)^\lambda (n-\gamma_2)^\lambda}x$ $(0 < \alpha < \lambda, \alpha \le \min\{1, 2-\lambda\})$, since for $x > 0$,

$$f(x, y) = \frac{x^\alpha}{1 + x^\lambda (y-\gamma_2)^\lambda} (y - \gamma_2)^{\alpha-1}$$

is strictly convex satisfying $f''_{y^2}(x, y) > 0$ with respect to $y \in \left(\frac{1}{2}, \infty\right)$, then, Condition (ii) is satisfied. In view of Example 4.1(i), the constant factor in (4.41) and (4.42) (for $u(x) = x - \gamma_1, v(n) = n - \gamma_2$) is the best possible, we have

$$\|T\| = \|\widetilde{T}\| = \frac{\pi}{\lambda \sin(\frac{\pi\alpha}{\lambda})}.$$

(ii) If $h((x-\gamma_1)(n-\gamma_2)) = \frac{1}{[1+(x-\gamma_1)(n-\gamma_2)]^\lambda}x$ $(0 < \alpha < \lambda, \alpha \le 1)$, by the same way and Example 4.1(ii), we have

$$\|T\| = \|\widetilde{T}\| = B(\alpha, \lambda - \alpha).$$

(iii) If $h((x-\gamma_1)(n-\gamma_2)) = \frac{\ln[(x-\gamma_1)(n-\gamma_2)]}{(x-\gamma_1)^\lambda (n-\gamma_2)^\lambda - 1}$ $(0 < \alpha < \lambda, \alpha \le 1)$, then by the same way and Example 4.1(iii), we have

$$\|T\| = \|\widetilde{T}\| = [\frac{\pi}{\lambda \sin \pi(\frac{\alpha}{\lambda})}]^2.$$

(iv) $h((x-\gamma_1)(n-\gamma_2)) = \prod_{k=1}^s \frac{1}{a_k[(x-\gamma_1)(n-\gamma_2)]^\lambda + 1}$ $(0 < a_1 < \cdots < a_s, \lambda > 0, 0 < \alpha < s, \alpha\lambda \le 1)$, then by the same way and Example 4.1(iv), we have

$$\|T\| = \|\widetilde{T}\| = \frac{\pi}{\lambda \sin(\pi\alpha)} \sum_{k=1}^s a_k^{s-\alpha-1} \prod_{j=1(j\ne k)}^s \frac{1}{a_k - a_j}.$$

(v) If $h((x-\gamma_1)(n-\gamma_2)) = \frac{1}{[(x-\gamma_1)(n-\gamma_2)]^\lambda + 2[(x-\gamma_1)(n-\gamma_2)]^{\lambda/2} \cos\beta + 1}$ $(\lambda > 0, 0 < \beta \le \frac{\pi}{2}, 0 < \alpha < 1, \alpha\lambda \le \min\{1, 2-\lambda\})$, then, by the same way and Example 4.1(v), we find

$$\|T\| = \|\widetilde{T}\| = \frac{2\pi \sin\beta(1 - 2\alpha)}{\lambda \sin\beta \sin(2\pi\alpha)} \in \mathbf{R}_+.$$

(vi) If $h((x-\gamma_1)(n-\gamma_2)) = \frac{1}{[(x-\gamma_1)(n-\gamma_2)]^\lambda + b[(x-\gamma_1)(n-\gamma_2)]^{\lambda/2} + c}$ $(c > 0, 0 \le b \le 2\sqrt{c}, 0 < \alpha = \frac{\lambda}{2} \le \frac{2}{3})$, then, by the same way and Example 4.1(vi), we have

$$\|T\| = \|\widetilde{T}\| = \frac{2}{\lambda} \begin{cases} \frac{\pi}{\sqrt{c}}, & b = 0, \\ \frac{4}{\sqrt{4c-b^2}} \arctan \frac{\sqrt{4c-b^2}}{b}, & 0 < b < 2\sqrt{c}, \\ \frac{2}{\sqrt{c}}, & b = 2\sqrt{c}. \end{cases}$$

(vii) If $h((x - \gamma_1)(n - \gamma_2)) = \frac{1}{e^{\gamma[(x-\gamma_1)(n-\gamma_2)]^\lambda} - Ae^{-\gamma[(x-\gamma_1)(n-\gamma_2)]^\lambda}}$ ($-1 \leq A < 1, \beta > 0$ ($A = 1, \beta > 1$), $\gamma, \lambda > 0, \alpha = \beta\lambda \leq 1$), then, by the same way and Example 4.1(vii), we have

$$||T|| = ||\widetilde{T}|| = \frac{\Gamma(\beta)}{\lambda\gamma^\beta} \sum_{k=0}^{\infty} \frac{A^k}{(2k+1)^\beta}.$$

(viii) If $h((x - \gamma_1)(n - \gamma_2)) = \ln\{\frac{b[(x-\gamma_1)(n-\gamma_2)]^\lambda + 1}{a[(x-\gamma_1)(n-\gamma_2)]^\lambda + 1}\}(0 \leq a < b, \lambda > 0, \alpha = -\beta\lambda, 0 < \beta < 1)$, then, by the same way and Example 4.1(viii), we have

$$||T|| = ||\widetilde{T}|| = \frac{(b^\beta - a^\beta)\pi}{\beta\lambda\sin(\beta\pi)}.$$

(ix) If $h((x - \gamma_1)(n - \gamma_2)) = \arctan[\frac{1}{(x-\gamma_1)(n-\gamma_2)}]^\lambda$ ($0 < \alpha < \lambda, \alpha \leq \min\{1, 2 - \lambda\}$), then, by the same way and Example 4.1(ix), we have

$$||T|| = ||\widetilde{T}|| = \frac{\pi}{2\alpha\cos(\frac{\pi\alpha}{2\lambda})}.$$

(3) For $a \geq \frac{2}{3}$, setting $u(x) = \ln ax (x \in (\frac{1}{a}, \infty)), v(n) = \ln an, n \geq n_0 = 2$, then $\frac{v'(y)}{v(y)} = \frac{1}{y\ln ay}(> 0)$ is decreasing for $y > \frac{3}{2}$.

(i) If $h(\ln ax \ln an) = \frac{1}{1+(\ln ax \ln an)^\lambda}(0 < \alpha < \lambda, \alpha \leq \min\{1, 2 - \lambda\})$, since for $x > 0$,

$$f(x,y) = \frac{x^\alpha}{1 + x^\lambda(\ln ay)^\lambda}(\ln ay)^{\alpha-1}$$

is strictly convex satisfying $f''_{y^2}(x, y) > 0$ with respect to $y \in (\frac{3}{2}, \infty)$, then, Condition (ii) is satisfied. In view of Example 4.1(i), the constant factor in (4.41) and (4.42) (for $u(x) = \ln ax, v(n) = \ln an$) is the best possible, we have

$$||T|| = ||\widetilde{T}|| = \frac{\pi}{\lambda\sin\pi(\frac{\alpha}{\lambda})}.$$

(ii) If $h(\ln ax \ln an) = \frac{1}{(1+\ln ax \ln an)^\lambda}$ ($0 < \alpha < \lambda, \alpha \leq 1$), by the same way and Example 4.1(ii), we have

$$||T|| = ||\widetilde{T}|| = B(\alpha, \lambda - \alpha).$$

(iii) If $h(\ln ax \ln an) = \frac{\ln(\ln ax \ln an)}{(x-\gamma_1)^\lambda(n-\gamma_2)^\lambda - 1}(0 < \alpha < \lambda, \alpha \leq 1)$, then by the same way and Example 4.1(iii), we have

$$||T|| = ||\widetilde{T}|| = [\frac{\pi}{\lambda\sin(\frac{\pi\alpha}{\lambda})}]^2.$$

(iv) If $h(\ln ax \ln an) = \prod_{k=1}^{s} \frac{1}{a_k(\ln ax \ln an)^\lambda + 1}(0 < a_1 < \cdots < a_s, \lambda > 0, 0 < \alpha < s, \alpha\lambda \le 1)$, then, by the same way and Example 4.1(iv), we have

$$||T|| = ||\widetilde{T}|| = \frac{\pi}{\lambda \sin(\pi\alpha)} \sum_{k=1}^{s} a_k^{s-\alpha-1} \prod_{j=1(j\neq k)}^{s} \frac{1}{a_k - a_j}.$$

(v) If $h(\ln ax \ln an) = \frac{1}{(\ln ax \ln an)^\lambda + 2(\ln ax \ln an)^{\lambda/2}\cos\beta + 1}(\lambda > 0, 0 < \beta \le \frac{\pi}{2}, 0 < \alpha < 1, \alpha\lambda \le \min\{1, 2 - \lambda\})$, then, by the same way and Example 4.1(v), we find

$$||T|| = ||\widetilde{T}|| = \frac{2\pi \sin\beta(1 - 2\alpha)}{\lambda \sin\beta \sin(2\pi\alpha)} \in \mathbf{R}_+.$$

(vi) If $h(\ln ax \ln an) = \frac{1}{(\ln ax \ln an)^\lambda + b(\ln ax \ln an)^{\lambda/2} + c}(c > 0, 0 \le b \le 2\sqrt{c}, 0 < \alpha = \frac{\lambda}{2} \le \frac{2}{3})$, then, by the same way and Example 4.1(vi), we have

$$||T|| = ||\widetilde{T}|| = \frac{2}{\lambda} \begin{cases} \frac{\pi}{\sqrt{c}}, & b = 0, \\ \frac{4}{\sqrt{4c-b^2}} \arctan\frac{\sqrt{4c-b^2}}{b}, & 0 < b < 2\sqrt{c}, \\ \frac{2}{\sqrt{c}}, & b = 2\sqrt{c}. \end{cases}$$

(vii) If $h(\ln ax \ln an) = \frac{1}{e^{\gamma(\ln ax \ln an)^\lambda} - Ae^{-\gamma(\ln ax \ln an)^\lambda}}(-1 \le A < 1, \beta > 0$ $(A = 1, \beta > 1), \gamma, \lambda > 0, \alpha = \beta\lambda \le 1)$, then, by the same way and Example 4.1(vii), we have

$$||T|| = ||\widetilde{T}|| = \frac{\Gamma(\beta)}{\lambda\gamma^\beta} \sum_{k=0}^{\infty} \frac{A^k}{(2k+1)^\beta}.$$

(viii) If $h(\ln ax \ln an) = \ln\left\{\frac{b(\ln ax \ln an)^\lambda + 1}{a(\ln ax \ln an)^\lambda + 1}\right\}$ $(0 \le a < b, \lambda > 0, \alpha = -\beta\lambda, 0 < \beta < 1)$, then, by the same way and Example 4.1(viii), we have

$$||T|| = ||\widetilde{T}|| = \frac{(b^\beta - a^\beta)\pi}{\beta\lambda \sin(\beta\pi)}.$$

(ix) If $h(\ln ax \ln an) = \arctan[\frac{1}{(\ln ax \ln an)}]^\lambda, 0 < \alpha < \lambda, \alpha \le \min\{1, 2 - \lambda\}$, then, by the same way and Example 4.1(ix), we have

$$||T|| = ||\widetilde{T}|| = \frac{\pi}{2\alpha \cos(\frac{\pi\alpha}{2\lambda})}.$$

Note. For $a \le \frac{1}{2}$, setting $u(x) = \ln(x - a)(x \in (a + 1, \infty)), v(n) = \ln(n - a), n \ge n_0 = 2$, then $\frac{v'(y)}{v(y)} = \frac{1}{(y-a)\ln(y-a)}(> 0)$ is decreasing for $y > \frac{3}{2}$. In this case, still can obtain some similar results of Example 4.15, (1)-(3)(see Yang [137]).

By the same way, we still have

Corollary 4.3. *Let the assumptions of Theorem 4.2 be fulfilled and additionally, let $u(x)$ be a strictly increasing differentiable function in (b, c), with $u(b^+) = 0$ and $u(c^-) = \infty$. For $0 < p < 1$, $\frac{1}{p} + \frac{1}{q} = 1$, setting*

$$\widetilde{\Phi}(x) = \Phi(x)(1 - \theta(u(x)))(x \in (b, c)),$$

if $f(x)$, $a_n \geq 0$, $f \in L_{p,\widetilde{\Phi}}(b, c)$, $a = \{a_n\}_{n=n_0}^{\infty} \in l_{q,\Psi}, ||f||_{p,\widetilde{\Phi}} > 0, ||a||_{q,\Psi} > 0$, then we have the following equivalent inequalities:

$$\sum_{n=n_0}^{\infty} a_n \int_b^c h(u(x)v(n))f(x)dx$$

$$= \int_b^c f(x) \sum_{n=n_0}^{\infty} a_n h(u(x)v(n))dx > k(\alpha)||f||_{p,\widetilde{\Phi}}||a||_{q,\Psi}, \quad (4.44)$$

$$\left\{ \sum_{n=n_0}^{\infty} [\Psi(n)]^{1-p} \left[\int_b^c h(u(x)v(n))f(x)dx \right]^p \right\}^{\frac{1}{p}} > k(\alpha)||f||_{p,\widetilde{\Phi}}, \quad (4.45)$$

and

$$\left\{ \int_b^c [\widetilde{\Phi}(x)]^{1-q} \left[\sum_{n=n_0}^{\infty} h(u(x)v(n))a_n \right]^q dx \right\}^{\frac{1}{q}} > k(\alpha)||a||_{q,\Psi}. \quad (4.46)$$

Moreover, if $\frac{v'(y)}{v(y)}(> 0)$ is decreasing in $[n_0, \infty)$ and there exist constants $\eta, \delta_2 > 0$, such that $\theta(x) = O(x^\eta)(0 < x \leq 1)$ and $k(\alpha + \delta_2) \in \mathbf{R}_+$, then, the constant factor $k(\alpha)$ in the above inequalities is the best possible.

In view of Remark 4.1, Corollary 4.2 and Corollary 4.3, we still have

Corollary 4.4. *Suppose that $\alpha \in \mathbf{R}$, $h(t)$ is a non-negative finite measurable function in $\mathbf{R}_+$, $k(\alpha) = \int_0^\infty h(t)t^{\alpha-1}dt$, $v(y)$ is a strict increasing differential function in $[n_0, \infty)(n_0 \in \mathbf{N})$ with $v(n_0) > 0$ and $v(\infty) = \infty, u(x)$ is a strict increasing differential function in (b, c), with $u(b^+) = 0$ and $u(c^-) = \infty$, $\frac{v'(y)}{v(y)}(> 0)$ is decreasing in $[n_0, \infty)$, and there exist constants $\eta, \delta_0 > 0$, such that for any $\widetilde{\alpha} \in (\alpha - \delta_0, \alpha + \delta_0)$,*

$$0 < k(\widetilde{\alpha})(1 - \theta(x)) < \varpi_{\widetilde{\alpha}}(x) < k(\widetilde{\alpha}) < \infty \quad (x \in \mathbf{R}_+), \quad (4.47)$$

where $\theta(x) = O(x^\eta)$ $(x \in (0, 1])$.

(i) For $p > 1(p < 0)$, $\frac{1}{p} + \frac{1}{q} = 1$, $f(x)$, $a_n \geq 0$, $f \in L_{p,\Phi}(b, c)$, $a = \{a_n\}_{n=n_0}^{\infty} \in l_{q,\Psi}$, $||f||_{p,\Phi} > 0$, $||a||_{q,\Psi} > 0$, if $p > 1$, then, we have the equivalent inequalities (4.40), (4.41) and (4.42), with the best constant factor $k(\alpha)$; if $p < 0$, then we have the equivalent reverses of (4.40), (4.41) and (4.42), with the same best constant factor $k(\alpha)$.

(ii) For $0 < p < 1$, $\frac{1}{p} + \frac{1}{q} = 1$, if $f(x)$, $a_n \geq 0$, $f \in L_{p,\widetilde{\Phi}}(b,c)$, $a = \{a_n\}_{n=n_0}^{\infty} \in l_{q,\Psi}$, $||f||_{p,\widetilde{\Phi}} > 0$, $||a||_{q,\Psi} > 0$, then, we have the equivalent inequalities (4.44), (4.45) and (4.46) with the same best constant factor $k(\alpha)$.

For $v(x) = u(x) = x(x \in (0,\infty))$, $n_0 = 1$, $k(\alpha) = \int_0^{\infty} h(t)t^{\alpha-1}dt$ in Corollary 4.4, setting $\varphi(x) = x^{p(1-\alpha)-1}$ $(x \in (0,\infty))$, $\psi(n) = n^{q(1-\alpha)-1}(n \in \mathbf{N})$, we have

$$\omega(\alpha,n) = n^{\alpha} \int_0^{\infty} h(xn)x^{\alpha-1}dx = k(\alpha), \quad n \in \mathbf{N}, \qquad (4.48)$$

$$\varpi(\alpha,x) = x^{\alpha} \sum_{n=1}^{\infty} h(xn)n^{\alpha-1}, \quad x \in \mathbf{R}_+, \qquad (4.49)$$

and the following corollary:

Corollary 4.5. *Suppose that $\alpha \in \mathbf{R}$, $h(t)$ is a non-negative finite measurable function in $\mathbf{R}_+$, and there exist constants $\eta, \delta_0 > 0$, such that for any $\widetilde{\alpha} \in (\alpha - \delta_0, \alpha + \delta_0)$,*

$$0 < k(\widetilde{\alpha})(1 - \theta_{\widetilde{\alpha}}(x)) < \varpi(\widetilde{\alpha},x) < k(\widetilde{\alpha}) < \infty \quad (x \in \mathbf{R}_+), \qquad (4.50)$$

where $\theta_{\widetilde{\alpha}}(x) = O(x^{\eta})$ $(x \in (0,1])$.

(i) For $p > 1(p < 0)$, $\frac{1}{p} + \frac{1}{q} = 1$, $f(x)$, $a_n \geq 0$, $f \in L_{p,\varphi}(\mathbf{R}_+)$, $a = \{a_n\}_{n=1}^{\infty} \in l_{q,\psi}$, $||f||_{p,\varphi} > 0$, $||a||_{q,\psi} > 0$, if $p > 1$, then we have the following equivalent inequalities:

$$\sum_{n=1}^{\infty} a_n \int_0^{\infty} h(xn)f(x)dx$$

$$= \int_0^{\infty} f(x) \sum_{n=1}^{\infty} a_n h(xn)dx < k(\alpha)||f||_{p,\varphi}||a||_{q,\psi}, \qquad (4.51)$$

$$\left\{ \sum_{n=1}^{\infty} [\psi(n)]^{1-p} \left[\int_0^{\infty} h(xn)f(x)dx \right]^p \right\}^{\frac{1}{p}} < k(\alpha)||f||_{p,\varphi}, \qquad (4.52)$$

and

$$\left\{ \int_0^{\infty} [\varphi(x)]^{1-q} \left[\sum_{n=1}^{\infty} h(xn)a_n \right]^q dx \right\}^{\frac{1}{q}} < k(\alpha)||a||_{q,\psi}, \qquad (4.53)$$

with the best constant factor $k(\alpha)$; if $p < 0$, then we have the equivalent reverses of (4.51), (4.52) and (4.53), with the same best constant factor $k(\alpha)$.

(ii) For $0 < p < 1$, $\frac{1}{p} + \frac{1}{q} = 1$, setting

$$\widetilde{\varphi}(x) = (1 - \theta_\alpha(x))\varphi(x) \quad (x \in \mathbf{R}_+),$$

if $f(x), a_n \geq 0$, $f \in L_{p,\widetilde{\varphi}}(\mathbf{R}_+)$, $a = \{a_n\}_{n=1}^\infty \in l_{q,\psi}$, $\|f\|_{p,\widetilde{\varphi}} > 0$, $\|a\|_{q,\psi} > 0$, then, we have the following equivalent inequalities:

$$\sum_{n=1}^\infty a_n \int_0^\infty h(xn) f(x) dx$$

$$= \int_0^\infty f(x) \sum_{n=1}^\infty a_n h(xn) dx > k(\alpha) \|f\|_{p,\widetilde{\varphi}} \|a\|_{q,\psi}, \qquad (4.54)$$

$$\left\{ \sum_{n=1}^\infty [\psi(n)]^{1-p} \left[\int_0^\infty h(xn) f(x) dx \right]^p \right\}^{\frac{1}{p}} > k(\alpha) \|f\|_{p,\widetilde{\varphi}}, \qquad (4.55)$$

and

$$\left\{ \int_0^\infty [\widetilde{\varphi}(x)]^{1-q} \left[\sum_{n=1}^\infty h(xn) a_n \right]^q dx \right\}^{\frac{1}{q}} > k(\alpha) \|a\|_{q,\psi}, \qquad (4.56)$$

with the same best constant factor $k(\alpha)$.

Note. (i) If we change the upper limit ∞ of the integral to $c > 0$, then, the results of Corollary 4.5 is still value for $p > 1$ and $0 < p < 1$. For $p < 0$, we may refer to Corollary 4.1 for $v(n) = n, n_0 = 1$. (ii) We referred to some reverses with the particular kernels in Yang (see [166], [167], [168]).

4.4 Some Particular Examples

4.4.1 *Applying Condition (i) and Corollary 4.5*

Example 4.4. We set $h(xy) = \frac{1}{A(\max\{1,xy\})^\lambda + B(\min\{1,xy\})^\lambda}((x,y) \in \mathbf{R}_+^2; 0 < B \leq A, \lambda > 0, 0 < \alpha < \min\{1,\lambda\})$. There exists a constant $0 < \delta_0 < \min\{\alpha, 1-\alpha, \lambda-\alpha\}$, such that for $\widetilde{\alpha} \in (\alpha - \delta_0, \alpha + \delta_0) \subset (0,1)$, $\widetilde{\alpha} < \lambda$. For $x > 0$,

$$f(x,y) = \frac{x^{\widetilde{\alpha}} y^{\widetilde{\alpha}-1}}{A(\max\{1,xy\})^\lambda + B(\min\{1,xy\})^\lambda}$$

$$= \begin{cases} \frac{x^{\widetilde{\alpha}} y^{\widetilde{\alpha}-1}}{A + B(xy)^\lambda}, & 0 < y \leq \frac{1}{x}, \\ \frac{x^{\widetilde{\alpha}} y^{\widetilde{\alpha}-1}}{A(xy)^\lambda + B}, & y > \frac{1}{x} \end{cases}$$

is strictly decreasing with respect to $y > 0$. We find

$$k(\widetilde{\alpha}) = \int_0^\infty \frac{1}{A(\max\{1,t\})^\lambda + B(\min\{1,t\})^\lambda} t^{\widetilde{\alpha}-1}\, dt$$

$$= \int_0^1 \frac{1}{A+Bt^\lambda} t^{\widetilde{\alpha}-1}\, dt + \int_1^\infty \frac{1}{At^\lambda + B} t^{\widetilde{\alpha}-1}\, dt$$

$$= \frac{1}{A}\int_0^1 \frac{1}{1+\frac{B}{A}t^\lambda}(t^{\widetilde{\alpha}-1} + t^{\lambda-\widetilde{\alpha}-1})\, dt$$

$$= \frac{1}{A}\int_0^1 \sum_{k=0}^\infty \left(-\frac{B}{A}\right)^k t^{\lambda k}(t^{\widetilde{\alpha}-1} + t^{\lambda-\widetilde{\alpha}-1})\, dt$$

$$= \frac{1}{A}\sum_{k=0}^\infty \left(-\frac{B}{A}\right)^k \int_0^1 (t^{\lambda k+\widetilde{\alpha}-1} + t^{\lambda k+\lambda-\widetilde{\alpha}-1})\, dt$$

$$= \frac{1}{A}\sum_{k=0}^\infty \left(-\frac{B}{A}\right)^k \left(\frac{1}{\lambda k+\widetilde{\alpha}} + \frac{1}{\lambda k+\lambda-\widetilde{\alpha}}\right) \in \mathbf{R}_+,$$

and obtain

$$\varpi(\widetilde{\alpha}, x) = x^{\widetilde{\alpha}} \sum_{n=1}^\infty \frac{n^{\widetilde{\alpha}-1}}{A(\max\{1,xn\})^\lambda + B(\min\{1,xn\})^\lambda}$$

$$> x^{\widetilde{\alpha}} \int_1^\infty \frac{y^{\widetilde{\alpha}-1}}{A(\max\{1,xy\})^\lambda + B(\min\{1,xy\})^\lambda}\, dy$$

$$= k(\widetilde{\alpha})(1 - \theta_{\widetilde{\alpha}}(x)) > 0,$$

where

$$\theta_{\widetilde{\alpha}}(x) = \frac{x^{\widetilde{\alpha}}}{k(\widetilde{\alpha})} \int_0^1 \frac{y^{\widetilde{\alpha}-1}\, dy}{A(\max\{1,xy\})^\lambda + B(\min\{1,xy\})^\lambda} > 0.$$

For $0 < x \le 1$, it follows that

$$0 < \theta_{\widetilde{\alpha}}(x) = \frac{x^{\widetilde{\alpha}}}{k(\widetilde{\alpha})} \int_0^1 \frac{y^{\widetilde{\alpha}-1}}{A + B(xy)^\lambda}\, dy$$

$$\le \frac{x^{\widetilde{\alpha}}}{k(\widetilde{\alpha})} \int_0^1 \frac{y^{\widetilde{\alpha}-1}}{A}\, dy = \frac{x^{\widetilde{\alpha}}}{Ak(\widetilde{\alpha})\widetilde{\alpha}} \le \frac{x^\eta}{Ak(\widetilde{\alpha})\widetilde{\alpha}},$$

namely, $\theta_{\widetilde{\alpha}}(x) = O(x^\eta)(\eta = \alpha - \delta_0 > 0)$. By Condition (i), it follows that

$$\varpi(\widetilde{\alpha}, x) = \sum_{n=1}^\infty \frac{x^{\widetilde{\alpha}} n^{\widetilde{\alpha}-1}}{A(\max\{1,xn\})^\lambda + B(\min\{1,xn\})^\lambda} < k(\widetilde{\alpha}).$$

Hence, we obtain (4.50).

By Corollary 4.5, for $h(xn) = \frac{1}{A(\max\{1,xn\})^\lambda + B(\min\{1,xn\})^\lambda}(0 < B \le A, \lambda > 0, 0 < \alpha < 1)$, if $p > 1$, we have equivalent inequalities (4.51), (4.52) and (4.53); if $p < 0$, we have equivalent reverses of (4.51), (4.52) and (4.53); if $0 < p < 1$, we have equivalent inequalities (4.54), (4.55) and (4.56). All the inequalities have the same best constant factor:

$$k(\alpha) = \frac{\lambda}{A} \sum_{k=0}^{\infty} \left(-\frac{B}{A}\right)^k \frac{2k+1}{(\lambda k + \alpha)(\lambda k + \lambda - \alpha)}. \qquad (4.57)$$

Remark 4.3. For $c > 0$, since for $\delta \in (\alpha, \lambda)$, $t^\delta h(t) \to 0$ $(t \to \infty)$, there exists a constant $L > 0$, such that

$$h(t) = \frac{1}{A(\max\{1, t\})^\lambda + B(\min\{1, t\})^\lambda} \leq L\left(\frac{1}{t^\delta}\right) \quad (t \in [c, \infty)).$$

By Corollary 4.1, for $p < 0$, $v(n) = n, n_0 = 1$,

$$\widetilde{\theta}_\alpha(n) = \frac{1}{k(\alpha)} \int_{cn}^{\infty} h(t) t^{\alpha-1} dt,$$

we still have (4.30), (4.31) and (4.32) with the same best constant factor.

Example 4.5. We set $h(xy) = \frac{(\min\{1, xy\})^\lambda}{A(\max\{1, xy\})^\lambda + B(\min\{1, xy\})^\lambda}((x, y) \in \mathbf{R}_+^2; 0 < B \leq A, \lambda > 0, -\lambda < \alpha < \min\{\lambda, 1 - \lambda\})$. There exists $0 < \delta_0 < \min\{\lambda, \lambda + \alpha, \lambda - \alpha, 1 - \lambda + \alpha\}$, such that for $\widetilde{\alpha} \in (\alpha - \delta_0, \alpha + \delta_0) \subset (-\lambda, \lambda)$, $\widetilde{\alpha} < \alpha + (1 - \lambda + \alpha) = 1 - \lambda$. For $x > 0$,

$$f(x, y) = \frac{(\min\{1, xy\})^\lambda x^{\widetilde{\alpha}} y^{\widetilde{\alpha}-1}}{A(\max\{1, xy\})^\lambda + B(\min\{1, xy\})^\lambda}$$

$$= \begin{cases} \frac{x^{\lambda+\widetilde{\alpha}} y^{\lambda+\widetilde{\alpha}-1}}{A + B(xy)^\lambda}, & 0 < y \leq \frac{1}{x}, \\ \frac{x^{\widetilde{\alpha}} y^{\widetilde{\alpha}-1}}{A(xy)^\lambda + B}, & y > \frac{1}{x} \end{cases}$$

is strictly decreasing with respect to $y > 0$. We find

$$k(\widetilde{\alpha}) = \int_0^{\infty} \frac{(\min\{1, t\})^\lambda}{A(\max\{1, t\})^\lambda + B(\min\{1, t\})^\lambda} t^{\widetilde{\alpha}-1} dt$$

$$= \int_0^1 \frac{1}{A + Bt^\lambda} t^{\lambda+\widetilde{\alpha}-1} dt + \int_1^{\infty} \frac{1}{At^\lambda + B} t^{\widetilde{\alpha}-1} dt$$

$$= \frac{1}{A} \int_0^1 \frac{1}{1 + \frac{B}{A} t^\lambda} (t^{\lambda+\widetilde{\alpha}-1} + t^{\lambda-\widetilde{\alpha}-1}) dt$$

$$= \frac{1}{A} \int_0^1 \sum_{k=0}^{\infty} \left(-\frac{B}{A}\right)^k t^{\lambda k} (t^{\lambda+\widetilde{\alpha}-1} + t^{\lambda-\widetilde{\alpha}-1}) dt$$

$$= \frac{1}{A} \sum_{k=0}^{\infty} \left(-\frac{B}{A}\right)^k \int_0^1 (t^{\lambda k+\lambda+\widetilde{\alpha}-1} + t^{\lambda k+\lambda-\widetilde{\alpha}-1}) dt$$

$$= \frac{1}{A} \sum_{k=1}^{\infty} \left(-\frac{B}{A}\right)^{k-1} \left(\frac{1}{\lambda k + \widetilde{\alpha}} + \frac{1}{\lambda k - \widetilde{\alpha}}\right) \in \mathbf{R}_+,$$

and then we obtain

$$\varpi(\widetilde{\alpha}, x) = x^{\widetilde{\alpha}} \sum_{n=1}^{\infty} \frac{(\min\{1, xn\})^{\lambda} n^{\widetilde{\alpha}-1}}{A(\max\{1, xn\})^{\lambda} + B(\min\{1, xn\})^{\lambda}}$$

$$> x^{\widetilde{\alpha}} \int_{1}^{\infty} \frac{(\min\{1, xy\})^{\lambda} y^{\widetilde{\alpha}-1}}{A(\max\{1, xy\})^{\lambda} + B(\min\{1, xy\})^{\lambda}} dy$$

$$= k(\widetilde{\alpha})(1 - \theta_{\widetilde{\alpha}}(x)) > 0,$$

where

$$\theta_{\widetilde{\alpha}}(x) = \frac{x^{\widetilde{\alpha}}}{k(\widetilde{\alpha})} \int_{0}^{1} \frac{(\min\{1, xy\})^{\lambda} y^{\widetilde{\alpha}-1} dy}{A(\max\{1, xy\})^{\lambda} + B(\min\{1, xy\})^{\lambda}} > 0.$$

For $0 < x \leq 1$, it follows that

$$0 < \theta_{\widetilde{\alpha}}(x) = \frac{x^{\widetilde{\alpha}}}{k(\widetilde{\alpha})} \int_{0}^{1} \frac{x^{\lambda} y^{\lambda+\widetilde{\alpha}-1}}{A + B(xy)^{\lambda}} dy$$

$$\leq \frac{x^{\lambda+\widetilde{\alpha}}}{k(\widetilde{\alpha})} \int_{0}^{1} \frac{y^{\lambda+\widetilde{\alpha}-1}}{A} dy$$

$$= \frac{x^{\lambda+\widetilde{\alpha}}}{Ak(\widetilde{\alpha})(\lambda + \widetilde{\alpha})} \leq \frac{x^{\eta}}{Ak(\widetilde{\alpha})(\lambda + \widetilde{\alpha})},$$

namely, $\theta_{\widetilde{\alpha}}(x) = O(x^{\eta})$ $(\eta = \lambda + \alpha - \delta_0 > 0)$.

By Condition (i), it follows that

$$\varpi(\widetilde{\alpha}, x) = x^{\widetilde{\alpha}} \sum_{n=1}^{\infty} \frac{(\min\{1, xn\})^{\lambda} n^{\widetilde{\alpha}-1}}{A(\max\{1, xn\})^{\lambda} + B(\min\{1, xn\})^{\lambda}} < k(\widetilde{\alpha}).$$

Hence, we obtain (4.50).

By Corollary 4.5, for $h(xn) = \frac{(\min\{1, xn\})^{\lambda}}{A(\max\{1, xn\})^{\lambda} + B(\min\{1, xn\})^{\lambda}}$ $(0 < B \leq A, \lambda > 0, -\lambda < \alpha < \min\{\lambda, 1-\lambda\})$, if $p > 1$, we have equivalent inequalities (4.51), (4.52) and (4.53); if $p < 0$, we have equivalent reverses of (4.51), (4.52) and (4.53); if $0 < p < 1$, we have equivalent inequalities (4.54), (4.55) and (4.56). All the inequalities have the same best possible constant factor

$$k(\alpha) = \frac{2\lambda}{A} \sum_{k=1}^{\infty} \left(-\frac{B}{A}\right)^{k-1} \frac{k}{\lambda^2 k^2 - \alpha^2}. \tag{4.58}$$

Remark 4.4. For $c > 0$, since for $\delta \in (\alpha, \lambda)$, $t^{\delta} h(t) \to 0 (t \to \infty)$, there exists $L > 0$, such that

$$h(t) = \frac{(\min\{1, t\})^{\lambda}}{A(\max\{1, t\})^{\lambda} + B(\min\{1, t\})^{\lambda}}$$

$$\leq L \left(\frac{1}{t^{\delta}}\right) (t \in [c, \infty)).$$

By Corollary 4.1, for $p < 0, v(n) = n, n_0 = 1$,

$$\widetilde{\theta}_\alpha(n) = \frac{1}{k(\alpha)} \int_{cn}^{\infty} h(t)t^{\alpha-1}dt,$$

we still have (4.30), (4.31) and (4.32) with the same best constant factor.

Example 4.6. We set $h(xy) = \frac{(\min\{1,xy\})^\beta}{(\max\{1,xy\})^{\lambda+\beta}}((x,y) \in \mathbf{R}_+^2; \lambda + 2\beta > 0,$
$-\beta < \alpha < \min\{\lambda + \beta, 1 - \beta\})$. There exists a constant $0 < \delta_0 < \min\{\beta + \alpha, \lambda + \beta - \alpha, 1 - \beta - \alpha\}$, such that for $\widetilde{\alpha} \in (\alpha - \delta_0, \alpha + \delta_0) \subset (-\beta, \lambda + \beta)$, $\widetilde{\alpha} < 1 - \beta$. For $x > 0$,

$$f(x,y) = \frac{(\min\{1, xy\})^\beta x^{\widetilde{\alpha}} y^{\widetilde{\alpha}-1}}{(\max\{1, xy\})^{\lambda+\beta}}$$

$$= \begin{cases} x^{\widetilde{\alpha}+\beta} y^{\widetilde{\alpha}+\beta-1}, & 0 < y \le \frac{1}{x}, \\ x^{\widetilde{\alpha}-\lambda-\beta} y^{\widetilde{\alpha}-\lambda-\beta-1}, & y > \frac{1}{x} \end{cases}$$

is strictly decreasing with respect to $y > 0$. We find

$$k(\widetilde{\alpha}) = \int_0^\infty \frac{(\min\{t, 1\})^\beta}{(\max\{t, 1\})^{\lambda+\beta}} t^{\widetilde{\alpha}-1}dt$$

$$= \int_0^1 t^{\beta+\widetilde{\alpha}-1}dt + \int_1^\infty \frac{1}{t^{\lambda+\beta}} t^{\widetilde{\alpha}-1}dt$$

$$= \frac{1}{\beta + \widetilde{\alpha}} + \frac{1}{\lambda + \beta - \widetilde{\alpha}}$$

$$= \frac{\lambda + 2\beta}{(\beta + \widetilde{\alpha})(\lambda + \beta - \widetilde{\alpha})} \in \mathbf{R}_+,$$

and then, we have

$$\varpi(\widetilde{\alpha}, x) = x^{\widetilde{\alpha}} \sum_{n=1}^\infty \frac{(\min\{1, xn\})^\beta n^{\widetilde{\alpha}-1}}{(\max\{1, xn\})^{\lambda+\beta}}$$

$$> x^{\widetilde{\alpha}} \int_1^\infty \frac{(\min\{1, xy\})^\beta y^{\widetilde{\alpha}-1}}{(\max\{1, xy\})^{\lambda+\beta}} dy$$

$$= k(\widetilde{\alpha})(1 - \theta_{\widetilde{\alpha}}(x)) > 0,$$

where

$$\theta_{\widetilde{\alpha}}(x) = \frac{x^{\widetilde{\alpha}}}{k(\widetilde{\alpha})} \int_0^1 \frac{(\min\{1, xy\})^\beta y^{\widetilde{\alpha}-1}}{(\max\{1, xy\})^{\lambda+\beta}} dy > 0.$$

For $0 < x \le 1$, it follows that

$$0 < \theta_{\widetilde{\alpha}}(x) = \frac{x^{\widetilde{\alpha}+\beta}}{k(\widetilde{\alpha})} \int_0^1 y^{\widetilde{\alpha}+\beta-1}dy$$

$$= \frac{x^{\widetilde{\alpha}+\beta}}{k(\widetilde{\alpha})(\widetilde{\alpha} + \beta)} \le \frac{x^\eta}{k(\widetilde{\alpha})(\widetilde{\alpha} + \beta)},$$

namely, $\theta_{\widetilde{\alpha}}(x) = O(x^\eta)(\eta = \alpha + \beta - \delta_0 > 0)$.

By Condition (i), it follows that

$$\varpi(\widetilde{\alpha}, x) = x^{\widetilde{\alpha}} \sum_{n=1}^\infty \frac{(\min\{1, xn\})^\beta n^{\widetilde{\alpha}-1}}{(\max\{1, xn\})^{\lambda+\beta}} < k(\widetilde{\alpha}).$$

Hence, we obtain (4.50).

By Corollary 4.5 with the Note, for $h(xn) = \frac{(\min\{1, xn\})^\beta}{(\max\{1, xn\})^{\lambda+\beta}} (\lambda + 2\beta > 0, -\beta < \alpha < \min\{\lambda + \beta, 1 - \beta\})$, if $p > 1$, we have equivalent inequalities (4.51), (4.52) and (4.53); if $p < 0$, we have equivalent reverses of (4.51), (4.52) and (4.53); if $0 < p < 1$, we have equivalent inequalities (4.54), (4.55) and (4.56). All the inequalities have the same best constant factor

$$k(\alpha) = \frac{\lambda + 2\beta}{(\beta + \alpha)(\lambda + \beta - \alpha)}. \tag{4.59}$$

Remark 4.5. (1) In particular, (i) for $\beta = 0, h(xn) = \frac{1}{(\max\{1, xn\})^\lambda}(\lambda > 0, 0 < \alpha < \min\{\lambda, 1\})$; (ii) for $\lambda = -\beta, h(xn) = (\min\{1, xn\})^\beta$ $(\beta > 0, -\beta < \alpha < \min\{0, 1 - \beta\})$; (iii) for $\lambda = 0, h(xn) = (\frac{\min\{1, xn\}}{\max\{1, xn\}})^\beta$ $(\beta > 0, -\beta < \alpha < \min\{\beta, 1 - \beta\})$, we imply respectively some results of Example 4.2 (x)-(xii).

(2) For $c > 0$, since for $\delta \in (\alpha, \lambda + \beta)$, $t^\delta h(t) \to 0(t \to \infty)$, there exists a constant $L > 0$, such that

$$h(t) = \frac{(\min\{t, 1\})^\beta}{(\max\{t, 1\})^{\lambda+\beta}} \leq L\left(\frac{1}{t^\delta}\right) \quad (t \in [c, \infty)).$$

By Corollary 4.1, for $p < 0$, $v(n) = n$, $n_0 = 1$,

$$\widetilde{\theta}_\alpha(n) = \frac{1}{k(\alpha)} \int_{cn}^\infty h(t) t^{\alpha-1} dt,$$

we still have (4.30), (4.31) and (4.32) with the same best constant factor.

4.4.2 *Applying Condition (iii) and Corollary 4.2*

Example 4.7. We set $h(xy) = \frac{1}{1+x^\lambda y^\lambda}((x, y) \in \mathbf{R}_+^2; \lambda > 0, 0 < \alpha < \min\{2, \lambda\})$. There exists a constant $0 < \delta_0 < \min\{\alpha, \lambda - \alpha, 2 - \alpha\}$, such that for $\widetilde{\alpha} \in (\alpha - \delta_0, \alpha + \delta_0) \subset (0, 2)$, $\widetilde{\alpha} < \lambda$. It follows

$$k(\widetilde{\alpha}) = \int_0^\infty \frac{t^{\widetilde{\alpha}-1}}{1+t^\lambda} dt = \frac{1}{\lambda} \int_0^\infty \frac{u^{(\widetilde{\alpha}/\lambda)-1}}{u+1} du$$

$$= \frac{\pi}{\lambda \sin(\pi\widetilde{\alpha}/\lambda)} \in \mathbf{R}_+.$$

For $\gamma < 1$, it is obvious that
$$\omega(\widetilde{\alpha}, n) = (n - \gamma)^{\widetilde{\alpha}} \int_0^\infty \frac{x^{\widetilde{\alpha}-1}dx}{1 + x^\lambda(n-\gamma)^\lambda} = k(\widetilde{\alpha}) \qquad (n \in \mathbf{N}).$$
In the following, we show that, for
$$\gamma \le \gamma(\widetilde{\alpha}) = 1 - \frac{1}{4}(\widetilde{\alpha} + \sqrt{2(\lambda+1)\widetilde{\alpha} - \widetilde{\alpha}^2}) < 1,$$
it follows that
$$0 < k(\widetilde{\alpha})(1 - \theta_{\widetilde{\alpha}}(x))$$
$$< \varpi(\widetilde{\alpha}, x) = \sum_{n=1}^\infty \frac{x^{\widetilde{\alpha}}(n-\gamma)^{\widetilde{\alpha}-1}}{1 + x^\lambda(n-\gamma)^\lambda} < k(\widetilde{\alpha}) \ (x > 0), \qquad (4.60)$$
where
$$0 < \theta_{\widetilde{\alpha}}(x) = O(x^\eta)(0 < x \le 1; \eta = \alpha - \delta_0 > 0). \qquad (4.61)$$
(1) Setting $g(x, y) = \frac{(y-\gamma)^{\widetilde{\alpha}-1}}{1+x^\lambda(y-\gamma)^\lambda}(x > 0, y > \gamma)$, we prove that
$$\widetilde{R}(x) = x^{\widetilde{\alpha}}\left[\int_\gamma^1 g(x,y)dy - \frac{1}{2}g(x,1) - \int_1^\infty P_1(y)g_y'(x,y)dy\right] > 0. \quad (4.62)$$
In fact, we obtain
$$\int_\gamma^1 g(x,y)dy = \int_\gamma^1 \frac{(y-\gamma)^{\widetilde{\alpha}-1}dy}{1+x^\lambda(y-\gamma)^\lambda} = \frac{1}{x^{\widetilde{\alpha}}}\int_0^{(1-\gamma)x} \frac{t^{\widetilde{\alpha}-1}dt}{1+t^\lambda}$$
$$-\frac{1}{2}g(x,1) = -\frac{1}{2}\frac{(1-\gamma)^{\widetilde{\alpha}-1}}{1+x^\lambda(1-\gamma)^\lambda}, \qquad (4.63)$$
and
$$g_y'(x,y) = -\frac{\lambda x^\lambda(y-\gamma)^{\widetilde{\alpha}+\lambda-2}}{[1+x^\lambda(y-\gamma)^\lambda]^2} + \frac{(\widetilde{\alpha}-1)(y-\gamma)^{\widetilde{\alpha}-2}}{1+x^\lambda(y-\gamma)^\lambda}$$
$$= -\frac{\lambda[1+x^\lambda(y-\gamma)^\lambda - 1](y-\gamma)^{\widetilde{\alpha}-2}}{[1+x^\lambda(y-\gamma)^\lambda]^2}$$
$$+ \frac{(\widetilde{\alpha}-1)(y-\gamma)^{\widetilde{\alpha}-2}}{1+x^\lambda(y-\gamma)^\lambda}$$
$$= -\frac{(\lambda-\widetilde{\alpha}+1)(y-\gamma)^{\widetilde{\alpha}-2}}{1+x^\lambda(y-\gamma)^\lambda} + \frac{\lambda(y-\gamma)^{\widetilde{\alpha}-2}}{[1+x^\lambda(y-\gamma)^\lambda]^2}.$$
By (2.37), for $\lambda > 0, \widetilde{\alpha} < 2$,
$$-\int_1^\infty P_1(y)g_y'(x,y)dy$$
$$= \int_1^\infty P_1(y)\frac{(\lambda-\widetilde{\alpha}+1)(y-\gamma)^{\widetilde{\alpha}-2}}{1+x^\lambda(y-\gamma)^\lambda}dy$$
$$- \int_1^\infty P_1(y)\frac{\lambda(y-\gamma)^{\widetilde{\alpha}-2}}{[1+x^\lambda(y-\gamma)^\lambda]^2}dy$$
$$> -\frac{1}{8}\frac{(\lambda-\widetilde{\alpha}+1)(1-\gamma)^{\widetilde{\alpha}-2}}{1+x^\lambda(1-\gamma)^\lambda} + 0. \qquad (4.64)$$

Then, by (4.62), (4.63) and (4.64), setting

$$A = [\frac{1-\gamma}{2} + \frac{1}{8}(\lambda - \widetilde{\alpha} + 1)](1-\gamma)^{\widetilde{\alpha}-2},$$

we have

$$\widetilde{R}(x) > h(x) = \int_0^{(1-\gamma)x} \frac{t^{\widetilde{\alpha}-1}dt}{1+t^\lambda} - \frac{Ax^{\widetilde{\alpha}}}{1 + x^\lambda(1-\gamma)^\lambda}. \qquad (4.65)$$

Since for $\gamma \le \gamma(\widetilde{\alpha}), 1 - \gamma \ge \frac{1}{4}[\widetilde{\alpha} + \sqrt{2(\lambda+1)\widetilde{\alpha} - \widetilde{\alpha}^2}]$, we have

$$B(\gamma) = (1-\gamma)^2 - \frac{\widetilde{\alpha}}{2}(1-\gamma) - \frac{\widetilde{\alpha}}{8}(\lambda - \widetilde{\alpha} + 1) \ge 0,$$

then, we find

$$h'(x) = \frac{(1-\gamma)^{\widetilde{\alpha}}x^{\widetilde{\alpha}-1}}{1 + x^\lambda(1-\gamma)^\lambda} + \frac{\lambda(1-\gamma)^\lambda Ax^{\widetilde{\alpha}+\lambda-1}}{[1 + x^\lambda(1-\gamma)^\lambda]^2}$$
$$- \frac{A\widetilde{\alpha}x^{\widetilde{\alpha}-1}}{1 + x^\lambda(1-\gamma)^\lambda}$$
$$= [(1-\gamma)^{\widetilde{\alpha}} - A\widetilde{\alpha}]\frac{x^{\widetilde{\alpha}-1}}{1 + x^\lambda(1-\gamma)^\lambda}$$
$$+ \frac{\lambda(1-\gamma)^\lambda Ax^{\widetilde{\alpha}+\lambda-1}}{[1 + x^\lambda(1-\gamma)^\lambda]^2}$$
$$> B(\gamma)\frac{(1-\gamma)^{\widetilde{\alpha}-2}x^{\widetilde{\alpha}-1}}{1 + x^\lambda(1-\gamma)^\lambda} > 0 \qquad (A > 0).$$

By (4.65), it follows that $\widetilde{R}(x) > h(0) = 0(x > 0)$. Hence, (4.62) holds.

(2) In view of $f(x, y) = x^{\widetilde{\alpha}}g(x, y)$, by (4.4) and Condition (iii), it follows that

$$\varpi(\widetilde{\alpha}, x) = \sum_{n=1}^\infty \frac{x^{\widetilde{\alpha}}(n-\gamma)^{\widetilde{\alpha}-1}}{1 + x^\lambda(n-\gamma)^\lambda}$$
$$= k(\widetilde{\alpha}) - \widetilde{R}(x) < k(\widetilde{\alpha})(x > 0).$$

By (4.64), we still obtain

$$-\int_1^\infty P_1(y)g_y'(x, y)dy < \frac{1}{8}\frac{\lambda(1-\gamma)^{\widetilde{\alpha}-2}}{[1 + x^\lambda(1-\gamma)^\lambda]^2},$$

and then, by (4.62), it follows that

$$\widetilde{R}(x) < \int_0^{(1-\gamma)x} \frac{t^{\widetilde{\alpha}-1}}{1+t^\lambda}dt - \frac{x^{\widetilde{\alpha}}(1-\gamma)^{\widetilde{\alpha}-1}}{2[1 + x^\lambda(1-\gamma)^\lambda]}$$
$$+ \frac{\lambda x^{\widetilde{\alpha}}(1-\gamma)^{\widetilde{\alpha}-2}}{8[1 + x^\lambda(1-\gamma)^\lambda]^2}. \qquad (4.66)$$

Setting $\theta_{\widetilde{\alpha}}(x) = \frac{1}{k(\widetilde{\alpha})}[\widetilde{R}(x) + \frac{x^{\widetilde{\alpha}}(1-\gamma)^{\widetilde{\alpha}-1}}{1+x^{\lambda}(1-\gamma)^{\lambda}}](>0)$, we find

$$\varpi(\widetilde{\alpha}, x) > \varpi(\widetilde{\alpha}, x) - \frac{x^{\widetilde{\alpha}}(1-\gamma)^{\widetilde{\alpha}-1}}{1+x^{\lambda}(1-\gamma)^{\lambda}}$$
$$= k(\widetilde{\alpha})(1 - \theta_{\widetilde{\alpha}}(x)) > 0.$$

For $\eta = \alpha - \delta_0 > 0, 0 < x \leq 1$, by (4.66), we obtain

$$0 < \frac{1}{x^{\eta}}\theta_{\widetilde{\alpha}}(x) \leq \frac{1}{x^{\widetilde{\alpha}}}\theta_{\widetilde{\alpha}}(x)$$
$$= \frac{1}{k(\widetilde{\alpha})}\left[\frac{1}{x^{\widetilde{\alpha}}}\widetilde{R}(x) + \frac{(1-\gamma)^{\widetilde{\alpha}-1}}{1+x^{\lambda}(1-\gamma)^{\lambda}}\right]$$
$$< \frac{1}{k(\widetilde{\alpha})}\left\{\frac{1}{x^{\widetilde{\alpha}}}\int_0^{(1-\gamma)x}\frac{t^{\widetilde{\alpha}-1}dt}{1+t^{\lambda}} - \frac{(1-\gamma)^{\widetilde{\alpha}-1}}{2[1+x^{\lambda}(1-\gamma)^{\lambda}]}\right.$$
$$\left. + \frac{\lambda(1-\gamma)^{\widetilde{\alpha}-2}}{8[1+x^{\lambda}(1-\gamma)^{\lambda}]^2} + \frac{(1-\gamma)^{\widetilde{\alpha}-1}}{1+x^{\lambda}(1-\gamma)^{\lambda}}\right\}$$
$$\to \text{ constant as } x \to 0,$$

namely, $\theta_{\widetilde{\alpha}}(x) = O(x^{\eta})(0 < x \leq 1; \eta = \alpha - \delta_0 > 0)$. Hence, we show that (4.60) and (4.61) are valid.

Setting $u(x) = x - \mu(x > \mu)$, by Corollary 4.2 for

$$\gamma \leq 1 - \frac{1}{4}[\alpha + \sqrt{2(\lambda+1)\alpha - \alpha^2}],$$

$$h(u(x)v(n)) = \frac{1}{1 + (x-\mu)^{\lambda}(n-\gamma)^{\lambda}}$$

$(x > \beta\mu, n \in \mathbf{N}; \lambda > 0, 0 < \alpha < \min\{2,\lambda\})$, if $p > 1$, we have equivalent inequalities (4.40), (4.41) and (4.42); if $p < 0$, we have equivalent reverses of (4.40), (4.41) and (4.42); if $0 < p < 1$, we have equivalent inequalities (4.44), (4.45) and (4.46). All the inequalities have the same best constant factor

$$k(\alpha) = \frac{\pi}{\lambda\sin(\pi\alpha/\lambda)}. \tag{4.67}$$

Example 4.8. We set $h(xy) = \frac{(\min\{1,xy\})^{\beta}}{(\max\{1,xy\})^{\lambda+\beta}}((x,y) \in \mathbf{R}_+^2; 2\beta + \lambda > 0,$ $-\beta < \alpha < \min\{\lambda+\beta, 2-\beta\})$. There exists $0 < \delta_0 < \min\{\alpha+\beta, \lambda+\beta-\alpha, 2-\beta-\alpha\}$, such that for $\widetilde{\alpha} \in (\alpha-\delta_0, \alpha+\delta_0) \subset (-\beta, \lambda+\beta), \widetilde{\alpha} < 2 - \beta$. It follows that

$$k(\widetilde{\alpha}) = \int_0^{\infty}\frac{(\min\{1,t\})^{\beta}t^{\widetilde{\alpha}-1}}{(\max\{1,t\})^{\lambda+\beta}}dt$$
$$= \frac{\lambda+2\beta}{(\beta+\widetilde{\alpha})(\lambda+\beta-\widetilde{\alpha})} \in \mathbf{R}_+.$$

For $0 \leq \gamma < 1$, it is obvious that

$$\omega(\widetilde{\alpha}, n) = (n - \gamma)^{\widetilde{\alpha}} \int_0^\infty \frac{(\min\{1, x(n - \gamma)\})^\beta x^{\widetilde{\alpha}-1}}{(\max\{1, x(n - \gamma)\})^{\lambda+\beta}} dx$$

$$= k(\widetilde{\alpha})(n \in \mathbf{N}).$$

In the following, we show that, for

$$\gamma \leq \gamma(\widetilde{\alpha}) = 1 - \frac{1}{4}[\beta + \widetilde{\alpha} + \sqrt{2(\beta + \widetilde{\alpha})(1 + \lambda + 2\beta) - (\beta + \widetilde{\alpha})^2}],$$

$$0 < k(\widetilde{\alpha})(1 - \theta_{\widetilde{\alpha}}(x)) < \varpi(\widetilde{\alpha}, x)$$

$$= \sum_{n=1}^\infty \frac{x^{\widetilde{\alpha}}(\min\{1, x(n - \gamma)\})^\beta}{(\max\{1, x(n - \gamma)\})^{\lambda+\beta}}(n - \gamma)^{\widetilde{\alpha}-1} < k(\widetilde{\alpha}) \ (x > 0), \quad (4.68)$$

where

$$0 < \theta_{\widetilde{\alpha}}(x) = O(x^\eta) \quad (0 < x \leq 1; \ \eta = \alpha + \beta - \delta_0 > 0). \quad (4.69)$$

(1) Setting

$$g(x, y) = \frac{(\min\{1, x(y - \gamma)\})^\beta}{(\max\{1, x(y - \gamma)\})^{\lambda+\beta}}(y - \gamma)^{\widetilde{\alpha}-1} \quad (x > 0, y > \gamma),$$

in the following, we prove that, for $\gamma \leq \gamma(\widetilde{\alpha})$,

$$\widetilde{R}(x) = x^{\widetilde{\alpha}} \left[\int_\gamma^1 g(x, y)dy - \frac{1}{2}g(x, 1) - \int_1^\infty P_1(y)g_y'(x, y)dy \right] > 0. \quad (4.70)$$

In fact, we obtain

$$g(x, y) = \begin{cases} x^\beta (y - \gamma)^{\widetilde{\alpha}+\beta-1}, & \gamma < y < \frac{1}{x} + \gamma, \\ x^{-\lambda-\beta}(y - \gamma)^{\widetilde{\alpha}-\lambda-\beta-1}, & y \geq \frac{1}{x} + \gamma, \end{cases}$$

$$-g_y'(x, y) = \begin{cases} h_1(x, y), \gamma < y < \frac{1}{x} + \gamma, \\ h_2(x, y), \quad y \geq \frac{1}{x} + \gamma, \end{cases}$$

$$h_1(x, y) = (1 - \widetilde{\alpha} - \beta)x^\beta (y - \gamma)^{\widetilde{\alpha}+\beta-2},$$

$$h_2(x, y) = (1 - \widetilde{\alpha} + \lambda + \beta)x^{-\lambda-\beta}(y - \gamma)^{\widetilde{\alpha}-\lambda-\beta-2}.$$

For $0 < \frac{1}{x} \leq 1 - \gamma, y \geq 1 \geq \frac{1}{x} + \gamma$, by equation (2.23) in Chapter 2, we have

$$-\int_1^\infty P_1(y)g_y'(x, y)dy = \int_1^\infty P_1(y)h_2(x, y)dy$$

$$> -\frac{1}{12x^{\lambda+\beta}}(1 - \widetilde{\alpha} + \lambda + \beta)(1 - \gamma)^{\widetilde{\alpha}-\lambda-\beta-2}$$

$$> -\frac{1}{8x^{\lambda+\beta}}(1 - \widetilde{\alpha} + \lambda + \beta)(1 - \gamma)^{\widetilde{\alpha}-\lambda-\beta-2}; \quad (4.71)$$

for $\frac{1}{x} > 1 - \gamma$, by (2.47) ($\varepsilon_0 \in (0,1), \varepsilon_1 \in [0,1]$), we find

$$
\begin{aligned}
&-\int_1^\infty P_1(y) g_y'(x,y) dy \\
&= \frac{\varepsilon_0}{8} \left(-h_1(x,1) + h_1(x, \frac{1}{x} + \gamma) - h_2(x, \frac{1}{x} + \gamma) \right) \\
&\quad - \frac{\varepsilon_1}{8} \left(h_1(x, \frac{1}{x} + \gamma) - h_2(x, \frac{1}{x} + \gamma) \right) \\
&= -\frac{\varepsilon_0}{8} \left[(1 - \widetilde{\alpha} - \beta) x^\beta (1 - \gamma)^{\widetilde{\alpha} + \beta - 2} + \frac{2\beta + \lambda}{x^{\widetilde{\alpha} - 2}} \right] + \frac{\varepsilon_1}{8} \frac{2\beta + \lambda}{x^{\widetilde{\alpha} - 2}} \\
&> -\frac{1}{8}(1 - \widetilde{\alpha} - \beta) x^\beta (1 - \gamma)^{\widetilde{\alpha} + \beta - 2} - \frac{2\beta + \lambda}{8 x^{\widetilde{\alpha} - 2}}. \quad\quad (4.72)
\end{aligned}
$$

We still find

$$
-\frac{1}{2} g(x,1) = \begin{cases} -\frac{1}{2} x^\beta (1 - \gamma)^{\widetilde{\alpha} + \beta - 1}, & \frac{1}{x} > 1 - \gamma, \\ -\frac{1}{2} x^{-\lambda - \beta} (1 - \gamma)^{\widetilde{\alpha} - \lambda - \beta - 1}, & 0 < \frac{1}{x} \le 1 - \gamma. \end{cases}
$$

For $0 < \frac{1}{x} \le 1 - \gamma$, we have

$$
\begin{aligned}
&\int_\gamma^1 g(x,y) dy \\
&= \int_\gamma^{\frac{1}{x} + \gamma} x^\beta (y - \gamma)^{\widetilde{\alpha} + \beta - 1} dy + \int_{\gamma + \frac{1}{x}}^1 \frac{(y - \gamma)^{\widetilde{\alpha} - \lambda - \beta - 1}}{x^{\lambda + \beta}} dy \\
&= \frac{1}{(\beta + \widetilde{\alpha}) x^{\widetilde{\alpha}}} + \frac{1}{(\beta + \lambda - \widetilde{\alpha}) x^{\widetilde{\alpha}}} - \frac{(1 - \gamma)^{\widetilde{\alpha} - \lambda - \beta}}{(\beta + \lambda - \widetilde{\alpha}) x^{\lambda + \beta}}; \quad\quad (4.73)
\end{aligned}
$$

for $\frac{1}{x} > 1 - \gamma$,

$$
\int_\gamma^1 g(x,y) dy = \int_\gamma^1 x^\beta (y - \gamma)^{\widetilde{\alpha} + \beta - 1} dy = \frac{(1 - \gamma)^{\widetilde{\alpha} + \beta} x^\beta}{\widetilde{\alpha} + \beta}. \quad\quad (4.74)
$$

(i) For $0 < \frac{1}{x} \le 1 - \gamma$, $\quad \gamma \le \gamma(\widetilde{\alpha})$,

$$
1 - \gamma \ge \frac{1}{4}[\beta + \widetilde{\alpha} + \sqrt{2(\beta + \widetilde{\alpha})(1 + \lambda + 2\beta) - (\beta + \widetilde{\alpha})^2}],
$$

it is obvious that

$$
A(\gamma) = (1 - \gamma)^2 - \frac{\beta + \widetilde{\alpha}}{2}(1 - \gamma) - \frac{1}{8}(\beta + \widetilde{\alpha})(1 + \lambda + \beta - \widetilde{\alpha}) \ge 0,
$$

and then, by (4.70), (4.71), (4.72) and (4.73), in view of $0 < \frac{1}{x} \le 1 - \gamma$, it follows that

$$
\begin{aligned}
\widetilde{R}(x) &> \frac{1}{\beta + \widetilde{\alpha}} + \frac{1}{\beta + \lambda - \widetilde{\alpha}} - \frac{(1-\gamma)^{\widetilde{\alpha}-\lambda-\beta}}{(\beta+\lambda-\widetilde{\alpha})x^{\lambda+\beta-\widetilde{\alpha}}} \\
&\quad - \frac{(1-\gamma)^{\widetilde{\alpha}-\lambda-\beta-1}}{2x^{\lambda+\beta-\widetilde{\alpha}}} - \frac{(1-\widetilde{\alpha}+\lambda+\beta)(1-\gamma)^{\widetilde{\alpha}-\lambda-\beta-2}}{8x^{\lambda+\beta-\widetilde{\alpha}}} \\
&> \frac{1}{\beta+\widetilde{\alpha}} + \frac{1}{\beta+\lambda-\widetilde{\alpha}} - \frac{(1-\gamma)^{\widetilde{\alpha}-\lambda-\beta}(1-\gamma)^{\lambda+\beta-\widetilde{\alpha}}}{\beta+\lambda-\widetilde{\alpha}} \\
&\quad - \frac{1}{2}(1-\gamma)^{\lambda+\beta-\widetilde{\alpha}}(1-\gamma)^{\widetilde{\alpha}-\lambda-\beta-1} \\
&\quad - \frac{1}{8}(1-\widetilde{\alpha}+\lambda+\beta)(1-\gamma)^{\widetilde{\alpha}-\lambda-\beta-2}(1-\gamma)^{\lambda+\beta-\widetilde{\alpha}} \\
&= \frac{A(\gamma)}{(\beta+\widetilde{\alpha})(1-\gamma)^2} \ge 0;
\end{aligned}
$$

(ii) for $\frac{1}{x} > 1 - \gamma$, $\gamma \le \gamma(\widetilde{\alpha})$, in view of $x^2 < x^{\beta+\widetilde{\alpha}}(1-\gamma)^{\widetilde{\alpha}+\beta-2}$, we still find

$$
\begin{aligned}
\widetilde{R}(x) &> \frac{(1-\gamma)^{\widetilde{\alpha}+\beta}x^{\beta+\widetilde{\alpha}}}{\widetilde{\alpha}+\beta} - \frac{1}{2}x^{\beta+\widetilde{\alpha}}(1-\gamma)^{\widetilde{\alpha}+\beta-1} \\
&\quad - \frac{1}{8}(1-\widetilde{\alpha}-\beta)x^{\beta+\widetilde{\alpha}}(1-\gamma)^{\widetilde{\alpha}+\beta-2} \\
&\quad - \frac{2\beta+\lambda}{8}x^{\beta+\widetilde{\alpha}}(1-\gamma)^{\widetilde{\alpha}+\beta-2} \\
&= A(\gamma)\frac{x^{\beta+\widetilde{\alpha}}(1-\gamma)^{\widetilde{\alpha}+\beta-2}}{\widetilde{\alpha}+\beta} \ge 0.
\end{aligned}
$$

Hence, (4.70) follows for $x > 0$.

(2) By (4.4) and Condition (iii), it follows that

$$
\varpi(\widetilde{\alpha}, x) = k(\widetilde{\alpha}) - \widetilde{R}(x) < k(\widetilde{\alpha}) \qquad (x > 0).
$$

Setting

$$
\theta_{\widetilde{\alpha}}(x) = \frac{1}{k(\widetilde{\alpha})}\left[\widetilde{R}(x) + \frac{x^{\widetilde{\alpha}}(\min\{1, x(1-\gamma)\})^{\beta}}{(\max\{1, x(1-\gamma)\})^{\lambda+\beta}}(1-\gamma)^{\widetilde{\alpha}-1}\right] > 0,
$$

we obtain

$$
\begin{aligned}
\varpi(\widetilde{\alpha}, x) &> \varpi(\widetilde{\alpha}, x) - \frac{x^{\widetilde{\alpha}}(\min\{1, x(1-\gamma)\})^{\beta}}{(\max\{1, x(1-\gamma)\})^{\lambda+\beta}}(1-\gamma)^{\widetilde{\alpha}-1} \\
&= k(\widetilde{\alpha})(1 - \theta_{\widetilde{\alpha}}(x)) > 0.
\end{aligned}
$$

For $0 < x \le 1$, it is obvious that $\frac{1}{x^{\tilde{\alpha}+\beta}}\theta_{\tilde{\alpha}}(x)$ is continuous in $(0,1]$. Since

$$\lim_{x\to 0^+} \frac{1}{x^{\tilde{\alpha}+\beta}} \frac{x^{\tilde{\alpha}}(\min\{1, x(1-\gamma)\})^{\beta}}{(\max\{1, x(1-\gamma)\})^{\lambda+\beta}}(1-\gamma)^{\tilde{\alpha}-1} = (1-\gamma)^{\tilde{\alpha}+\beta-1},$$

$$\lim_{x\to 0^+} \frac{x^{\tilde{\alpha}}}{x^{\tilde{\alpha}+\beta}} \int_{\gamma}^{1} g(x,y)dy$$

$$= \lim_{x\to 0^+} \frac{(1-\gamma)^{\tilde{\alpha}+\beta}x^{\beta}}{x^{\beta}(\tilde{\alpha}+\beta)} = \frac{(1-\gamma)^{\tilde{\alpha}+\beta}}{\tilde{\alpha}+\beta},$$

and

$$\lim_{x\to 0^+} \frac{x^{\tilde{\alpha}}}{x^{\tilde{\alpha}+\beta}}\left(-\int_{1}^{\infty} P_1(y)g_y'(x,y)dy\right)$$

$$= \lim_{x\to 0^+} \frac{1}{x^{\beta}}\left\{-\frac{\varepsilon_0}{8}\left[(1-\tilde{\alpha}-\beta)x^{\beta}(1-\gamma)^{\tilde{\alpha}+\beta-2} + \frac{2\beta+\lambda}{x^{\tilde{\alpha}-2}}\right]\right.$$

$$\left.+\frac{\varepsilon_1(2\beta+\lambda)}{8x^{\tilde{\alpha}-2}}\right\}$$

$$= -\frac{\varepsilon_0}{8}(1-\tilde{\alpha}-\beta)(1-\gamma)^{\tilde{\alpha}+\beta-2},$$

then, by (4.70), we find

$$0 < \frac{1}{x^{\eta}}\theta_{\tilde{\alpha}}(x) \le \frac{1}{x^{\tilde{\alpha}+\beta}}\theta_{\tilde{\alpha}}(x) \to \text{constant, as } x \to 0^+,$$

namely, $\theta_{\tilde{\alpha}}(x) = O(x^{\eta})$ $(0 < x \le 1; \eta = \alpha + \beta - \delta_0 > 0)$.

Hence, we show that (4.68) and (4.69) are valid.

Setting $u(x) = x - \mu (x > \mu)$, by Corollary 4.2, for

$$\gamma \le 1 - \frac{1}{4}\left[\beta + \alpha + \sqrt{2(\beta+\alpha)(1+\lambda+2\beta) - (\beta+\alpha)^2}\right],$$

$$h(u(x)v(n)) = \frac{(\min\{1, (x-\mu)(n-\gamma)\})^{\beta}}{(\max\{1, (x-\gamma)(n-\gamma)\})^{\lambda+\beta}}$$

$(x > \mu, n \in \mathbf{N}; 2\beta + \lambda > 0, -\beta < \alpha < \min\{\lambda+\beta, 2-\beta\})$, if $p > 1$, we have equivalent inequalities (4.40), (4.41) and (4.42); if $p < 0$, we have equivalent reverses of (4.40), (4.41) and (4.42); if $0 < p < 1$, we have equivalent inequalities (4.44), (4.45) and (4.46). All the inequalities have the same best constant factor

$$k(\alpha) = \frac{\lambda+2\beta}{(\beta+\alpha)(\lambda+\beta-\alpha)}. \tag{4.75}$$

Remark 4.6. If we add the condition $2\beta + \lambda \le 1$ in the above example, then it follows that $\gamma(\tilde{\alpha}) > 0$. Setting $y = \beta + \tilde{\alpha} \in (0,2)$, then, $\gamma(\tilde{\alpha}) > 0$ is equivalent to

$$G(y) = y^2 - (5 + \lambda + 2\beta)y + 8 > 0.$$

Since $G'(y) = 2y - (5+\lambda+2\beta)$ is increasing with $G'(2) = 4 - (5+\lambda+2\beta) < 0$, and $G'(y) < 0 (y \in [0,2])$, then, we find

$$G(y) > G(2) = 2[1 - (\lambda + 2\beta)] \geq 0.$$

In this case, for $\gamma = 0 \ (< \gamma(\widetilde{\alpha}))$, we can get some results of Example 4.7.

Chapter 5

Multi-dimensional Half-Discrete Hilbert-Type Inequalities

"... in a subject (inequalities) like this, which has applications in every part of mathematics but never been developed systematically."

G. H. Hardy

"A great discovery solves a great problem but there is a grain of discovery in the solution of any problem. Your problem may be modest; but if it challenges your curiosity and brings into play your incentive faculties, and if you solve it by your own means, you may experience the tension and enjoy the triumph of discovery."

George Polya

5.1 Introduction

This chapter deals with two kinds of multi-dimensional half-discrete Hilbert-type inequalities using the way of weight functions and techniques of real analysis. These inequalities are extensions of the two-dimensional cases studied in Chapters 3 and 4. The best possible constant factors are proved. Included are equivalent forms, the operator expressions, the reverses and many particular examples.

5.2   Some Preliminary Results and Lemmas

5.2.1   *Some Related Lemmas*

Lemma 5.1. *(see Wang and Guo [73]) If $m \in \mathbf{N}$, α, $M > 0$, $\Psi(u)$ is a non-negative measurable function in $(0, 1]$, and*

$$D_M = \left\{ x \in \mathbf{R}_+^m; \sum_{i=1}^m x_i^\alpha \le M^\alpha \right\},$$

then, we have

$$\int \cdots \int_{D_M} \Psi\left(\sum_{i=1}^m \left(\frac{x_i}{M} \right)^\alpha \right) dx_1 \cdots dx_m = \frac{M^m \Gamma^m \left(\frac{1}{\alpha} \right)}{\alpha^m \Gamma \left(\frac{m}{\alpha} \right)} \int_0^1 \Psi(u) u^{\frac{m}{\alpha} - 1} du. \tag{5.1}$$

Lemma 5.2. *Suppose that $m, s, n^{(0)} \in \mathbf{N}, \alpha, \beta > 0$, $v(t)$ is an increasing differentiable function in $[n^{(0)}, \infty)$ with $v(n^{(0)}) > 0$ and $v'(t)$ is decreasing in $[n^{(0)}, \infty)$. We set*

$$||x||_\alpha = \left(\sum_{k=1}^m |x_k|^\alpha \right)^{\frac{1}{\alpha}} \quad (x = (x_1, \cdots, x_m) \in \mathbf{R}^m),$$

$$||v(y)||_\beta = \left(\sum_{k=1}^s (v(y_k))^\beta \right)^{\frac{1}{\beta}} \quad (y = (y_1, \cdots, y_s) \in [n^{(0)}, \infty)^s),$$

where

$$[n^{(0)}, \infty)^s = \overbrace{[n^{(0)}, \infty) \times \cdots \times [n^{(0)}, \infty)}^{s}.$$

Then, for $\varepsilon > 0$, we have

$$J(\varepsilon) = \int_{\{x \in \mathbf{R}_+^m; ||x||_\alpha \ge 1\}} ||x||_\alpha^{-m-\varepsilon} dx = \frac{\Gamma^m \left(\frac{1}{\alpha} \right)}{\varepsilon \alpha^{m-1} \Gamma \left(\frac{m}{\alpha} \right)}, \tag{5.2}$$

$$\widetilde{J}(\varepsilon) = \int_{\{x \in \mathbf{R}_+^m; ||x||_\alpha \le 1\}} ||x||_\alpha^{-m+\varepsilon} dx = \frac{\Gamma^m \left(\frac{1}{\alpha} \right)}{\varepsilon \alpha^{m-1} \Gamma \left(\frac{m}{\alpha} \right)}, \tag{5.3}$$

and

$$H(\varepsilon) = \sum_{n \in \mathbf{N}_{n^{(0)}}^s} ||v(n)||_\beta^{-s-\varepsilon} \prod_{k=1}^s v'(n_k) = \frac{\Gamma^s \left(\frac{1}{\beta} \right)}{\varepsilon \beta^{s-1} \Gamma(\frac{s}{\beta})} + O(1), \tag{5.4}$$

where $n_i \in \mathbf{N}_{n^{(0)}} = \{n^{(0)}, n^{(0)} + 1, \cdots\}$ $(i = 1, 2, \cdots, s)$, $n = (n_1, n_2, \cdots, n_s)$.

Proof.   For $M > 1$, setting $\Psi(u)$ as follows:

$$\Psi(u) = 0(u \in (0, M^{-\alpha}));$$

$$\Psi(u) = \frac{1}{(Mu^{1/\alpha})^{m+\varepsilon}}(u \in [M^{-\alpha}, 1]),$$

by (5.1), since $||x||_\alpha \geq 1$ means that

$$\sum_{i=1}^{m}(\frac{x_i}{M})^\alpha \geq M^{-\alpha},$$

we find

$$J(\varepsilon) = \lim_{M\to\infty} \int \cdots \int_{D_M} \Psi\left(\sum_{i=1}^{m}(\frac{x_i}{M})^\alpha\right) dx_1 \cdots dx_m$$

$$= \lim_{M\to\infty} \frac{M^m\Gamma^m(\frac{1}{\alpha})}{\alpha^m\Gamma(\frac{m}{\alpha})} \int_0^1 \Psi(u)u^{\frac{m}{\alpha}-1}du$$

$$= \lim_{M\to\infty} \frac{M^m\Gamma^m(\frac{1}{\alpha})}{\alpha^m\Gamma(\frac{m}{\alpha})} \int_{M^{-\alpha}}^1 \frac{1}{(Mu^{1/\alpha})^{m+\varepsilon}}u^{\frac{m}{\alpha}-1}du$$

$$= \frac{\Gamma^m(\frac{1}{\alpha})}{\alpha^m\Gamma(\frac{m}{\alpha})} \lim_{M\to\infty} \frac{1}{M^\varepsilon} \int_{M^{-\alpha}}^1 u^{\frac{-\varepsilon}{\alpha}-1}du$$

$$= \frac{\Gamma^m(\frac{1}{\alpha})}{\alpha^m\Gamma(\frac{m}{\alpha})} \lim_{M\to\infty} \frac{\alpha}{\varepsilon}\left(1 - \frac{1}{M^\varepsilon}\right),$$

and then (5.2) is valid.

In view of (5.1) (for $M = 1$), it follows that

$$\tilde{J}(\varepsilon) = \int_{\{x\in\mathbf{R}_+^m;||x||_\alpha\leq 1\}} \left(\sum_{i=1}^{m}x_i^\alpha\right)^{\frac{1}{\alpha}(-m+\varepsilon)} dx$$

$$= \frac{\Gamma^m(\frac{1}{\alpha})}{\alpha^m\Gamma(\frac{m}{\alpha})} \int_0^1 u^{\frac{1}{\alpha}(-m+\varepsilon)}u^{\frac{m}{\alpha}-1}du = \frac{\Gamma^m(\frac{1}{\alpha})}{\varepsilon\alpha^{m-1}\Gamma(\frac{m}{\alpha})}.$$

Hence, (5.3) is valid.

By the decreasing property of $[v(t)]^{-1}$ and $v'(t)$, we have

$$H(\varepsilon) = c_0 + \sum_{n\in\mathbf{N}^s_{n^{(0)}+1}} ||v(n)||_\beta^{-s-\varepsilon} \prod_{k=1}^{s} v'(n_k)$$

$$\leq c_0 + \int_{\mathbf{n}^{(0)}}^\infty \cdots \int_{\mathbf{n}^{(0)}}^\infty ||v(y)||_\beta^{-s-\varepsilon} \prod_{k=1}^{s} v'(y_k)dy_1 \cdots dy_s,$$

$$(u_k = v(y_k)(k = 1, \cdots, s))$$

$$= c_0 + \int_{v(\mathbf{n}^{(0)})}^{\infty} \cdots \int_{v(\mathbf{n}^{(0)})}^{\infty} ||u||_\beta^{-s-\varepsilon} du_1 \cdots du_s$$

$$= c_0 + c_1 + \int_1^{\infty} \cdots \int_1^{\infty} ||u||_\beta^{-s-\varepsilon} du_1 \cdots du_s$$

$$= \frac{\Gamma^s(\frac{1}{\beta})}{\varepsilon \beta^{s-1} \Gamma(\frac{s}{\beta})} + c_0 + c_1,$$

$$H(\varepsilon) \geq \int_{\mathbf{n}^{(0)}}^{\infty} \cdots \int_{\mathbf{n}^{(0)}}^{\infty} ||v(y)||_\beta^{-s-\varepsilon} \prod_{k=1}^{s} v'(y_k) dy_1 \cdots dy_s,$$

$$(u_k = v(y_k)(k = 1, \cdots, s))$$

$$= \int_{v(\mathbf{n}^{(0)})}^{\infty} \cdots \int_{v(\mathbf{n}^{(0)})}^{\infty} ||u||_\beta^{-s-\varepsilon} du_1 \cdots du_s$$

$$= \frac{\Gamma^s(\frac{1}{\beta})}{\varepsilon \, \beta^{s-1} \Gamma(\frac{s}{\beta})} + c_1,$$

and then, we obtain (5.4). $\qquad\qquad\qquad\qquad\qquad\qquad\qquad\qquad\square$

5.2.2   *Some Results about the Weight Functions*

Definition 5.1. Suppose that $m, s, n^{(0)} \in \mathbf{N}, \alpha, \beta > 0, \lambda_1, \lambda_2 \in \mathbf{R}, H(t, u)$ is a non-negative finite measurable function in R_+^2, $H(t, u)u^{\lambda_2 - s}$ is decreasing with respect to $u \in \mathbf{R}_+$ and strictly decreasing in an interval $I \subset (n^{(0)}, \infty), v'(t) > 0, v''(t) \leq 0$ $(t \in (n^{(0)} - 1, \infty))$ with $v(n^{(0)} - 1) \geq 0$. Define two weight functions $\omega(n)$ and $\varpi(x)$ as follows:

$$\omega(n) = \int_{R_+^m} H(||x||_\alpha, ||v(n)||_\beta) \frac{||v(n)||_\beta^{\lambda_2}}{||x||_\alpha^{m-\lambda_1}} dx \ (n \in \mathbf{N}_{n^{(0)}}^s), \quad (5.5)$$

$$\varpi(x) = \sum_{n \in \mathbf{N}_{n^{(0)}}^s} H(||x||_\alpha, ||v(n)||_\beta) \frac{||x||_\alpha^{\lambda_1}}{||v(n)||_\beta^{s-\lambda_2}} \prod_{k=1}^{s} v'(n_k) \ (x \in \mathbf{R}_+^m). \quad (5.6)$$

If the expression

$$\lim_{M \to \infty} M^{\lambda_2} \int_0^1 H(||x||_\alpha, M v^{\frac{1}{\beta}}) v^{\frac{\lambda_2}{\beta} - 1} dv = a \in \mathbf{R}_+,$$

then applying (5.1), we get the following results:

$$\omega(n) = \lim_{M \to \infty} \int_{D_M} H\left(M[\sum_{k=1}^{m} (\frac{x_k}{M})^\alpha]^{\frac{1}{\alpha}}, ||v(n)||_\beta\right)$$

$$\times \frac{||v(n)||_\beta^{\lambda_2}}{\{M[\sum_{k=1}^{m} (x_k/M)^\alpha]^{\frac{1}{\alpha}}\}^{m-\lambda_1}} dx$$

$$= \lim_{M \to \infty} \frac{M^m \Gamma^m(\frac{1}{\alpha})}{\alpha^m \Gamma(\frac{m}{\alpha})} \int_0^1 H(Mu^{\frac{1}{\alpha}}, ||v(n)||_\beta) \frac{||v(n)||_\beta^{\lambda_2}}{(Mu^{\frac{1}{\alpha}})^{m-\lambda_1}} u^{\frac{m}{\alpha}-1} du$$

$$= \frac{\Gamma^m(\frac{1}{\alpha})}{\alpha^m \Gamma(\frac{m}{\alpha})} ||v(n)||_\beta^{\lambda_2} \lim_{M \to \infty} M^{\lambda_1} \int_0^1 H(Mu^{\frac{1}{\alpha}}, ||v(n)||_\beta) u^{\frac{\lambda_1}{\alpha}-1} du; \quad (5.7)$$

$$\varpi(x) < \int_{\mathbf{n}^{(0)}-1}^{\infty} \cdots \int_{\mathbf{n}^{(0)}-1}^{\infty} H(||x||_\alpha, ||v(y)||_\beta) \frac{||x||_\alpha^{\lambda_1}}{||v(y)||_\beta^{s-\lambda_2}} \prod_{k=1}^{s} v'(y_k) dy_1 \cdots dy_s,$$

$$(u_k = v(y_k)(k=1, \cdots, s))$$

$$= \int_{v(\mathbf{n}^{(0)}-1)}^{\infty} \cdots \int_{v(\mathbf{n}^{(0)}-1)}^{\infty} H(||x||_\alpha, ||u||_\beta) \frac{||x||_\alpha^{\lambda_1}}{||u||_\beta^{s-\lambda_2}} du_1 \cdots du_s$$

$$\leq \int_{\mathbf{R}_+^s} H(||x||_\alpha, ||u||_\beta) \frac{||x||_\alpha^{\lambda_1}}{||u||_\beta^{s-\lambda_2}} du$$

$$= \frac{\Gamma^s(\frac{1}{\beta})}{\beta^s \Gamma\left(\frac{s}{\beta}\right)} ||x||_\alpha^{\lambda_1} \lim_{M \to \infty} M^{\lambda_2} \int_0^1 H(||x||_\alpha, Mv^{\frac{1}{\beta}}) v^{\frac{\lambda_2}{\beta}-1} dv, \quad (5.8)$$

$$\varpi(x) > \int_{\mathbf{n}^{(0)}}^{\infty} \cdots \int_{\mathbf{n}^{(0)}}^{\infty} H(||x||_\alpha, ||v(y)||_\beta) \frac{||x||_\alpha^{\lambda_1}}{||v(y)||_\beta^{s-\lambda_2}} \prod_{k=1}^{s} v'(y_k) dy_1 \cdots dy_s,$$

$$(u_k = v(y_k)(k=1, \cdots, s))$$

$$= \int_{v(\mathbf{n}^{(0)})}^{\infty} \cdots \int_{v(\mathbf{n}^{(0)})}^{\infty} H(||x||_\alpha, ||u||_\beta) \frac{||x||_\alpha^{\lambda_1}}{||u||_\beta^{s-\lambda_2}} du_1 \cdots du_s$$

$$= \int_{R_+^s} H(||x||_\alpha, ||u||_\beta) \frac{||x||_\alpha^{\lambda_1}}{||u||_\beta^{s-\lambda_2}} du$$

$$- \delta_0 \int_{\{u \in \mathbf{R}_+^s; 0 < ||u||_\alpha \leq 1\}} H(||x||_\alpha, ||u||_\beta) \frac{||x||_\alpha^{\lambda_1}}{||u||_\beta^{s-\lambda_2}} du$$

$$= \frac{\Gamma^s(\frac{1}{\beta})}{\beta^s \Gamma\left(\frac{s}{\beta}\right)} ||x||_\alpha^{\lambda_1} \left[\lim_{M \to \infty} M^{\lambda_2} \int_0^1 H(||x||_\alpha, Mv^{\frac{1}{\beta}}) v^{\frac{\lambda_2}{\beta}-1} dv \right.$$

$$\left. - \delta_0 \int_0^1 H(||x||_\alpha, v^{\frac{1}{\beta}}) v^{\frac{\lambda_2}{\beta}-1} dv \right] \quad (0 < \delta_0 \leq 1). \quad (5.9)$$

In particular, for $\beta = 1$, if

$$\frac{\partial}{\partial u}(H(t,u)u^{\lambda_2-s}) < 0, \quad \frac{\partial^2}{\partial u^2}(H(t,u)u^{\lambda_2-s}) > 0 \quad (u \in \mathbf{R}^+),$$

and $v'(t) > 0, v''(t) \leq 0, v'''(t) \geq 0$ $(t \in (n^{(0)} - \frac{1}{2}, \infty))$, with $v(n^{(0)} - \frac{1}{2}) \geq 0$, then we can find that

$$\frac{\partial}{\partial y_i}||v(y)||_1 > 0, \quad \frac{\partial^2}{\partial y_i^2}||v(y)||_1 \leq 0,$$

and

$$H(||x||_\alpha, ||v(y)||_1)\frac{||x||_\alpha^{\lambda_1}}{||v(y)||_1^{s-\lambda_2}} \prod_{k=1}^{s} v'(y_k)$$

$$= \frac{H(||x||_\alpha, v(y_1) + \cdots + v(y_s))||x||_\alpha^{\lambda_1}}{(v(y_1) + \cdots + v(y_s))^{s-\lambda_2}} \prod_{k=1}^{s} v'(y_k)$$

is strictly decreasing and strict convex with respect to any $y_i \in (n^{(0)} - \frac{1}{2}, \infty)(i = 1, \cdots, s)$. By Hermite-Hadamard's inequality (see Kuang [47]), it follows that

$$\varpi_1(x) = \sum_{n \in \mathbf{N}_{n^{(0)}}^s} H(||x||_\alpha, ||v(n)||_1)\frac{||x||_\alpha^{\lambda_1}}{||v(n)||_1^{s-\lambda_2}} \prod_{k=1}^{s} v'(n_k)$$

$$< \int_{\mathbf{n}^{(0)}-\frac{1}{2}}^{\infty} \cdots \int_{\mathbf{n}^{(0)}-\frac{1}{2}}^{\infty} H(||x||_\alpha, ||v(y)||_1)\frac{||x||_\alpha^{\lambda_1}}{||v(y)||_1^{s-\lambda_2}} \prod_{k=1}^{s} v'(y_k)dy_1 \cdots dy_s,$$

$$(u_k = v(y_k)(k = 1, \cdots, s))$$

$$= \int_{v(\mathbf{n}^{(0)}-\frac{1}{2})}^{\infty} \cdots \int_{v(\mathbf{n}^{(0)}-\frac{1}{2})}^{\infty} H(||x||_\alpha, ||u||_1)\frac{||x||_\alpha^{\lambda_1}}{||u||_1^{s-\lambda_2}} du_1 \cdots du_s$$

$$\leq \int_{\mathbf{R}_+^s} H(||x||_\alpha, ||u||_1)\frac{||x||_\alpha^{\lambda_1}}{||u||_1^{s-\lambda_2}} du$$

$$= \frac{||x||_\alpha^{\lambda_1}}{(s-1)!} \lim_{M \to \infty} M^{\lambda_2} \int_0^1 H(||x||_\alpha, Mv)v^{\lambda_2-1}dv, \quad (5.10)$$

$$\varpi_1(x) > \int_{\mathbf{n}^{(0)}}^{\infty} \cdots \int_{\mathbf{n}^{(0)}}^{\infty} H(||x||_\alpha, ||v(y)||_1)\frac{||x||_\alpha^{\lambda_1}}{||v(y)||_1^{s-\lambda_2}} \prod_{k=1}^{s} v'(y_k)dy_1 \cdots dy_s,$$

$$(u_k = v(y_k)(k = 1, \cdots, s))$$

$$= \int_{v(\mathbf{n}^{(0)})}^{\infty} \cdots \int_{v(\mathbf{n}^{(0)})}^{\infty} H(||x||_\alpha, ||u||_1) \frac{||x||_\alpha^{\lambda_1}}{||u||_1^{s-\lambda_2}} du_1 \cdots du_s$$

$$= \int_{\mathbf{R}_+^s} H(||x||_\alpha, ||u||_1) \frac{||x||_\alpha^{\lambda_1}}{||u||_1^{s-\lambda_2}} du$$

$$-\delta_0 \int_{\{u \in \mathbf{R}_+^s; 0<||u||_\alpha \leq 1\}} H(||x||_\alpha, ||u||_1) \frac{||x||_\alpha^{\lambda_1}}{||u||_1^{s-\lambda_2}} du$$

$$= \frac{||x||_\alpha^{\lambda_1}}{(s-1)!} \left[\lim_{M \to \infty} M^{\lambda_2} \int_0^1 H(||x||_\alpha, Mv) v^{\lambda_2-1} dv \right.$$

$$\left. -\delta_0 \int_0^1 H(||x||_\alpha, v) v^{\lambda_2-1} dv \right] \quad (0 < \delta_0 \leq 1). \tag{5.11}$$

5.2.3   *Two Preliminary Inequalities*

Lemma 5.3. *Let the assumptions of Definition 5.1 be fulfilled and additionally, let $p \in \mathbf{R}\backslash\{0,1\}, \frac{1}{p} + \frac{1}{q} = 1$, $a(n) = a(n_1, \cdots, n_s) \geq 0 (n \in \mathbf{N}_{n^{(0)}}^s)$, $f(x) = f(x_1,\cdots,x_m)$ be a non-negative measurable function in $\mathbf{R}_+^m$. Then*

(i) *for $p > 1$, we have the following inequalities:*

$$\widetilde{J}_1 = \left\{ \sum_{n \in \mathbf{N}_{n^{(0)}}^s} \frac{||v(n)||_\beta^{p\lambda_2-s}}{[\omega(n)]^{p-1}} \prod_{k=1}^s v'(n_k) \left(\int_{R_+^m} H(||x||_\alpha, ||v(n)||_\beta) f(x) dx \right)^p \right\}^{\frac{1}{p}}$$

$$\leq \left\{ \int_{R_+^m} \varpi(x) ||x||_\alpha^{p(m-\lambda_1)-m} f^p(x) dx \right\}^{\frac{1}{p}}, \tag{5.12}$$

and

$$\widetilde{J}_2 = \left\{ \int_{R_+^m} \frac{||x||_\alpha^{q\lambda_1-m}}{[\varpi(x)]^{q-1}} \left(\sum_{n \in \mathbf{N}_{n^{(0)}}^s} H(||x||_\alpha, ||v(n)||_\beta) a(n) \right)^q dx \right\}^{\frac{1}{q}}$$

$$\leq \left\{ \sum_{n \in \mathbf{N}_{n^{(0)}}^s} \omega(n) ||v(n)||_\beta^{q(s-\lambda_2)-s} \left(\prod_{k=1}^s v'(n_k) \right)^{1-q} a^q(n) \right\}^{\frac{1}{q}}; \tag{5.13}$$

(ii) *for $p < 0$, or $0 < p < 1$, we have the reverses of (5.12) and (5.13).*

Proof.   (i) For $p > 1$, by Hölder's inequality (see Kuang [47]), we have

$$
\left(\int_{R_+^m} H(||x||_\alpha, ||v(n)||_\beta) f(x) dx \right)^p
$$

$$
= \left\{ \int_{R_+^m} H(||x||_\alpha, ||v(n)||_\beta) \left[\frac{||x||_\alpha^{(m-\lambda_1)/q}}{||v(n)||_\beta^{(s-\lambda_2)/p}} \left(\prod_{k=1}^s v'(n_k) \right)^{\frac{1}{p}} f(x) \right] \right.
$$

$$
\left. \times \left[\frac{||v(n)||_\beta^{(s-\lambda_2)/p}}{||x||_\alpha^{(m-\lambda_1)/q}} \left(\prod_{k=1}^s v'(n_k) \right)^{\frac{-1}{p}} \right] dx \right\}^p
$$

$$
\leq \int_{R_+^m} H(||x||_\alpha, ||v(n)||_\beta) \frac{||x||_\alpha^{(m-\lambda_1)(p-1)}}{||v(n)||_\beta^{s-\lambda_2}} \prod_{k=1}^s v'(n_k) f^p(x) dx
$$

$$
\times \left\{ \int_{R_+^m} H(||x||_\alpha, ||v(n)||_\beta) \frac{||v(n)||_\beta^{(s-\lambda_2)(q-1)}}{||x||_\alpha^{m-\lambda_1}} \left(\prod_{k=1}^s v'(n_k) \right)^{1-q} dx \right\}^{p-1}
$$

$$
= [\omega(n)]^{p-1} ||v(n)||_\beta^{s-p\lambda_2} \left(\prod_{k=1}^s v'(n_k) \right)^{-1}
$$

$$
\times \int_{R_+^m} H(||x||_\alpha, ||v(n)||_\beta) \frac{||x||_\alpha^{(m-\lambda_1)(p-1)}}{||v(n)||_\beta^{s-\lambda_2}} \prod_{k=1}^s v'(n_k) f^p(x) dx.
$$

Then by the Lebesgue term by term integration theorem (see Kuang [49]), it follows that

$$
\tilde{J}_1 \leq \left\{ \sum_{n \in \mathbf{N}_{n(0)}^s} \int_{R_+^m} H(||x||_\alpha, ||v(n)||_\beta) \right.
$$

$$
\left. \times \frac{||x||_\alpha^{(m-\lambda_1)(p-1)}}{||v(n)||_\beta^{s-\lambda_2}} \prod_{k=1}^s v'(n_k) f^p(x) dx \right\}^{\frac{1}{p}}
$$

$$
= \left\{ \int_{R_+^m} \sum_{n \in \mathbf{N}_{n(0)}^s} H(||x||_\alpha, ||v(n)||_\beta) \frac{||x||_\alpha^{(m-\lambda_1)(p-1)}}{||v(n)||_\beta^{s-\lambda_2}} \prod_{k=1}^s v'(n_k) f^p(x) dx \right\}^{\frac{1}{p}}
$$

$$
= \left\{ \int_{R_+^m} \varpi(x) ||x||_\alpha^{p(m-\lambda_1)-m} f^p(x) dx \right\}^{\frac{1}{p}},
$$

and hence, we obtain inequality (5.12).

Still by Hölder's inequality, we have

$$
\left(\sum_{n \in \mathbf{N}_{n(0)}^s} H(||x||_\alpha, ||v(n)||_\beta) a(n) \right)^q
$$

$$
= \left\{ \sum_{n \in \mathbf{N}_{n(0)}^s} H(||x||_\alpha, ||v(n)||_\beta) \left[\frac{||x||_\alpha^{(m-\lambda_1)/q}}{||v(n)||_\beta^{(s-\lambda_2)/p}} \left(\prod_{k=1}^s v'(n_k) \right)^{\frac{1}{p}} \right] \right.
$$

$$
\left. \times \left[\frac{||v(n)||_\beta^{(s-\lambda_2)/p}}{||x||_\alpha^{(m-\lambda_1)/q}} \left(\prod_{k=1}^s v'(n_k) \right)^{\frac{-1}{p}} a(n) \right] \right\}^q
$$

$$
\leq \left\{ \sum_{n \in \mathbf{N}_{n(0)}^s} H(||x||_\alpha, ||v(n)||_\beta) \frac{||x||_\alpha^{(m-\lambda_1)(p-1)}}{||v(n)||_\beta^{s-\lambda_2}} \prod_{k=1}^s v'(n_k) \right\}^{q-1}
$$

$$
\times \sum_{n \in \mathbf{N}_{n(0)}^s} H(||x||_\alpha, ||v(n)||_\beta) \frac{||v(n)||_\beta^{(s-\lambda_2)(q-1)}}{||x||_\alpha^{m-\lambda_1}} \left(\prod_{k=1}^s v'(n_k) \right)^{1-q} a^q(n)
$$

$$
= \frac{[\omega(x)]^{q-1}}{||x||_\alpha^{q\lambda_1 - m}} \sum_{n \in \mathbf{N}_{n(0)}^s} H(||x||_\alpha, ||v(n)||_\beta)
$$

$$
\times \frac{||v(n)||_\beta^{(s-\lambda_2)(q-1)}}{||x||_\alpha^{m-\lambda_1}} \left(\prod_{k=1}^s v'(n_k) \right)^{1-q} a^q(n).
$$

Then, by the Lebesgue term by term integration theorem, it follows that

$$
\widetilde{J}_2 \leq \left\{ \int_{R_+^m} \sum_{n \in \mathbf{N}_{n(0)}^s} H(||x||_\alpha, ||v(n)||_\beta) \right.
$$

$$
\left. \times \frac{||v(n)||_\beta^{(s-\lambda_2)(q-1)}}{||x||_\alpha^{m-\lambda_1}} \left(\prod_{k=1}^s v'(n_k) \right)^{1-q} a^q(n) dx \right\}^{\frac{1}{q}}
$$

$$
= \left\{ \sum_{n \in \mathbf{N}_{n(0)}^s} \left[\int_{R_+^m} H(||x||_\alpha, ||v(n)||_\beta) \frac{||v(n)||_\beta^{(s-\lambda_2)(q-1)}}{||x||_\alpha^{m-\lambda_1}} dx \right] \right.
$$

$$
\left. \times \left(\prod_{k=1}^s v'(n_k) \right)^{1-q} a^q(n) \right\}^{\frac{1}{q}}
$$

$$= \left\{ \sum_{n \in \mathbf{N}_{n(0)}^s} \omega(n) ||v(n)||_\beta^{q(s-\lambda_2)-s} \left(\prod_{k=1}^{s} v'(n_k) \right)^{1-q} a^q(n) \right\}^{\frac{1}{q}},$$

and we have proved inequality (5.13).

(ii) For $p < 0$, or $0 < p < 1$, by the reverse Hölder's inequality with weight and in the same way, we obtain the reverses of (5.12) and (5.13). $\square$

5.3 Some Inequalities Related to a General Homogeneous Kernel

5.3.1 *Several Lemmas*

As the assumptions of Definition 5.1, if $\lambda \in \mathbf{R}$, $k_\lambda(t, u)$ is a finite homogeneous function of degree $-\lambda$ in $\mathbf{R}_+^2$, $H(t, u) = k_\lambda(t, u)$, $\lambda_1 + \lambda_2 = \lambda$, then (5.5) and (5.6) reduce to the following weight functions:

$$\omega(\lambda_2, n) = \int_{R_+^m} k_\lambda(||x||_\alpha, ||v(n)||_\beta) \frac{||v(n)||_\beta^{\lambda_2}}{||x||_\alpha^{m-\lambda_1}} dx (n \in \mathbf{N}_{n(0)}^s), \qquad (5.14)$$

$$\varpi(\lambda_1, x) = \sum_{n \in \mathbf{N}_{n(0)}^s} k_\lambda(||x||_\alpha, ||v(n)||_\beta) \frac{||x||_\alpha^{\lambda_1}}{||v(n)||_\beta^{s-\lambda_2}} \prod_{k=1}^{s} v'(n_k), \ (x \in \mathbf{R}_+^m). \qquad (5.15)$$

For $k(\lambda_1) = \int_0^\infty k_\lambda(t, 1) t^{\lambda_1 - 1} dt \in \mathbf{R}_+$, by (5.7), (5.8) and (5.9), we find

$$\omega(\lambda_2, n) = \frac{\Gamma^m(\frac{1}{\alpha})}{\alpha^m \Gamma(\frac{m}{\alpha})} ||v(n)||_\beta^{\lambda_2}$$

$$\times \lim_{M \to \infty} M^{\lambda_1} \int_0^1 k_\lambda(Mu^{\frac{1}{\alpha}}, ||v(n)||_\beta) u^{\frac{\lambda_1}{\alpha} - 1} du,$$

$$(u = (||v(n)||_\beta t/M)^\alpha)$$

$$= \frac{\Gamma^m(\frac{1}{\alpha})}{\alpha^{m-1} \Gamma(\frac{m}{\alpha})} \int_0^\infty k_\lambda(t, 1) t^{\lambda_1 - 1} dt$$

$$= K(\alpha, m, \lambda_1) = \frac{\Gamma^m(\frac{1}{\alpha})}{\alpha^{m-1} \Gamma(\frac{m}{\alpha})} k(\lambda_1); \qquad (5.16)$$

$$\varpi(\lambda_1, x) < \frac{\Gamma^s(\frac{1}{\beta})}{\beta^s \Gamma\left(\frac{s}{\beta}\right)} ||x||_\alpha^{\lambda_1} \lim_{M \to \infty} M^{\lambda_2} \int_0^1 k_\lambda(||x||_\alpha, M v^{\frac{1}{\beta}}) v^{\frac{\lambda_2}{\beta}-1} dv,$$

$$(v = (||x||_\alpha / M t)^\beta)$$

$$= \frac{\Gamma^s(\frac{1}{\beta})}{\beta^{s-1} \Gamma\left(\frac{s}{\beta}\right)} \int_0^\infty k_\lambda(t, 1) t^{\lambda_1-1} dt$$

$$= K(\beta, s, \lambda_1) = \frac{\Gamma^s(\frac{1}{\beta})}{\beta^{s-1} \Gamma\left(\frac{s}{\beta}\right)} k(\lambda_1), \tag{5.17}$$

$$\varpi(\lambda_1, x) > \frac{\Gamma^s(\frac{1}{\beta})}{\beta^s \Gamma\left(\frac{s}{\beta}\right)} ||x||_\alpha^{\lambda_1} \left[\lim_{M \to \infty} M^{\lambda_2} \int_0^1 k_\lambda(||x||_\alpha, M v^{\frac{1}{\beta}}) v^{\frac{\lambda_2}{\beta}-1} dv \right.$$

$$\left. - \delta_0 \int_0^1 k_\lambda(||x||_\alpha, v^{\frac{1}{\beta}}) v^{\frac{\lambda_2}{\beta}-1} dv \right]$$

$$= K(\beta, s, \lambda_1) - \delta_0 \int_{||x||_\alpha}^\infty k_\lambda(t, 1) t^{\lambda_1-1} dt$$

$$= K(\beta, s, \lambda_1)(1 - \theta(||x||_\alpha)) > 0, \tag{5.18}$$

where

$$\theta(||x||_\alpha) = \frac{\delta_0}{K(\beta, s, \lambda_1)} \int_{||x||_\alpha}^\infty k_\lambda(t, 1) t^{\lambda_1-1} dt > 0 \qquad (0 < \delta_0 \le 1).$$

Lemma 5.4. *Suppose that* $m, s, n^{(0)} \in \mathbf{N}, \alpha, \beta > 0, p \in \mathbf{R} \backslash \{0, 1\}, \frac{1}{p} + \frac{1}{q} = 1,$ $\lambda_1, \lambda_2 \in \mathbf{R}, \lambda_1 + \lambda_2 = \lambda, k_\lambda(t, u)$ *is a finite homogeneous function of degree* $-\lambda$ *in* $\mathbf{R}_+^2$, $k_\lambda(t, u) u^{\lambda_2-s}$ *is decreasing with respect to* $u \in \mathbf{R}_+$, *and strict decreasing in an interval* $I \subset (n^{(0)}, \infty), v'(t) > 0, v''(t) \le 0$ $(t \in (n^{(0)} - 1, \infty))$, *with* $v(n^{(0)} - 1) \ge 0$. *If there exist constants* $\widetilde{\delta}_0 > 0$ *and* $\eta > \lambda_1$, *such that for any* $\delta \in [0, \widetilde{\delta}_0]$,

$$k(\lambda_1 \pm \delta) = \int_0^\infty k_\lambda(t, 1) t^{(\lambda_1 \pm \delta)-1} dt \in \mathbf{R}_+,$$

and

$$k_\lambda(t, 1) \le L\left(\frac{1}{t^\eta}\right) \qquad (t \ge 1; L > 0),$$

then, for any $0 < \varepsilon < |q|\widetilde{\delta}_0$, we have

$$\widetilde{I}(\varepsilon) = \int_{\{x \in \mathbf{R}_+^m; ||x||_\alpha \geq 1\}} \sum_{n \in \mathbf{N}_{n^{(0)}}^s} k_\lambda(||x||_\alpha, ||v(n)||_\beta) \frac{||x||_\alpha^{\lambda_1 - m - \frac{\varepsilon}{p}}}{||v(n)||_\beta^{s - (\lambda_2 - \frac{\varepsilon}{q})}} \prod_{k=1}^s v'(n_k) dx$$

$$= \frac{\Gamma^m(\frac{1}{\alpha})}{\varepsilon \alpha^{m-1} \Gamma(\frac{m}{\alpha})} \frac{\Gamma^s(\frac{1}{\beta})}{\beta^{s-1} \Gamma(\frac{s}{\beta})} (k(\lambda_1) + o(1))(1 - \varepsilon O_1(1))(\varepsilon \to 0^+).$$

(5.19)

Proof. In view of (5.15), we find

$$\widetilde{I}(\varepsilon) = \int_{\{x \in \mathbf{R}_+^m; ||x||_\alpha \geq 1\}} ||x||_\alpha^{-m-\varepsilon} \sum_{n \in \mathbf{N}_{n^{(0)}}^s} k_\lambda(||x||_\alpha, ||v(n)||_\beta)$$

$$\times \frac{||x||_\alpha^{\lambda_1 + \frac{\varepsilon}{q}}}{||v(n)||_\beta^{s - (\lambda_2 - \frac{\varepsilon}{q})}} \prod_{k=1}^s v'(n_k) dx$$

$$= \int_{\{x \in \mathbf{R}_+^m; ||x||_\alpha \geq 1\}} ||x||_\alpha^{-m-\varepsilon} \varpi(\lambda_1 + \frac{\varepsilon}{q}, x) dx$$

$$\geq K(\beta, s, \lambda_1 + \frac{\varepsilon}{q}) \int_{\{x \in \mathbf{R}_+^m; ||x||_\alpha \geq 1\}} ||x||_\alpha^{-m-\varepsilon}(1 - \widetilde{\theta}(||x||_\alpha)) dx$$

$$= \frac{\Gamma^m(\frac{1}{\alpha})}{\alpha^{m-1} \Gamma(\frac{m}{\alpha})} k(\lambda_1 + \frac{\varepsilon}{q}) \int_{\{x \in \mathbf{R}_+^m; ||x||_\alpha \geq 1\}} ||x||_\alpha^{-m-\varepsilon}(1 - \widetilde{\theta}(||x||_\alpha)) dx,$$

where

$$\widetilde{\theta}(||x||_\alpha) = \frac{\delta_0}{K(\beta, s, \lambda_1 + \frac{\varepsilon}{q})} \int_{||x||_\alpha}^\infty k_\lambda(t, 1) t^{\lambda_1 - 1} dt \qquad (0 < \delta_0 \leq 1).$$

Since $Mu^{1/\alpha} = ||x||_\alpha \geq 1$, we find

$$0 < \widetilde{\theta}(Mu^{1/\alpha}) = \frac{\delta_0}{K(\beta, s, \lambda_1 + \frac{\varepsilon}{q})} \int_{Mu^{1/\alpha}}^\infty k_\lambda(t, 1) t^{\lambda_1 - 1} dt$$

$$\leq \frac{\delta_0 L}{K(\beta, s, \lambda_1 + \frac{\varepsilon}{q})} \int_{Mu^{1/\alpha}}^\infty t^{\lambda_1 - \eta - 1} dt$$

$$= \frac{\delta_0 L M^{\lambda_1 - \eta} u^{\frac{\lambda_1 - \eta}{\alpha}}}{(\eta - \lambda_1) K(\beta, s, \lambda_1 + \frac{\varepsilon}{q})},$$

$$\int_{\{x \in \mathbf{R}_+^m; ||x||_\alpha \geq 1\}} ||x||_\alpha^{-m-\varepsilon}(1 - \widetilde{\theta}(||x||_\alpha)) dx$$

$$= \lim_{M \to \infty} \int \cdots \int_{D_M} \Psi\left(\sum_{i=1}^m \left(\frac{x_i}{M}\right)^\alpha\right) dx_1 \cdots dx_m$$

$$= \lim_{M \to \infty} \frac{M^m \Gamma^m(\frac{1}{\alpha})}{\alpha^m \Gamma(\frac{m}{\alpha})} \int_0^1 \Psi(u) u^{\frac{m}{\alpha}-1} du$$

$$= \lim_{M \to \infty} \frac{M^m \Gamma^m(\frac{1}{\alpha})}{\alpha^m \Gamma(\frac{m}{\alpha})} \int_{M^{-\alpha}}^1 \frac{(1 - \widetilde{\theta}(Mu^{1/\alpha}))}{(Mu^{1/\alpha})^{m+\varepsilon}} u^{\frac{m}{\alpha}-1} du$$

$$= \frac{\Gamma^m(\frac{1}{\alpha})}{\alpha^m \Gamma(\frac{m}{\alpha})} \lim_{M \to \infty} \frac{1}{M^\varepsilon} \int_{M^{-\alpha}}^1 (1 - \widetilde{\theta}(Mu^{1/\alpha})) u^{\frac{-\varepsilon}{\alpha}-1} du$$

$$= \frac{\Gamma^m(\frac{1}{\alpha})}{\alpha^m \Gamma(\frac{m}{\alpha})} \lim_{M \to \infty} \frac{1}{M^\varepsilon} \int_{M^{-\alpha}}^1 \left[1 - M^{\lambda_1 - \eta} O(u^{\frac{\lambda_1 - \eta}{\alpha}})\right] u^{\frac{-\varepsilon}{\alpha}-1} du$$

$$= \frac{\Gamma^m(\frac{1}{\alpha})}{\varepsilon \alpha^{m-1} \Gamma(\frac{m}{\alpha})} (1 - \varepsilon O_1(1)) \quad (\varepsilon \to 0^+). \tag{5.20}$$

By Lemma 3.6, we have

$$k\left(\lambda_1 + \frac{\varepsilon}{q}\right) = k(\lambda_1) + o(1) \quad (\varepsilon \to 0^+),$$

and then it follows that

$$\widetilde{I}(\varepsilon) \geq \frac{\Gamma^m(\frac{1}{\alpha})}{\varepsilon \alpha^{m-1} \Gamma(\frac{m}{\alpha})} \frac{\Gamma^s(\frac{1}{\beta})}{\beta^{s-1} \Gamma(\frac{s}{\beta})} k\left(\lambda_1 + \frac{\varepsilon}{q}\right) (1 - \varepsilon O_1(1))$$

$$= \frac{\Gamma^m(\frac{1}{\alpha})}{\varepsilon \alpha^{m-1} \Gamma(\frac{m}{\alpha})} \frac{\Gamma^s(\frac{1}{\beta})}{\beta^{s-1} \Gamma(\frac{s}{\beta})} (k(\lambda_1) + o(1))(1 - \varepsilon O_1(1)).$$

By (5.2), we still have

$$\widetilde{I}(\varepsilon) \leq \frac{\Gamma^m(\frac{1}{\alpha})}{\alpha^{m-1} \Gamma(\frac{m}{\alpha})} k\left(\lambda_1 + \frac{\varepsilon}{q}\right) \int_{\{x \in \mathbf{R}_+^m; ||x||_\alpha \geq 1\}} ||x||_\alpha^{-m-\varepsilon} dx$$

$$= \frac{\Gamma^m(\frac{1}{\alpha})}{\varepsilon \alpha^{m-1} \Gamma(\frac{m}{\alpha})} \frac{\Gamma^s(\frac{1}{\beta})}{\beta^{s-1} \Gamma(\frac{s}{\beta})} (k(\lambda_1) + o(1)).$$

Thus, we have (5.19). □

In particular, for $\beta = 1$, if

$$\frac{\partial}{\partial u}(k_\lambda(t, u) u^{\lambda_2 - s}) < 0, \quad \frac{\partial^2}{\partial u^2}(k_\lambda(t, u) u^{\lambda_2 - s}) > 0 \quad (u \in \mathbf{R}^+),$$

$v'(t) > 0, v''(t) \leq 0, v'''(t) \geq 0 \ (t \in (n^{(0)} - \frac{1}{2}, \infty))$, with $v(n^{(0)} - \frac{1}{2}) \geq 0$, then, in view of (5.14) and (5.15), we set the following weight functions:

$$\omega_1(\lambda_2, n) = \int_{\mathbf{R}_+^m} k_\lambda(||x||_\alpha, ||v(n)||_1) \frac{||v(n)||_1^{\lambda_2}}{||x||_\alpha^{m-\lambda_1}} dx \ (n \in \mathbf{N}_{n^{(0)}}^s), \tag{5.21}$$

$$\varpi_1(\lambda_1, x) = \sum_{n \in \mathbf{N}_{n^{(0)}}^s} k_\lambda(||x||_\alpha, ||v(n)||_1) \frac{||x||_\alpha^{\lambda_1}}{||v(n)||_1^{s-\lambda_2}} \prod_{k=1}^s v'(n_k) \ (x \in \mathbf{R}_+^m),$$

$$\tag{5.22}$$

and by (5.16), (5.10) and (5.11), we find

$$\omega_1(\lambda_2, n) = K(\alpha, m, \lambda_1) = \frac{\Gamma^m(\frac{1}{\alpha})}{\alpha^{m-1}\Gamma(\frac{m}{\alpha})} k(\lambda_1); \qquad (5.23)$$

$$\varpi_1(\lambda_1, x) < \frac{||x||_\alpha^{\lambda_1}}{(s-1)!} \lim_{M \to \infty} M^{\lambda_2} \int_0^1 k_\lambda(||x||_\alpha, Mv) v^{\lambda_2 - 1} dv$$

$$= K(1, s, \lambda_1) = \frac{1}{(s-1)!} k(\lambda_1), \qquad (5.24)$$

and

$$\varpi(\lambda_1, x) > \frac{||x||_\alpha^{\lambda_1}}{(s-1)!} \left[\lim_{M \to \infty} M^{\lambda_2} \int_0^1 k_\lambda(||x||_\alpha, Mv) v^{\lambda_2 - 1} dv \right.$$

$$\left. -\delta_0 \int_0^1 k_\lambda(||x||_\alpha, v) v^{\lambda_2 - 1} dv \right]$$

$$= \frac{1}{(s-1)!} k(\lambda_1)(1 - \theta_1(||x||_\alpha)) > 0, \qquad (5.25)$$

where

$$\theta_1(||x||_\alpha) = \frac{\delta_0(s-1)!}{k(\lambda_1)} \int_{||x||_\alpha}^\infty k_\lambda(t, 1) t^{\lambda_1 - 1} dt > 0 \qquad (0 < \delta_0 \le 1).$$

By the same way of Lemma 5.4, we still have

Lemma 5.5. *Suppose that* $m, s, n^{(0)} \in \mathbf{N}, \alpha > 0, p \in \mathbf{R} \backslash \{0, 1\}, \frac{1}{p} + \frac{1}{q} = 1,$ $\lambda_1, \lambda_2 \in \mathbf{R}, \lambda_1 + \lambda_2 = \lambda, k_\lambda(t, u)$ *is a finite homogeneous function of degree* $-\lambda$ *in* $\mathbf{R}_+^2$, *satisfying*

$$\frac{\partial}{\partial u}(k_\lambda(t, u)u^{\lambda_2 - s}) < 0, \frac{\partial^2}{\partial u^2}(k_\lambda(t, u)u^{\lambda_2 - s}) > 0(u \in \mathbf{R}^+),$$

$v'(t) > 0, v''(t) \le 0, v'''(t) \ge 0$ *(*$t \in (n^{(0)} - \frac{1}{2}, \infty)$*), with* $v(n^{(0)} - \frac{1}{2}) \ge 0$. *If there exist constants* $\widetilde{\delta}_0 > 0$ *and* $\eta > \lambda_1$, *such that, for any* $\delta \in [0, \widetilde{\delta}_0]$,

$$k(\lambda_1 \pm \delta) = \int_0^\infty k_\lambda(t, 1) t^{(\lambda_1 \pm \delta) - 1} dt \in \mathbf{R}_+,$$

and

$$k_\lambda(t, 1) \le L\left(\frac{1}{t^\eta}\right) \qquad (t \ge 1; L > 0),$$

then, for any $0 < \varepsilon < |q|\widetilde{\delta}_0$, *we have*

$$\widetilde{I}(\varepsilon) = \int_{\{x \in \mathbf{R}_+^m; ||x||_\alpha \ge 1\}} \sum_{n \in \mathbf{N}_{n^{(0)}}^s} k_\lambda(||x||_\alpha, ||v(n)||_1)$$

$$\times \frac{||x||_\alpha^{\lambda_1 - m - \frac{\varepsilon}{p}}}{||v(n)||_1^{s - (\lambda_2 - \frac{\varepsilon}{q})}} \prod_{k=1}^s v'(n_k) dx$$

$$= \frac{\Gamma^m(\frac{1}{\alpha})}{\varepsilon(s-1)!\alpha^{m-1}\Gamma(\frac{m}{\alpha})}(k(\lambda_1) + o(1))(1 - \varepsilon O_1(1)) \quad (\varepsilon \to 0^+). \quad (5.26)$$

5.3.2 Main Results

For $p \in \mathbf{R}\backslash\{0,1\}, \frac{1}{p} + \frac{1}{q} = 1$, we set two functions

$$\varphi(x) = ||x||_{\alpha}^{p(m-\lambda_1)-m} \quad (x \in \mathbf{R}_+^m),$$

and

$$\Psi(n) = ||v(n)||_{\beta}^{q(s-\lambda_2)-s} \left(\prod_{k=1}^{s} v'(n_k)\right)^{1-q} \quad (n \in \mathbf{N}_{n^{(0)}}^s),$$

wherefrom

$$[\varphi(x)]^{1-q} = ||x||_{\alpha}^{q\lambda_1-m},$$

and

$$[\Psi(n)]^{1-p} = ||v(n)||_{\beta}^{p\lambda_2-s} \prod_{k=1}^{s} v'(n_k).$$

We also set

$$\widetilde{\varphi}(x) = (1 - \theta(||x||_{\alpha}))||x||_{\alpha}^{p(m-\lambda_1)-m},$$

and

$$[\widetilde{\varphi}(x)]^{1-q} = (1 - \theta(||x||_{\alpha}))^{1-q}||x||_{\alpha}^{q\lambda_1-m},$$

where

$$\theta(||x||_{\alpha}) = \frac{\delta_0}{K(\beta, s, \lambda_1)} \int_{||x||_{\alpha}}^{\infty} k_{\lambda}(t,1)t^{\lambda_1-1}dt > 0 \quad (0 < \delta_0 \leq 1).$$

We define two sets as follows:

$$\mathbf{L}_{p,\varphi}(\mathbf{R}_+^m) = \left\{f; ||f||_{p,\varphi} = \left\{\int_{\mathbf{R}_+^m} ||x||_{\alpha}^{p(m-\lambda_1)-m}|f(x)|^p dx\right\}^{\frac{1}{p}} < \infty\right\},$$

$$l_{q,\Psi} = \left\{a; ||a||_{q,\Psi} = \left\{\sum_{n \in \mathbf{N}_{n^{(0)}}^s} ||v(n)||_{\beta}^{q(s-\lambda_2)-s} \right. \right.$$

$$\left. \left. \times \left(\prod_{k=1}^{s} v'(n_k)\right)^{1-q} |a(n)|^q\right\}^{\frac{1}{q}} < \infty\right\}.$$

Note. For $p > 1(q > 1)$, the above two sets with the norms are normed linear spaces. For $0 < p < 1(q < 0)$ or $p < 0(0 < q < 1)$, we still use the sets with $||f||_{p,\varphi}$ and $||a||_{q,\Psi}$ as the formal symbols.

We have the following theorem:

Theorem 5.1. *Suppose that* $m, s, n^{(0)} \in \mathbf{N}, \alpha, \beta > 0, p \in \mathbf{R}\backslash\{0,1\}, \frac{1}{p} + \frac{1}{q} = 1, \ \lambda, \lambda_1, \lambda_2 \in \mathbf{R}, \ \lambda_1 + \lambda_2 = \lambda, k_\lambda(t,u)$ *is a finite homogeneous function of degree* $-\lambda$ *in* $\mathbf{R}_+^2, \ k_\lambda(t,u)t^{\lambda_2-s}$ *is decreasing with respect to* $u \in \mathbf{R}_+$ *and strictly decreasing in an interval* $I \subset (n^{(0)}, \infty), \ v'(t) > 0, v''(t) \le 0$ $(t \in (n^{(0)} - 1, \infty)),$ *with* $v(n^{(0)} - 1) \ge 0,$

$$K(\lambda_1) = \left(\frac{\Gamma^m(\frac{1}{\alpha})}{\alpha^{m-1}\Gamma(\frac{m}{\alpha})}\right)^{\frac{1}{q}} \left(\frac{\Gamma^s(\frac{1}{\beta})}{\beta^{s-1}\Gamma\left(\frac{s}{\beta}\right)}\right)^{\frac{1}{p}} k(\lambda_1), \qquad (5.27)$$

where $k(\lambda_1) = \int_0^\infty k_\lambda(t,1)t^{\lambda_1-1}dt \in \mathbf{R}_+.$ *If* $f(x) \ge 0, a(n) \ge 0, \ f \in \mathbf{L}_{p,\varphi}(\mathbf{R}_+^m), a = \{a(n)\} \in l_{q,\Psi}$ *such that* $\|f\|_{p,\varphi} > 0, \|a\|_{q,\Psi} > 0,$ *then*

(i) *for* $p > 1$, *we have the following equivalent inequalities:*

$$I = \int_{\mathbf{R}_+^m} \sum_{n \in \mathbf{N}_{n^{(0)}}^s} k_\lambda(\|x\|_\alpha, \|v(n)\|_\beta)a(n)f(x)dx$$

$$< K(\lambda_1)\|f\|_{p,\varphi}\|a\|_{q,\Psi}, \qquad (5.28)$$

$$J_1 = \left\{\sum_{n \in \mathbf{N}_{n^{(0)}}^s} [\Psi(n)]^{1-p} \left(\int_{\mathbf{R}_+^m} k_\lambda(\|x\|_\alpha, \|v(n)\|_\beta)f(x)dx\right)^p\right\}^{\frac{1}{p}}$$

$$< K(\lambda_1)\|f\|_{p,\varphi}, \qquad (5.29)$$

and

$$J_2 = \left\{\int_{\mathbf{R}_+^m} [\varphi(x)]^{1-q} \left(\sum_{n \in \mathbf{N}_{n^{(0)}}^s} k_\lambda(\|x\|_\alpha, \|v(n)\|_\beta)a(n)\right)^q dx\right\}^{\frac{1}{q}}$$

$$< K(\lambda_1)\|a\|_{q,\Psi}; \qquad (5.30)$$

(ii) *for* $p < 0$, *we have the reverses of* (5.28), (5.29) *and* (5.30);

(iii) *for* $0 < p < 1$, *if* $f(x) \ge 0, \ a(n) \ge 0, \ f \in \mathbf{L}_{p,\widetilde{\varphi}}(\mathbf{R}_+^m), \ a = \{a(n)\} \in l_{q,\Psi}$ *such that* $\|f\|_{p,\widetilde{\varphi}} > 0, \|a\|_{q,\Psi} > 0,$ *then we have the following reverse equivalent inequalities:*

$$I = \int_{\mathbf{R}_+^m} \sum_{n \in \mathbf{N}_{n^{(0)}}^s} k_\lambda(\|x\|_\alpha, \|v(n)\|_\beta)a(n)f(x)dx$$

$$> K(\lambda_1)\|f\|_{p,\widetilde{\varphi}}\|a\|_{q,\Psi}, \qquad (5.31)$$

$$J_1 = \left\{\sum_{n \in \mathbf{N}_{n^{(0)}}^s} [\Psi(n)]^{1-p} \left(\int_{\mathbf{R}_+^m} k_\lambda(\|x\|_\alpha, \|v(n)\|_\beta)f(x)dx\right)^p\right\}^{\frac{1}{p}}$$

$$> K(\lambda_1)\|f\|_{p,\widetilde{\varphi}}, \qquad (5.32)$$

and

$$\tilde{J}_2 = \left\{ \int_{\mathbf{R}_+^m} [\tilde{\varphi}(x)]^{1-q} \left(\sum_{n \in \mathbf{N}_{n(0)}^s} k_\lambda(||x||_\alpha, ||v(n)||_\beta) a(n) \right)^q dx \right\}^{\frac{1}{q}}$$
$$> K(\lambda_1) ||a||_{q,\Psi}.$$

(5.33)

Proof. (i) For $p > 1$, by (5.12) and (5.13), for

$$H(||x||_\alpha, ||v(n)||_\beta) = k_\lambda(||x||_\alpha, ||v(n)||_\beta),$$

$\omega(n) = \omega(\lambda_2, n)$ and $\varpi(x) = \varpi(\lambda_1, x)$, in view of (5.16), (5.17) and the assumptions, we have (5.29) and (5.30).

By Hölder's inequality, we have

$$I = \sum_{n \in \mathbf{N}_{n(0)}^s} \int_{\mathbf{R}_+^m} k_\lambda(||x||_\alpha, ||v(n)||_\beta) a(n) f(x) dx$$
$$= \sum_{n \in \mathbf{N}_{n(0)}^s} \left\{ [\Psi(n)]^{\frac{-1}{q}} \int_{\mathbf{R}_+^m} k_\lambda(||x||_\alpha, ||v(n)||_\beta) f(x) dx \right\} \{ [\Psi(n)]^{\frac{1}{q}} a(n) \}$$
$$\le J_1 ||a||_{q,\Psi}.$$

(5.34)

Then, by (5.29), we have inequality (5.28).

On the other hand, assuming that (5.28) is valid, setting

$$a(n) = [\Psi(n)]^{1-p} \left(\int_{\mathbf{R}_+^m} k_\lambda(||x||_\alpha, ||v(n)||_\beta) f(x) dx \right)^{p-1} \qquad (n \in \mathbf{N}_{n(0)}^s),$$

then, we have $||a||_{q,\Psi} = J_1^{p-1}$. By (5.12) and the assumption of $0 < ||f||_{p,\varphi} < \infty$, we have $J_1 < \infty$. If $J_1 = 0$, then (5.29) is trivially valid; if $J_1 > 0$, then by (5.28), it follows that

$$||a||_{q,\Psi}^q = J_1^p = I < K(\lambda_1) ||f||_{p,\varphi} ||a||_{q,\Psi},$$
$$||a||_{q,\Psi}^{q-1} = J_1 < K(\lambda_1) ||f||_{p,\varphi},$$

then, (5.29) follows. Hence, (5.28) and (5.29) are equivalent.

By Hölder's inequality, we still have

$$I = \int_{\mathbf{R}_+^m} \{ [\varphi(x)]^{\frac{1}{p}} f(x) \} \left\{ [\varphi(x)]^{\frac{-1}{p}} \sum_{n \in \mathbf{N}_{n(0)}^s} k_\lambda(||x||_\alpha, ||v(n)||_\beta) a(n) \right\} dx$$
$$\le ||f||_{p,\varphi} J_2.$$

(5.35)

Then, by (5.30), we have (5.28).

On the other hand, assuming that (5.28) is valid, setting

$$f(x) = [\varphi(x)]^{1-q} \left(\sum_{n \in \mathbf{N}_{n^{(0)}}^s} k_\lambda(||x||_\alpha, ||v(n)||_\beta) a(n) \right)^{q-1} \qquad (x \in \mathbf{R}_+^m),$$

then, we have $||f||_{p,\varphi} = J_2^{q-1}$. By (5.13) and the assumption of $0 < ||a||_{q,\Psi} < \infty$, we have $J_2 < \infty$. If $J_2 = 0$, then (5.30) is valid trivially; if $J_2 > 0$, then by (5.28), it follows that

$$||f||_{p,\varphi}^p = J_2^q = I < K(\lambda_1)||f||_{p,\varphi}||a||_{q,\Psi},$$
$$||f||_{p,\varphi}^{p-1} = J_2 < K(\lambda_1)||a||_{q,\Psi}.$$

Thus, (5.30) follows, and (5.28) and (5.30) are equivalent.

Hence, (5.28), (5.29) and (5.30) are equivalent.

(ii) For $p < 0$ $(0 < q < 1)$, by the reverses of (5.12) and (5.13), for

$$H(||x||_\alpha, ||v(n)||_\beta) = k_\lambda(||x||_\alpha, ||v(n)||_\beta),$$

$\omega(n) = \omega(\lambda_2, n)$ and $\varpi(x) = \varpi(\lambda_1, x)$, in view of (5.16), (5.17) and the assumptions, we have the reverses of (5.29) and (5.30).

By the reverse Hölder's inequality, we have

$$I \geq J_1 ||a||_{q,\Psi}. \tag{5.36}$$

Then, by the reverse of (5.29), we have the reverse of (5.28).

On the other hand, assuming that the reverse of (5.28) is valid, setting $a(n)$ as (i), then we have $||a||_{q,\Psi} = J_1^{p-1}$. By the reverse of (5.12) and the assumption of $0 < ||f||_{p,\varphi} < \infty$, we have $J_1 > 0$. If $J_1 = \infty$, then the reverse of (5.29) is valid trivially; if $J_1 < \infty$, then by the reverse of (5.28), it follows that

$$||a||_{q,\Psi}^q = J_1^p = I > K(\lambda_1)||f||_{p,\varphi}||a||_{q,\Psi},$$
$$||a||_{q,\Psi}^{q-1} = J_1 > K(\lambda_1)||f||_{p,\varphi}.$$

Thus, the reverse of (5.29) follows. Hence, the reverses of (5.28) and (5.29) are equivalent.

By the reverse Hölder's inequality, we still have

$$I \geq ||f||_{p,\varphi} J_2. \tag{5.37}$$

Then, by the reverse of inequality (5.30), we have the reverse of (5.28).

On the other hand, assuming that the reverse of (5.28) is valid, setting $f(x)$ as (i), then we have $||f||_{p,\varphi} = J_2^{q-1}$. By the reverse of (5.13) and the assumption of $0 < ||a||_{q,\Psi} < \infty$, we have $J_2 > 0$. If $J_2 = \infty$, then the reverse of (5.30) is trivially valid; if $J_2 < \infty$, then by the reverse of (5.28), it follows that

$$||f||_{p,\varphi}^p = J_2^q = I > K(\lambda_1)||f||_{p,\varphi}||a||_{q,\Psi},$$
$$||f||_{p,\varphi}^{p-1} = J_2 > K(\lambda_1)||a||_{q,\Psi}.$$

Thus, the reverse of (5.30) follows and then, the reverses of (5.28) and (5.30) are equivalent.

Hence, the reverses of (5.28), (5.29) and (5.30) are equivalent.

(iii) For $0 < p < 1(q < 0)$, by the reverses of (5.12) and (5.13), for

$$H(||x||_\alpha, ||v(n)||_\beta) = k_\lambda(||x||_\alpha, ||v(n)||_\beta),$$

$\omega(n) = \omega(\lambda_2, n)$ and $\varpi(x) = \varpi(\lambda_1, x)$, in view of (5.16), (5.18) and the assumptions, we have (5.32) and (5.33).

By the reverse Hölder's inequality, we have

$$I \geq J_1||a||_{q,\Psi}. \tag{5.38}$$

Then, by (5.32), we have (5.30).

On the other hand, assuming that (5.30) is valid, setting $a(n)$ as (i), then we have $||a||_{q,\Psi} = J_1^{p-1}$. By the reverse of (5.12) and the assumption of $0 < ||f||_{p,\widetilde{\varphi}} < \infty$, we have $J_1 > 0$. If $J_1 = \infty$, then (5.32) is trivially valid; if $J_1 < \infty$, then by (5.31), it follows that

$$||a||_{q,\Psi}^q = J_1^p = I > K(\lambda_1)||f||_{p,\widetilde{\varphi}}||a||_{q,\Psi},$$
$$||a||_{q,\Psi}^{q-1} = J_1 > K(\lambda_1)||f||_{p,\widetilde{\varphi}}.$$

Thus, (5.31) follows. Hence, (5.30) and (5.31) are equivalent.

By the reverse Hölder's inequality, we still have

$$I \geq ||f||_{p,\widetilde{\varphi}}\widetilde{J}_2. \tag{5.39}$$

Then, by (5.33), we have (5.31).

On the other hand, assuming that (5.31) is valid, setting $f(x)$ as follows:

$$f(x) = [\widetilde{\varphi}(x)]^{1-q} \left(\sum_{n \in \mathbf{N}_n^s(0)} k_\lambda(||x||_\alpha, ||v(n)||_\beta)a(n) \right)^{q-1}, \qquad x \in \mathbf{R}_+^m,$$

then, we have $||f||_{p,\widetilde{\varphi}} = \widetilde{J}_2^{q-1}$. By the reverse of (5.13) and the assumption of $0 < ||a||_{q,\Psi} < \infty$, we have $\widetilde{J}_2 > 0$. If $\widetilde{J}_2 = \infty$, then, (5.33) is trivially valid; if $\widetilde{J}_2 < \infty$, and then, by (5.31), it follows that

$$||f||_{p,\widetilde{\varphi}}^p = \widetilde{J}_2^q = I > K(\lambda_1)||f||_{p,\widetilde{\varphi}}||a||_{q,\Psi},$$
$$||f||_{p,\widetilde{\varphi}}^{p-1} = \widetilde{J}_2 > K(\lambda_1)||a||_{q,\Psi}.$$

Thus, (5.33) follows and then (5.31) and (5.33) are equivalent.

Hence, (5.31), (5.32) and (5.33) are equivalent. $\qquad \square$

Theorem 5.2. *As the assumptions of Theorem 5.1, if there exist $\widetilde{\delta}_0 > 0$, and $\eta > \lambda_1$, such that for any $\delta \in (0, \widetilde{\delta}_0]$,*

$$k(\lambda_1 \pm \delta) = \int_0^\infty k_\lambda(t, 1) t^{(\lambda_1 \pm \delta) - 1} dt \in \mathbf{R}_+,$$

and

$$k_\lambda(t, 1) \le L\left(\frac{1}{t^\eta}\right) \qquad (t \ge 1; L > 0),$$

then, the inequalities in Theorem 5.1 have the same best possible constant factor $K(\lambda_1)$ given by (5.27).

Proof. For $0 < \varepsilon < |q|\widetilde{\delta}_0$, we set $\widetilde{f}(x), \widetilde{a}(n)$ as follows:

$$\widetilde{f}(x) = \begin{cases} 0, & 0 < ||x||_\alpha < 1, \\ ||x||_\alpha^{\lambda_1 - m - \frac{\varepsilon}{p}}, & ||x||_\alpha \ge 1, \end{cases}$$

$$\widetilde{a}(n) = ||v(n)||_\beta^{\lambda_2 - \frac{\varepsilon}{q} - s} \prod_{k=1}^s v'(n_k), n \in \mathbf{N}_{n^{(0)}}^s.$$

(i) For $p > 1$, if there exists a constant $k(\le K(\lambda_1))$, such that (5.28) is still valid as we replace $K(\lambda_1)$ by k, then, in particular, it follows that

$$\widetilde{I}(\varepsilon) = \int_{\mathbf{R}_+^m} \sum_{n \in \mathbf{N}_{n^{(0)}}^s} k_\lambda(||x||_\alpha, ||v(n)||_\beta) \widetilde{a}(n) \widetilde{f}(x) dx < k||\widetilde{f}||_{p,\varphi} ||\widetilde{a}||_{q,\Psi}.$$

$$(5.40)$$

Since by (5.2) and (5.4), we have

$$||\widetilde{f}||_{p,\varphi} = \left\{ \int_{\{x \in \mathbf{R}_+^m ; ||x||_\alpha \ge 1\}} ||x||_\alpha^{-m - \varepsilon} dx \right\}^{\frac{1}{p}} = \left\{ \frac{\Gamma^m(\frac{1}{\alpha})}{\varepsilon \alpha^{m-1} \Gamma(\frac{m}{\alpha})} \right\}^{\frac{1}{p}},$$

$$||\widetilde{a}||_{q,\Psi} = \left\{ \sum_{n \in \mathbf{N}_{n^{(0)}}^s} ||v(n)||_\beta^{-s - \varepsilon} \prod_{k=1}^s v'(n_k) \right\}^{\frac{1}{q}} = \left\{ \frac{\Gamma^s(\frac{1}{\beta})}{\varepsilon \beta^{s-1} \Gamma(\frac{s}{\beta})} + O(1) \right\}^{\frac{1}{q}},$$

then, by (5.19) and (5.40), it follows that

$$\frac{\Gamma^m(\frac{1}{\alpha})}{\alpha^{m-1} \Gamma(\frac{m}{\alpha})} \frac{\Gamma^s(\frac{1}{\beta})}{\beta^{s-1} \Gamma(\frac{s}{\beta})} (k(\lambda_1) + o(1)) (1 - \varepsilon O_1(1))$$

$$< k \left\{ \frac{\Gamma^m(\frac{1}{\alpha})}{\alpha^{m-1} \Gamma(\frac{m}{\alpha})} \right\}^{\frac{1}{p}} \left\{ \frac{\Gamma^s(\frac{1}{\beta})}{\beta^{s-1} \Gamma(\frac{s}{\beta})} + \varepsilon O(1) \right\}^{\frac{1}{q}},$$

and then $K(\lambda_1) \leq k$ $(\varepsilon \to 0^+)$. Hence, $k = K(\lambda_1)$ is the best possible constant factor of (5.28).

By the equivalency, the constant factor in (5.29) and (5.30) is the best possible. Otherwise, we can get a contradiction by (5.34) and (5.35) that the constant factor in (5.28) is not the best possible.

(ii) For $p < 0$, if there exists a constant $k \geq K(\lambda_1)$, such that the reverse of (5.28) is still valid as we replace $K(\lambda_1)$ by k, then, in particular, it follows by (i) that

$$\frac{\Gamma^m(\frac{1}{\alpha})}{\alpha^{m-1}\Gamma(\frac{m}{\alpha})}\frac{\Gamma^s(\frac{1}{\beta})}{\beta^{s-1}\Gamma\left(\frac{s}{\beta}\right)}(k(\lambda_1)+o(1))(1-\varepsilon O_1(1))$$

$$> k\left\{\frac{\Gamma^m(\frac{1}{\alpha})}{\alpha^{m-1}\Gamma(\frac{m}{\alpha})}\right\}^{\frac{1}{p}}\left\{\frac{\Gamma^s(\frac{1}{\beta})}{\beta^{s-1}\Gamma\left(\frac{s}{\beta}\right)}+\varepsilon O(1)\right\}^{\frac{1}{q}\frac{1}{q}},$$

and then $K(\lambda_1) \geq k$ $(\varepsilon \to 0^+)$. Hence, $k = K(\lambda_1)$ is the best possible constant factor of the reverse of (5.28).

By the equivalency, the constant factor in the reverse of (5.29) (the reverse of (5.30)) is the best possible. Otherwise, we can get a contradiction by (5.36) ((5.37)) that the constant factor in the reverse of (5.28) is not the best possible.

(iii) For $0 < p < 1$, if there exists a constant $k \geq K(\lambda_1)$, such that (5.31) is still valid as we replace $K(\lambda_1)$ by k, then in particular, it follows that

$$\widetilde{I}(\varepsilon) = \int_{\mathbf{R}_+^m}\sum_{n\in\mathbf{N}_n^s{}_{(0)}}k_\lambda(||x||_\alpha,||v(n)||_\beta)\widetilde{a}(n)\widetilde{f}(x)dx$$

$$> k||\widetilde{f}||_{p,\widetilde{\varphi}}||\widetilde{a}||_{q,\Psi}.$$

Since, by (5.20), we have

$$||\widetilde{f}||_{p,\widetilde{\varphi}} = \left\{\int_{\{x\in\mathbf{R}_+^m;||x||_\alpha\geq1\}}||x||_\alpha^{-m-\varepsilon}(1-\theta(||x||_\alpha))dx\right\}^{\frac{1}{p}}$$

$$= \left\{\frac{\Gamma^m(\frac{1}{\alpha})}{\varepsilon\alpha^{m-1}\Gamma(\frac{m}{\alpha})}(1-\varepsilon O_1(1))\right\}^{\frac{1}{p}},$$

then it follows that

$$\frac{\Gamma^m(\frac{1}{\alpha})}{\alpha^{m-1}\Gamma(\frac{m}{\alpha})}\frac{\Gamma^s(\frac{1}{\beta})}{\beta^{s-1}\Gamma\left(\frac{s}{\beta}\right)}(k(\lambda_1)+o(1))\,(1-\varepsilon O_1(1))$$

$$> k\left\{\frac{\Gamma^m(\frac{1}{\alpha})}{\alpha^{m-1}\Gamma(\frac{m}{\alpha})}(1-\varepsilon O_1(1))\right\}^{\frac{1}{p}}\left\{\frac{\Gamma^s(\frac{1}{\beta})}{\beta^{s-1}\Gamma\left(\frac{s}{\beta}\right)}+\varepsilon O(1)\right\}^{\frac{1}{q}},$$

and then $K(\lambda_1) \geq k$ ($\varepsilon \to 0^+$). Hence, $k = K(\lambda_1)$ is the best value of (5.31).

By the equivalency, the constant factor in (5.32), (5.33) is the best possible. Otherwise, we can get a contradiction by (5.38) ((5.39)) that the constant factor in (5.31) is not the best possible. □

For $\beta = 1$, we set

$$\Psi_1(n) = ||v(n)||_1^{q(s-\lambda_2)-s}\left(\prod_{k=1}^{s}v'(n_k)\right)^{1-q}\quad(n\in\mathbf{N}_{n^{(0)}}^s),$$

wherefrom

$$[\Psi_1(n)]^{1-p} = ||v(n)||_1^{q\lambda_2-s}\prod_{k=1}^{s}v'(n_k).$$

We also set

$$\widetilde{\varphi}_1(x) = (1-\theta_1(||x||_\alpha))||x||_\alpha^{p(m-\lambda_1)-m},$$

and

$$[\widetilde{\varphi}_1(x)]^{1-q} = (1-\theta_1(||x||_\alpha))^{1-q}||x||_\alpha^{q\lambda_1-m},$$

where

$$\theta_1(||x||_\alpha) = \frac{\delta_0(s-1)!}{k(\lambda_1)}\int_{||x||_\alpha}^{\infty}k_\lambda(t,1)t^{\lambda_1-1}dt\quad(0<\delta_0\leq 1).$$

By (5.21)-(5.25), in the same way of Theorem 5.1 and Theorem 5.2, we have

Theorem 5.3. *Suppose that* $m, s \in \mathbf{N}$, $\alpha > 0$, $p \in \mathbf{R}\backslash\{0,1\}$, $\frac{1}{p}+\frac{1}{q}=1$, λ, $\lambda_1, \lambda_2 \in \mathbf{R}, \lambda_1+\lambda_2=\lambda$, $k_\lambda(t,u)$ *is a finite homogeneous function of degree* $-\lambda$ *in* $\mathbf{R}_+^2$, *satisfying*

$$\frac{\partial}{\partial u}(k_\lambda(t,u)u^{\lambda_2-s}) < 0,\quad\frac{\partial^2}{\partial u^2}(k_\lambda(t,u)u^{\lambda_2-s})>0\quad(u\in\mathbf{R}^+),$$

$$v'(t) > 0, v''(t) \leq 0, v'''(t)\geq 0(t\in(n^{(0)}-\frac{1}{2},\infty)), \text{ with } v\left(n^{(0)}-\frac{1}{2}\right)\geq 0,$$

$$K_1(\lambda_1) = \left(\frac{\Gamma^m(\frac{1}{\alpha})}{\alpha^{m-1}\Gamma(\frac{m}{\alpha})}\right)^{\frac{1}{q}}\left(\frac{1}{(s-1)!}\right)^{\frac{1}{p}}k(\lambda_1),\qquad(5.41)$$

where $k(\lambda_1) = \int_0^\infty k_\lambda(t,1)t^{\lambda_1-1}du \in \mathbf{R}_+.$ *If* $f(x) \geq 0, a(n) \geq 0,$ $f \in$ $\mathbf{L}_{p,\varphi}(\mathbf{R}_+^m),$ $a = \{a(n)\} \in l_{q,\Psi_1}$ *such that* $||f||_{p,\varphi} > 0, ||a||_{q,\Psi_1} > 0,$ *then,*

(i) for $p > 1$, we have the following equivalent inequalities:

$$\int_{\mathbf{R}_+^m} \sum_{n \in \mathbf{N}_{n(0)}^s} k_\lambda(||x||_\alpha, ||v(n)||_1)a(n)f(x)dx$$

$$< K_1(\lambda_1)||f||_{p,\varphi}||a||_{q,\Psi_1}, \tag{5.42}$$

$$\left\{ \sum_{n \in \mathbf{N}_{n(0)}^s} [\Psi_1(n)]^{1-p} \left(\int_{\mathbf{R}_+^m} k_\lambda(||x||_\alpha, ||v(n)||_1)f(x)dx \right)^p \right\}^{\frac{1}{p}}$$

$$< K_1(\lambda_1)||f||_{p,\varphi}, \tag{5.43}$$

and

$$\left\{ \int_{\mathbf{R}_+^m} [\varphi(x)]^{1-q} \left(\sum_{n \in \mathbf{N}_{n(0)}^s} k_\lambda(||x||_\alpha, ||v(n)||_1)a(n) \right)^q dx \right\}^{\frac{1}{q}}$$

$$< K_1(\lambda_1)||a||_{q,\Psi_1}; \tag{5.44}$$

(ii) for $p < 0$, we have the reverses of (5.42), (5.43) and (5.44);

(iii) for $0 < p < 1$, if $f(x) \geq 0$, $a(n) \geq 0$, $f \in \mathbf{L}_{p,\widetilde{\varphi}}(\mathbf{R}_+^m)$, $a = \{a(n)\} \in$ l_{q,Ψ_1} such that $||f||_{p,\widetilde{\varphi}_1} > 0, ||a||_{q,\Psi_1} > 0$, then we have the following reverse equivalent inequalities:

$$\int_{\mathbf{R}_+^m} \sum_{n \in \mathbf{N}_{n(0)}^s} k_\lambda(||x||_\alpha, ||v(n)||_1)a(n)f(x)dx$$

$$> K_1(\lambda_1)||f||_{p,\widetilde{\varphi}_1}||a||_{q,\Psi_1}, \tag{5.45}$$

$$\left\{ \sum_{n \in \mathbf{N}_{n(0)}^s} [\Psi_1(n)]^{1-p} \left(\int_{\mathbf{R}_+^m} k_\lambda(||x||_\alpha, ||v(n)||_1)f(x)dx \right)^p \right\}^{\frac{1}{p}}$$

$$> K_1(\lambda_1)||f||_{p,\widetilde{\varphi}_1}, \tag{5.46}$$

and

$$\left\{ \int_{\mathbf{R}_+^m} [\widetilde{\varphi}_1(x)]^{1-q} \left(\sum_{n \in \mathbf{N}_{n(0)}^s} k_\lambda(||x||_\alpha, ||v(n)||_1)a(n) \right)^q dx \right\}^{\frac{1}{q}}$$

$$> K_1(\lambda_1)||a||_{q,\Psi_1}. \tag{5.47}$$

Moreover, if there exist $\widetilde{\delta}_0 > 0$ and $\eta > \lambda_1$, such that for any $\delta \in (0, \widetilde{\delta}_0]$,

$$k(\lambda_1 \pm \delta) = \int_0^\infty k_\lambda(t,1)t^{(\lambda_1 \pm \delta)-1}dt \in \mathbf{R}_+,$$

and

$$k_\lambda(t,1) \leq L\left(\frac{1}{t^\eta}\right) \ (t \geq 1; L > 0),$$

then, the above inequalities have the same best possible constant factor $K_1(\lambda_1)$.

5.3.3   *Some Corollaries*

If $w(t)$ is a strictly increasing differentiable function in (b,c) $(-\infty \leq b < c \leq \infty)$, with $w(b+) = 0$, $w(c-) = \infty$,

$$\Phi(x) = ||w(x)||_\alpha^{p(m-\lambda_1)-m}(\prod_{i=1}^m w'(x_i))^{1-p},$$

and $\widetilde{\Phi}(x) = (1 - \theta(||w(x)||_\alpha))\Phi(x)$, setting $x = w(X) = (w(X_1), \cdots, w(X_m))$ in the inequalities of Theorem 5.1, after simplification, replacing $f(w(X))\prod_{i=1}^m w'(X_i)$ by $f(x)$, we have:

Corollary 5.1. *Suppose that m, s, $n^{(0)} \in \mathbf{N}$, α, $\beta > 0$, $p \in \mathbf{R}\backslash\{0,1\}$, $\frac{1}{p} + \frac{1}{q} = 1$, λ, λ_1, $\lambda_2 \in \mathbf{R}$, $\lambda_1 + \lambda_2 = \lambda$, $k_\lambda(t,u)$ is a finite homogeneous function of degree $-\lambda$ in $\mathbf{R}_+^2$, $k_\lambda(t,u)u^{\lambda_2-s}$ is decreasing with respect to $u \in \mathbf{R}_+$ and strictly decreasing in an interval $I \subset (n^{(0)}, \infty)$, $w(t)$ is strictly increasing in (b,c), with $w(b+) = 0$, $w(c-) = \infty$, and $v'(t) > 0$, $v''(t) \leq 0$ $(t \in (n^{(0)} - 1, \infty))$, with $v(n^{(0)} - 1) \geq 0$,*

$$K(\lambda_1) = \left(\frac{\Gamma^m(\frac{1}{\alpha})}{\alpha^{m-1}\Gamma(\frac{m}{\alpha})}\right)^{\frac{1}{q}}\left(\frac{\Gamma^s(\frac{1}{\beta})}{\beta^{s-1}\Gamma\left(\frac{s}{\beta}\right)}\right)^{\frac{1}{p}}k(\lambda_1), \qquad (5.48)$$

where $k(\lambda_1) = \int_0^\infty k_\lambda(t,1)t^{\lambda_1-1}dt \in \mathbf{R}_+$. If $f(x) \geq 0$, $a(n) \geq 0$, $f \in \mathbf{L}_{p,\Phi}((\mathbf{b},\mathbf{c})^m)$, $a = \{a(n)\} \in l_{q,\Psi}$, such that $||f||_{p,\Phi} > 0$, $||a||_{q,\Psi} > 0$, then,

(i) for $p > 1$, we have the following equivalent inequalities:

$$\int_{(\mathbf{b},\mathbf{c})^m} \sum_{n \in \mathbf{N}_{n^{(0)}}^s} k_\lambda(||w(x)||_\alpha, ||v(n)||_\beta)a(n)f(x)dx$$

$$< K(\lambda_1)||f||_{p,\Phi}||a||_{q,\Psi}, \qquad (5.49)$$

$$\left\{\sum_{n \in \mathbf{N}_{n^{(0)}}^s}[\Psi(n)]^{1-p}\left(\int_{(\mathbf{b},\mathbf{c})^m} k_\lambda(||w(x)||_\alpha, ||v(n)||_\beta)f(x)dx\right)^p\right\}^{\frac{1}{p}}$$

$$< K(\lambda_1)||f||_{p,\Phi}, \qquad (5.50)$$

and

$$\left\{ \int_{(\mathbf{b},\mathbf{c})^m} [\Phi(x)]^{1-q} \left(\sum_{n \in \mathbf{N}^s_{n(0)}} k_\lambda(||w(x)||_\alpha, ||v(n)||_\beta) a(n) \right)^q dx \right\}^{\frac{1}{q}}$$
$$< K(\lambda_1)||a||_{q,\Psi}; \qquad (5.51)$$

(ii) for $p < 0$, we have the reverses of (5.49), (5.50) and (5.51);

(iii) for $0 < p < 1$, if $f(x) \geq 0$, $a(n) \geq 0$, $f \in \mathbf{L}_{p,\widetilde{\Phi}}(\mathbf{R}^m_+)$, $a = \{a(n)\} \in l_{q,\Psi}$, such that $||f||_{p,\widetilde{\Phi}} > 0$, $||a||_{q,\Psi} > 0$, then we have the following reverse equivalent inequalities:

$$\int_{(\mathbf{b},\mathbf{c})^m} \sum_{n \in \mathbf{N}^s_{n(0)}} k_\lambda(||w(x)||_\alpha, ||v(n)||_\beta) a(n) f(x) dx$$
$$< K(\lambda_1)||f||_{p,\widetilde{\Phi}}||a||_{q,\Psi}, \qquad (5.52)$$

$$\left\{ \sum_{n \in \mathbf{N}^s_{n(0)}} [\Psi(n)]^{1-p} \left(\int_{(\mathbf{b},\mathbf{c})^m} k_\lambda(||w(x)||_\alpha, ||v(n)||_\beta) f(x) dx \right)^p \right\}^{\frac{1}{p}}$$
$$> K(\lambda_1)||f||_{p,\widetilde{\Phi}}, \qquad (5.53)$$

and

$$\left\{ \int_{(\mathbf{b},\mathbf{c})^m} [\widetilde{\varphi}(x)]^{1-q} \left(\sum_{n \in \mathbf{N}^s_{n(0)}} k_\lambda(||w(x)||_\alpha, ||v(n)||_\beta) a(n) \right)^q dx \right\}^{\frac{1}{q}}$$
$$> K(\lambda_1)||a||_{q,\Psi}. \qquad (5.54)$$

For $w(x) = x$ in (5.49)-(5.54), we reduce (5.28)-(5.33). Hence, Theorem 5.1 and Corollary 5.1 are equivalent. We can conclude that the constant factor in the above inequalities is the best possible by adding some conditions as follows:

Corollary 5.2. *As the assumptions of Corollary 5.1, if there exist $\widetilde{\delta}_0 > 0$ and $\eta > \lambda_1$, such that for any $\delta \in (0, \widetilde{\delta}_0]$,*

$$k(\lambda_1 \pm \delta) = \int_0^\infty k_\lambda(t, 1) t^{(\lambda_1 \pm \delta) - 1} dt \in \mathbf{R}_+,$$

and

$$k_\lambda(t, 1) \leq L\frac{1}{t^\eta} \quad (t \geq 1; L > 0),$$

then the inequalities in Corollary 5.1 are all with the same best possible constant factor $K(\lambda_1)$.

For $v(t) = t, n^{(0)} = 1$ in Theorem 5.1 and Theorem 5.2, setting

$$\psi(n) = ||n||_\beta^{q(s-\lambda_2)-s} (n \in \mathbf{N}^s),$$

we have

Corollary 5.3. *Suppose that $m, s \in \mathbf{N}$, $\alpha, \beta > 0$, $p \in \mathbf{R}\backslash\{0,1\}$, $\frac{1}{p} + \frac{1}{q} = 1$, $\lambda, \lambda_1, \lambda_2 \in \mathbf{R}$, $\lambda_1 + \lambda_2 = \lambda$, $k_\lambda(t,u)$ is a finite homogeneous function of degree $-\lambda$ in $\mathbf{R}_+^2$, $k_\lambda(t,u)u^{\lambda_2-s}$ is decreasing with respect to $u \in \mathbf{R}_+$ and strict decreasing in an interval $I \subset (1,\infty)$,*

$$K(\lambda_1) = \left(\frac{\Gamma^m(\frac{1}{\alpha})}{\alpha^{m-1}\Gamma(\frac{m}{\alpha})}\right)^{\frac{1}{q}} \left(\frac{\Gamma^s(\frac{1}{\beta})}{\beta^{s-1}\Gamma\left(\frac{s}{\beta}\right)}\right)^{\frac{1}{p}} k(\lambda_1), \qquad (5.55)$$

where, $k(\lambda_1) = \int_0^\infty k_\lambda(t,1)t^{\lambda_1-1}dt \in \mathbf{R}_+$. If $f(x) \geq 0$, $a(n) \geq 0$, $f \in \mathbf{L}_{p,\varphi}(\mathbf{R}_+^m)$, $a = \{a(n)\} \in \mathbf{l}_{q,\psi}$, such that $||f||_{p,\varphi} > 0$, $||a||_{q,\psi} > 0$, then

(i) for $p > 1$, we have the following equivalent inequalities:

$$\int_{\mathbf{R}_+^m} \sum_{n \in \mathbf{N}^s} k_\lambda(||x||_\alpha, ||n||_\beta)a(n)f(x)dx < K(\lambda_1)||f||_{p,\varphi}||a||_{q,\psi}, \qquad (5.56)$$

$$\left\{\sum_{n \in \mathbf{N}^s} [\psi(n)]^{1-p} \left(\int_{\mathbf{R}_+^m} k_\lambda(||x||_\alpha, ||n||_\beta)f(x)dx\right)^p\right\}^{\frac{1}{p}} < K(\lambda_1)||f||_{p,\varphi},$$
$$(5.57)$$

and

$$\left\{\int_{\mathbf{R}_+^m} [\varphi(x)]^{1-q} \left(\sum_{n \in \mathbf{N}^s} k_\lambda(||x||_\alpha, ||n||_\beta)a(n)\right)^q dx\right\}^{\frac{1}{q}} < K(\lambda_1)||a||_{q,\psi};$$
$$(5.58)$$

(ii) for $p < 0$, we have the reverses of (5.56), (5.57) and (5.58);

(iii) for $0 < p < 1$, if $f(x) \geq 0, a(n) \geq 0$, $f \in \mathbf{L}_{p,\widetilde{\varphi}}(\mathbf{R}_+^m)$, $a = \{a(n)\} \in \mathbf{l}_{q,\psi}$, such that $||f||_{p,\widetilde{\varphi}} > 0, ||a||_{q,\psi} > 0$, then we have the following reverse equivalent inequalities:

$$\int_{\mathbf{R}_+^m} \sum_{n \in \mathbf{N}^s} k_\lambda(||x||_\alpha, ||n||_\beta)a(n)f(x)dx > K(\lambda_1)||f||_{p,\widetilde{\varphi}}||a||_{q,\psi},$$
$$(5.59)$$

$$\left\{\sum_{n \in \mathbf{N}^s} [\psi(n)]^{1-p} \left(\int_{\mathbf{R}_+^m} k_\lambda(||x||_\alpha, ||n||_\beta)f(x)dx\right)^p\right\}^{\frac{1}{p}} > K(\lambda_1)||f||_{p,\widetilde{\varphi}}, \quad (5.60)$$

and

$$\left\{ \int_{\mathbf{R}_+^m} [\widetilde{\varphi}(x)]^{1-q} \left(\sum_{n \in \mathbf{N}^s} k_\lambda(||x||_\alpha, ||n||_\beta) a(n) \right)^q dx \right\}^{\frac{1}{q}} > K(\lambda_1) ||a||_{q,\Psi}.$$

(5.61)

Moreover, if there exist $\widetilde{\delta}_0 > 0$ and $\eta > \lambda_1$, such that for any $\delta \in (0, \widetilde{\delta}_0]$,

$$k(\lambda_1 \pm \delta) = \int_0^\infty k_\lambda(t, 1) t^{(\lambda_1 \pm \delta) - 1} dt \in \mathbf{R}_+,$$

and

$$k_\lambda(t, 1) \le L \frac{1}{t^\eta} \quad (t \ge 1; L > 0),$$

then, the above inequalities are all with the best constant factor $K(\lambda_1)$.

For $v(t) = (t - \xi)^\gamma (0 < \gamma \le 1, 0 \le \xi \le \frac{1}{2}), n^{(0)} = 1$ in Theorem 5.3, setting

$$\Psi_\xi(n) = ||(n - \xi)^\gamma||_1^{q(s-\lambda_2)-s} (n \in \mathbf{N}^s),$$

we have some more accurate inequalities of Corollary 5.3 (for $\beta = 1$) as follows:

Corollary 5.4. *Suppose that* $m, s \in \mathbf{N}, \alpha > 0, p \in \mathbf{R}\backslash\{0, 1\}, \frac{1}{p} + \frac{1}{q} = 1,$ $\lambda, \lambda_1, \lambda_2 \in \mathbf{R}, \lambda_1 + \lambda_2 = \lambda, k_\lambda(t, u)$ *is a finite homogeneous function of degree* $-\lambda$ *in* $\mathbf{R}_+^2$, *satisfying*

$$\frac{\partial}{\partial u}(k_\lambda(t, u) u^{\lambda_2 - s}) < 0, \quad \frac{\partial^2}{\partial u^2}(k_\lambda(t, u) u^{\lambda_2 - s}) > 0 \quad (u \in \mathbf{R}^+),$$

$$K_1(\lambda_1) = \left(\frac{\Gamma^m(\frac{1}{\alpha})}{\alpha^{m-1}\Gamma(\frac{m}{\alpha})} \right)^{\frac{1}{q}} \left(\frac{1}{(s-1)!} \right)^{\frac{1}{p}} k(\lambda_1), \quad (5.62)$$

where $k(\lambda_1) = \int_0^\infty k_\lambda(t, 1) t^{\lambda_1 - 1} dt \in \mathbf{R}_+$. *If* $f(x) \ge 0, a(n) \ge 0, f \in$ $\mathbf{L}_{p,\varphi}(\mathbf{R}_+^m), a = \{a(n)\} \in l_{q,\Psi_\xi}$ *such that* $||f||_{p,\varphi} > 0, ||a||_{q,\Psi_\xi} > 0$, *then*

(i) *for* $p > 1$, *we have the following equivalent inequalities:*

$$\int_{\mathbf{R}_+^m} \sum_{n \in \mathbf{N}^s} k_\lambda(||x||_\alpha, ||(n - \xi)^\gamma||_1) a(n) f(x) dx$$
$$< K_1(\lambda_1) ||f||_{p,\varphi} ||a||_{q,\Psi_\xi}, \quad (5.63)$$

$$\left\{ \sum_{n \in \mathbf{N}^s} [\Psi_\xi(n)]^{1-p} \left(\int_{\mathbf{R}_+^m} k_\lambda(||x||_\alpha, ||(n - \xi)^\gamma||_1) f(x) dx \right)^p \right\}^{\frac{1}{p}}$$
$$< K_1(\lambda_1) ||f||_{p,\varphi}, \quad (5.64)$$

and

$$\left\{\int_{\mathbf{R}_+^m} [\varphi(x)]^{1-q} \left(\sum_{n\in\mathbf{N}^s} k_\lambda(||x||_\alpha, ||(n-\xi)^\gamma||_1)a(n)\right)^q dx\right\}^{\frac{1}{q}}$$
$$< K_1(\lambda_1)||a||_{q,\Psi_\xi}; \tag{5.65}$$

(ii) for $p < 0$, we have the reverses of (5.63), (5.64) and (5.65);

(iii) for $0 < p < 1$, if $f(x) \geq 0, a(n) \geq 0$, $f \in \mathbf{L}_{p,\widetilde{\varphi}}(\mathbf{R}_+^m)$, $a = \{a(n)\} \in l_{q,\Psi_\xi}$, such that $||f||_{p,\widetilde{\varphi}} > 0, ||a||_{q,\Psi_\xi} > 0$, then we have the following reverse equivalent inequalities:

$$\int_{\mathbf{R}_+^m} \sum_{n\in\mathbf{N}^s} k_\lambda(||x||_\alpha, ||(n-\xi)^\gamma||_1)a(n)f(x)dx$$
$$> K_1(\lambda_1)||f||_{p,\widetilde{\varphi}_1}||a||_{q,\Psi_\xi}, \tag{5.66}$$

$$\left\{\sum_{n\in\mathbf{N}^s} [\Psi_1(n)]^{1-p} \left(\int_{\mathbf{R}_+^m} k_\lambda(||x||_\alpha, ||(n-\xi)^\gamma||_1)f(x)dx\right)^p\right\}^{\frac{1}{p}}$$
$$> K_1(\lambda_1)||f||_{p,\widetilde{\varphi}_1}, \tag{5.67}$$

and

$$\left\{\int_{\mathbf{R}_+^m} [\widetilde{\varphi}_1(x)]^{1-q} \left(\sum_{n\in\mathbf{N}^s} k_\lambda(||x||_\alpha, ||(n-\xi)^\gamma||_1)a(n)\right)^q dx\right\}^{\frac{1}{q}}$$
$$> K_1(\lambda_1)||a||_{q,\Psi_\xi}. \tag{5.68}$$

Moreover, if there exist $\widetilde{\delta}_0 > 0$ and $\eta > \lambda_1$, such that for any $\delta \in (0, \widetilde{\delta}_0]$,

$$k(\lambda_1 \pm \delta) = \int_0^\infty k_\lambda(t,1)t^{(\lambda_1\pm\delta)-1}dt \in \mathbf{R}_+,$$

and

$$k_\lambda(t,1) \leq L\left(\frac{1}{t^\eta}\right) \quad (t \geq 1; L > 0),$$

then, the above inequalities are all with the same best possible constant factor $K_1(\lambda_1)$.

5.3.4 *Operator Expressions and Some Particular Examples*

As the assumptions of Theorem 5.1, for $p > 1$, setting

$$\varphi(x) = ||x||_\alpha^{p(m-\lambda_1)-m} \quad (x \in \mathbf{R}_+^m),$$

and

$$\Psi(n) = ||v(n)||_\beta^{q(s-\lambda_2)-s} \left(\prod_{k=1}^{s} v'(n_k) \right)^{1-q} \quad (n \in \mathbf{N}_{n^{(0)}}^s),$$

we define a first kind of multi-dimensional half-discrete Hilbert-type operator with the homogeneous kernel $T_1 : \mathbf{L}_{p,\varphi}(\mathbf{R}_+^m) \to l_{p,\Psi^{1-p}}$ as follows: For any $f \in \mathbf{L}_{p,\varphi}(\mathbf{R}_+^m)$, there exists a $T_1 f$, satisfying

$$T_1 f(n) = \int_{\mathbf{R}_+^m} k_\lambda(||x||_\alpha, ||v(n)||_\beta) f(x) dx (n \in \mathbf{N}_{n^{(0)}}^s). \tag{5.69}$$

By (5.29), we can write

$$||T_1 f||_{p,\Psi^{1-p}} < K(\lambda_1) ||f||_{p,\varphi},$$

and then $T_1 f \in l_{p,\Psi^{1-p}}$. Hence, T_1 is a bounded linear operator with the norm (see Tailor and Lay [72])

$$||T_1|| = \sup_{f(\neq\theta)\in\mathbf{L}_{p,\varphi}(\mathbf{R}_+^m)} \frac{||T_1 f||_{p,\Psi^{1-p}}}{||f||_{p,\varphi}} \leq K(\lambda_1).$$

Also we define a second kind of multi-dimensional half-discrete Hilbert-type operator with the homogeneous kernel $T_2 : l_{q,\Psi} \to \mathbf{L}_{q,\varphi^{1-q}}(\mathbf{R}_+^m)$ as follows: For any $a \in l_{q,\Psi}$, there exists a $T_2 a$, satisfying

$$T_2 a(x) = \sum_{n\in\mathbf{N}_{n^{(0)}}^s} k_\lambda(||x||_\alpha, ||v(n)||_\beta) a(n) (x \in \mathbf{R}_+^m). \tag{5.70}$$

By (5.30), we can write

$$||T_2 a||_{q,\varphi^{1-q}} < K(\lambda_1) ||a||_{q,\Psi},$$

and then $T_2 a \in \mathbf{L}_{q,\varphi^{1-q}}(\mathbf{R}_+^m)$. Hence, T_2 is a bounded linear operator with

$$||T_2|| = \sup_{a(\neq\theta)\in l_{q,\Psi}} \frac{||T_2 a||_{q,\varphi^{1-q}}}{||a||_{q,\Psi}} \leq K(\lambda_1).$$

By Theorem 5.2, it follows that

Theorem 5.4. *With the assumptions of Theorem 5.1, T_1 and T_2 are defined by (5.69) and (5.70). If there exist $\widetilde{\delta}_0 > 0$ and $\eta > \lambda_1$, such that, for any $\delta \in (0, \widetilde{\delta}_0]$,*

$$k(\lambda_1 \pm \delta) = \int_0^\infty k_\lambda(t,1) t^{(\lambda_1\pm\delta)-1} dt \in \mathbf{R}_+,$$

and

$$k_\lambda(t,1) \leq L\frac{1}{t^\eta} \quad (t \geq 1; L > 0),$$

then, we have

$$||T_1|| = ||T_2|| = K(\lambda_1) = \left(\frac{\Gamma^m(\frac{1}{\alpha})}{\alpha^{m-1}\Gamma(\frac{m}{\alpha})} \right)^{\frac{1}{q}} \left(\frac{\Gamma^s(\frac{1}{\beta})}{\beta^{s-1}\Gamma\left(\frac{s}{\beta}\right)} \right)^{\frac{1}{p}} k(\lambda_1).$$

With the assumptions of Theorem 5.3, for $p > 1$, setting
$$\varphi(x) = ||x||_\alpha^{p(m-\lambda_1)-m} \quad (x \in \mathbf{R}_+^m),$$

and

$$\Psi_1(n) = ||v(n)||_1^{q(s-\lambda_2)-s} \left(\prod_{k=1}^s v'(n_k) \right)^{1-q} \quad (n \in \mathbf{N}_{n^{(0)}}^s),$$

we define a first kind half-discrete operator with the homogeneous kernel and multi-variables $\widetilde{T}_1 : \mathbf{L}_{p,\varphi}(\mathbf{R}_+^m) \to \mathbf{l}_{p,\Psi_1^{1-p}}$ as follows:

For any $f \in \mathbf{L}_{p,\varphi}(\mathbf{R}_+^m)$, there exists a $\widetilde{T}_1 f$, satisfying

$$\widetilde{T}_1 f(n) = \int_{R_+^m} k_\lambda(||x||_\alpha, ||v(n)||_1) f(x) dx (n \in \mathbf{N}_{n^{(0)}}^s). \tag{5.71}$$

By (5.43), we can write
$$||\widetilde{T}_1 f||_{p,\Psi_1^{1-p}} < K_1(\lambda_1) ||f||_{p,\varphi},$$

and then, $\widetilde{T}_1 f \in \mathbf{l}_{p,\Psi_1^{1-p}}$. Hence, $\widetilde{T}_1$ is a bounded linear operator with the norm

$$||\widetilde{T}_1|| = \sup_{f(\neq\theta)\in\mathbf{L}_{p,\varphi}(\mathbf{R}_+^m)} \frac{||\widetilde{T}_1 f||_{p,\Psi_1^{1-p}}}{||f||_{p,\varphi}} \leq K_1(\lambda_1).$$

Also we define a second kind half-discrete operator with the homogeneous kernel and multi-variables $\widetilde{T}_2 : \mathbf{l}_{q,\Psi_1} \to \mathbf{L}_{q,\varphi^{1-q}}(\mathbf{R}_+^m)$ as follows:

For any $a \in \mathbf{l}_{q,\Psi_1}$, there exists a $\widetilde{T}_2 a$, satisfying

$$\widetilde{T}_2 a(x) = \sum_{n \in \mathbf{N}_{n^{(0)}}^s} k_\lambda(||x||_\alpha, ||v(n)||_1) a(n) \quad (x \in \mathbf{R}_+^m). \tag{5.72}$$

By (5.44), we can write
$$||\widetilde{T}_2 a||_{q,\varphi^{1-q}} < K(\lambda_1) ||a||_{q,\Psi_1},$$

and then $\widetilde{T}_2 a \in \mathbf{L}_{q,\varphi^{1-q}}(\mathbf{R}_+^m)$. Hence, $\widetilde{T}_2$ is a bounded linear operator with

$$||\widetilde{T}_2|| = \sup_{a(\neq\theta)\in\mathbf{l}_{q,\Psi}} \frac{||\widetilde{T}_2 a||_{q,\varphi^{1-q}}}{||a||_{q,\Psi_1}} \leq K_1(\lambda_1).$$

By Theorem 5.3, it follows that

Theorem 5.5. *As the assumptions of Theorem 5.3, $\widetilde{T}_1$ and $\widetilde{T}_2$ are defined by (5.71) and (5.72). If there exist $\tilde{\delta}_0 > 0$ and $\eta > \lambda_1$, such that for any $\delta \in (0, \tilde{\delta}_0]$,*

$$k(\lambda_1 \pm \delta) = \int_0^\infty k_\lambda(t, 1) t^{(\lambda_1\pm\delta)-1} dt \in \mathbf{R}_+,$$

and

$$k_\lambda(t, 1) \le L\left(\frac{1}{t^\eta}\right) \quad (t \ge 1; L > 0),$$

then, we have

$$\|\widetilde{T_1}\| = \|\widetilde{T_2}\| = K_1(\lambda_1) = \left(\frac{\Gamma^m(\frac{1}{\alpha})}{\alpha^{m-1}\Gamma(\frac{m}{\alpha})}\right)^{\frac{1}{q}} \left(\frac{1}{(s-1)!}\right)^{\frac{1}{p}} k(\lambda_1).$$

Example 5.1. We set $k_\lambda(t, u) = \frac{1}{(t+u)^\lambda}$ ($\lambda > 0$, $\lambda_1 > 0$, $\lambda_2 \le s$), then, we find that

$$\frac{\partial}{\partial u}(k_\lambda(t, u)u^{\lambda_2-s}) < 0, \quad \frac{\partial^2}{\partial u^2}(k_\lambda(t, u)u^{\lambda_2-s}) > 0 \quad (u \in \mathbf{R}^+).$$

For $\widetilde{\delta}_0 = \frac{1}{2}\min\{\lambda_1, \lambda_2\}, \delta \in (0, \widetilde{\delta}_0]$, we find

$$k(\lambda_1 \pm \delta) = \int_0^\infty \frac{t^{(\lambda_1 \pm \delta)-1}}{(t+1)^\lambda} du = B(\lambda_1 \pm \delta, \lambda_2 \mp \delta) \in \mathbf{R}_+,$$

and

$$k_\lambda(t, 1) = \frac{1}{(t+1)^\lambda} \le \frac{1}{t^\eta} \quad (t \ge 1; \eta = \frac{\lambda+\lambda_1}{2} \in (\lambda_1, \lambda)).$$

(i) For $v(t) = \ln t (t \ge 2 = n^{(0)}), v(n^{(0)} - 1) = 0$, setting

$$\varphi(x) = \|x\|_\alpha^{p(m-\lambda_1)-m} \quad (x \in \mathbf{R}_+^m),$$

and

$$\Psi(n) = \|\ln n\|_\beta^{q(s-\lambda_2)-s}\left(\prod_{k=1}^s \frac{1}{n_k}\right)^{1-q} \quad (n \in \mathbf{N}_2^s),$$

we define an operator $T_1 : \mathbf{L}_{p,\varphi}(\mathbf{R}_+^m) \to \mathbf{l}_{p,\Psi^{1-p}}$ as follows:
For any $f \in \mathbf{L}_{p,\varphi}(\mathbf{R}_+^m)$, there exists a T_1f, satisfying

$$\widetilde{T}_1 f(n) = \int_{R_+^m} \frac{1}{(\|x\|_\alpha + \|\ln n\|_\beta)^\lambda} f(x) dx \quad (n \in \mathbf{N}_2^s).$$

We also define an operator $T_2 : \mathbf{l}_{q,\Psi} \to \mathbf{L}_{q,\varphi^{1-q}}(\mathbf{R}_+^m)$ as follows:
For any $a \in \mathbf{l}_{q,\Psi}$, there exists a T_2a, satisfying

$$T_2 a(x) = \sum_{n \in \mathbf{N}_2^s} \frac{1}{(\|x\|_\alpha + \|\ln n\|_\beta)^\lambda} a(n) \quad (x \in \mathbf{R}_+^m).$$

Then by Theorem 5.4, we have

$$\|T_1\| = \|T_2\| = \left(\frac{\Gamma^m(\frac{1}{\alpha})}{\alpha^{m-1}\Gamma(\frac{m}{\alpha})}\right)^{\frac{1}{q}} \left(\frac{\Gamma^s(\frac{1}{\beta})}{\beta^{s-1}\Gamma\left(\frac{s}{\beta}\right)}\right)^{\frac{1}{p}} B(\lambda_1, \lambda_2).$$

(ii) For $v(t) = \ln(t - \gamma)(t \geq 2 = n^{(0)}; \gamma \leq \frac{1}{2})$, $v(n^{(0)} - \frac{1}{2}) = \ln(\frac{3}{2} - \gamma) \geq 0$, setting

$$\varphi(x) = ||x||_\alpha^{p(m-\sigma)-m} \quad (x \in \mathbf{R}_+^m),$$

and

$$\Psi_\gamma(n) = ||\ln(n-\gamma)||_1^{q(s-\sigma)-s} \left(\prod_{k=1}^s \frac{1}{n_k - \gamma}\right)^{1-q} \quad (n \in \mathbf{N}_2^s),$$

we define an operator $\widetilde{T}_1 : \mathbf{L}_{p,\varphi}(\mathbf{R}_+^m) \to \mathbf{l}_{p,\Psi_\gamma^{1-p}}$ as follows:

For any $f \in \mathbf{L}_{p,\varphi}(\mathbf{R}_+^m)$, there exists a $\widetilde{T}_1 f$, satisfying

$$\widetilde{T}_1 f(n) = \int_{R_+^m} \frac{1}{(||x||_\alpha + ||\ln(n-\gamma)||_1)^\lambda} f(x)dx \quad (n \in \mathbf{N}_2^s).$$

Also we define an operator $\widetilde{T}_2 : \mathbf{l}_{q,\Psi_\gamma} \to \mathbf{L}_{q,\varphi^{1-q}}(\mathbf{R}_+^m)$ as follows:

For any $a \in \mathbf{l}_{q,\Psi_\gamma}$, there exists $T_2 a$, satisfying

$$\widetilde{T}_2 a(x) = \sum_{n \in \mathbf{N}_2^s} \frac{1}{(||x||_\alpha + ||\ln(n-\gamma)||_1)^\lambda} a(n)(x \in \mathbf{R}_+^m).$$

Then, by Theorem 5.5, we have

$$||\widetilde{T}_1|| = ||\widetilde{T}_2|| = \left(\frac{\Gamma^m(\frac{1}{\alpha})}{\alpha^{m-1}\Gamma(\frac{m}{\alpha})}\right)^{\frac{1}{q}} \left(\frac{1}{(s-1)!}\right)^{\frac{1}{p}} B(\lambda_1, \lambda_2).$$

(iii) For $v(t) = \ln \kappa t (t \geq 2 = n^{(0)}; \kappa \geq \frac{2}{3})$, $v(n^{(0)} - \frac{1}{2}) = \ln \kappa(\frac{3}{2}) \geq 0$, setting

$$\varphi(x) = ||x||_\alpha^{p(m-\lambda_1)-m} \quad (x \in \mathbf{R}_+^m),$$

and

$$\Psi_\kappa(n) = ||\ln \kappa n||_1^{q(s-\lambda_2)-s} \left(\prod_{k=1}^s \frac{1}{n_k}\right)^{1-q} \quad (n \in \mathbf{N}_2^s),$$

we define an operator $\widetilde{T}_1 : \mathbf{L}_{p,\varphi}(\mathbf{R}_+^m) \to \mathbf{l}_{p,\Psi_\kappa^{1-p}}$ as follows:

For any $f \in \mathbf{L}_{p,\varphi}(\mathbf{R}_+^m)$, there exists $\widetilde{T}_1 f$, satisfying

$$\widetilde{T}_1 f(n) = \int_{R_+^m} \frac{1}{(||x||_\alpha + ||\ln \kappa n||_1)^\lambda} f(x)dx \quad (n \in \mathbf{N}_2^s).$$

We also define an operator $\widetilde{T}_2 : \mathbf{l}_{q,\Psi_\kappa} \to \mathbf{L}_{q,\varphi^{1-q}}(\mathbf{R}_+^m)$ as follows:

For any $a \in \mathbf{l}_{q,\Psi_\kappa}$, there exists a $\widetilde{T}_2 a$, satisfying

$$\widetilde{T}_2 a(x) = \sum_{n \in \mathbf{N}_2^s} \frac{1}{(||x||_\alpha + ||\ln \kappa n||_1)^\lambda} a(n) \quad (x \in \mathbf{R}_+^m).$$

Then, by Theorem 5.5, we have

$$||\widetilde{T_1}|| = ||\widetilde{T_2}|| = \left(\frac{\Gamma^m(\frac{1}{\alpha})}{\alpha^{m-1}\Gamma(\frac{m}{\alpha})}\right)^{\frac{1}{q}}\left(\frac{1}{(s-1)!}\right)^{\frac{1}{p}}B(\lambda_1, \lambda_2).$$

Example 5.2. We set $k_\lambda(t, u) = \frac{\ln(t/u)}{t^\lambda - u^\lambda}(0 < \lambda \le 1, \lambda_1 > 0, \lambda_2 \le s)$, then we find (see Yang [134])

$$\frac{\partial}{\partial u}(k_\lambda(t, u)u^{\lambda_2 - s}) < 0, \quad \frac{\partial^2}{\partial u^2}(k_\lambda(t, u)u^{\lambda_2 - s}) > 0 \quad (u \in \mathbf{R}^+).$$

For $\widetilde{\delta}_0 = \frac{1}{2}\min\{\lambda_1, \lambda_2\}, \delta \in (0, \widetilde{\delta}_0]$, we obtain

$$\begin{aligned}
k(\lambda_1 \pm \delta) &= \int_0^\infty \frac{(\ln t)t^{(\lambda_1 \pm \delta) - 1}}{t^\lambda - 1}dt \\
&= \frac{1}{\lambda^2}\int_0^\infty \frac{(\ln v)v^{(\lambda_1 \pm \delta)/\lambda - 1}}{v - 1}dv \\
&= \left[\frac{\pi}{\lambda\sin\pi(\frac{\lambda_1 \pm \delta}{\lambda})}\right]^2 \in \mathbf{R}_+,
\end{aligned}$$

and

$$k_\lambda(t, 1) = \frac{\ln t}{t^\lambda - 1} \le L\frac{1}{t^\eta},$$

where $t \ge 1$; $\eta = \frac{\lambda + \lambda_1}{2} \in (\lambda_1, \lambda), L > 0$.

(i) For $v(t) = \ln t(t \ge 2 = n^{(0)}), v(n^{(0)} - 1) = 0$, setting $\varphi(x)$ and $\Psi(n)$ as Example 5.1(i), we define an operator $T_1 : \mathbf{L}_{p,\varphi}(\mathbf{R}^m_+) \to l_{p,\Psi^{1-p}}$ as follows:

For any $f \in \mathbf{L}_{p,\varphi}(\mathbf{R}^m_+)$, there exists a T_1f, satisfying

$$T_1f(n) = \int_{R^m_+} \frac{\ln(||x||_\alpha/||\ln n||_\beta)}{||x||_\alpha^\lambda - ||\ln n||_\beta^\lambda}f(x)dx \quad (n \in \mathbf{N}_2^s).$$

We also define an operator $T_2 : l_{q,\Psi} \to \mathbf{L}_{q,\varphi^{1-q}}(\mathbf{R}^m_+)$ as follows:

For any $a \in l_{q,\Psi}$, there exists a T_2a, satisfying

$$T_2a(x) = \sum_{n \in \mathbf{N}_2^s} \frac{\ln(||x||_\alpha/||\ln n||_\beta)}{||x||_\alpha^\lambda - ||\ln n||_\beta^\lambda}a(n) \quad (x \in \mathbf{R}^m_+).$$

Then, by Theorem 5.4, we have

$$||T_1|| = ||T_2|| = \left(\frac{\Gamma^m(\frac{1}{\alpha})}{\alpha^{m-1}\Gamma(\frac{m}{\alpha})}\right)^{\frac{1}{q}}\left(\frac{\Gamma^s(\frac{1}{\beta})}{\beta^{s-1}\Gamma(\frac{s}{\beta})}\right)^{\frac{1}{p}}\left[\frac{\pi}{\lambda\sin\pi(\frac{\lambda_1}{\lambda})}\right]^2.$$

(ii) For $v(t) = \ln(t-\gamma)(t \geq 2 = n^{(0)}; \gamma \leq \frac{1}{2})$, $v(n^{(0)} - \frac{1}{2}) = \ln(\frac{3}{2} - \gamma) \geq 0$, setting $\varphi(x)$ and $\Psi_\gamma(n)$ as Example 5.1(ii), we define an operator $\widetilde{T}_1$: $\mathbf{L}_{p,\varphi}(\mathbf{R}_+^m) \to \mathbf{l}_{p,\Psi_\gamma^{1-p}}$ as follows: For any $f \in \mathbf{L}_{p,\varphi}(\mathbf{R}_+^m)$, there exists a $\widetilde{T}_1 f$, satisfying

$$\widetilde{T}_1 f(n) = \int_{R_+^m} \frac{\ln(||x||_\alpha / ||\ln(n-\gamma)||_1)}{||x||_\alpha^\lambda - ||\ln(n-\gamma)||_1^\lambda} f(x) dx \quad (n \in \mathbf{N}_2^s).$$

Also we define an operator $\widetilde{T}_2 : \mathbf{l}_{q,\Psi_\gamma} \to \mathbf{L}_{q,\varphi^{1-q}}(\mathbf{R}_+^m)$ as follows:

For any $a \in \mathbf{l}_{q,\Psi_\gamma}$, there exists $\widetilde{T}_2 a$, satisfying

$$\widetilde{T}_2 a(x) = \sum_{n \in \mathbf{N}_2^s} \frac{\ln(||x||_\alpha / ||\ln(n-\gamma)||_1)}{||x||_\alpha^\lambda - ||\ln(n-\gamma)||_1^\lambda} a(n) \quad (x \in \mathbf{R}_+^m).$$

Then, by Theorem 5.5, we have

$$||\widetilde{T}_1|| = ||\widetilde{T}_2|| = \left(\frac{\Gamma^m(\frac{1}{\alpha})}{\alpha^{m-1}\Gamma(\frac{m}{\alpha})} \right)^{\frac{1}{q}} \left(\frac{1}{(s-1)!} \right)^{\frac{1}{p}} \left[\frac{\pi}{\lambda \sin \pi(\frac{\lambda_1}{\lambda})} \right]^2.$$

(iii) For $v(t) = \ln \kappa t (t \geq 2 = n^{(0)}; \kappa \geq \frac{2}{3})$, $v(n^{(0)} - \frac{1}{2}) = \ln \kappa(\frac{3}{2}) \geq 0$, setting $\varphi(x)$ and $\Psi_\kappa(n)$ as Example 5.1(iii), we define $\widetilde{T}_1 : \mathbf{L}_{p,\varphi}(\mathbf{R}_+^m) \to \mathbf{l}_{p,\Psi_\kappa^{1-p}}$ as follows: For any $f \in \mathbf{L}_{p,\varphi}(\mathbf{R}_+^m)$, there exists a $\widetilde{T}_1 f$, satisfying

$$\widetilde{T}_1 f(n) = \int_{R_+^m} \frac{\ln(||x||_\alpha / ||\ln \kappa n||_1)}{||x||_\alpha^\lambda - ||\ln \kappa n||_1^\lambda} f(x) dx \quad (n \in \mathbf{N}_2^s).$$

We also define an operator $\widetilde{T}_2 : \mathbf{l}_{q,\Psi_\kappa} \to \mathbf{L}_{q,\varphi^{1-q}}(\mathbf{R}_+^m)$ as follows:

For any $a \in \mathbf{l}_{q,\Psi_\kappa}$, there exists $\widetilde{T}_2 a$, satisfying

$$\widetilde{T}_2 a(x) = \sum_{n \in \mathbf{N}_2^s} \frac{\ln(||x||_\alpha / ||\ln \kappa n||_1)}{||x||_\alpha^\lambda - ||\ln \kappa n||_1^\lambda} a(n) \quad (x \in \mathbf{R}_+^m).$$

Then by Theorem 5.5, we have

$$||\widetilde{T}_1|| = ||\widetilde{T}_2|| = \left(\frac{\Gamma^m(\frac{1}{\alpha})}{\alpha^{m-1}\Gamma(\frac{m}{\alpha})} \right)^{\frac{1}{q}} \left(\frac{1}{(s-1)!} \right)^{\frac{1}{p}} \left[\frac{\pi}{\lambda \sin \pi(\frac{\lambda_1}{\lambda})} \right]^2.$$

Example 5.3. We set $k_0(t, u) = e^{-\eta(u/t)}(\eta > 0, \lambda_1 = -\sigma < 0, \lambda_2 = \sigma \leq s)$, then, we find that

$$\frac{\partial}{\partial u}(k_0(t, u)u^{\sigma-s}) < 0, \quad \frac{\partial^2}{\partial u^2}(k_0(t, u)u^{\sigma-s}) > 0 \quad (u \in \mathbf{R}^+).$$

For $\widetilde{\delta}_0 = \frac{1}{2}\sigma, \delta \in (0, \widetilde{\delta}_0]$, we have

$$k(-\sigma \pm \delta) = \int_0^\infty e^{-\eta(1/t)} t^{-\sigma \pm \delta - 1} dt = \frac{1}{\eta^{\sigma \pm \delta}} \int_0^\infty e^{-v} v^{\sigma \pm \delta - 1} dv$$

$$= \frac{1}{\eta^{\sigma \pm \delta}} \Gamma(\sigma \pm \delta) \in \mathbf{R}_+,$$

and

$$k_0(t, 1) = e^{-\eta(1/t)} \le L \frac{1}{t^\eta}$$

where $t \ge 1; L > 0$, $\eta = -\frac{1}{2}\sigma > -\sigma$.

(i) For $v(t) = \ln t (t \ge 2 = n^{(0)})$, $v(n^{(0)} - 1) = 0$, setting $\varphi(x)$ and $\Psi(n)$ as Example 5.1(i), we define an operator $T_1 : \mathbf{L}_{p,\varphi}(\mathbf{R}_+^m) \to \mathbf{l}_{p,\Psi^{1-p}}$ as follows:

For any $f \in \mathbf{L}_{p,\varphi}(\mathbf{R}_+^m)$, there exists a $T_1 f$, satisfying

$$T_1 f(n) = \int_{R_+^m} e^{-\eta(\|\ln n\|_\beta / \|x\|_\alpha)} f(x) dx (n \in \mathbf{N}_2^s).$$

We also define an operator $T_2 : \mathbf{l}_{q,\Psi} \to \mathbf{L}_{q,\varphi^{1-q}}(\mathbf{R}_+^m)$ as follows:

For any $a \in \mathbf{l}_{q,\Psi}$, there exists a $T_2 a$, satisfying

$$T_2 a(x) = \sum_{n \in \mathbf{N}_2^s} e^{-\eta(\|\ln n\|_\beta / \|x\|_\alpha)} a(n) \quad (x \in \mathbf{R}_+^m).$$

Then, by Theorem 5.5, we have

$$\|T_1\| = \|T_2\| = \left(\frac{\Gamma^m(\frac{1}{\alpha})}{\alpha^{m-1} \Gamma(\frac{m}{\alpha})} \right)^{\frac{1}{q}} \left(\frac{\Gamma^s(\frac{1}{\beta})}{\beta^{s-1} \Gamma\left(\frac{s}{\beta}\right)} \right)^{\frac{1}{p}} \frac{1}{\eta^\sigma} \Gamma(\sigma).$$

(ii) For $v(t) = \ln(t - \gamma)(t \ge 2 = n^{(0)}; \gamma \le \frac{1}{2})$, $v(n^{(0)} - \frac{1}{2}) = \ln(\frac{3}{2} - \gamma) \ge 0$, setting $\varphi(x)$ and $\Psi_\gamma(n)$ as Example 5.1(ii), we define an operator $\widetilde{T}_1 : \mathbf{L}_{p,\varphi}(\mathbf{R}_+^m) \to \mathbf{l}_{p,\Psi_\gamma^{1-p}}$ as follows:

For any $f \in \mathbf{L}_{p,\varphi}(\mathbf{R}_+^m)$, there exists a $\widetilde{T}_1 f$, satisfying

$$\widetilde{T}_1 f(n) = \int_{R_+^m} e^{-\eta(\|\ln(n-\gamma)\|_1 / \|x\|_\alpha)} f(x) dx (n \in \mathbf{N}_2^s).$$

Also we define an operator $\widetilde{T}_2 : \mathbf{l}_{q,\Psi_\gamma} \to \mathbf{L}_{q,\varphi^{1-q}}(\mathbf{R}_+^m)$ as follows:

For any $a \in \mathbf{l}_{q,\Psi_\gamma}$, there exists $\widetilde{T}_2 a$, satisfying

$$\widetilde{T}_2 a(x) = \sum_{n \in \mathbf{N}_2^s} e^{-\eta(\|\ln(n-\gamma)\|_1 / \|x\|_\alpha)} a(n) \quad (x \in \mathbf{R}_+^m).$$

Then, by Theorem 5.5, we have

$$||\widetilde{T}_1|| = ||\widetilde{T}_2|| = \left(\frac{\Gamma^m(\frac{1}{\alpha})}{\alpha^{m-1}\Gamma(\frac{m}{\alpha})}\right)^{\frac{1}{q}} \left(\frac{1}{(s-1)!}\right)^{\frac{1}{p}} \frac{1}{\eta^\sigma}\Gamma(\sigma).$$

(iii) For $v(t) = \ln \kappa t (t \geq 2 = n^{(0)}; \kappa \geq \frac{2}{3})$, $v(n^{(0)} - \frac{1}{2}) = \ln \kappa(\frac{3}{2}) \geq 0$, setting $\varphi(x)$ and $\Psi_\kappa(n)$ as Example 5.1(iii), we define an operator $\widetilde{T}_1$: $\mathbf{L}_{p,\varphi}(\mathbf{R}_+^m) \to \mathbf{l}_{p,\Psi_\kappa^{1-p}}$ as follows:

For any $f \in \mathbf{L}_{p,\varphi}(\mathbf{R}_+^m)$, there exists a $\widetilde{T}_1 f$, satisfying

$$\widetilde{T}_1 f(n) = \int_{R_+^m} e^{-\eta(||\ln \kappa n||_1/||x||_\alpha)} f(x) dx \qquad (n \in \mathbf{N}_2^s).$$

We also define an operator $\widetilde{T}_2 : \mathbf{l}_{q,\Psi_\kappa} \to \mathbf{L}_{q,\varphi^{1-q}}(\mathbf{R}_+^m)$ as follows: For any $a \in \mathbf{l}_{q,\Psi_\kappa}$, there exists a $\widetilde{T}_2 a$, satisfying

$$\widetilde{T}_2 a(x) = \sum_{n \in \mathbf{N}_2^s} e^{-\eta(||\ln \kappa n||_1/||x||_\alpha)} a(n) \qquad (x \in \mathbf{R}_+^m).$$

Then by Theorem 5.5, we have

$$||\widetilde{T}_1|| = ||\widetilde{T}_2|| = \left(\frac{\Gamma^m(\frac{1}{\alpha})}{\alpha^{m-1}\Gamma(\frac{m}{\alpha})}\right)^{\frac{1}{q}} \left(\frac{1}{(s-1)!}\right)^{\frac{1}{p}} \frac{1}{\eta^\sigma}\Gamma(\sigma).$$

Example 5.4. We set $k_\lambda(t,u) = \frac{(\min\{t,u\})^\gamma}{(\max\{t,u\})^{\lambda+\gamma}}(\lambda \geq 0, \gamma \geq 0, \lambda_1 + \lambda_2 = \lambda$, $\lambda_1 + \gamma > 0, \lambda_2 + \gamma > 0)$. For $\lambda_2 + \gamma \leq s$, we find

$$k_\lambda(t,u)u^{\lambda_2-s} = \frac{(\min\{t,u\})^\gamma}{(\max\{t,u\})^{\lambda+\gamma}} u^{\lambda_2-s}$$

$$= \begin{cases} \frac{u^{\gamma+\lambda_2-s}}{t^{\lambda+\gamma}}, & 0 < u \leq t, \\ \frac{t^\gamma}{u^{\lambda+\gamma-\lambda_2+s}}, & u > t \end{cases}$$

is decreasing with respect to $u \in \mathbf{R}_+$ and strictly decreasing in an interval of $(1, \infty)$. For $\widetilde{\delta}_0 = \frac{1}{2}\min\{\lambda_1 + \gamma, \lambda_2 + \gamma\}, \delta \in (0, \widetilde{\delta}_0]$, we obtain

$$k(\lambda_1 \pm \delta) = \int_0^\infty \frac{(\min\{t,1\})^\gamma}{(\max\{t,1\})^{\lambda+\gamma}} t^{\lambda_1-1} dt$$

$$= \frac{1}{\gamma + \lambda_1 \pm \delta} + \frac{1}{\gamma + \lambda_2 \mp \delta} \in \mathbf{R}_+,$$

and

$$k_\lambda(t,1) = \frac{(\min\{t,1\})^\gamma}{(\max\{t,1\})^{\lambda+\gamma}} = \frac{1}{t^{\lambda+\gamma}} \leq \frac{1}{t^\eta},$$

where $t \geq 1; \eta = \frac{1}{2}(\lambda_1 + \lambda + \gamma) \in (\lambda_1, \lambda + \gamma)$.

For $v(t) = \ln t (t \geq 2 = n^{(0)})$, $v(n^{(0)} - 1) = 0$, setting $\varphi(x)$ and $\Psi(n)$ as Example 5.1(i), we define an operator $T_1 : \mathbf{L}_{p,\varphi}(\mathbf{R}_+^m) \to l_{p,\Psi^{1-p}}$ as follows:

For any $f \in \mathbf{L}_{p,\varphi}(\mathbf{R}_+^m)$, there exists a $T_1 f$, satisfying

$$T_1 f(n) = \int_{R_+^m} \frac{(\min\{||x||_\alpha, ||\ln n||_\beta\})^\gamma}{(\max\{||x||_\alpha, ||\ln n||_\beta\})^{\lambda+\gamma}} f(x) dx \qquad (n \in \mathbf{N}_2^s).$$

Also we define an operator $T_2 : l_{q,\Psi} \to \mathbf{L}_{q,\varphi^{1-q}}(\mathbf{R}_+^m)$ as follows: For any $a \in l_{q,\Psi}$, there exists a $T_2 a$, satisfying

$$T_2 a(x) = \sum_{n \in \mathbf{N}_2^s} \frac{(\min\{||x||_\alpha, ||\ln n||_\beta\})^\gamma}{(\max\{||x||_\alpha, ||\ln n||_\beta\})^{\lambda+\gamma}} a(n) \qquad (x \in \mathbf{R}_+^m).$$

Then, by Theorem 5.4, we have

$$||T_1|| = ||T_2||$$

$$= \left(\frac{\Gamma^m(\frac{1}{\alpha})}{\alpha^{m-1}\Gamma(\frac{m}{\alpha})} \right)^{\frac{1}{q}} \left(\frac{\Gamma^s(\frac{1}{\beta})}{\beta^{s-1}\Gamma\left(\frac{s}{\beta}\right)} \right)^{\frac{1}{p}} \frac{2\gamma + \lambda}{(\gamma + \lambda_1)(\gamma + \lambda_2)}.$$

Remark 5.1. (i) We can still write some particular inequalities with the norms as the best possible constant factors in the above examples by using Theorem 5.1 and Theorem 5.3.

(ii) In particular, for $m = s = 1$ in the theorems and corollaries of this chapter, we get

$$\left(\frac{\Gamma^m(\frac{1}{\alpha})}{\alpha^{m-1}\Gamma(\frac{m}{\alpha})} \right)^{\frac{1}{q}} \left(\frac{\Gamma^s(\frac{1}{\beta})}{\beta^{s-1}\Gamma\left(\frac{s}{\beta}\right)} \right)^{\frac{1}{p}} = 1,$$

and imply some corresponding results of Chapter 3.

5.4 Some Inequalities Relating a General Non-Homogeneous Kernel

5.4.1 *Some Lemmas*

With the assumptions of Definition 5.1, if $h(u)$ is a non-negative finite measurable function in $\mathbf{R}_+, H(t, u) = h(tu)$, $\lambda_1 = \lambda_2 = \sigma \in \mathbf{R}$, then (5.5) and (5.6) reduce to

$$\omega(\sigma, n) = \int_{\mathbf{R}_+^m} h(||x||_\alpha ||v(n)||_\beta) \frac{||v(n)||_\beta^\sigma}{||x||_\alpha^{m-\sigma}} dx \quad (n \in \mathbf{N}_{n^{(0)}}^s), \tag{5.73}$$

$$\varpi(\sigma, x) = \sum_{n \in \mathbf{N}_{n^{(0)}}^s} h(||x||_\alpha ||v(n)||_\beta) \frac{||x||_\alpha^\sigma}{||v(n)||_\beta^{s-\sigma}} \prod_{k=1}^s v'(n_k) \ (x \in \mathbf{R}_+^m). \tag{5.74}$$

For $k(\sigma) = \int_0^\infty h(t)t^{\sigma-1}dt \in \mathbf{R}_+$, by (5.7), (5.8) and (5.9), we find

$$\omega(\sigma, n) = \frac{\Gamma^m(\frac{1}{\alpha})}{\alpha^m \Gamma(\frac{m}{\alpha})}||v(n)||_\beta^\sigma \lim_{M\to\infty} M^\sigma \int_0^1 h(Mu^{\frac{1}{\alpha}}||v(n)||_\beta)u^{\frac{\sigma}{\alpha}-1}du,$$

$$(u = (t/||v(n)||_\beta M)^\alpha)$$

$$= \frac{\Gamma^m(\frac{1}{\alpha})}{\alpha^{m-1}\Gamma(\frac{m}{\alpha})} \int_0^\infty h(t)t^{\sigma-1}dt$$

$$= K(\alpha, m, \sigma) = \frac{\Gamma^m(\frac{1}{\alpha})}{\alpha^{m-1}\Gamma(\frac{m}{\alpha})}k(\sigma); \qquad (5.75)$$

$$\varpi(\sigma, x) < \frac{\Gamma^s(\frac{1}{\beta})}{\beta^s \Gamma(\frac{s}{\beta})}||x||_\alpha^\sigma \lim_{M\to\infty} M^\sigma \int_0^1 h(||x||_\alpha M v^{\frac{1}{\beta}})v^{\frac{\sigma}{\beta}-1}dv,$$

$$(v = (t/||x||_\alpha M)^\beta)$$

$$= \frac{\Gamma^s(\frac{1}{\beta})}{\beta^{s-1}\Gamma\left(\frac{s}{\beta}\right)} \int_0^\infty h(t)t^{\sigma-1}dt$$

$$= K(\beta, s, \sigma) = \frac{\Gamma^s(\frac{1}{\beta})}{\beta^{s-1}\Gamma\left(\frac{s}{\beta}\right)} k(\sigma), \qquad (5.76)$$

and

$$\varpi(\sigma, x) > \frac{\Gamma^s(\frac{1}{\beta})}{\beta^s \Gamma(\frac{s}{\beta})}||x||_\alpha^\sigma \left[\lim_{M\to\infty} M^\sigma \int_0^1 h(||x||_\alpha M v^{\frac{1}{\beta}})v^{\frac{\sigma}{\beta}-1}dv \right.$$

$$\left. - \delta_0 \int_0^1 h(||x||_\alpha v^{\frac{1}{\beta}})v^{\frac{\sigma}{\beta}-1}dv \right]$$

$$= K(\beta, s, \sigma) - \delta_0 \int_0^{||x||_\alpha} h(t)t^{\sigma-1}dt$$

$$= K(\beta, s, \sigma)(1 - \theta(||x||_\alpha)) > 0, \qquad (5.77)$$

where

$$\theta(||x||_\alpha) = \frac{\delta_0}{K(\beta, s, \sigma)} \int_0^{||x||_\alpha} h(t)t^{\sigma-1}dt > 0 \quad (0 < \delta_0 \le 1).$$

Lemma 5.6. *Suppose that m, s, $n^{(0)} \in \mathbf{N}$, α, $\beta > 0$, $p \in \mathbf{R}\backslash\{0,1\}$, $\frac{1}{p} + \frac{1}{q} = 1$, $\sigma \in \mathbf{R}$, $h(u)$ is a non-negative finite measurable function in $\mathbf{R}_+$, $h(u)u^{\sigma-s}$ is decreasing with respect to $u \in \mathbf{R}_+$ and strict decreasing in an interval $I \subset (n^{(0)}, \infty), v'(t) > 0, v''(t) \le 0 (t \in (n^{(0)} - 1, \infty))$, with $v(n^{(0)} - 1) \ge 0$. If there exist constants $\widetilde{\delta}_0 > 0$ and $\eta < \sigma$, such that for any $\delta \in [0, \widetilde{\delta}_0]$,*

$$k(\sigma \pm \delta) = \int_0^\infty h(t)t^{(\sigma\pm\delta)-1}dt \in \mathbf{R}_+,$$

and

$$h(t) \leq L\left(\frac{1}{t^\eta}\right) \quad (0 < t \leq 1; L > 0),$$

then, for any $0 < \varepsilon < |q|\widetilde{\delta}_0$, we have

$$\widetilde{I}(\varepsilon) = \int_{\{x \in \mathbf{R}_+^m; ||x||_\alpha \leq 1\}} \sum_{n \in \mathbf{N}_{n^{(0)}}^s} h(||x||_\alpha ||v(n)||_\beta) \frac{||x||_\alpha^{\sigma - m + \frac{\varepsilon}{p}}}{||v(n)||_\beta^{s - (\sigma - \frac{\varepsilon}{q})}} \prod_{k=1}^s v'(n_k) dx$$

$$= \frac{\Gamma^m(\frac{1}{\alpha})}{\varepsilon \alpha^{m-1} \Gamma(\frac{m}{\alpha})} \frac{\Gamma^s(\frac{1}{\beta})}{\beta^{s-1} \Gamma(\frac{s}{\beta})} (k(\sigma) + o(1))(1 - \varepsilon O_2(1))(\varepsilon \to 0^+). \quad (5.78)$$

Proof. In view of (5.77), we have

$$\widetilde{I}(\varepsilon) = \int_{\{x \in \mathbf{R}_+^m; ||x||_\alpha \leq 1\}} ||x||_\alpha^{-m+\varepsilon} \sum_{n \in \mathbf{N}_{n^{(0)}}^s} h(||x||_\alpha ||v(n)||_\beta) \frac{||x||_\alpha^{\sigma - \frac{\varepsilon}{q}}}{||v(n)||_\beta^{s - (\sigma - \frac{\varepsilon}{q})}}$$

$$\times \prod_{k=1}^s v'(n_k) dx$$

$$= \int_{\{x \in \mathbf{R}_+^m; ||x||_\alpha \leq 1\}} ||x||_\alpha^{-m+\varepsilon} \varpi(\sigma - \frac{\varepsilon}{q}, x) dx$$

$$\geq K(\beta, s, \sigma - \frac{\varepsilon}{q}) \int_{\{x \in \mathbf{R}_+^m; ||x||_\alpha \leq 1\}} ||x||_\alpha^{-m+\varepsilon} (1 - \widetilde{\theta}(||x||_\alpha)) dx$$

$$= \frac{\Gamma^s(\frac{1}{\beta})}{\beta^{s-1} \Gamma\left(\frac{s}{\beta}\right)} k(\sigma - \frac{\varepsilon}{q}) \int_{\{x \in \mathbf{R}_+^m; ||x||_\alpha \leq 1\}} ||x||_\alpha^{-m+\varepsilon} (1 - \widetilde{\theta}(||x||_\alpha)) dx,$$

where

$$\widetilde{\theta}(||x||_\alpha) = \frac{\delta_0}{K(\beta, s, \sigma - \frac{\varepsilon}{q})} \int_0^{||x||_\alpha} h(t) t^{\sigma-1} dt \quad (0 < \delta_0 \leq 1).$$

Since $u^{1/\alpha} = ||x||_\alpha \leq 1$, we find

$$0 < \widetilde{\theta}(u^{1/\alpha}) = \frac{\delta_0}{K(\beta, s, \sigma - \frac{\varepsilon}{q})} \int_0^{u^{1/\alpha}} h(t) t^{\sigma-1} dt$$

$$\leq \frac{\delta_0 L}{K(\beta, s, \sigma - \frac{\varepsilon}{q})} \int_0^{u^{1/\alpha}} t^{\sigma-\eta-1} dt$$

$$= \frac{\delta_0 L u^{\frac{\sigma-\eta}{\alpha}}}{(\sigma - \eta) K(\beta, s, \sigma - \frac{\varepsilon}{q})},$$

$$\int_{\{x \in \mathbf{R}_+^m; ||x||_\alpha \leq 1\}} ||x||_\alpha^{-m+\varepsilon} (1 - \widetilde{\theta}(||x||_\alpha)) dx$$

$$= \int \cdots \int_{D_M} \Psi \left(\sum_{i=1}^{m} x_i{}^{\alpha} \right) dx_1 \cdots dx_m$$

$$= \frac{\Gamma^m(\frac{1}{\alpha})}{\alpha^m \Gamma(\frac{m}{\alpha})} \int_0^1 \Psi(u) u^{\frac{m}{\alpha}-1} du$$

$$= \frac{\Gamma^m(\frac{1}{\alpha})}{\alpha^m \Gamma(\frac{m}{\alpha})} \int_0^1 \frac{(1 - \widetilde{\theta}(u^{1/\alpha}))}{(u^{1/\alpha})^{m-\varepsilon}} u^{\frac{m}{\alpha}-1} du$$

$$= \frac{\Gamma^m(\frac{1}{\alpha})}{\alpha^m \Gamma(\frac{m}{\alpha})} \int_0^1 [1 - \widetilde{\theta}(u^{1/\alpha})] u^{\frac{\varepsilon}{\alpha}-1} du$$

$$= \frac{\Gamma^m(\frac{1}{\alpha})}{\alpha^m \Gamma(\frac{m}{\alpha})} \int_0^1 [1 - O(u^{\frac{\sigma-\eta}{\alpha}})] u^{\frac{\varepsilon}{\alpha}-1} du$$

$$= \frac{\Gamma^m(\frac{1}{\alpha})}{\varepsilon \alpha^{m-1} \Gamma(\frac{m}{\alpha})} (1 - \varepsilon O_1(1)), \qquad (\varepsilon \to 0^+). \qquad (5.79)$$

By Lemma 4.5 in Chapter 4, we have

$$k(\sigma - \frac{\varepsilon}{q}) = k(\sigma) + o(1) \quad (\varepsilon \to 0^+),$$

and then, it follows that

$$\widetilde{I}(\varepsilon) \geq \frac{\Gamma^m(\frac{1}{\alpha})}{\varepsilon \alpha^{m-1} \Gamma(\frac{m}{\alpha})} \frac{\Gamma^s(\frac{1}{\beta})}{\beta^{s-1} \Gamma\left(\frac{s}{\beta}\right)} k(\sigma - \frac{\varepsilon}{q})(1 - \varepsilon O_2(1))$$

$$= \frac{\Gamma^m(\frac{1}{\alpha})}{\varepsilon \alpha^{m-1} \Gamma(\frac{m}{\alpha})} \frac{\Gamma^s(\frac{1}{\beta})}{\beta^{s-1} \Gamma(\frac{s}{\beta})} (k(\sigma) + o(1))(1 - \varepsilon O_2(1)).$$

By (5.76) and (5.3), we still have

$$\widetilde{I}(\varepsilon) \leq \frac{\Gamma^m(\frac{1}{\alpha})}{\alpha^{m-1} \Gamma(\frac{m}{\alpha})} k(\sigma - \frac{\varepsilon}{q}) \int_{\{x \in \mathbf{R}_+^m; ||x||_\alpha \leq 1\}} ||x||_\alpha^{-m+\varepsilon} dx$$

$$= \frac{\Gamma^m(\frac{1}{\alpha})}{\varepsilon \alpha^{m-1} \Gamma(\frac{m}{\alpha})} \frac{\Gamma^s(\frac{1}{\beta})}{\beta^{s-1} \Gamma(\frac{s}{\beta})} (k(\alpha) + o(1)).$$

Then we have (5.78).   □

In particular, for $\beta = 1$, if

$$\frac{d}{du}(h(u)u^{\sigma-s}) < 0, \quad \frac{d^2}{du^2}(h(u)u^{\sigma-s}) > 0 \quad (u \in \mathbf{R}_+)$$

and $v'(t) > 0, v''(t) \leq 0, v'''(t) \geq 0$ $(t \in (n^{(0)} - \frac{1}{2}, \infty))$, with $v(n^{(0)} - \frac{1}{2}) \geq 0$, in view of (5.73) and (5.74), we set the following weight functions:

$$\omega_1(\sigma, n) = \int_{\mathbf{R}_+^m} h(||x||_\alpha ||v(n)||_1) \frac{||v(n)||_1^\sigma}{||x||_\alpha^{m-\sigma}} dx \ (n \in \mathbf{N}_{n^{(0)}}^s), \qquad (5.80)$$

$$\varpi_1(\sigma, x) = \sum_{n \in \mathbf{N}_{n^{(0)}}^s} h(||x||_\alpha ||v(n)||_1) \frac{||x||_\alpha^\sigma}{||v(n)||_1^{s-\sigma}} \prod_{k=1}^{s} v'(n_k)(x \in \mathbf{R}_+^m), \ (5.81)$$

and by (5.75), (5.10) and (5.11), we find

$$\omega_1(\sigma, n) = K(\alpha, m, \sigma) = \frac{\Gamma^m(\frac{1}{\alpha})}{\alpha^{m-1}\Gamma(\frac{m}{\alpha})}k(\sigma); \tag{5.82}$$

$$\varpi_1(\sigma, x) < \frac{||x||_\alpha^\sigma}{(s-1)!} \lim_{M\to\infty} M^\sigma \int_0^1 h(||x||_\alpha Mv)v^{\sigma-1}dv$$

$$= K(1, s, \sigma) = \frac{k(\sigma)}{(s-1)!}, \tag{5.83}$$

and

$$\varpi(\sigma, x) > \frac{||x||_\alpha^\sigma}{(s-1)!}$$

$$\left[\lim_{M\to\infty} M^\sigma \int_0^1 h(||x||_\alpha Mv)v^{\sigma-1}dv - \delta_0 \int_0^1 h(||x||_\alpha v)v^{\sigma-1}dv\right]$$

$$= \frac{k(\sigma)}{(s-1)!}(1 - \theta_1(||x||_\alpha)) > 0, \tag{5.84}$$

where

$$\theta_1(||x||_\alpha) = \frac{\delta_0(s-1)!}{k(\sigma)} \int_0^{||x||_\alpha} h(t)t^{\sigma-1}dt > 0 \quad (0 < \delta_0 \leq 1).$$

By the same way of Lemma 5.6, we still have

Lemma 5.7. *Suppose that* $m, s \in \mathbf{N}, \alpha > 0, p \in \mathbf{R}\backslash\{0, 1\}, \frac{1}{p} + \frac{1}{q} = 1, h(u)$ *is a non-negative finite measurable function in* $\mathbf{R}_+$, $\sigma \in \mathbf{R}$, *satisfying*

$$\frac{d}{du}(h(u)u^{\sigma-s}) < 0, \frac{d^2}{du^2}(h(u)u^{\sigma-s}) > 0 \quad (u \in \mathbf{R}^+),$$

$v'(t) > 0, v''(t) \leq 0, v'''(t) \geq 0 \ (t \in (n^{(0)} - \frac{1}{2}, \infty))$, *with* $v(n^{(0)} - \frac{1}{2}) \geq 0$.
If there exist constants $\widetilde{\delta}_0 > 0$ *and* $\eta < \sigma$, *such that for any* $\delta \in [0, \widetilde{\delta}_0]$,

$$k(\sigma \pm \delta) = \int_0^\infty h(t)t^{(\sigma\pm\delta)-1}dt \in \mathbf{R}_+,$$

and

$$h(t) \leq L\left(\frac{1}{t^\eta}\right) \quad (0 < t \leq 1; L > 0),$$

then, for any $0 < \varepsilon < |q|\widetilde{\delta}_0$, *we have*

$$\widetilde{I}(\varepsilon) = \int_{\{x\in\mathbf{R}_+^m;||x||_\alpha\leq1\}} \sum_{n\in\mathbf{N}_{n^{(0)}}^s} h(||x||_\alpha||v(n)||_1)\frac{||x||_\alpha^{\sigma-m+\frac{\varepsilon}{p}}}{||v(n)||_1^{s-(\sigma-\frac{\varepsilon}{q})}} \prod_{k=1}^s v'(n_k)dx$$

$$= \frac{\Gamma^m(\frac{1}{\alpha})}{\varepsilon(s-1)!\alpha^{m-1}\Gamma(\frac{m}{\alpha})}(k(\sigma) + o(1))(1 - \varepsilon O_2(1))(\varepsilon \to 0^+). \tag{5.85}$$

5.4.2 **Main Results**

For $p \in \mathbf{R}\backslash\{0,1\}$, $\frac{1}{p} + \frac{1}{q} = 1$, we set two functions

$$\varphi(x) = ||x||_\alpha^{p(m-\sigma)-m} \quad (x \in \mathbf{R}_+^m),$$

and

$$\Psi(n) = ||v(n)||_\beta^{q(s-\sigma)-s} \left(\prod_{k=1}^s v'(n_k)\right)^{1-q} \quad (n \in \mathbf{N}_{n^{(0)}}^s),$$

wherefrom $[\varphi(x)]^{1-q} = ||x||_\alpha^{q\sigma-m}$, and $[\Psi(n)]^{1-p} = ||v(n)||_\beta^{q\sigma-s} \prod_{k=1}^s v'(n_k)$. We also set

$$\widetilde{\varphi}(x) = (1 - \theta(||x||_\alpha))||x||_\alpha^{p(m-\sigma)-m},$$

and

$$[\widetilde{\varphi}(x)]^{1-q} = (1 - \theta(||x||_\alpha))^{1-q}||x||_\alpha^{q\sigma-m},$$

where

$$\theta(||x||_\alpha) = \frac{\delta_0}{K(\beta,s,\sigma)} \int_0^{||x||_\alpha} h(t)t^{\sigma-1}dt > 0 \quad (0 < \delta_0 \le 1).$$

Then we define two sets as follows:

$$\mathbf{L}_{p,\varphi}(\mathbf{R}_+^m) = \left\{ f; ||f||_{p,\varphi} = \left\{ \int_{\mathbf{R}_+^m} ||x||_\alpha^{p(m-\alpha)-m}|f(x)|^p dx \right\}^{\frac{1}{p}} < \infty \right\},$$

$$\mathbf{l}_{q,\Psi} = \left\{ a; ||a||_{q,\Psi} = \left\{ \sum_{n \in \mathbf{N}_{n^{(0)}}^s} ||v(n)||_\beta^{q(s-\sigma)-s} \right. \right.$$

$$\left. \left. \times \left(\prod_{k=1}^s v'(n_k)\right)^{1-q} |a(n)|^q \right\}^{\frac{1}{q}} < \infty \right\}.$$

Note. For p, $q > 1$, the above two sets with the norms are normed linear spaces. For the other cases, we still use them as the formal symbols.

We have the following theorem:

Theorem 5.6. *Suppose that* m, s, $n^{(0)} \in \mathbf{N}$, α, $\beta > 0$, $p \in \mathbf{R}\backslash\{0,1\}$, $\frac{1}{p} + \frac{1}{q} = 1$, $\sigma \in \mathbf{R}$, $h(u)$ *is a non-negative finite measurable function in* $\mathbf{R}_+$, $h(u)u^{\sigma-s}$ *is decreasing with respect to* $u \in \mathbf{R}_+$ *and strict decreasing* *in an interval* $I \subset (n^{(0)}, \infty)$, $v'(t) > 0, v''(t) \le 0 (t \in (n^{(0)} - 1, \infty))$, *with* $v(n^{(0)} - 1) \ge 0$,

$$K(\sigma) = \left(\frac{\Gamma^m(\frac{1}{\alpha})}{\alpha^{m-1}\Gamma(\frac{m}{\alpha})} \right)^{\frac{1}{q}} \left(\frac{\Gamma^s(\frac{1}{\beta})}{\beta^{s-1}\Gamma(\frac{s}{\beta})} \right)^{\frac{1}{p}} k(\sigma), \qquad (5.86)$$

where $k(\sigma) = \int_0^\infty h(t)t^{\sigma-1}dt \in \mathbf{R}_+$. *If* $f(x) \ge 0$, $a(n) \ge 0$, $f \in \mathbf{L}_{p,\varphi}(\mathbf{R}_+^m), a = \{a(n)\} \in l_{q,\Psi}$ *such that* $||f||_{p,\varphi} > 0, ||a||_{q,\Psi} > 0$, *then*

(i) *for* $p > 1$, *we have the following equivalent inequalities:*

$$I = \int_{\mathbf{R}_+^m} \sum_{n \in \mathbf{N}_{n^{(0)}}^s} h(||x||_\alpha ||v(n)||_\beta)a(n)f(x)dx < K(\sigma)||f||_{p,\varphi}||a||_{q,\Psi}, \quad (5.87)$$

$$J_1 = \left\{ \sum_{n \in \mathbf{N}_{n^{(0)}}^s} [\Psi(n)]^{1-p} \left(\int_{\mathbf{R}_+^m} h(||x||_\alpha ||v(n)||_\beta)f(x)dx \right)^p \right\}^{\frac{1}{p}}$$
$$< K(\sigma)||f||_{p,\varphi}, \qquad (5.88)$$

and

$$J_2 = \left\{ \int_{\mathbf{R}_+^m} [\varphi(x)]^{1-q} \left(\sum_{n \in \mathbf{N}_{n^{(0)}}^s} h(||x||_\alpha ||v(n)||_\beta)a(n) \right)^q dx \right\}^{\frac{1}{q}}$$
$$< K(\sigma)||a||_{q,\Psi}; \qquad (5.89)$$

(ii) *for* $p < 0$, *we have the reverses of* (5.87), (5.88) *and* (5.89);

(iii) *for* $0 < p < 1$, *if* $f(x) \ge 0, a(n) \ge 0$, $f \in \mathbf{L}_{p,\widetilde{\varphi}}(\mathbf{R}_+^m)$, $a = \{a(n)\} \in l_{q,\Psi}$, *such that* $||f||_{p,\widetilde{\varphi}} > 0, ||a||_{q,\Psi} > 0$, *then, we have the following reverse equivalent inequalities:*

$$I = \int_{\mathbf{R}_+^m} \sum_{n \in \mathbf{N}_{n^{(0)}}^s} h(||x||_\alpha ||v(n)||_\beta)a(n)f(x)dx > K(\sigma)||f||_{p,\widetilde{\varphi}}||a||_{q,\Psi}, \quad (5.90)$$

$$J_1 = \left\{ \sum_{n \in \mathbf{N}_{n^{(0)}}^s} [\Psi(n)]^{1-p} \left(\int_{\mathbf{R}_+^m} h(||x||_\alpha ||v(n)||_\beta)f(x)dx \right)^p \right\}^{\frac{1}{p}}$$
$$> K(\sigma)||f||_{p,\widetilde{\varphi}}, \qquad (5.91)$$

and

$$\widetilde{J}_2 = \left\{ \int_{\mathbf{R}_+^m} [\widetilde{\varphi}(x)]^{1-q} \left(\sum_{n \in \mathbf{N}_{n^{(0)}}^s} h(||x||_\alpha ||v(n)||_\beta)a(n) \right)^q dx \right\}^{\frac{1}{q}}$$
$$> K(\sigma)||a||_{q,\Psi}. \qquad (5.92)$$

Proof.   (i) For $p > 1$, by (5.12) and (5.13), for

$$H(||x||_\alpha, ||v(n)||_\beta) = h(||x||_\alpha ||v(n)||_\beta),$$

$\omega(n) = \omega(\sigma, n)$ and $\varpi(x) = \varpi(\sigma, x)$, in view of (5.75), (5.76) and the assumptions, we have (5.88) and (5.89).

By Hölder's inequality, we have

$$
\begin{aligned}
I &= \sum_{n \in \mathbf{N}_{n(0)}^s} \int_{\mathbf{R}_+^m} h(||x||_\alpha ||v(n)||_\beta) a(n) f(x) dx \\
&= \sum_{n \in \mathbf{N}_{n(0)}^s} \left\{ [\Psi(n)]^{\frac{-1}{q}} \int_{\mathbf{R}_+^m} h(||x||_\alpha ||v(n)||_\beta) f(x) dx \right\} \{ [\Psi(n)]^{\frac{1}{q}} a(n) \} \\
&\leq J_1 ||a||_{q,\Psi}.
\end{aligned}
\tag{5.93}
$$

Then by (5.88), we have (5.87).

On the other hand, assuming that (5.87) is valid, and setting

$$
a(n) = [\Psi(n)]^{1-p} \left(\int_{\mathbf{R}_+^m} h(||x||_\alpha ||v(n)||_\beta) f(x) dx \right)^{p-1} \quad (n \in \mathbf{N}_{n(0)}^s),
$$

then, we have $||a||_{q,\Psi} = J_1^{p-1}$. By (5.12) and the assumption of $0 < ||f||_{p,\varphi} < \infty$, we have $J_1 < \infty$. If $J_1 = 0$, then (5.88) is trivially valid; if $J_1 > 0$, then by (5.87), it follows that

$$
\begin{aligned}
||a||_{q,\Psi}^q &= J_1^p = I < K(\sigma) ||f||_{p,\varphi} ||a||_{q,\Psi}, \text{ that is,} \\
||a||_{q,\Psi}^{q-1} &= J_1 < K(\sigma) ||f||_{p,\varphi}.
\end{aligned}
$$

Thus, (5.88) follows. Hence, (5.87) and (5.88) are equivalent.

By Hölder's inequality, we still have

$$
\begin{aligned}
I &= \int_{\mathbf{R}_+^m} \{ [\varphi(x)]^{\frac{1}{p}} f(x) \} \left\{ [\varphi(x)]^{\frac{-1}{p}} \sum_{n \in \mathbf{N}_{n(0)}^s} h(||x||_\alpha ||v(n)||_\beta) a(n) \right\} dx \\
&\leq ||f||_{p,\varphi} J_2.
\end{aligned}
\tag{5.94}
$$

Then, by (5.89), we have (5.87).

On the other hand, assuming that (5.87) is valid, and setting

$$
f(x) = [\varphi(x)]^{1-q} \left(\sum_{n \in \mathbf{N}_{n(0)}^s} h(||x||_\alpha ||v(n)||_\beta) a(n) \right)^{q-1} \quad (x \in \mathbf{R}_+^m),
$$

then, we have $||f||_{p,\varphi} = J_2^{q-1}$. By (5.13) and the assumption of $0 < ||a||_{q,\Psi} < \infty$, we have $J_2 < \infty$. If $J_2 = 0$, then (5.89) is trivially valid; if $J_2 > 0$, then by (5.87), it follows that

$$||f||_{p,\varphi}^p = J_2^q = I < K(\sigma)||f||_{p,\varphi}||a||_{q,\Psi}, \text{ that is,}$$
$$||f||_{p,\varphi}^{p-1} = J_2 < K(\sigma)||a||_{q,\Psi}.$$

Thus, (5.89) follows, and then (5.87) and (5.89) are equivalent.

Hence, (5.87), (5.88) and (5.89) are equivalent.

(ii) For $p < 0$ $(0 < q < 1)$, by the reverses of (5.12) and (5.13), for

$$H(||x||_\alpha, ||v(n)||_\beta) = h(||x||_\alpha ||v(n)||_\beta),$$

$\omega(n) = \omega(\sigma, n)$ and $\varpi(x) = \varpi(\sigma, x)$, in view of (5.75), (5.76) and the assumptions, we have the reverses of (5.88) and (5.89).

By the reverse Hölder's inequality, we have

$$I \geq J_1 ||a||_{q,\Psi}. \tag{5.95}$$

Then, by the reverse of (5.88), we have the reverse of (5.87).

On the other hand, assuming that the reverse of (5.87) is valid, setting $a(n)$ as (i), then we have $||a||_{q,\Psi} = J_1^{p-1}$. By the reverse of (5.12) and the assumption of $0 < ||f||_{p,\varphi} < \infty$, we have $J_1 > 0$. If $J_1 = \infty$, then the reverse of (5.88) is valid trivially; if $J_1 < \infty$, then by the reverse of (5.87), it follows that

$$||a||_{q,\Psi}^q = J_1^p = I > K(\sigma)||f||_{p,\varphi}||a||_{q,\Psi}, \text{ that is,}$$
$$||a||_{q,\Psi}^{q-1} = J_1 > K(\sigma)||f||_{p,\varphi}.$$

Thus, the reverse of (5.88) follows, and then the reverses of (5.87) and (5.88) are equivalent.

By the reverse Hölder's inequality, we still have

$$I \geq ||f||_{p,\varphi} J_2. \tag{5.96}$$

Then, by the reverse of (5.89), we have the reverse of (5.87).

On the other hand, assuming that the reverse of (5.87) is valid, and setting $f(x)$ as in (i), then we have $||f||_{p,\varphi} = J_2^{q-1}$. By the reverse of (5.13) and the assumption of $0 < ||a||_{q,\Psi} < \infty$, we have $J_2 > 0$. If $J_2 = \infty$, then the reverse of (5.89) is valid trivially; if $J_2 < \infty$, then, by the reverse of (5.87), it follows that

$$||f||_{p,\varphi}^p = J_2^q = I > K(\sigma)||f||_{p,\varphi}||a||_{q,\Psi}, \text{ that is,}$$
$$||f||_{p,\varphi}^{p-1} = J_2 > K(\sigma)||a||_{q,\Psi}.$$

Hence, the reverse of (5.89) follows, and then, the reverses of (5.87) and (5.89) are equivalent.

Hence, the reverses of (5.87), (5.88) and (5.89) are equivalent.

(iii) For $0 < p < 1$ $(q < 0)$, by the reverses of (5.12) and (5.13), for

$$H(||x||_\alpha, ||v(n)||_\beta) = h(||x||_\alpha ||v(n)||_\beta),$$

$\omega(n) = \omega(\sigma, n)$ and $\varpi(x) = \varpi(\sigma, x)$, in view of (5.75), (5.77) and the assumptions, we have (5.91) and (5.92).

By the reverse Hölder's inequality, we have

$$I \geq J_1 ||a||_{q,\Psi}. \tag{5.97}$$

Then, by (5.91), we have (5.90).

On the other hand, assuming that (5.90) is valid, and setting $a(n)$ as in (i), then, we have $||a||_{q,\Psi} = J_1^{p-1}$. By the reverse of (5.12) and the assumption of $0 < ||f||_{p,\widetilde\varphi} < \infty$, we have $J_1 > 0$. If $J_1 = \infty$, then, (5.91) is valid trivially; if $J_1 < \infty$, then, by (5.90), it follows that

$$||a||_{q,\Psi}^q = J_1^p = I > K(\sigma)||f||_{p,\widetilde\varphi}||a||_{q,\Psi}, \text{that is,}$$
$$||a||_{q,\Psi}^{q-1} = J_1 > K(\sigma)||f||_{p,\widetilde\varphi}.$$

Thus, (5.91) follows, and then (5.90) and (5.91) are equivalent.

By the reverse Hölder's inequality, we still have

$$I \geq ||f||_{p,\widetilde\varphi}\, \widetilde{J_2}. \tag{5.98}$$

Then, by (5.92), we have inequality (5.90).

On the other hand, assuming that (5.90) is valid, and setting $f(x)$ as follows:

$$f(x) = [\widetilde\varphi(x)]^{1-q} \left(\sum_{n \in \mathbf{N}_{n}^{s}(0)} h(||x||_\alpha ||v(n)||_\beta)a(n) \right)^{q-1} \quad (x \in \mathbf{R}_+^m),$$

then, we have $||f||_{p,\widetilde\varphi} = \widetilde{J_2}^{q-1}$. By the reverse of (5.13) and the assumption of $0 < ||a||_{q,\Psi} < \infty$, we have $\widetilde{J_2} > 0$. If $\widetilde{J_2} = \infty$, then (5.92) is valid trivially; if $\widetilde{J_2} < \infty$, then, by (5.90), it follows that

$$||f||_{p,\widetilde\varphi}^p = \widetilde{J_2}^q = I > K(\sigma)||f||_{p,\widetilde\varphi}||a||_{q,\Psi}, \text{ that is,}$$
$$||f||_{p,\widetilde\varphi}^{p-1} = \widetilde{J_2} > K(\sigma)||a||_{q,\Psi}.$$

Thus, (5.92) follows, and then (5.90) and (5.92) are equivalent.

Hence, (5.90), (5.91) and (5.92) are equivalent. $\square$

Theorem 5.7. *With the assumptions of Theorem 5.6, if there exist $\widetilde{\delta}_0 > 0$ and $\eta < \sigma$, such that, for any $\delta \in (0, \widetilde{\delta}_0]$,*

$$k(\sigma \pm \delta) = \int_0^\infty h(t) t^{(\sigma \pm \delta) - 1} dt \in \mathbf{R}_+,$$

and

$$h(t) \le L\left(\frac{1}{t^\eta}\right) \quad (0 < t \le 1; L > 0),$$

then, the inequalities in Theorem 5.6 have the same best possible constant factor $K(\sigma)$.

Proof. For $0 < \varepsilon < |q|\widetilde{\delta}_0$, we set $\widetilde{f}(x), \widetilde{a}(n)$ as follows:

$$\widetilde{f}(x) = \begin{cases} ||x||_\alpha^{\sigma - m + \frac{\varepsilon}{p}}, & 0 < ||x||_\alpha < 1, \\ 0, & ||x||_\alpha \ge 1, \end{cases}$$

$$\widetilde{a}(n) = ||v(n)||_\beta^{\sigma - \frac{\varepsilon}{q} - s} \prod_{k=1}^s v'(n_k), \quad n \in \mathbf{N}_{n(0)}^s.$$

(i) For $p > 1$, if there exists a constant $k(\le K(\sigma))$, such that (5.87) is still valid as we replace $K(\sigma)$ by k, then, in particular, it follows that

$$\widetilde{I}(\varepsilon) = \int_{\mathbf{R}_+^m} \sum_{n \in \mathbf{N}_{n(0)}^s} h(||x||_\alpha ||v(n)||_\beta) \widetilde{a}(n) \widetilde{f}(x) dx < k ||\widetilde{f}||_{p,\varphi} ||\widetilde{a}||_{q,\Psi}. \quad (5.99)$$

Since, by (5.3) and (5.4), we have

$$||\widetilde{f}||_{p,\varphi} = \left\{ \int_{\{x; ||x||_\alpha \le 1\}} ||x||_\alpha^{-m + \varepsilon} dx \right\}^{\frac{1}{p}} = \left\{ \frac{\Gamma^m(\frac{1}{\alpha})}{\varepsilon \alpha^{m-1} \Gamma(\frac{m}{\alpha})} \right\}^{\frac{1}{p}},$$

$$||\widetilde{a}||_{q,\Psi} = \left\{ \sum_{n \in \mathbf{N}_{n(0)}^s} ||v(n)||_\beta^{-s - \varepsilon} \prod_{k=1}^s v'(n_k) \right\}^{\frac{1}{q}}$$

$$= \left\{ \frac{\Gamma^s(\frac{1}{\beta})}{\varepsilon \beta^{s-1} \Gamma\left(\frac{s}{\beta}\right)} + O(1) \right\}^{\frac{1}{q}},$$

then, by (5.78) and (5.99), it follows that

$$\frac{\Gamma^m(\frac{1}{\alpha})}{\alpha^{m-1} \Gamma(\frac{m}{\alpha})} \frac{\Gamma^s(\frac{1}{\beta})}{\beta^{s-1} \Gamma\left(\frac{s}{\beta}\right)} (k(\sigma) + o(1))(1 - \varepsilon O_2(1))$$

$$< k \left\{ \frac{\Gamma^m(\frac{1}{\alpha})}{\alpha^{m-1} \Gamma(\frac{m}{\alpha})} \right\}^{\frac{1}{p}} \left\{ \frac{\Gamma^s(\frac{1}{\beta})}{\beta^{s-1} \Gamma\left(\frac{s}{\beta}\right)} + \varepsilon O(1) \right\}^{\frac{1}{q}},$$

and then $K(\sigma) \leq k$ ($\varepsilon \to 0^+$). Hence, $k = K(\sigma)$ is the best value of (5.87).

By the equivalency, the constant factor in (5.88) ((5.89)) is the best possible. Otherwise, we can get a contradiction by (5.93) ((5.94)) that the constant factor in (5.87) is not the best possible.

(ii) For $p < 0$, if there exists a constant $k(\geq K(\sigma))$, such that the reverse of (5.87) is still valid as we replace $K(\sigma)$ by k, then, in particular, by (i), it follows that

$$\frac{\Gamma^m(\frac{1}{\alpha})}{\alpha^{m-1}\Gamma(\frac{m}{\alpha})}\frac{\Gamma^s(\frac{1}{\beta})}{\beta^{s-1}\Gamma\left(\frac{s}{\beta}\right)}(k(\sigma)+o(1))(1-\varepsilon O_2(1))$$

$$> k\left\{\frac{\Gamma^m(\frac{1}{\alpha})}{\alpha^{m-1}\Gamma(\frac{m}{\alpha})}\right\}^{\frac{1}{p}}\left\{\frac{\Gamma^s(\frac{1}{\beta})}{\beta^{s-1}\Gamma\left(\frac{s}{\beta}\right)}+\varepsilon O(1)\right\}^{\frac{1}{q}},$$

and then, $K(\sigma) \geq k(\varepsilon \to 0^+)$. Hence, $k = K(\sigma)$ is the best value of the reverse of (5.87).

By the equivalency, the constant factor in the reverse of (5.88) (the reverse of (5.89)) is the best possible. Otherwise, we can get a contradiction by (5.95) ((5.96)) that the constant factor in the reverse of (5.87) is not the best possible.

(iii) For $0 < p < 1$, if there exists a constant $k(\geq K(\sigma))$, such that (5.90) is still valid as we replace $K(\sigma)$ by k, then in particular, it follows that

$$\widetilde{I}(\varepsilon) = \int_{\mathbf{R}_+^m}\sum_{n\in\mathbf{N}_{n}^{s}(0)} h(||x||_\alpha||v(n)||_\beta)\widetilde{a}(n)\widetilde{f}(x)dx > k||\widetilde{f}||_{p,\widetilde{\varphi}}||\widetilde{a}||_{q,\Psi}.$$

By (5.79), we have

$$||\widetilde{f}||_{p,\widetilde{\varphi}} = \left\{\int_{\{x\in\mathbf{R}_+^m;||x||_\alpha\leq 1\}}||x||_\alpha^{-m+\varepsilon}(1-\theta(||x||_\alpha))dx\right\}^{\frac{1}{p}}$$

$$= \left\{\frac{\Gamma^m(\frac{1}{\alpha})}{\varepsilon\alpha^{m-1}\Gamma(\frac{m}{\alpha})}(1-\varepsilon O_2(1))\right\}^{\frac{1}{p}}.$$

Then, we obtain

$$\frac{\Gamma^m(\frac{1}{\alpha})}{\alpha^{m-1}\Gamma(\frac{m}{\alpha})}\frac{\Gamma^s(\frac{1}{\beta})}{\beta^{s-1}\Gamma\left(\frac{s}{\beta}\right)}(k(\sigma)+o(1))(1-\varepsilon O_2(1))$$

$$> k\left\{\frac{\Gamma^m(\frac{1}{\alpha})}{\alpha^{m-1}\Gamma(\frac{m}{\alpha})}(1-\varepsilon O_1(1))\right\}^{\frac{1}{p}}\left\{\frac{\Gamma^s(\frac{1}{\beta})}{\beta^{s-1}\Gamma\left(\frac{s}{\beta}\right)}+\varepsilon O(1)\right\}^{\frac{1}{q}},$$

and then $K(\sigma) \geq k(\varepsilon \to 0^+)$. Hence, $k = K(\sigma)$ is the best value of (5.90).

By the equivalency, the constant factor in (5.91) and (5.92) is the best possible. Otherwise, we can get a contradiction by (5.97) and (5.98) that the constant factor in (5.90) is not the best possible. □

For $\beta = 1$, we set

$$\Psi_1(n) = ||v(n)||_1^{q(s-\sigma)-s} \left(\prod_{k=1}^{s} v'(n_k)\right)^{1-q} \quad (n \in \mathbf{N}_{n^{(0)}}^s),$$

wherefrom

$$[\Psi_1(n)]^{1-p} = ||v(n)||_1^{q\sigma-s} \prod_{k=1}^{s} v'(n_k).$$

We also set

$$\widetilde{\varphi}_1(x) = (1 - \theta_1(||x||_\alpha))||x||_\alpha^{p(m-\sigma)-m},$$

and

$$[\widetilde{\varphi}_1(x)]^{1-q} = (1 - \theta_1(||x||_\alpha))^{1-q}||x||_\alpha^{q\sigma-m},$$

where

$$\theta_1(||x||_\alpha) = \frac{\delta_0(s-1)!}{k(\sigma)} \int_{||x||_\alpha}^{\infty} h(t)t^{\sigma-1}dt \quad (0 < \delta_0 \leq 1).$$

By (5.87)-(5.93), in the same way of Theorem 5.6 and Theorem 5.7, we have

Theorem 5.8. *Suppose that* $m, s \in \mathbf{N}$, $\alpha > 0$, $p \in \mathbf{R}\backslash\{0,1\}$, $\frac{1}{p} + \frac{1}{q} = 1$, $\sigma \in \mathbf{R}$, $h(u)$ *is a non-negative finite measurable function in* $\mathbf{R}_+$, *satisfying*

$$\frac{d}{du}(h(u)u^{\sigma-s}) < 0, \quad \frac{d^2}{du^2}(h(u)u^{\sigma-s}) > 0 \quad (u \in \mathbf{R}_+),$$

$v'(t) > 0$, $v''(t) \leq 0$, $v'''(t) \geq 0$ $(t \in (n^{(0)} - \frac{1}{2}, \infty))$ *with* $v(n^{(0)} - \frac{1}{2}) \geq 0$,

$$K_1(\sigma) = \left(\frac{\Gamma^m(\frac{1}{\alpha})}{\alpha^{m-1}\Gamma(\frac{m}{\alpha})}\right)^{\frac{1}{q}} \left(\frac{1}{(s-1)!}\right)^{\frac{1}{p}} k(\sigma), \qquad (5.100)$$

where $k(\sigma) = \int_0^{\infty} h(t)t^{\sigma-1}du \in \mathbf{R}_+$. *If* $f(x) \geq 0$, $a(n) \geq 0$, $f \in \mathbf{L}_{p,\varphi}(\mathbf{R}_+^m)$, $a = \{a(n)\} \in l_{q,\Psi_1}$, *such that* $||f||_{p,\varphi} > 0$, $||a||_{q,\Psi_1} > 0$, *then*

(i) for $p > 1$, we have the following equivalent inequalities:

$$\int_{\mathbf{R}_+^m} \sum_{n \in \mathbf{N}_{n(0)}^s} h(||x||_\alpha ||v(n)||_1) a(n) f(x) dx$$

$$< K_1(\sigma) ||f||_{p,\varphi} ||a||_{q,\Psi_1}, \tag{5.101}$$

$$\left\{ \sum_{n \in \mathbf{N}_{n(0)}^s} [\Psi_1(n)]^{1-p} \left(\int_{\mathbf{R}_+^m} h(||x||_\alpha ||v(n)||_1) f(x) dx \right)^p \right\}^{\frac{1}{p}}$$

$$< K_1(\sigma) ||f||_{p,\varphi}, \tag{5.102}$$

and

$$\left\{ \int_{\mathbf{R}_+^m} [\varphi(x)]^{1-q} \left(\sum_{n \in \mathbf{N}_{n(0)}^s} h(||x||_\alpha ||v(n)||_1) a(n) \right)^q dx \right\}^{\frac{1}{q}} < K_1(\sigma) ||a||_{q,\Psi_1}; \tag{5.103}$$

(ii) for $p < 0$, we have the reverses of (5.101), (5.102) and (5.103);

(iii) For $0 < p < 1$, if $f(x) \geq 0$, $a(n) \geq 0$, $f \in \mathbf{L}_{p,\widetilde{\varphi}}(\mathbf{R}_+^m)$, $a = \{a(n)\} \in l_{q,\Psi_1}$ such that $||f||_{p,\widetilde{\varphi}_1} > 0$, $||a||_{q,\Psi_1} > 0$, then we have the following reverse equivalent inequalities:

$$\int_{\mathbf{R}_+^m} \sum_{n \in \mathbf{N}_{n(0)}^s} h(||x||_\alpha ||v(n)||_1) a(n) f(x) dx$$

$$> K_1(\sigma) ||f||_{p,\widetilde{\varphi}_1} ||a||_{q,\Psi_1}, \tag{5.104}$$

$$\left\{ \sum_{n \in \mathbf{N}_{n(0)}^s} [\Psi_1(n)]^{1-p} \left(\int_{\mathbf{R}_+^m} h(||x||_\alpha ||v(n)||_1) f(x) dx \right)^p \right\}^{\frac{1}{p}}$$

$$> K_1(\sigma) ||f||_{p,\widetilde{\varphi}_1}, \tag{5.105}$$

and

$$\left\{ \int_{\mathbf{R}_+^m} [\widetilde{\varphi}_1(x)]^{1-q} \left(\sum_{n \in \mathbf{N}_{n(0)}^s} h(||x||_\alpha ||v(n)||_1) a(n) \right)^q dx \right\}^{\frac{1}{q}} > K_1(\sigma) ||a||_{q,\Psi_1}. \tag{5.106}$$

Moreover, if there exist $\widetilde{\delta}_0 > 0$ and $\eta < \sigma$, such that for any $\delta \in (0, \widetilde{\delta}_0]$,

$$k(\sigma \pm \delta) = \int_0^\infty h(t) t^{(\sigma \pm \delta)-1} dt \in \mathbf{R}_+,$$

and

$$h(t) \leq L \left(\frac{1}{t^\eta} \right) (0 < t \leq 1; L > 0),$$

then, the above inequalities are all with the same best possible constant factor $K_1(\sigma)$.

5.4.3 Some Corollaries

If $w(t)$ is a strictly increasing differentiable function in $(b,c)(-\infty \leq b < c \leq \infty)$, with $w(b+) = 0$, $w(c-) = \infty$, then

$$\Phi(x) = ||w(x)||_\alpha^{p(m-\sigma)-m} \left(\prod_{i=1}^{m} w'(x_i) \right)^{1-p},$$

and $\widetilde{\Phi}(x) = (1 - \theta(||w(x)||_\alpha))\Phi(x)$, setting $x = w(X) = (w(X_1), \cdots, w(X_m))$ in the inequalities of Theorem 5.7, after simplification, replacing $f(w(X)) \prod_{i=1}^{m} w'(X_i)$ by $f(x)$, we have

Corollary 5.5. *Suppose that* $m, s, n^{(0)} \in \mathbf{N}, \alpha, \beta > 0, p \in \mathbf{R}\backslash\{0,1\}, \frac{1}{p}+\frac{1}{q} = 1, \sigma \in \mathbf{R}, h(u)$ *is a non-negative finite measurable function in* $\mathbf{R}_+, h(u)u^{\sigma-s}$ *is decreasing with respect to* $u \in \mathbf{R}_+$ *and strict decreasing in an interval* $I \subset (n^{(0)}, \infty), w(t)$ *is strict increasing in* (b,c)*, with* $w(b+) = 0, w(c-) = \infty, v'(t) > 0, v''(t) \leq 0$ $(t \in (n^{(0)} - 1, \infty))$*, with* $v(n^{(0)} - 1) \geq 0$*,*

$$K(\sigma) = \left(\frac{\Gamma^m(\frac{1}{\alpha})}{\alpha^{m-1}\Gamma(\frac{m}{\alpha})} \right)^{\frac{1}{q}} \left(\frac{\Gamma^s(\frac{1}{\beta})}{\beta^{s-1}\Gamma\left(\frac{s}{\beta}\right)} \right)^{\frac{1}{p}} k(\sigma), \tag{5.107}$$

where $k(\sigma) = \int_0^\infty h(t)t^{\sigma-1}dt \in \mathbf{R}_+$. *If* $f(x) \geq 0, a(n) \geq 0, f \in \mathbf{L}_{p,\Phi}((\mathbf{b},\mathbf{c})^m), a = \{a(n)\} \in \mathbf{l}_{q,\Psi}$*, such that* $||f||_{p,\Phi} > 0, ||a||_{q,\Psi} > 0$*, then*

(i) *for* $p > 1$*, we have the following equivalent inequalities:*

$$\int_{(\mathbf{b},\mathbf{c})^m} \sum_{n \in \mathbf{N}_{n^{(0)}}^s} h(||w(x)||_\alpha ||v(n)||_\beta)a(n)f(x)dx$$
$$< K(\sigma)||f||_{p,\Phi}||a||_{q,\Psi}, \tag{5.108}$$

$$\left\{ \sum_{n \in \mathbf{N}_{n^{(0)}}^s} [\Psi(n)]^{1-p} \left(\int_{(\mathbf{b},\mathbf{c})^m} h(||w(x)||_\alpha ||v(n)||_\beta)f(x)dx \right)^p \right\}^{\frac{1}{p}}$$
$$< K(\sigma)||f||_{p,\Phi}, \tag{5.109}$$

and

$$\left\{ \int_{(\mathbf{b},\mathbf{c})^m} [\Phi(x)]^{1-q} \left(\sum_{n \in \mathbf{N}_{n^{(0)}}^s} h(||w(x)||_\alpha ||v(n)||_\beta)a(n) \right)^q dx \right\}^{\frac{1}{q}}$$
$$< K(\sigma)||a||_{q,\Psi}; \tag{5.110}$$

(ii) for $p < 0$, we have the reverses of (5.108), (5.109) and (5.110);

(iii) for $0 < p < 1$, if $f(x) \geq 0, a(n) \geq 0$, $f \in \mathbf{L}_{p,\widetilde{\Phi}}(\mathbf{R}_+^m), a = \{a(n)\} \in l_{q,\Psi}$, such that $||f||_{p,\widetilde{\Phi}} > 0, ||a||_{q,\Psi} > 0$, then we have the following reverse equivalent inequalities:

$$\int_{(\mathbf{b},\mathbf{c})^m} \sum_{n \in \mathbf{N}_{n^{(0)}}^s} h(||w(x)||_\alpha ||v(n)||_\beta) a(n) f(x) dx$$
$$> K(\sigma) ||f||_{p,\widetilde{\Phi}} ||a||_{q,\Psi}, \tag{5.111}$$

$$\left\{ \sum_{n \in \mathbf{N}_{n^{(0)}}^s} [\Psi(n)]^{1-p} \left(\int_{(\mathbf{b},\mathbf{c})^m} h(||w(x)||_\alpha ||v(n)||_\beta) f(x) dx \right)^p \right\}^{\frac{1}{p}}$$
$$> K(\sigma) ||f||_{p,\widetilde{\Phi}}, \tag{5.112}$$

and

$$\left\{ \int_{(\mathbf{b},\mathbf{c})^m} [\widetilde{\varphi}(x)]^{1-q} \left(\sum_{n \in \mathbf{N}_{n^{(0)}}^s} h(||w(x)||_\alpha ||v(n)||_\beta) a(n) \right)^q dx \right\}^{\frac{1}{q}}$$
$$> K(\sigma) ||a||_{q,\Psi}. \tag{5.113}$$

If $w(x) = x$, (5.108)-(5.113) reduce to (5.87)-(5.92). Hence, Theorem 5.7 and Corollary 5.5 are equivalent. We can conclude that the constant factor in the above inequalities is the best possible by adding some conditions as follows:

Corollary 5.6. *With the assumptions of Corollary 5.5, if there exist $\widetilde{\delta}_0 > 0$ and $\eta < \sigma$, such that, for any $\delta \in (0, \widetilde{\delta}_0]$,*

$$k(\sigma \pm \delta) = \int_0^\infty h(t) t^{(\sigma \pm \delta) - 1} dt \in \mathbf{R}_+,$$

and

$$h(t) \leq L \left(\frac{1}{t^\eta} \right) \quad (0 < t \leq 1; L > 0),$$

then, the inequalities in Corollary 5.5 have the same best possible constant factor $K(\sigma)$.

For $v(t) = t, n^{(0)} = 1$ in Theorem 5.6 and Theorem 5.7, setting

$$\psi(n) = ||n||_\beta^{q(s-\sigma)-s} \quad (n \in \mathbf{N}^s),$$

we have the following corollary:

Corollary 5.7. *Suppose that* $m, s \in \mathbf{N}, \alpha, \beta > 0, p \in R\backslash\{0, 1\}, \frac{1}{p} + \frac{1}{q} = 1, \sigma \in \mathbf{R}$, $h(u)$ *is a non-negative finite measurable function in* $\mathbf{R}_+$, $h(u)u^{\sigma-s}$ *is decreasing with respect to* $u \in \mathbf{R}_+$ *and strictly decreasing in an interval* $I \subset (1, \infty)$,

$$K(\sigma) = \left(\frac{\Gamma^m(\frac{1}{\alpha})}{\alpha^{m-1}\Gamma(\frac{m}{\alpha})} \right)^{\frac{1}{q}} \left(\frac{\Gamma^s(\frac{1}{\beta})}{\beta^{s-1}\Gamma\left(\frac{s}{\beta}\right)} \right)^{\frac{1}{p}} k(\sigma), \tag{5.114}$$

where $k(\sigma) = \int_0^\infty h(t)t^{\sigma-1}dt \in \mathbf{R}_+$. *If* $f(x) \geq 0, a(n) \geq 0$, $f \in \mathbf{L}_{p,\varphi}(\mathbf{R}_+^m), a = \{a(n)\} \in l_{q,\psi}$ *such that* $||f||_{p,\varphi} > 0, ||a||_{q,\psi} > 0$, *then*

(i) *for* $p > 1$, *we have the following equivalent inequalities:*

$$\int_{\mathbf{R}_+^m} \sum_{n \in \mathbf{N}^s} h(||x||_\alpha ||n||_\beta)a(n)f(x)dx < K(\sigma)||f||_{p,\varphi}||a||_{q,\psi},$$
$$\tag{5.115}$$

$$\left\{ \sum_{n \in \mathbf{N}^s} [\psi(n)]^{1-p} \left(\int_{\mathbf{R}_+^m} h(||x||_\alpha ||n||_\beta)f(x)dx \right)^p \right\}^{\frac{1}{p}} < K(\sigma)||f||_{p,\varphi},$$
$$\tag{5.116}$$

and

$$\left\{ \int_{\mathbf{R}_+^m} [\varphi(x)]^{1-q} \left(\sum_{n \in \mathbf{N}^s} h(||x||_\alpha ||n||_\beta)a(n) \right)^q dx \right\}^{\frac{1}{q}} < K(\sigma)||a||_{q,\psi};$$
$$\tag{5.117}$$

(ii) *for* $p < 0$, *we have the reverses of* (5.115), (5.116) *and* (5.117);

(iii) *for* $0 < p < 1$, *if* $f(x) \geq 0, a(n) \geq 0$, $f \in \mathbf{L}_{p,\widetilde{\varphi}}(\mathbf{R}_+^m)$, $a = \{a(n)\} \in l_{q,\psi}$, *such that* $||f||_{p,\widetilde{\varphi}} > 0$, $||a||_{q,\psi} > 0$, *then, we have the following reverse equivalent inequalities:*

$$\int_{\mathbf{R}_+^m} \sum_{n \in \mathbf{N}^s} h(||x||_\alpha ||n||_\beta)a(n)f(x)dx > K(\sigma)||f||_{p,\widetilde{\varphi}}||a||_{q,\psi},$$
$$\tag{5.118}$$

$$\left\{ \sum_{n \in \mathbf{N}^s} [\psi(n)]^{1-p} \left(\int_{\mathbf{R}_+^m} h(||x||_\alpha ||n||_\beta)f(x)dx \right)^p \right\}^{\frac{1}{p}} > K(\sigma)||f||_{p,\widetilde{\varphi}}, \tag{5.119}$$

and

$$\left\{ \int_{\mathbf{R}_+^m} [\widetilde{\varphi}(x)]^{1-q} \left(\sum_{n \in \mathbf{N}^s} h(||x||_\alpha ||n||_\beta)a(n) \right)^q dx \right\}^{\frac{1}{q}} > K(\sigma)||a||_{q,\Psi}.$$
$$\tag{5.120}$$

Moreover, if there exist $\widetilde{\delta}_0 > 0$ and $\eta < \sigma$, such that for any $\delta \in (0, \widetilde{\delta}_0]$,

$$k(\sigma \pm \delta) = \int_0^\infty h(t)t^{(\sigma \pm \delta)-1}dt \in \mathbf{R}_+,$$

and

$$h(t) \leq L\left(\frac{1}{t^\eta}\right) \quad (0 < t \leq 1; L > 0),$$

then, the above inequalities hold with the same best possible constant factor $K(\sigma)$.

For $v(t) = (t - \xi)^\gamma (0 < \gamma \leq 1, 0 \leq \xi \leq \frac{1}{2})$, $n^{(0)} = 1$ in Theorem 5.8, setting

$$\psi_\xi(n) = ||(n - \xi)^\gamma||_1^{q(s-\sigma)-s}(n \in \mathbf{N}^s),$$

we have some more accurate inequalities of Corollary 5.7 (for $\beta = 1$) as follows:

Corollary 5.8. *Suppose that* m, $s \in \mathbf{N}$, $\alpha > 0$, $p \in \mathbf{R}\backslash\{0,1\}$, $\frac{1}{p} + \frac{1}{q} = 1$, $\sigma \in \mathbf{R}$, $h(u)$ *is a non-negative finite measurable function in* $\mathbf{R}_+$, *satisfying*

$$\frac{d}{du}(h(u)u^{\sigma-s}) < 0, \quad \frac{d^2}{du^2}(h(u)u^{\sigma-s}) > 0 \quad (u \in \mathbf{R}_+),$$

$$K_1(\sigma) = \left(\frac{\Gamma^m(\frac{1}{\alpha})}{\alpha^{m-1}\Gamma(\frac{m}{\alpha})}\right)^{\frac{1}{q}}\left(\frac{1}{(s-1)!}\right)^{\frac{1}{p}}k(\sigma), \qquad (5.121)$$

where $k(\sigma) = \int_0^\infty h(t)t^{\sigma-1}dt \in \mathbf{R}_+$. *If* $f(x) \geq 0$, $a(n) \geq 0$, $f \in \mathbf{L}_{p,\varphi}(\mathbf{R}_+^m)$, $a = \{a(n)\} \in l_{q,\Psi_\xi}$, *such that* $||f||_{p,\varphi} > 0$, $||a||_{q,\Psi_\xi} > 0$, *then*

(i) *for* $p > 1$, *we have the following equivalent inequalities:*

$$\int_{\mathbf{R}_+^m}\sum_{n\in\mathbf{N}^s}h(||x||_\alpha||(n-\xi)^\gamma||_1)a(n)f(x)dx$$

$$< K_1(\sigma)||f||_{p,\varphi}||a||_{q,\Psi_\xi}, \qquad (5.122)$$

$$\left\{\sum_{n\in\mathbf{N}^s}[\Psi_\xi(n)]^{1-p}\left(\int_{\mathbf{R}_+^m}h(||x||_\alpha||(n-\xi)^\gamma||_1)f(x)dx\right)^p\right\}^{\frac{1}{p}}$$

$$< K_1(\sigma)||f||_{p,\varphi}, \qquad (5.123)$$

and

$$\left\{\int_{\mathbf{R}_+^m}[\varphi(x)]^{1-q}\left(\sum_{n\in\mathbf{N}^s}h(||x||_\alpha||(n-\xi)^\gamma||_1)a(n)\right)^q dx\right\}^{\frac{1}{q}}$$

$$< K_1(\sigma)||a||_{q,\Psi_\xi}; \qquad (5.124)$$

(ii) for $p < 0$, we have the reverses of (5.122), (5.123) and (5.124);

(iii) for $0 < p < 1$, if $f(x) \geq 0$, $a(n) \geq 0$, $f \in \mathbf{L}_{p,\widetilde{\varphi}}(\mathbf{R}_+^m)$, $a = \{a(n)\} \in l_{q,\Psi_\xi}$ such that $||f||_{p,\widetilde{\varphi}} > 0$, $||a||_{q,\Psi_\xi} > 0$, then, we have the following reverse equivalent inequalities:

$$\int_{\mathbf{R}_+^m} \sum_{n \in \mathbf{N}^s} h(||x||_\alpha||(n-\xi)^\gamma||_1)a(n)f(x)dx$$
$$> K_1(\sigma)||f||_{p,\widetilde{\varphi}_1}||a||_{q,\Psi_\xi}, \tag{5.125}$$

$$\left\{ \sum_{n \in \mathbf{N}^s} [\Psi_1(n)]^{1-p} \left(\int_{\mathbf{R}_+^m} h(||x||_\alpha||(n-\xi)^\gamma||_1)f(x)dx \right)^p \right\}^{\frac{1}{p}}$$
$$> K_1(\sigma)||f||_{p,\widetilde{\varphi}_1}, \tag{5.126}$$

$$\left\{ \int_{\mathbf{R}_+^m} [\widetilde{\varphi}_1(x)]^{1-q} \left(\sum_{n \in \mathbf{N}^s} h(||x||_\alpha||(n-\xi)^\gamma||_1)a(n) \right)^q dx \right\}^{\frac{1}{q}}$$
$$> K_1(\sigma)||a||_{q,\Psi_\xi}. \tag{5.127}$$

Moreover, if there exist $\widetilde{\delta}_0 > 0$ and $\eta < \sigma$, such that, for any $\delta \in (0, \widetilde{\delta}_0]$,

$$k(\sigma \pm \delta) = \int_0^\infty h(t)t^{(\sigma \pm \delta)-1}dt \in \mathbf{R}_+,$$

and

$$h(t) \leq L\left(\frac{1}{t^\eta}\right) \quad (0 < t \leq 1; L > 0),$$

then the above inequalities have the same best possible constant factor $K_1(\sigma)$.

5.4.4 *Operator Expressions and Some Particular Examples*

With the assumptions of Theorem 5.6, for $p > 1$, we set

$$\varphi(x) = ||x||_\alpha^{p(m-\sigma)-m} \quad (x \in \mathbf{R}_+^m),$$

and

$$\Psi(n) = ||v(n)||_\beta^{q(s-\sigma)-s} \left(\prod_{k=1}^s v'(n_k) \right)^{1-q} \quad (n \in \mathbf{N}_{n^{(0)}}^s),$$

and we define the first kind of multi-dimensional half-discrete Hilbert-type operator with the non-homogeneous kernel $T_1 : \mathbf{L}_{p,\varphi}(\mathbf{R}_+^m) \to l_{p,\Psi^{1-p}}$ as follows:

For any $f \in \mathbf{L}_{p,\varphi}(\mathbf{R}_+^m)$, there exists a $T_1 f$, satisfying

$$T_1 f(n) = \int_{\mathbf{R}_+^m} h(||x||_\alpha ||v(n)||_\beta) f(x) dx \quad (n \in \mathbf{N}_{n^{(0)}}^s). \tag{5.128}$$

By (5.88), we can write

$$||T_1 f||_{p,\Psi^{1-p}} < K(\sigma)||f||_{p,\varphi},$$

and then, $T_1 f \in l_{p,\Psi^{1-p}}$. Hence, T_1 is a bounded linear operator with the norm

$$||T_1|| = \sup_{f(\neq\theta)\in\mathbf{L}_{p,\varphi}(\mathbf{R}_+^m)} \frac{||T_1 f||_{p,\Psi^{1-p}}}{||f||_{p,\varphi}} \leq K(\sigma).$$

Also we define the second kind of multi-dimensional half-discrete Hilbert-type operator with the non-homogeneous kernel $T_2 : l_{q,\Psi} \to \mathbf{L}_{q,\varphi^{1-q}}(\mathbf{R}_+^m)$ as follows:

For any $a \in l_{q,\Psi}$, there exists a $T_2 a$, satisfying

$$T_2 a(x) = \sum_{n\in\mathbf{N}_{n^{(0)}}^s} h(||x||_\alpha ||v(n)||_\beta) a(n) \quad (x \in \mathbf{R}_+^m). \tag{5.129}$$

By (5.89), we can write

$$||T_2 a||_{q,\varphi^{1-q}} < K(\sigma)||a||_{q,\Psi},$$

and then, $T_2 a \in \mathbf{L}_{q,\varphi^{1-q}}(\mathbf{R}_+^m)$. Hence, T_2 is a bounded linear operator with the norm

$$||T_2|| = \sup_{a(\neq\theta)\in l_{q,\Psi}} \frac{||T_2 a||_{q,\varphi^{1-q}}}{||a||_{q,\Psi}} \leq K(\sigma).$$

By Theorem 5.7, it follows that

Theorem 5.9. *With the assumptions of Theorem 5.6, T_1 and T_2 are defined by (5.128) and (5.129). If there exist $\widetilde{\delta}_0 > 0$ and $\eta < \sigma$, such that for any $\delta \in (0, \widetilde{\delta}_0]$,*

$$k(\sigma \pm \delta) = \int_0^\infty h(t) t^{(\sigma\pm\delta)-1} dt \in \mathbf{R}_+,$$

and

$$h(t) \leq L\left(\frac{1}{t^\eta}\right) \ (0 < t \leq 1; L > 0),$$

then, we have

$$||T_1|| = ||T_2|| = K(\sigma)$$

$$= \left(\frac{\Gamma^m(\frac{1}{\alpha})}{\alpha^{m-1}\Gamma(\frac{m}{\alpha})}\right)^{\frac{1}{q}} \left(\frac{\Gamma^s(\frac{1}{\beta})}{\beta^{s-1}\Gamma(\frac{s}{\beta})}\right)^{\frac{1}{p}} k(\sigma).$$

With the assumptions of Theorem 5.7, for $p > 1$, we set

$$\varphi(x) = ||x||_\alpha^{p(m-\sigma)-m} \quad (x \in \mathbf{R}_+^m),$$

and

$$\Psi_1(n) = ||v(n)||_1^{q(s-\sigma)-s} \left(\prod_{k=1}^s v'(n_k)\right)^{1-q} \quad (n \in \mathbf{N}_{n^{(0)}}^s),$$

and we define the first kind half-discrete Hilbert-type operator with the non-homogeneous kernel and multi-variables $\widetilde{T}_1 : \mathbf{L}_{p,\varphi}(\mathbf{R}_+^m) \to \mathbf{l}_{p,\Psi_1^{1-p}}$ as follows:

For any $f \in \mathbf{L}_{p,\varphi}(\mathbf{R}_+^m)$, there exists a $\widetilde{T}_1 f$, satisfying

$$\widetilde{T}_1 f(n) = \int_{\mathbf{R}_+^m} h(||x||_\alpha ||v(n)||_1) f(x) dx \quad (n \in \mathbf{N}_{n^{(0)}}^s). \tag{5.130}$$

By (5.102), we can write

$$||\widetilde{T}_1 f||_{p,\Psi_1^{1-p}} < K_1(\sigma) ||f||_{p,\varphi},$$

and then, $\widetilde{T}_1 f \in \mathbf{l}_{p,\Psi_1^{1-p}}$. Hence, $\widetilde{T}_1$ is a bounded linear operator with the norm

$$||\widetilde{T}_1|| = \sup_{f(\neq\theta)\in\mathbf{L}_{p,\varphi}(\mathbf{R}_+^m)} \frac{||\widetilde{T}_1 f||_{p,\Psi_1^{1-p}}}{||f||_{p,\varphi}} \leq K_1(\sigma).$$

Also we define the second kind half-discrete Hilbert-type operator with the non-homogeneous kernel and multi-variables $\widetilde{T}_2 : \mathbf{l}_{q,\Psi_1} \to \mathbf{L}_{q,\varphi^{1-q}}(\mathbf{R}_+^m)$ as follows:

For any $a \in \mathbf{l}_{q,\Psi_1}$, there exists a $\widetilde{T}_2 a$, satisfying

$$\widetilde{T}_2 a(x) = \sum_{n\in\mathbf{N}_{n^{(0)}}^s} h(||x||_\alpha ||v(n)||_1) a(n) \quad (x \in \mathbf{R}_+^m). \tag{5.131}$$

By (5.103), we can write

$$||\widetilde{T}_2 a||_{q,\varphi^{1-q}} < K(\sigma) ||a||_{q,\Psi_1},$$

and then, $\widetilde{T}_2 a \in \mathbf{L}_{q,\varphi^{1-q}}(\mathbf{R}_+^m)$. Hence, $\widetilde{T}_2$ is a bounded linear operator with the norm

$$||\widetilde{T}_2|| = \sup_{a(\neq\theta)\in\mathbf{l}_{q,\Psi}} \frac{||\widetilde{T}_2 a||_{q,\varphi^{1-q}}}{||a||_{q,\Psi_1}} \leq K_1(\sigma).$$

By Theorem 5.9, it follows that

Theorem 5.10. *With the assumptions of Theorem 5.9, $\widetilde{T}_1$ and $\widetilde{T}_2$ are defined by (5.130) and (5.131). If there exist $\widetilde{\delta}_0 > 0$ and $\eta < \sigma$, such that, for any $\delta \in (0, \widetilde{\delta}_0]$,*

$$k(\sigma \pm \delta) = \int_0^\infty h(t) t^{(\sigma \pm \delta)-1} dt \in \mathbf{R}_+,$$

and

$$h(t) \le L\left(\frac{1}{t^\eta}\right) \quad (0 < t \le 1; L > 0),$$

then, we have

$$\|\widetilde{T}_1\| = \|\widetilde{T}_2\| = K_1(\sigma) = \left(\frac{\Gamma^m(\frac{1}{\alpha})}{\alpha^{m-1}\Gamma(\frac{m}{\alpha})}\right)^{\frac{1}{q}} \left(\frac{1}{(s-1)!}\right)^{\frac{1}{p}} k(\sigma).$$

Example 5.5. We set $h(u) = \frac{1}{(1+u)^\lambda}$ $(0 < \sigma < \lambda, \sigma \le s)$, then we find

$$\frac{d}{du}(h(u)u^{\sigma-s}) < 0, \quad \frac{d^2}{du^2}(h(u)u^{\sigma-s}) > 0 \quad (u \in \mathbf{R}_+).$$

For $\widetilde{\delta}_0 = \frac{1}{2}\min\{\sigma, \lambda - \sigma\}, \delta \in (0, \widetilde{\delta}_0]$,

$$\begin{aligned}
k(\sigma \pm \delta) &= \int_0^\infty \frac{t^{(\sigma \pm \delta)-1}}{(1+t)^\lambda} du \\
&= B(\sigma \pm \delta, \lambda - \sigma \mp \delta) \in \mathbf{R}_+,
\end{aligned}$$

and

$$h(t) = \frac{1}{(1+t)^\lambda} \le \frac{1}{t^\eta} \quad \left(0 < t \le 1; \eta = \frac{\sigma}{2} < \sigma\right).$$

(i) For $v(t) = \ln t$ $(t \ge 2 = n^{(0)})$, $v(n^{(0)} - 1) = 0$, setting

$$\varphi(x) = \|x\|_\alpha^{p(m-\sigma)-m} \quad (x \in \mathbf{R}_+^m),$$

and

$$\Psi(n) = \|\ln n\|_\beta^{q(s-\sigma)-s} \left(\prod_{k=1}^s \frac{1}{n_k}\right)^{1-q} \quad (n \in \mathbf{N}_2^s),$$

we define an operator $T_1 : \mathbf{L}_{p,\varphi}(\mathbf{R}_+^m) \to l_{p,\Psi^{1-p}}$ as follows:
For any $f \in \mathbf{L}_{p,\varphi}(\mathbf{R}_+^m)$, there exists a $T_1 f$, satisfying

$$T_1 f(n) = \int_{R_+^m} \frac{1}{(1 + \|x\|_\alpha \|\ln n\|_\beta)^\lambda} f(x) dx \quad (n \in \mathbf{N}_2^s).$$

Also we define an operator $T_2 : l_{q,\Psi} \to \mathbf{L}_{q,\varphi^{1-q}}(\mathbf{R}_+^m)$ as follows:

For any $a \in l_{q,\Psi}$, there exists a $T_2 a$, satisfying

$$T_2 a(x) = \sum_{n \in \mathbf{N}_2^s} \frac{1}{(1 + ||x||_\alpha|| \ln n||_\beta)^\lambda} a(n) \quad (x \in \mathbf{R}_+^m).$$

Then, by Theorem 5.7, we have

$$||T_1|| = ||T_2|| = \left(\frac{\Gamma^m(\frac{1}{\alpha})}{\alpha^{m-1}\Gamma(\frac{m}{\alpha})} \right)^{\frac{1}{q}} \left(\frac{\Gamma^s(\frac{1}{\beta})}{\beta^{s-1}\Gamma\left(\frac{s}{\beta}\right)} \right)^{\frac{1}{p}} B(\sigma, \lambda - \sigma).$$

(ii) For $v(t) = \ln(t - \gamma)$ $(t \geq 2 = n^{(0)}; \gamma \leq \frac{1}{2})$, $v(n^{(0)} - \frac{1}{2}) = \ln(\frac{3}{2} - \gamma) \geq 0$, setting $\varphi(x)$ as in (i) and

$$\Psi_\gamma(n) = || \ln(n - \gamma)||_1^{q(s-\sigma)-s} \left(\prod_{k=1}^s \frac{1}{n_k - \gamma} \right)^{1-q} \quad (n \in \mathbf{N}_2^s),$$

we define an operator $\widetilde{T}_1 : \mathbf{L}_{p,\varphi}(\mathbf{R}_+^m) \to l_{p,\Psi_\gamma^{1-p}}$ as follows:

For any $f \in \mathbf{L}_{p,\varphi}(\mathbf{R}_+^m)$, there exists a $\widetilde{T}_1 f$, satisfying

$$\widetilde{T}_1 f(n) = \int_{R_+^m} \frac{1}{(1 + ||x||_\alpha|| \ln(n - \gamma)||_1)^\lambda} f(x) dx \quad (n \in \mathbf{N}_2^s).$$

Also we define an operator $\widetilde{T}_2 : l_{q,\Psi_\gamma} \to \mathbf{L}_{q,\varphi^{1-q}}(\mathbf{R}_+^m)$ as follows:

For any $a \in l_{q,\Psi_\gamma}$, there exists a $T_2 a$, satisfying

$$\widetilde{T}_2 a(x) = \sum_{n \in \mathbf{N}_2^s} \frac{1}{(1 + ||x||_\alpha|| \ln(n - \gamma)||_1)^\lambda} a(n) \quad (x \in \mathbf{R}_+^m).$$

Then, by Theorem 5.8, we have

$$||\widetilde{T}_1|| = ||\widetilde{T}_2|| = \left(\frac{\Gamma^m(\frac{1}{\alpha})}{\alpha^{m-1}\Gamma(\frac{m}{\alpha})} \right)^{\frac{1}{q}} \left(\frac{1}{(s-1)!} \right)^{\frac{1}{p}} B(\sigma, \lambda - \sigma).$$

(iii) For $v(t) = \ln \kappa t$ $(t \geq 2 = n^{(0)}; \kappa \geq \frac{2}{3})$, $v(n^{(0)} - \frac{1}{2}) = \ln \kappa(\frac{3}{2}) \geq 0$, setting $\varphi(x)$ as in (i) and

$$\Psi_\kappa(n) = || \ln \kappa n||_1^{q(s-\sigma)-s} \left(\prod_{k=1}^s \frac{1}{n_k} \right)^{1-q} \quad (n \in \mathbf{N}_2^s),$$

we define an operator $\widetilde{T}_1 : \mathbf{L}_{p,\varphi}(\mathbf{R}_+^m) \to l_{p,\Psi_\kappa^{1-p}}$ as follows:

For any $f \in \mathbf{L}_{p,\varphi}(\mathbf{R}_+^m)$, there exists a $\widetilde{T}_1 f$, satisfying

$$\widetilde{T}_1 f(n) = \int_{R_+^m} \frac{1}{(1 + ||x||_\alpha|| \ln \kappa n||_1)^\lambda} f(x) dx \quad (n \in \mathbf{N}_2^s).$$

Also we define an operator $\widetilde{T}_2 : l_{q,\Psi_\kappa} \to \mathbf{L}_{q,\varphi^{1-q}}(\mathbf{R}_+^m)$ as follows:

For any $a \in l_{q,\Psi_\kappa}$, there exists a $\widetilde{T}_2 a$, satisfying

$$\widetilde{T}_2 a(x) = \sum_{n \in \mathbf{N}_2^s} \frac{1}{(1 + ||x||_\alpha || \ln \kappa n||_1)^\lambda} a(n) \quad (x \in \mathbf{R}_+^m).$$

Then, by Theorem 5.8, we have

$$||\widetilde{T}_1|| = ||\widetilde{T}_2|| = \left(\frac{\Gamma^m(\frac{1}{\alpha})}{\alpha^{m-1}\Gamma(\frac{m}{\alpha})} \right)^{\frac{1}{q}} \left(\frac{1}{(s-1)!} \right)^{\frac{1}{p}} B(\sigma, \lambda - \sigma).$$

Example 5.6. We set $h(u) = \frac{\ln u}{u^\lambda - 1} (0 < \sigma < \lambda \le 1, \sigma \le s)$, then we find

$$\frac{d}{du}(h(u)u^{\sigma-s}) < 0, \quad \frac{d^2}{du^2}(h(u)u^{\sigma-s}) > 0 \quad (u \in \mathbf{R}_+).$$

For $\widetilde{\delta}_0 = \frac{1}{2}\min\{\sigma, \lambda - \sigma\}, \delta \in (0, \widetilde{\delta}_0]$,

$$\begin{aligned}
k(\sigma \pm \delta) &= \int_0^\infty \frac{(\ln t)t^{(\sigma \pm \delta)-1}}{t^\lambda - 1} dt \\
&= \frac{1}{\lambda^2} \int_0^\infty \frac{(\ln v)v^{(\sigma \pm \delta)/\lambda - 1}}{v - 1} dv \\
&= \left[\frac{\pi}{\lambda \sin \pi(\frac{\sigma \pm \delta}{\lambda})} \right]^2 \in \mathbf{R}_+,
\end{aligned}$$

and

$$h(t) = \frac{\ln t}{t^\lambda - 1} \le L \frac{1}{t^\eta} \quad \left(0 < t \le 1; \eta = \frac{\sigma}{2} < \sigma, L > 0 \right).$$

(i) For $v(t) = \ln t$ $(t \ge 2 = n^{(0)})$, $v(n^{(0)} - 1) = 0$, setting $\varphi(x)$ and $\Psi(n)$ as in Example 5.5(i), we define an operator $T_1 : \mathbf{L}_{p,\varphi}(\mathbf{R}_+^m) \to l_{p,\Psi^{1-p}}$ as follows:

For any $f \in \mathbf{L}_{p,\varphi}(\mathbf{R}_+^m)$, there exists a $T_1 f$, satisfying

$$T_1 f(n) = \int_{R_+^m} \frac{\ln(||x||_\alpha || \ln n||_\beta)}{(||x||_\alpha || \ln n||_\beta)^\lambda - 1} f(x) dx \quad (n \in \mathbf{N}_2^s).$$

Also we define an operator $T_2 : l_{q,\Psi} \to \mathbf{L}_{q,\varphi^{1-q}}(\mathbf{R}_+^m)$ as follows:

For any $a \in l_{q,\Psi}$, there exists a $T_2 a$, satisfying

$$T_2 a(x) = \sum_{n \in \mathbf{N}_2^s} \frac{\ln(||x||_\alpha || \ln n||_\beta)}{(||x||_\alpha || \ln n||_\beta)^\lambda - 1} a(n) \quad (x \in \mathbf{R}_+^m).$$

Then, by Theorem 5.10, we have

$$||T_1|| = ||T_2|| = \left(\frac{\Gamma^m(\frac{1}{\alpha})}{\alpha^{m-1}\Gamma(\frac{m}{\alpha})} \right)^{\frac{1}{q}} \left(\frac{\Gamma^s(\frac{1}{\beta})}{\beta^{s-1}\Gamma\left(\frac{s}{\beta}\right)} \right)^{\frac{1}{p}} \left[\frac{\pi}{\lambda \sin \pi(\frac{\sigma}{\lambda})} \right]^2.$$

(ii) For $v(t) = \ln(t - \gamma)$ $(t \geq 2 = n^{(0)}; \gamma \leq \frac{1}{2})$, $v(n^{(0)} - \frac{1}{2})^{\cdot} = \ln(\frac{3}{2} - \gamma) \geq 0$, setting $\varphi(x)$ and $\Psi_\gamma(n)$ as in Example 5.5(ii), we define an operator $\widetilde{T}_1 : \mathbf{L}_{p,\varphi}(\mathbf{R}_+^m) \to \mathbf{l}_{p,\Psi_\gamma^{1-p}}$ as follows:

For any $f \in \mathbf{L}_{p,\varphi}(\mathbf{R}_+^m)$, there exists a $\widetilde{T}_1 f$, satisfying

$$\widetilde{T}_1 f(n) = \int_{R_+^m} \frac{\ln(||x||_\alpha || \ln(n - \gamma)||_1)}{(||x||_\alpha || \ln(n - \gamma)||_1)^\lambda - 1} f(x) dx \quad (n \in \mathbf{N}_2^s).$$

Also we define an operator $\widetilde{T}_2 : \mathbf{l}_{q,\Psi_\gamma} \to \mathbf{L}_{q,\varphi^{1-q}}(\mathbf{R}_+^m)$ as follows:

For any $a \in \mathbf{l}_{q,\Psi_\gamma}$, there exists a $\widetilde{T}_2 a$, satisfying

$$\widetilde{T}_2 a(x) = \sum_{n \in \mathbf{N}_2^s} \frac{\ln(||x||_\alpha || \ln(n - \gamma)||_1)}{(||x||_\alpha || \ln(n - \gamma)||_1)^\lambda - 1} a(n) \quad (x \in \mathbf{R}_+^m).$$

Then, by Theorem 5.8, we have

$$||\widetilde{T}_1|| = ||\widetilde{T}_2|| = \left(\frac{\Gamma^m(\frac{1}{\alpha})}{\alpha^{m-1}\Gamma(\frac{m}{\alpha})} \right)^{\frac{1}{q}} \left(\frac{1}{(s-1)!} \right)^{\frac{1}{p}} \left[\frac{\pi}{\lambda \sin \pi(\frac{\sigma}{\lambda})} \right]^2.$$

(iii) For $v(t) = \ln \kappa t$ $(t \geq 2 = n^{(0)}; \kappa \geq \frac{2}{3})$, $v(n^{(0)} - \frac{1}{2}) = \ln \kappa(\frac{3}{2}) \geq 0$, setting $\varphi(x)$ and $\Psi_\kappa(n)$ as Example in 5.5(iii), we define an operator $\widetilde{T}_1 : \mathbf{L}_{p,\varphi}(\mathbf{R}_+^m) \to \mathbf{l}_{p,\Psi_\kappa^{1-p}}$ as follows:

For any $f \in \mathbf{L}_{p,\varphi}(\mathbf{R}_+^m)$, there exists a $\widetilde{T}_1 f$, satisfying

$$\widetilde{T}_1 f(n) = \int_{R_+^m} \frac{\ln(||x||_\alpha || \ln \kappa n||_1)}{(||x||_\alpha || \ln \kappa n||_1) - 1} f(x) dx \quad (n \in \mathbf{N}_2^s).$$

Also we define an operator $\widetilde{T}_2 : \mathbf{l}_{q,\Psi_\kappa} \to \mathbf{L}_{q,\varphi^{1-q}}(\mathbf{R}_+^m)$ as follows:

For any $a \in \mathbf{l}_{q,\Psi_\kappa}$, there exists a $\widetilde{T}_2 a$, satisfying

$$\widetilde{T}_2 a(x) = \sum_{n \in \mathbf{N}_2^s} \frac{\ln(||x||_\alpha || \ln \kappa n||_1)}{(||x||_\alpha || \ln \kappa n||_1) - 1} a(n) \quad (x \in \mathbf{R}_+^m).$$

Then, by Theorem 5.10, we have

$$||\widetilde{T}_1|| = ||\widetilde{T}_2|| = \left(\frac{\Gamma^m(\frac{1}{\alpha})}{\alpha^{m-1}\Gamma(\frac{m}{\alpha})} \right)^{\frac{1}{q}} \left(\frac{1}{(s-1)!} \right)^{\frac{1}{p}} \left[\frac{\pi}{\lambda \sin \pi(\frac{\sigma}{\lambda})} \right]^2.$$

Example 5.7. We set $h(u) = e^{-\eta u}(\eta > 0, 0 < \sigma \leq s)$, then we find

$$\frac{d}{du}(h(u)u^{\sigma-s}) < 0, \quad \frac{d^2}{du^2}(h(u)u^{\sigma-s}) > 0 \quad (u \in \mathbf{R}_+).$$

For $\tilde{\delta}_0 = \frac{1}{2}\sigma$, $\delta \in (0, \tilde{\delta}_0]$, we have

$$k(-\sigma \pm \delta) = \int_0^\infty e^{-\eta t} t^{\sigma \pm \delta - 1} dt = \frac{1}{\eta^{\sigma \pm \delta}} \int_0^\infty e^{-v} v^{\sigma \pm \delta - 1} dv$$

$$= \frac{1}{\eta^{\sigma \pm \delta}} \Gamma(\sigma \pm \delta) \in \mathbf{R}_+,$$

and

$$h(t) = e^{-\eta t} \leq L \frac{1}{t^\eta} \quad (0 < t \leq 1; L > 0, \ \eta = \frac{1}{2}\sigma < \sigma).$$

(i) For $v(t) = \ln t$ $(t \geq 2 = n^{(0)})$, $v(n^{(0)} - 1) = 0$, setting $\varphi(x)$ and $\Psi(n)$ as in Example 5.5(i), we define an operator $T_1 : \mathbf{L}_{p,\varphi}(\mathbf{R}_+^m) \to l_{p,\Psi^{1-p}}$ as follows:

For any $f \in \mathbf{L}_{p,\varphi}(\mathbf{R}_+^m)$, there exists a $T_1 f$, satisfying

$$T_1 f(n) = \int_{R_+^m} e^{-\eta \|x\|_\alpha \| \ln n\|_\beta} f(x) dx \quad (n \in \mathbf{N}_2^s).$$

Also we define an operator $T_2 : l_{q,\Psi} \to \mathbf{L}_{q,\varphi^{1-q}}(\mathbf{R}_+^m)$ as follows:

For any $a \in l_{q,\Psi}$, there exists a $T_2 a$, satisfying

$$T_2 a(x) = \sum_{n \in \mathbf{N}_2^s} e^{-\eta \|x\|_\alpha \| \ln n\|_\beta} a(n) \quad (x \in \mathbf{R}_+^m).$$

Then, by Theorem 5.10, we have

$$\|T_1\| = \|T_2\| = \left(\frac{\Gamma^m(\frac{1}{\alpha})}{\alpha^{m-1}\Gamma(\frac{m}{\alpha})} \right)^{\frac{1}{q}} \left(\frac{\Gamma^s(\frac{1}{\beta})}{\beta^{s-1}\Gamma\left(\frac{s}{\beta}\right)} \right)^{\frac{1}{p}} \frac{1}{\eta^\sigma} \Gamma(\sigma).$$

(ii) For $v(t) = \ln(t - \gamma)$ $(t \geq 2 = n^{(0)}; \ \gamma \leq \frac{1}{2})$, $v(n^{(0)} - \frac{1}{2}) = \ln(\frac{3}{2} - \gamma) \geq 0$, setting $\varphi(x)$ and $\Psi_\gamma(n)$ as in Example 5.5(ii), we define an operator $\tilde{T}_1 : \mathbf{L}_{p,\varphi}(\mathbf{R}_+^m) \to l_{p,\Psi_\gamma^{1-p}}$ as follows:

For any $f \in \mathbf{L}_{p,\varphi}(\mathbf{R}_+^m)$, there exists a $\tilde{T}_1 f$, satisfying

$$\tilde{T}_1 f(n) = \int_{R_+^m} e^{-\eta \|x\|_\alpha \| \ln(n-\gamma)\|_1} f(x) dx \quad (n \in \mathbf{N}_2^s).$$

Also we define an operator $\tilde{T}_2 : l_{q,\Psi_\gamma} \to \mathbf{L}_{q,\varphi^{1-q}}(\mathbf{R}_+^m)$ as follows:

For any $a \in l_{q,\Psi_\gamma}$, there exists $\widetilde{T}_2 a$, satisfying

$$\widetilde{T}_2 \, a(x) = \sum_{n \in \mathbf{N}_2^s} e^{-\eta ||x||_\alpha || \ln(n-\gamma)||_1} a(n) \quad (x \in \mathbf{R}_+^m).$$

Then, by Theorem 5.8, we have

$$||\widetilde{T}_1|| = ||\widetilde{T}_2|| = \left(\frac{\Gamma^m(\frac{1}{\alpha})}{\alpha^{m-1}\Gamma(\frac{m}{\alpha})}\right)^{\frac{1}{q}} \left(\frac{1}{(s-1)!}\right)^{\frac{1}{p}} \frac{1}{\eta^\sigma} \Gamma(\sigma).$$

(iii) For $v(t) = \ln \kappa t$ $(t \geq 2 = n^{(0)}; \kappa \geq \frac{2}{3})$, $v(n^{(0)} - \frac{1}{2}) = \ln \kappa(\frac{3}{2}) \geq 0$, setting $\varphi(x)$ and $\Psi_\kappa(n)$ as in Example 5.5(iii), we define an operator $\widetilde{T}_1 \mathbf{L}_{p,\varphi}(\mathbf{R}_+^m) \to l_{p,\Psi_\kappa^{1-p}}$ as follows:

For any $f \in \mathbf{L}_{p,\varphi}(\mathbf{R}_+^m)$, there exists a $\widetilde{T}_1 f$, satisfying

$$\widetilde{T}_1 f(n) = \int_{R_+^m} e^{-\eta ||x||_\alpha || \ln \kappa n||_1} f(x) dx \quad (n \in \mathbf{N}_2^s).$$

Also we define an operator $\widetilde{T}_2 : l_{q,\Psi_\kappa} \to \mathbf{L}_{q,\varphi^{1-q}}(\mathbf{R}_+^m)$ as follows:

For any $a \in l_{q,\Psi_\kappa}$, there exists a $\widetilde{T}_2 a$, satisfying

$$\widetilde{T}_2 a(x) = \sum_{n \in \mathbf{N}_2^s} e^{-\eta ||x||_\alpha || \ln \kappa n||_1} a(n) \quad (x \in \mathbf{R}_+^m).$$

Then, by Theorem 5.10, we have

$$||\widetilde{T}_1|| = ||\widetilde{T}_2|| = \left(\frac{\Gamma^m(\frac{1}{\alpha})}{\alpha^{m-1}\Gamma(\frac{m}{\alpha})}\right)^{\frac{1}{q}} \left(\frac{1}{(s-1)!}\right)^{\frac{1}{p}} \frac{1}{\eta^\sigma} \Gamma(\sigma).$$

Example 5.8. We set $h(u) = \frac{(\min\{1,u\})^\gamma}{(\max\{1,u\})^{\lambda+\gamma}}$ $(\lambda \geq 0, \ \gamma \geq 0, \ \lambda + \gamma - \sigma > 0, \ \sigma + \gamma > 0)$.

For $\sigma + \gamma \leq s$, we find

$$k_\lambda(t,u)u^{\lambda_2 - s} = \frac{(\min\{t,u\})^\gamma}{(\max\{t,u\})^{\lambda+\gamma}}u^{\sigma - s}$$

$$= \begin{cases} \frac{u^{\gamma+\sigma-s}}{t^{\lambda+\gamma}}, & 0 < u \leq t \\ \frac{t^\gamma}{u^{\lambda+\gamma-\sigma+s}}, & u > t \end{cases}$$

is decreasing with respect to $u \in \mathbf{R}_+$, and strictly decreasing in an interval $I \subset (1, \infty)$. For $\widetilde{\delta}_0 = \frac{1}{2}(\sigma + \gamma), \delta \in (0, \widetilde{\delta}_0]$, we have

$$k(\sigma \pm \delta) = \int_0^\infty \frac{(\min\{t,1\})^\gamma}{(\max\{t,1\})^{\lambda+\gamma}} t^{\sigma \pm \delta - 1} dt$$

$$= \frac{1}{\gamma + \sigma \pm \delta} + \frac{1}{\lambda + \gamma - \sigma \mp \delta} \in \mathbf{R}_+,$$

and

$$h(t) = \frac{(\min\{1,t\})^\gamma}{(\max\{1,t\})^{\lambda+\gamma}} = t^\gamma \leq \frac{1}{t^\eta},$$

where $0 < t \leq 1$; $\eta = \frac{\sigma}{2} < \sigma$.

For $v(t) = \ln t$ $(t \geq 2 = n^{(0)})$, $v(n^{(0)} - 1) = 0$, setting $\varphi(x)$ and $\Psi(n)$ as in Example 5.5(i), we define an operator $T_1 : \mathbf{L}_{p,\varphi}(\mathbf{R}_+^m) \to l_{p,\Psi^{1-p}}$ as follows:

For any $f \in \mathbf{L}_{p,\varphi}(\mathbf{R}_+^m)$, there exists a $T_1 f$ satisfying

$$\widetilde{T}_1 f(n) = \int_{R_+^m} \frac{(\min\{1, ||x||_\alpha|| \ln n||_\beta\})^\gamma}{(\max\{1, ||x||_\alpha|| \ln n||_\beta\})^{\lambda+\gamma}} f(x)\, dx \quad (n \in \mathbf{N}_2^s).$$

Also we define an operator $T_2 : l_{q,\Psi} \to \mathbf{L}_{q,\varphi^{1-q}}(\mathbf{R}_+^m)$ as follows:

For any $a \in l_{q,\Psi}$, there exists a $T_2 a$, satisfying

$$T_2 a(x) = \sum_{n \in \mathbf{N}_2^s} \frac{(\min\{1, ||x||_\alpha|| \ln n||_\beta\})^\gamma}{(\max\{1, ||x||_\alpha|| \ln n||_\beta\})^{\lambda+\gamma}} a(n) \quad (x \in \mathbf{R}_+^m).$$

Then, by Theorem 5.9, we have

$$||T_1|| = ||T_2|| = \left(\frac{\Gamma^m(\frac{1}{\alpha})}{\alpha^{m-1}\Gamma(\frac{m}{\alpha})} \right)^{\frac{1}{q}} \left(\frac{\Gamma^s(\frac{1}{\beta})}{\beta^{s-1}\Gamma(\frac{s}{\beta})} \right)^{\frac{1}{p}} \frac{2\gamma + \lambda}{(\gamma + \sigma)(\lambda + \gamma - \sigma)}.$$

Remark 5.2. (i) We can still write some particular inequalities with the norms as the best possible constant factors in the above examples by using Theorem 5.6 and Theorem 5.8.

(ii) In particular, for $m = s = 1$ in the theorems and corollaries of this chapter, we can obtain corresponding results of Chapter 4.

Chapter 6

Multiple Half-Discrete Hilbert-Type Inequalities

"As long as a branch of science affords an abundance of problems, it is full of life; want of problems means death or cessation of independent development. Just as every human enterprise prosecutes final aims, so mathematical research needs problems. Their solution steels the force of the investigator; thus he discovers new methods and view points and widens his horizon."

David Hilbert

"... we have always found, even with the most famous inequalities, that we have a little, new to add."

G. H. Hardy

6.1 Introduction

This chapter is devoted to two kinds of multiple half-discrete Hilbert-type inequalities which are derived by using the weight functions and techniques of real analysis. These inequalities are generalizations of the double cases mentioned in Chapters 3 and 4. The best possible constant factors involved in the inequalities are proved. Included are the equivalent forms, the operator expressions, some kinds of reverses, many theorems and corollaries, and many examples with particular kernels.

We also consider many lemmas related to weight functions in subsections 6.2.1 and 6.3.1. A large number of corollaries dealing with equivalent inequalities is presented in Section 6.4.

6.2 First Kind of Multiple Hilbert-type Inequalities

6.2.1 *Lemmas Related to the Weight Functions*

Lemma 6.1. *If* $m \in \mathbf{N}$, $p_i \in \mathbf{R}\backslash\{0,1\}$, $\lambda_i \in \mathbf{R}(i = 1, \cdots, m + 1)$, $\sum_{i=1}^{m+1} \frac{1}{p_i} = 1$, *then we have the following expression:*

$$A = \prod_{i=1}^{m+1} \left[x_i^{(\lambda_i-1)(1-p_i)} \prod_{j=1(j\neq i)}^{m+1} x_j^{\lambda_j-1} \right]^{\frac{1}{p_i}} = 1. \tag{6.1}$$

Proof. We find

$$A = \prod_{i=1}^{m+1} \left[x_i^{(\lambda_i-1)(1-p_i)+1-\lambda_i} \prod_{j=1}^{m+1} x_j^{\lambda_j-1} \right]^{\frac{1}{p_i}}$$

$$= \prod_{i=1}^{m+1} [x_i^{(\lambda_i-1)(-p_i)}]^{\frac{1}{p_i}} \left(\prod_{j=1}^{m+1} x_j^{\lambda_j-1} \right)^{\frac{1}{p_i}}$$

$$= \prod_{i=1}^{m+1} x_i^{1-\lambda_i} \left(\prod_{j=1}^{m+1} x_j^{\lambda_j-1} \right)^{\sum_{i=1}^{m+1} \frac{1}{p_i}}$$

$$= \prod_{i=1}^{m+1} x_i^{1-\lambda_i} \prod_{j=1}^{m+1} x_j^{\lambda_j-1} = 1.$$

Thus, (6.1) follows. □

Definition 6.1. *If* $m \in \mathbf{N}$, $\lambda \in \mathbf{R}$, $k_\lambda(x_1, \cdots, x_{m+1})$ *is a measurable function in* $\mathbf{R}_+^{m+1}$, *satisfying for any* $u > 0$ *and* $(x_1, \cdots, x_{m+1}) \in \mathbf{R}_+^{m+1}$,

$$k_\lambda(ux_1, \cdots, ux_{m+1}) = u^{-\lambda} k_\lambda(x_1, \cdots, x_{m+1}),$$

then, we call $k_\lambda(x_1, \cdots, x_{m+1})$ *the homogeneous function of degree* $-\lambda$ *in* $\mathbf{R}_+^{m+1}$.

Lemma 6.2. *If* $m \in \mathbf{N}$, $\lambda_i \in \mathbf{R}(i = 1, \cdots, m + 1)$, $\sum_{i=1}^{m+1} \lambda_i = \lambda$, $k_\lambda(x_1, \cdots, x_{m+1})(\geq 0)$ *is a homogeneous function of degree* $-\lambda$ *in* $\mathbf{R}_+^{m+1}$ *such that*

$$k(\lambda_{m+1}) = H(m + 1)$$

$$= \int_0^\infty \cdots \int_0^\infty k_\lambda(u_1, \cdots, u_m, 1) \prod_{j=1}^m u_j^{\lambda_j-1} du_1 \cdots du_m, \tag{6.2}$$

then, for $i = 1, \cdots, m$, we still have

$$H(i) = \int_0^\infty \cdots \int_0^\infty k_\lambda(u_1, \cdots, u_{i-1}, 1, u_{i+1}, \cdots, u_{m+1})$$

$$\times \prod_{j=1(j\neq i)}^{m+1} u_j^{\lambda_j - 1} du_1 \cdots du_{i-1} du_{i+1} \cdots du_{m+1} = k(\lambda_{m+1}). \quad (6.3)$$

Proof. Setting $u_j = u_{m+1} v_j$ $(j \neq i, m+1)$ in $H(i)$, we find

$$H(i) = \int_0^\infty \cdots \int_0^\infty k_\lambda(v_1, \cdots, v_{i-1}, u_{m+1}^{-1}, v_{i+1}, \cdots, v_m, 1)$$

$$\times u_{m+1}^{-1-\lambda_i} \prod_{j=1(j\neq i)}^m v_j^{\lambda_j - 1} dv_1 \cdots dv_{i-1} dv_{i+1} \cdots dv_m du_{m+1}.$$

Setting $v_i = u_{m+1}^{-1}$ in the above integral, we obtain $H(i) = H(m+1) = k(\lambda_{m+1})$ $(i = 1, \cdots, m)$. $\qquad \square$

In the following of this chapter, we agree on that $m \in \mathbf{N}, p_i \in \mathbf{R} \backslash \{0, 1\}, \lambda_i \in \mathbf{R}(i = 1, \cdots, m+1), \frac{1}{p} = \sum_{i=1}^m \frac{1}{p_i} = 1 - \frac{1}{p_{m+1}}, \sum_{i=1}^{m+1} \lambda_i = \lambda,$ and $k_\lambda(x_1, \cdots, x_{m+1})(\geq 0)$ is a finite homogeneous function of degree $-\lambda$ in $\mathbf{R}_+^{m+1}$.

Lemma 6.3. *If $k(\lambda_{m+1}) \in \mathbf{R}_+$, there exists a constant $\delta_0 > 0$, such that for any $m \in \mathbf{N}$ and $0 < \varepsilon < \delta_0 \min_{1 \leq j \leq m+1} \{|p_j|, |p|\}$,*

$$k(\lambda_{m+1} + \frac{\varepsilon}{p})$$

$$= \int_0^\infty \cdots \int_0^\infty k_\lambda(u_1, \cdots, u_m, 1) \prod_{j=1}^m u_j^{(\lambda_j - \frac{\varepsilon}{p_j})-1} du_1 \cdots du_m \in \mathbf{R}_+,$$

then, we have

$$k\left(\lambda_{m+1} + \frac{\varepsilon}{p}\right) = k(\lambda_{m+1}) + o(1) \ (\varepsilon \to 0^+). \quad (6.4)$$

Proof. We prove (6.4) by mathematical induction.

For $m = 1$, by Lemma 3.7, we have

$$k\left(\lambda_2 + \frac{\varepsilon}{p}\right) = \int_0^\infty k_\lambda(u_1, 1) u_1^{(\lambda_1 - \frac{\varepsilon}{p_1})-1} du_1$$

$$= \int_0^\infty k_\lambda(u_1, 1) u_1^{\lambda_1 - 1} du_1 + o(1)$$

$$= k(\lambda_2) + o(1) \ (\varepsilon \to 0^+).$$

Assuming that (6.4) is valid for $m = n - 1$, then, for $m = n$, since

$$\lambda_{n+1} + \frac{\varepsilon}{p} + \sum_{j=1}^{n} \left(\lambda_j - \frac{\varepsilon}{p_j} \right) = \lambda,$$

by the Fubini theorem (see Kuang [49]), we have

$$k\left(\lambda_{n+1} + \frac{\varepsilon}{p}\right) = \int_0^\infty \left[\int_0^\infty \cdots \int_0^\infty k_\lambda(u_1, \cdots, u_n, 1) \right.$$
$$\times \left. \prod_{j=1}^{n-1} u_j^{(\lambda_j - \frac{\varepsilon}{p_j})-1} du_1 \cdots du_{n-1} \right] u_n^{(\lambda_n - \frac{\varepsilon}{p_n})-1} du_n$$

$$= \int_0^\infty \left[\int_0^\infty \cdots \int_0^\infty k_\lambda(u_1, \cdots, u_n, 1) \right.$$
$$\times \left. \prod_{j=1}^{n-1} u_j^{(\lambda_j - \frac{\varepsilon}{p_j})-1} du_1 \cdots du_{n-1} \right] u_n^{\lambda_n - 1} du_n + o_1(1)$$

$$= \int_0^\infty \left[\int_0^\infty \cdots \int_0^\infty k_\lambda(u_1, \cdots, u_n, 1) u_n^{\lambda_n - 1} du_n \right]$$
$$\times \prod_{j=1}^{n-1} u_j^{(\lambda_j - \frac{\varepsilon}{p_j})-1} du_1 \cdots du_{n-1} + o_1(1)$$

$$= \int_0^\infty \left[\int_0^\infty \cdots \int_0^\infty k_\lambda(u_1, \cdots, u_n, 1) u_n^{\lambda_n - 1} du_n \right]$$
$$\times \prod_{j=1}^{n-1} u_j^{\lambda_j - 1} du_1 \cdots du_{n-1} + o_2(1) + o_1(1)$$

$$= H(n + 1) + o(1) = k(\lambda_{n+1}) + o(1) (\varepsilon \to 0^+).$$

Hence, we prove that (6.4) is valid for any $m \in \mathbf{N}$. $\square$

Definition 6.2. If $n \in \mathbf{N}$, $x_i \in \mathbf{R}_+$ $(i = 1, \cdots, m)$, we define weight functions $\omega_i(x_i)$ and $\varpi_{m+1}(n)$ as follows:

$$\omega_i(x_i) = x_i^{\lambda_i} \sum_{n=1}^\infty n^{\lambda_{m+1}-1} \int_0^\infty \cdots \int_0^\infty k_\lambda(x_1, \cdots, x_m, n)$$
$$\times \prod_{j=1(j \neq i)}^{m} x_j^{\lambda_j - 1} dx_1 \cdots dx_{i-1} dx_{i+1} \cdots dx_m, \tag{6.5}$$

$$\varpi_{m+1}(n) = n^{\lambda_{m+1}} \int_0^\infty \cdots \int_0^\infty k_\lambda(x_1, \cdots, x_m, n) \prod_{j=1}^{m} x_j^{\lambda_j - 1} dx_1 \cdots dx_m. \tag{6.6}$$

In particular, for $\lambda_{m+1} = \frac{\lambda}{2}$, replacing the kernel $k_\lambda(x_1, \cdots, x_m, n)$ by $k_\lambda(nx_1, \cdots, nx_m, 1)$ in (6.5) and (6.6), we can still define another weight functions $w_i(x_i)$ ($x_i \in \mathbf{R}_+; i = 1, \cdots, m$) and $\widetilde{w}_{m+1}(n)$ ($n \in \mathbf{N}$) as follows:

$$w_i(x_i) = x_i^{\lambda_i} \sum_{n=1}^{\infty} n^{\frac{\lambda}{2}-1} \int_0^\infty \cdots \int_0^\infty k_\lambda(nx_1, \cdots, nx_m, 1)$$

$$\times \prod_{j=1(j\neq i)}^{m} x_j^{\lambda_j - 1} dx_1 \cdots dx_{i-1} dx_{i+1} \cdots dx_m,$$

$$\widetilde{w}_{m+1}(n) = n^{\frac{\lambda}{2}} \int_0^\infty \cdots \int_0^\infty k_\lambda(nx_1, \cdots, nx_m, 1) \prod_{j=1}^{m} x_j^{\lambda_j - 1} dx_1 \cdots dx_m.$$

Lemma 6.4. *For $n \in \mathbf{N}$, we have $\varpi_{m+1}(n) = k(\lambda_{m+1})$. If $k(\lambda_{m+1}) \in \mathbf{R}_+$,*

$$k_\lambda(x_1, \cdots, x_m, y)y^{\lambda_{m+1}-1}$$

is decreasing with respect to $y \in \mathbf{R}_+$ and strictly decreasing in an interval $I \subset (1, \infty)$, then, for any $i = 1, \cdots, m$, we have

$$0 < k(\lambda_{m+1})(1 - \theta_i(x_i)) < \omega_i(x_i) < k(\lambda_{m+1}) \quad (x_i > 0), \tag{6.7}$$

where

$$\theta_i(x_i) = \frac{1}{k(\lambda_{m+1})}$$

$$\times \int_0^{1/x_i} u_{m+1}^{\lambda_{m+1}-1} \left[\int_0^\infty \cdots \int_0^\infty k_\lambda(u_1, \cdots, u_{i-1}, 1, u_{i+1}, \cdots, u_{m+1}) \right.$$

$$\left. \times \prod_{j=1(j\neq i)}^{m} u_j^{\lambda_j - 1} du_1 \cdots du_{i-1} du_{i+1} \cdots du_m \right] du_{m+1} > 0. \tag{6.8}$$

Note. *For $m = i = 1$, $\theta_i(x_i) = \frac{1}{k(\lambda_{m+1})} \int_0^{1/x_i} u_{m+1}^{\lambda_{m+1}-1} k_\lambda(1, u_{m+1}) du_{m+1}$.*

Moreover, if there exist constants $\alpha, L > 0$, such that

$$A_i(u_{m+1}) = \int_0^\infty \cdots \int_0^\infty k_\lambda(u_1, \cdots, u_{i-1}, 1, u_{i+1}, \cdots, u_{m+1})$$

$$\times \prod_{j=1(j\neq i)}^{m} u_j^{\lambda_j - 1} du_1 \cdots du_{i-1} du_{i+1} \cdots du_m \leq L u_{m+1}^{\alpha - \lambda_{m+1}},$$

$$\tag{6.9}$$

then, we still have

$$\theta_i(x_i) = O\left(\frac{1}{x_i^\alpha}\right) \quad (x_i > 0; i = 1, \cdots, m).$$

Note. For $m = i = 1$, $A_i(u_{m+1}) = k_\lambda(1, u_{m+1})$.

Proof. Setting $x_j = nu_j$ $(j = 1, \cdots, m)$ in (6.6) gives

$$\varpi_{m+1}(n) = n^{\lambda_{m+1}} n^m \int_0^\infty \cdots \int_0^\infty k_\lambda(nu_1, \cdots, nu_m, n)$$
$$\times \prod_{j=1}^m (nu_j)^{\lambda_j - 1} du_1 \cdots du_m$$
$$= H(m+1) = k(\lambda_{m+1}).$$

Using the decreasing property and Lemma 6.2, for $i = 1, \cdots, m$, we have

$$\omega_i(x_i) < x_i^{\lambda_i} \int_0^\infty x_{m+1}^{\lambda_{m+1}-1} \int_0^\infty \cdots \int_0^\infty k_\lambda(x_1, \cdots, x_m, x_{m+1}) dx_{m+1}$$
$$\times \prod_{j=1(j \neq i)}^m x_j^{\lambda_j - 1} dx_1 \cdots dx_{i-1} dx_{i+1} \cdots dx_m$$
$$= H(i) = k(\lambda_{m+1}) \ (u_j = x_j / x_i (j \neq i)),$$

$$\omega_i(x_i) > x_i^{\lambda_i} \int_1^\infty x_{m+1}^{\lambda_{m+1}-1} \left[\int_0^\infty \cdots \int_0^\infty k_\lambda(x_1, \cdots, x_m, x_{m+1}) \right.$$
$$\times \prod_{j=1(j \neq i)}^m x_j^{\lambda_j - 1} dx_1 \cdots dx_{i-1} dx_{i+1} \cdots dx_m \bigg] dx_{m+1},$$
$$(u_j = x_j / x_i (j \neq i))$$

$$= \int_{1/x_i}^\infty u_{m+1}^{\lambda_{m+1}-1} \left[\int_0^\infty \cdots \int_0^\infty k_\lambda(u_1, \cdots, u_{i-1}, 1, u_{i+1}, \cdots, u_{m+1}) \right.$$
$$\times \prod_{j=1(j \neq i)}^m u_j^{\lambda_j - 1} du_1 \cdots du_{i-1} du_{i+1} \cdots du_m \bigg] du_{m+1}$$
$$= k(\lambda_{m+1})(1 - \theta_i(x_i)) > 0,$$

where $\theta_i(x_i)$ is given by (6.8).

Moreover, by (6.9), we have

$$0 < \theta_i(x_i) = \frac{1}{k(\lambda_{m+1})} \int_0^{1/x_i} A_i(u_{m+1}) u_{m+1}^{\lambda_{m+1}-1} du_{m+1}$$
$$\leq \frac{L}{k(\lambda_{m+1})} \int_0^{1/x_i} u_{m+1}^{\alpha-1} du_{m+1} = \frac{L}{\alpha k(\lambda_{m+1}) x_i^\alpha}.$$

Thus, $\theta_i(x_i) = O\left(\frac{1}{x_i^\alpha}\right)$ $(x_i > 0; i = 1, \cdots, m)$. $\qquad\square$

Lemma 6.5. *Suppose that* $\lambda_{m+1} = \frac{\lambda}{2}$, $k(\lambda_{m+1}) \in \mathbf{R}_+$, *and* $k_\lambda(yx_1, \cdots, yx_m, 1)y^{\frac{\lambda}{2}-1}$ *is decreasing with respect to* $y \in \mathbf{R}_+$ *and strictly decreasing in an interval* $I \subset (1, \infty)$. *Then, we have* $\widetilde{w}_{m+1}(n) = k(\lambda_{m+1})$, *and for any* $i = 1, \cdots, m$,

$$0 < k(\lambda_{m+1})(1 - \widetilde{\theta}_i(x_i)) < w_i(x_i) < k(\lambda_{m+1}) \ (x_i > 0), \tag{6.10}$$

where

$$\widetilde{\theta}_i(x_i) = \frac{1}{k(\lambda_{m+1})} \int_0^{x_i} u_i^{\lambda_i - 1} \left[\int_0^\infty \cdots \int_0^\infty k_\lambda(u_1, \cdots, u_m, 1) \right.$$

$$\left. \times \prod_{j=1(j\neq i)}^m u_j^{\lambda_j - 1} du_1 \cdots du_{i-1} du_{i+1} \cdots du_m \right] du_i > 0. \tag{6.11}$$

Note. For $m = i = 1$, $\widetilde{\theta}_i(x_i) = \frac{1}{k(\lambda_{m+1})} \int_0^{x_i} u_i^{\lambda_i - 1} k_\lambda(u_i, 1) du_i$.

Moreover, if there exist constants α, $L > 0$ such that

$$B_1(u_i) = \int_0^\infty \cdots \int_0^\infty k_\lambda(u_1, \cdots, u_m, 1)$$

$$\times \prod_{j=1(j\neq i)}^m u_j^{\lambda_j - 1} du_1 \cdots du_{i-1} du_{i+1} \cdots du_m \leq L u_i^{\alpha - \lambda_i}, \tag{6.12}$$

then, we have

$$\widetilde{\theta}_i(x_i) = O(x_i^\alpha) \ (x_i > 0; i = 1, \cdots, m).$$

Note. For $m = i = 1$, $B_1(u_i) = k_\lambda(u_i, 1)$.

Proof. Setting $u_j = nx_j(j = 1, \cdots, m)$ in $\widetilde{w}_{m+1}(n)$ (see Definition 6.2), simplifying, we find $\widetilde{w}_{m+1}(n) = k(\lambda_{m+1})$.

Using the decreasing property and Lemma 6.4, we have

$$w_i(x_i) < x_i^{\lambda_i} \int_0^\infty x_{m+1}^{\frac{\lambda}{2}-1} \left[\int_0^\infty \cdots \int_0^\infty k_\lambda(x_{m+1}x_1, \cdots, x_{m+1}x_m, 1) \right.$$

$$\left. \times \prod_{j=1(j\neq i)}^m x_j^{\lambda_j - 1} dx_1 \cdots dx_{i-1} dx_{i+1} \cdots dx_m \right] dx_{m+1},$$

$$(x'_{m+1} = x_{m+1}^{-1})$$

$$= x_i^{\lambda_i} \int_0^\infty x_{m+1}^{\prime(-\frac{\lambda}{2}-1)} \left[\int_0^\infty \cdots \int_0^\infty k_\lambda(x_{m+1}^{\prime-1}x_1, \cdots, x_{m+1}^{\prime-1}x_m, 1) \right.$$

$$\left. \times \prod_{j=1(j\neq i)}^m x_j^{\lambda_j - 1} dx_1 \cdots dx_{i-1} dx_{i+1} \cdots dx_m \right] dx'_{m+1}$$

$$= x_i^{\lambda_i} \int_0^\infty \cdots \int_0^\infty k_\lambda(x_1, \cdots, x_m, x_{m+1})$$

$$\times \prod_{j=1(j\neq i)}^{m+1} x_j^{\lambda_j-1} dx_1 \cdots dx_{i-1} dx_{i+1} \cdots dx_{m+1}$$

$$= H(i) = k(\lambda_{m+1}) \ (u_j = x_j/x_i (j \neq i)),$$

$$w_i(x_i) > x_i^{\lambda_i} \int_1^\infty x_{m+1}^{\frac{\lambda}{2}-1} \left[\int_0^\infty \cdots \int_0^\infty k_\lambda(x_{m+1}x_1, \cdots, x_{m+1}x_m, 1) \right.$$

$$\left. \times \prod_{j=1(j\neq i)}^m x_j^{\lambda_j-1} dx_1 \cdots dx_{i-1} dx_{i+1} \cdots dx_m \right] dx_{m+1}.$$

Setting $u_j = x_{m+1}x_j (j \neq i)$ in the above integral, simplifying, and putting $u_i = x_{m+1}x_i$, we find

$$w_i(x_i) > \int_{x_i}^\infty u_i^{\lambda_i-1} \left[\int_0^\infty \cdots \int_0^\infty k_\lambda(u_1, \cdots, u_m, 1) \right.$$

$$\left. \times \prod_{j=1(j\neq i)}^m u_j^{\lambda_j-1} du_1 \cdots du_{i-1} du_{i+1} \cdots du_m \right] du_i$$

$$= k(\lambda_{m+1})(1 - \widetilde{\theta}_i(x_i)) > 0,$$

where $\widetilde{\theta}_i(x_i)$ is given by (6.11).

Moreover, by (6.11) and (6.12), we have

$$0 < \widetilde{\theta}_i(x_i) = \frac{1}{k(\lambda_{m+1})} \int_0^{x_i} B_1(u_i) u_i^{\lambda_i-1} du_i$$

$$\leq \frac{L}{k(\lambda_{m+1})} \int_0^{x_i} u_i^{\lambda_i-1} u_i^{\alpha-\lambda_i} du_i = \frac{L x_i^\alpha}{\alpha k(\lambda_{m+1})},$$

that is, $\widetilde{\theta}_i(x_i) = O(x_i^\alpha) \ (x_i > 0; i = 1, \cdots, m).$ $\qquad\square$

Lemma 6.6. *Suppose that there exists a constant $\delta_0 > 0$, such that, for any $\widetilde{\lambda}_i \in (\lambda_i - \delta_0, \lambda_i + \delta_0) \ (i = 1, \cdots, m+1)$, $\sum_{i=1}^{m+1} \widetilde{\lambda}_i = \lambda$, $k(\widetilde{\lambda}_{m+1}) \in \mathbf{R}_+$. If $k_\lambda(x_1, \cdots, x_m, y)y^{\widetilde{\lambda}_{m+1}-1}$ is decreasing with respect to $y \in \mathbf{R}_+$, then, for $0 < \varepsilon < \delta_0 \min_{1 \leq j \leq m+1}\{|p_j|, |p|\}$, we have*

$$I(\varepsilon) = \varepsilon \sum_{n=1}^\infty n^{(\lambda_{m+1}-\frac{\varepsilon}{p_{m+1}})-1} \int_1^\infty \cdots \int_1^\infty k_\lambda(x_1, \cdots, x_m, n)$$

$$\times \prod_{j=1}^m x_j^{(\lambda_j-\frac{\varepsilon}{p_j})-1} dx_1 \cdots dx_m = k(\lambda_{m+1}) + o(1)(\varepsilon \to 0^+). \quad (6.13)$$

Proof. For $\widetilde{\lambda}_i = \lambda_i - \frac{\varepsilon}{p_i} \ (i = 1, \cdots, m), \widetilde{\lambda}_{m+1} = \lambda_{m+1} + \frac{\varepsilon}{p} \in (\lambda_i - \delta_0, \lambda_i + \delta_0),$

$$k_\lambda(x_1, \cdots, x_m, y) \ y^{\lambda_{m+1}-\frac{\varepsilon}{p_{m+1}}-1} \ (= k_\lambda(x_1, \cdots, x_m, y)y^{\widetilde{\lambda}_{m+1}-1}y^{-\varepsilon})$$

is decreasing with respect to $y \in \mathbf{R}_+$. By the decreasing property, we find

$$I(\varepsilon) \geq \varepsilon \int_1^\infty x_{m+1}^{\lambda_{m+1}-\frac{\varepsilon}{p_{m+1}}-1} \int_1^\infty \cdots \int_1^\infty k_\lambda(x_1, \cdots, x_m, x_{m+1})$$
$$\times \prod_{j=1}^m x_j^{(\lambda_j-\frac{\varepsilon}{p_j})-1} dx_1 \cdots dx_m dx_{m+1}, \ (u_j = x_j/x_{m+1})$$

$$= \varepsilon \int_1^\infty x_{m+1}^{-\varepsilon-1} \left[\int_{\frac{1}{x_{m+1}}}^\infty \cdots \int_{\frac{1}{x_{m+1}}}^\infty k_\lambda(u_1, \cdots, u_m, 1) \right.$$
$$\left. \times \prod_{j=1}^m u_j^{(\lambda_j-\frac{\varepsilon}{p_j})-1} du_1 \cdots du_m \right] dx_{m+1}$$

$$\geq \varepsilon \int_1^\infty x_{m+1}^{-\varepsilon-1} \left[\int_0^\infty \cdots \int_0^\infty k_\lambda(u_1, \cdots, u_m, 1) \right.$$
$$\left. \times \prod_{j=1}^m u_j^{(\lambda_j-\frac{\varepsilon}{p_j})-1} du_1 \cdots du_m \right] dx_{m+1}$$
$$- \varepsilon \sum_{i=1}^m \int_1^\infty x_{m+1}^{-1} A_i(x_{m+1}) dx_{m+1}$$

$$= k(\lambda_{m+1}+\frac{\varepsilon}{p}) - \varepsilon \sum_{i=1}^m \int_1^\infty x_{m+1}^{-1} A_i(x_{m+1}) dx_{m+1},$$

where

$$A_i(x_{m+1}) = \int_0^\infty \cdots \int_0^\infty \int_0^{1/x_{m+1}} \int_0^\infty \cdots \int_0^\infty k_\lambda(u_1, \cdots, u_m, 1)$$
$$\times \prod_{j=1}^m u_j^{(\lambda_j-\frac{\varepsilon}{p_j})-1} du_1 \cdots du_{i-1} du_i du_{i+1} \cdots du_m.$$

Without loss of generality, we estimate $\int_1^\infty x_{m+1}^{-1} A_m(x_{m+1}) dx_{m+1}$ as follows:

$$0 < \int_1^\infty x_{m+1}^{-1} A_m(x_{m+1}) dx_{m+1}$$

$$= \int_1^\infty x_{m+1}^{-1} \left[\int_0^{1/x_{m+1}} \int_0^\infty \cdots \int_0^\infty k_\lambda(u_1, \cdots, u_m, 1) \right.$$
$$\left. \times \prod_{j=1}^m u_j^{(\lambda_j-\frac{\varepsilon}{p_j})-1} du_1 \cdots du_{m-1} du_m \right] dx_{m+1}$$

$$= \int_0^1 \left(\int_1^{1/u_m} x_{m+1}^{-1} dx_{m+1} \right) \int_0^\infty \cdots \int_0^\infty k_\lambda(u_1, \cdots, u_m, 1)$$

$$\times \prod_{j=1}^m u_j^{(\lambda_j - \frac{\varepsilon}{p_j})-1} du_1 \cdots du_{m-1} du_m$$

$$= \int_0^1 (-\ln u_m) \int_0^\infty \cdots \int_0^\infty k_\lambda(u_1, \cdots, u_m, 1)$$

$$\times \prod_{j=1}^m u_j^{(\lambda_j - \frac{\varepsilon}{p_j})-1} du_1 \cdots du_{m-1} du_m. \tag{6.14}$$

Setting $\alpha > 0$, such that $\max\{\alpha + \frac{\varepsilon}{|p_m|}, \alpha + \varepsilon \frac{1}{|p|}\} < \delta_0$, since $\lim_{u_m \to 0^+} u_m^\alpha(-\ln u_m) = 0$, there exists a constant $M_m > 0$, such that

$$0 < u_m^\alpha(-\ln u_m) \le M_m \ (u_m \in (0,1]).$$

In view of (6.14) and (6.2), we find

$$0 < \int_1^\infty x_{m+1}^{-1} A_m(x_{m+1}) \, dx_{m+1}$$

$$\le M_m \int_0^\infty \int_0^\infty \cdots \int_0^\infty k_\lambda(u_1, \cdots, u_m, 1)$$

$$\times \prod_{j=1}^{m-1} u_j^{(\lambda_j - \frac{\varepsilon}{p_j})-1} u_m^{(\lambda_m - \frac{\alpha p_m + \varepsilon}{p_m})-1} du_1 \cdots du_{m-1} du_m$$

$$= M_m k \left(\lambda_{m+1} + \alpha + \frac{\varepsilon}{p} \right) < \infty.$$

Thus,

$$\sum_{i=1}^m \int_1^\infty x_i^{-1} A_m(x_{m+1}) dx_{m+1} = O(1).$$

Hence, we have

$$I(\varepsilon) \ge k \left(\lambda_{m+1} + \varepsilon \left(1 - \frac{1}{p_{m+1}} \right) \right) - \varepsilon O(1).$$

We still obtain

$$I(\varepsilon) \le \varepsilon \sum_{n=1}^\infty n^{-\varepsilon-1} \left[n^{\lambda_{m+1}+\varepsilon - \frac{\varepsilon}{p_{m+1}}} \int_0^\infty \cdots \int_0^\infty k_\lambda(x_1, \cdots, x_m, n) \right.$$

$$\left. \times \prod_{j=1}^m x_j^{(\lambda_j - \frac{\varepsilon}{p_j})-1} dx_1 \cdots dx_m \right]$$

$$= \varepsilon \left(1 + \sum_{n=2}^{\infty} n^{-\varepsilon-1} \right) k \left(\lambda_{m+1} + \frac{\varepsilon}{p} \right)$$

$$\leq \varepsilon \left(1 + \int_1^{\infty} y^{-\varepsilon-1} dy \right) k \left(\lambda_{m+1} + \frac{\varepsilon}{p} \right)$$

$$= (\varepsilon + 1) \, k \left(\lambda_{m+1} + \frac{\varepsilon}{p} \right).$$

Hence, by (6.4), we have (6.13). $\qquad\square$

Lemma 6.7. *Suppose that $\lambda_{m+1} = \frac{\lambda}{2}$, there exists a constant $\delta_0 > 0$, such that, for any $\widetilde{\lambda}_i \in (\lambda_i - \delta_0, \lambda_i + \delta_0)$ $(i = 1, \cdots, m+1)$, $\sum_{i=1}^{m+1} \widetilde{\lambda}_i = \lambda$, $k(\widetilde{\lambda}_{m+1}) \in \mathbf{R}_+$, and*

$$k_\lambda(yx_1, \cdots, yx_m, 1) \, y^{\widetilde{\lambda}_{m+1}-1}$$

is decreasing with respect to $y \in \mathbf{R}_+$. Then, for $0 < \varepsilon < \delta_0 \min_{1 \leq j \leq m+1}\{|p_j|, |p|\}$, we have

$$\widetilde{I}(\varepsilon) = \varepsilon \sum_{n=1}^{\infty} n^{\frac{\lambda}{2} - \frac{\varepsilon}{p_{m+1}} - 1} \int_0^1 \cdots \int_0^1 k_\lambda(nx_1, \cdots, nx_m, 1)$$

$$\times \prod_{j=1}^{m} x_j^{(\lambda_j + \frac{\varepsilon}{p_j})-1} dx_1 \cdots dx_m = k(\lambda_{m+1}) + o(1)(\varepsilon \to 0^+). \quad (6.15)$$

Proof. By the decreasing property, we find

$$\widetilde{I}(\varepsilon) \geq \varepsilon \int_1^{\infty} x_{m+1}^{\frac{\lambda}{2} - \frac{\varepsilon}{p_{m+1}} - 1} \left[\int_0^1 \cdots \int_0^1 k_\lambda(x_{m+1}x_1, \cdots, x_{m+1}x_m, 1) \right.$$

$$\times \left. \prod_{j=1}^{m} x_j^{(\lambda_j - \frac{\varepsilon}{p_j})-1} dx_1 \cdots dx_m \right] dx_{m+1} \; u_j = x_{m+1}x_j (j = 1, \cdots, m)$$

$$= \varepsilon \int_1^{\infty} x_{m+1}^{-\varepsilon-1} \left[\int_0^{x_{m+1}} \cdots \int_0^{x_{m+1}} k_\lambda(u_1, \cdots, u_m, 1) \right.$$

$$\times \left. \prod_{j=1}^{m} u_j^{(\lambda_j + \frac{\varepsilon}{p_j})-1} du_1 \cdots du_m \right] dx_{m+1}$$

$$\geq \varepsilon \int_1^{\infty} x_{m+1}^{-\varepsilon-1} \left[\int_0^{\infty} \cdots \int_0^{\infty} k_\lambda(u_1, \cdots, u_m, 1) \right.$$

$$\times \left. \prod_{j=1}^{m} u_j^{(\lambda_j + \frac{\varepsilon}{p_j})-1} du_1 \cdots du_m \right] dx_{m+1}$$

$$-\varepsilon \sum_{i=1}^{m} \int_{1}^{\infty} x_{m+1}^{-1} B_i(x_{m+1}) dx_{m+1}$$

$$= k(\lambda_{m+1} - \varepsilon + \frac{\varepsilon}{p_{m+1}}) - \varepsilon \sum_{i=1}^{m} \int_{1}^{\infty} x_{m+1}^{-1} B_i(x_{m+1}) dx_{m+1}, \quad (6.16)$$

where

$$B_i(x_{m+1}) = \int_{0}^{\infty} \cdots \int_{0}^{\infty} \int_{x_{m+1}}^{\infty} \int_{0}^{\infty} \cdots \int_{0}^{\infty} k_{\lambda}(u_1, \cdots, u_m, 1)$$

$$\times \prod_{j=1}^{m} u_j^{(\lambda_j + \frac{\varepsilon}{p_j})-1} du_1 \cdots du_{i-1} du_i du_{i+1} \cdots du_m (1 \le i \le m).$$

Without loss of generality, we estimate $\int_{1}^{\infty} x_{m+1}^{-1} B_m(x_{m+1}) dx_{m+1}$ as follows:

$$0 < \int_{1}^{\infty} x_{m+1}^{-1} B_m(x_{m+1}) dx_{m+1}$$

$$= \int_{1}^{\infty} x_{m+1}^{-1} \left[\int_{x_{m+1}}^{\infty} \int_{0}^{\infty} \cdots \int_{0}^{\infty} k_{\lambda}(u_1, \cdots, u_m, 1) \right.$$

$$\left. \times \prod_{j=1}^{m} u_j^{(\lambda_j + \frac{\varepsilon}{p_j})-1} du_1 \cdots du_{m-1} du_m \right] dx_{m+1}$$

$$= \int_{1}^{\infty} \left(\int_{1}^{u_m} x_{m+1}^{-1} dx_{m+1} \right) \int_{0}^{\infty} \cdots \int_{0}^{\infty} k_{\lambda}(u_1, \cdots, u_m, 1)$$

$$\times \prod_{j=1}^{m} u_j^{(\lambda_j + \frac{\varepsilon}{p_j})-1} du_1 \cdots du_{m-1} du_m$$

$$= \int_{1}^{\infty} \ln u_m \int_{0}^{\infty} \cdots \int_{0}^{\infty} k_{\lambda}(u_1, \cdots, u_m, 1)$$

$$\times \prod_{j=1}^{m} u_j^{(\lambda_j + \frac{\varepsilon}{p_j})-1} du_1 \cdots du_{m-1} du_m.$$

Setting $\alpha > 0$, such that $\max\{\alpha + \frac{\varepsilon}{|p_m|}, \alpha + \frac{\varepsilon}{|p|}\} < \delta_0$, since

$$\lim_{u_m \to \infty} u_m^{-\alpha} \ln u_m = 0,$$

there exists a constant $M_m > 0$, such that

$$0 < u_m^{-\alpha} \ln u_m \le M_m \ (u_m \in [1, \infty)).$$

In view of (6.2), we obtain

$$
0 < \int_1^\infty x_{m+1}^{-1} B_m(x_{m+1}) dx_{m+1}
$$

$$
\leq M_m \int_0^\infty \int_0^\infty \cdots \int_0^\infty k_\lambda(u_1, \cdots, u_m, 1)
$$

$$
\times \prod_{j=1}^{m-1} u_j^{(\lambda_j + \frac{\varepsilon}{p_j})-1} u_m^{(\lambda_m + \frac{\alpha p_m + \varepsilon}{p_m})-1} du_1 \cdots du_{m-1} du_m
$$

$$
= M_m k \left(\lambda_{m+1} - \alpha - \frac{\varepsilon}{p} \right) < \infty,
$$

and then, by (6.16), we have

$$
\widetilde{I}(\varepsilon) \geq k \left(\lambda_{m+1} - \frac{\varepsilon}{p} \right) - \varepsilon O(1).
$$

By Lemma 6.4, we still obtain

$$
\widetilde{I}(\varepsilon) \leq \varepsilon \sum_{n=1}^\infty n^{-\varepsilon-1} \left[n^{\frac{\lambda}{2} + \varepsilon - \frac{\varepsilon}{p_{m+1}}} \int_0^\infty \cdots \int_0^\infty k_\lambda(nx_1, \cdots, nx_m, 1) \right.
$$

$$
\left. \times \prod_{j=1}^m x_j^{(\lambda_j + \frac{\varepsilon}{p_j})-1} dx_1 \cdots dx_m \right], \quad (u_j = nx_j (j = 1, \cdots, m))
$$

$$
= \varepsilon \left(1 + \sum_{n=2}^\infty n^{-\varepsilon-1} \right) \int_0^\infty \cdots \int_0^\infty k_\lambda(u_1, \cdots, u_m, 1)
$$

$$
\times \prod_{j=1}^m u_j^{(\lambda_j + \frac{\varepsilon}{p_j})-1} du_1 \cdots du_m
$$

$$
\leq \varepsilon \left(1 + \int_1^\infty y^{-\varepsilon-1} dy \right) k(\lambda_{m+1} - \frac{\varepsilon}{p})
$$

$$
= (\varepsilon + 1) k(\lambda_{m+1} - \frac{\varepsilon}{p}).
$$

Then, by (6.4), we have (6.15). □

6.2.2 *Two Preliminary Inequalities*

Lemma 6.8. *If $f_i(x_i)(i = 1, \cdots, m)$ are non-negative measurable functions in $\mathbf{R}_+$, then*

(i) for $p_i > 1(i = 1, \cdots, m+1)$, we have the following inequality:

$$J = \left\{ \sum_{n=1}^{\infty} n^{p\lambda_{m+1}-1} \left(\int_0^{\infty} \cdots \int_0^{\infty} k_\lambda(x_1, \cdots, x_m, n) \right.\right.$$

$$\left.\left. \times \prod_{j=1}^{m} f_j(x_j) dx_1 \cdots dx_m \right)^p \right\}^{\frac{1}{p}}$$

$$\leq [k(\lambda_{m+1})]^{\frac{1}{p_{m+1}}} \prod_{i=1}^{m} \left\{ \int_0^{\infty} \omega_i(x_i) x_i^{p_i(1-\lambda_i)-1} f_i(x_i) dx_i \right\}^{\frac{1}{p_i}} \; ; \quad (6.17)$$

(ii) for $0 < p_1 < 1$, $p_i < 0$ $(i = 2, \cdots, m+1)$, or for $p_i < 0$ $(i = 1, \cdots, m)$, $0 < p_{m+1} < 1$, we have the reverse of (6.17).

Proof.   (i) For $p_i > 1(i = 1, \cdots, m+1)(p > 1)$, by Hölder's inequality with weight (see Kuang [47]) and (6.1), we have

$$\left(\int_0^{\infty} \cdots \int_0^{\infty} k_\lambda(x_1, \cdots, x_m, n) \prod_{i=1}^{m} f_i(x_i) dx_1 \cdots dx_m \right)^p$$

$$= \left\{ \int_0^{\infty} \cdots \int_0^{\infty} k_\lambda(x_1, \cdots, x_m, n) \right.$$

$$\times \prod_{i=1}^{m} \left[x_i^{(\lambda_i-1)(1-p_i)} n^{\lambda_{m+1}-1} \prod_{j=1(j\neq i)}^{m} x_j^{\lambda_j-1} \right]^{\frac{1}{p_i}} f_i(x_i)$$

$$\times \left[n^{(\lambda_{m+1}-1)(1-p_{m+1})} \prod_{j=1}^{m} x_j^{\lambda_j-1} \right]^{\frac{1}{p_{m+1}}} dx_1 \cdots dx_m \right\}^p$$

$$\leq \int_0^{\infty} \cdots \int_0^{\infty} k_\lambda(x_1, \cdots, x_m, n)$$

$$\times \prod_{i=1}^{m} \left[x_i^{(\lambda_i-1)(1-p_i)} n^{\lambda_{m+1}-1} \prod_{j=1(j\neq i)}^{m} x_j^{\lambda_j-1} \right]^{\frac{p}{p_i}} f_i^p(x_i) dx_1 \cdots dx_m$$

$$\times \left\{ n^{p_{m+1}(1-\lambda_{m+1})-1} \left[n^{\lambda_{m+1}} \int_0^{\infty} \cdots \int_0^{\infty} k_\lambda(x_1, \cdots, x_m, n) \right.\right.$$

$$\times \prod_{j=1}^{m} x_j^{\lambda_j - 1} dx_1 \cdots dx_m \Bigg]\Bigg\}^{\frac{p}{p_{m+1}}}$$

$$= \frac{[\omega_{m+1}(n)]^{\frac{p}{p_{m+1}}}}{n^{p\lambda_{m+1}-1}} \int_0^\infty \cdots \int_0^\infty k_\lambda(x_1,\cdots,x_m,n) n^{\lambda_{m+1}-1}$$

$$\times \prod_{i=1}^{m} \left[x_i^{(\lambda_i-1)(1-p_i)} \prod_{j=1(j\neq i)}^{m} x_j^{\lambda_j-1} \right]^{\frac{p}{p_i}} f_i^p(x_i) dx_1 \cdots dx_m.$$

Since $\omega_{m+1}(n) = k(\lambda_{m+1})$, by the Lebesgue term by term integration theorem (see Kuang [49]), we find

$$J \leq [k(\lambda_{m+1})]^{\frac{1}{p_{m+1}}} \Bigg\{ \sum_{n=1}^{\infty} \int_0^\infty \cdots \int_0^\infty k_\lambda(x_1,\cdots,x_m,n) n^{\lambda_{m+1}-1}$$

$$\times \prod_{i=1}^{m} \left[x_i^{(\lambda_i-1)(1-p_i)} \prod_{j=1(j\neq i)}^{m} x_j^{\lambda_j-1} \right]^{\frac{p}{p_i}} f_i^p(x_i) dx_1 \cdots dx_m \Bigg\}^{\frac{1}{p}}$$

$$= [k(\lambda_{m+1})]^{\frac{1}{p_{m+1}}} \Bigg\{ \int_0^\infty \cdots \int_0^\infty \sum_{n=1}^{\infty} k_\lambda(x_1,\cdots,x_m,n) n^{\lambda_{m+1}-1}$$

$$\times \prod_{i=1}^{m} \left[x_i^{(\lambda_i-1)(1-p_i)} \prod_{j=1(j\neq i)}^{m} x_j^{\lambda_j-1} \right]^{\frac{p}{p_i}} f_i^p(x_i) dx_1 \cdots dx_m \Bigg\}^{\frac{1}{p}}.$$

In view of $\sum_{i=1}^{m} \frac{1}{(p_i/p)} = 1$, still by Hölder's inequality, we obtain

$$J \leq [k(\lambda_{m+1})]^{\frac{1}{p_{m+1}}} \prod_{i=1}^{m} \Bigg\{ \int_0^\infty \left[x_i^{\lambda_i} \sum_{n=1}^{\infty} n^{\lambda_{m+1}-1} \right. $$

$$\times \int_0^\infty \cdots \int_0^\infty k_\lambda(x_1,\cdots,x_m,n) \prod_{j=1(j\neq i)}^{m} x_j^{\lambda_j-1}$$

$$\times dx_1 \cdots dx_{i-1} dx_{i+1} \cdots dx_m \Bigg] x_i^{p_i(1-\lambda_i)-1} f_i^{p_i}(x_i) dx_i \Bigg\}^{\frac{p}{pp_i}}$$

$$= [k(\lambda_{m+1})]^{\frac{1}{p_{m+1}}} \prod_{i=1}^{m} \Bigg\{ \int_0^\infty \omega_i(x_i) x_i^{p_i(1-\lambda_i)-1} f_i^{p_i}(x_i) dx_i \Bigg\}^{\frac{1}{p_i}},$$

and then, inequality (6.17) follows.

(ii) For $0 < p_1 < 1$, $p_i < 0$ $(i = 2, \cdots, m+1)$, or for $p_i < 0$ $(i = 1, \cdots, m)$, $0 < p_{m+1} < 1$, by the reverse Hölder's inequality (see Kuang [47]) and the same way, we obtain the reverse of (6.17). $\square$

For $\lambda_{m+1} = \frac{\lambda}{2}$, replacing $k_\lambda(x_1, \cdots, x_m, n)$ by $k_\lambda(nx_1, \cdots, nx_m, 1)$ in Lemma 6.8, we still have

Lemma 6.9. *Suppose that* $\lambda_{m+1} = \frac{\lambda}{2}$, $f_i(x_i)$ $(i = 1, \cdots, m)$ *are non-negative measurable functions in* $\mathbf{R}_+$. *Then*

(i) *if* $p_i > 1$ $(i = 1, \cdots, m+1)$, *we have the following inequality:*

$$J_1 = \left\{ \sum_{n=1}^{\infty} n^{\frac{p\lambda}{2}-1} \left(\int_0^\infty \cdots \int_0^\infty k_\lambda(nx_1, \cdots, nx_m, 1) \right. \right.$$
$$\left. \left. \times \prod_{j=1}^m f_j(x_j) dx_1 \cdots dx_m \right)^p \right\}^{\frac{1}{p}}$$
$$\leq [k(\lambda_{m+1})]^{\frac{1}{p_{m+1}}} \prod_{i=1}^m \left\{ \int_0^\infty w_i(x_i) x_i^{p_i(1-\lambda_i)-1} f_i(x_i) dx_i \right\}^{\frac{1}{p_i}}; \quad (6.18)$$

(ii) *if* $0 < p_1 < 1$, $p_i < 0$ $(i = 2, \cdots, m+1)$, *or* $p_i < 0$ $(i = 1, \cdots, m)$, $0 < p_{m+1} < 1$, *then, we have the reverse of* (6.18).

6.2.3 *Main Results and Operator Expressions*

In the following, we set two functions:
$$\varphi_i(x_i) = x_i^{p_i(1-\lambda_i)-1} \quad (x_i \in \mathbf{R}_+; i = 1, \cdots, m),$$
and $\psi(n) = n^{p_{m+1}(1-\lambda_{m+1})-1}$ $(n \in \mathbf{N})$.

The spaces $L_{p_i,\varphi_i}(\mathbf{R}_+)$ and $l_{p_{m+1},\psi}$ with the norms $||f_i||_{p_i,\varphi_i}$ and $||a||_{p_{m+1},\psi}$ are defined by

$$L_{p_i,\varphi_i}(\mathbf{R}_+) = \left\{ f_i; ||f_i||_{p_i,\varphi_i} = \left\{ \int_0^\infty \varphi_i(x_i)|f_i(x_i)|^{p_i} dx_i \right\}^{\frac{1}{p_i}} < \infty \right\},$$

$$l_{p_{m+1},\psi} = \left\{ a = \{a_n\}; ||a||_{p_{m+1},\psi} = \left\{ \sum_{n=1}^{\infty} \psi(n)|a_n|^{p_{m+1}} \right\}^{\frac{1}{p_{m+1}}} < \infty \right\}.$$

Theorem 6.1. *Suppose that* $p_i > 1$ $(i = 1, \cdots, m+1)$, *there exists a constant* $\delta_0 > 0$ *such that, for any* $\tilde{\lambda}_i \in (\lambda_i - \delta_0, \lambda_i + \delta_0)$ $(i = 1, \cdots, m+1)$, $\sum_{i=1}^{m+1} \tilde{\lambda}_i = \lambda$,

$$k(\tilde{\lambda}_{m+1}) = \int_0^\infty \cdots \int_0^\infty k_\lambda(u_1, \cdots, u_m, 1) \prod_{j=1}^m u_j^{\tilde{\lambda}_j-1} du_1 \cdots du_m \in \mathbf{R}_+,$$

and $k_\lambda(x_1, \cdots, x_m, y) \, y^{\widetilde{\lambda}_{m+1}-1}$ is decreasing with respect to $y \in \mathbf{R}_+$ and strictly decreasing in an interval $I \subset (1, \infty)$. If $f_i(x_i) \geq 0, f_i \in L_{p_i, \varphi_i}(\mathbf{R}_+)$ $(i = 1, \cdots, m), a_n \geq 0, a = \{a_n\}_{n=1}^\infty \in l_{p_{m+1}, \psi}, ||f_i||_{p_i, \varphi_i} > 0, ||a||_{p_{m+1}, \psi} > 0$, then, we have the following equivalent inequalities:

$$I = \sum_{n=1}^\infty a_n \int_0^\infty \cdots \int_0^\infty k_\lambda(x_1, \cdots, x_m, n) \prod_{j=1}^m f_j(x_j) dx_1 \cdots dx_m$$

$$< k(\lambda_{m+1}) ||a||_{p_{m+1}, \psi} \prod_{i=1}^m ||f_i||_{p_i, \varphi_i}, \tag{6.19}$$

and

$$J = \left\{ \sum_{n=1}^\infty n^{p\lambda_{m+1}-1} \left(\int_0^\infty \cdots \int_0^\infty k_\lambda(x_1, \cdots, x_m, n) \right.\right.$$

$$\left.\left. \times \prod_{j=1}^m f_j(x_j) dx_1 \cdots dx_m \right)^p \right\}^{\frac{1}{p}}$$

$$< k(\lambda_{m+1}) \prod_{i=1}^m ||f_i||_{p_i, \varphi_i}, \tag{6.20}$$

where the constant factor $k(\lambda_{m+1})$ in the above inequalities is the best possible.

Proof. By (6.17) and the assumptions, since $\varpi_i(x_i) < k(\lambda_{m+1})$, we have (6.20).

By Hölder's inequality (see Kuang [47]), we obtain

$$I = \sum_{n=1}^\infty \left[(\psi(n))^{\frac{-1}{p_{m+1}}} \int_0^\infty \cdots \int_0^\infty k_\lambda(x_1, \cdots, x_m, n) \right.$$

$$\left. \times \prod_{j=1}^m f_j(x_j) dx_1 \cdots dx_m \right] [(\psi(n))^{\frac{1}{p_{m+1}}} a_n]$$

$$\leq J ||a||_{p_{m+1}, \psi}. \tag{6.21}$$

Then, by (6.20), we have inequality (6.19).

Assuming that (6.19) is valid and setting

$$a_n = n^{p\lambda_{m+1}-1} \left(\int_0^\infty \cdots \int_0^\infty k_\lambda(x_1, \cdots, x_m, n) \right.$$

$$\left. \times \prod_{j=1}^m f_j(x_j) dx_1 \cdots dx_m \right)^{p-1} \quad (n \in \mathbf{N}),$$

then, we have $J^{p-1} = ||a||_{p_{m+1},\psi}$. By (6.17), we have $J < \infty$. If $J = 0$, then (6.20) is trivially valid; if $J > 0$, then, by (6.19), it follows that

$$||a||_{p_{m+1},\psi}^{p_{m+1}} = J^p = I < k(\lambda_{m+1})||a||_{p_{m+1},\psi} \prod_{i=1}^{m} ||f_i||_{p_i,\varphi_i}, \quad \text{that is,}$$

$$||a||_{p_{m+1},\psi}^{p_{m+1}-1} = J < k(\lambda_{m+1}) \prod_{i=1}^{m} ||f_i||_{p_i,\varphi_i},$$

and then, inequality (6.20) follows.

Hence, (6.19) and (6.20) are equivalent.

For $0 < \varepsilon < \delta_0 \min_{1 \leq j \leq m+1}\{p_j, p\}$, we set $\widetilde{f}_i(x_i)$ and $\widetilde{a}_n$ as follows:

$$\widetilde{f}_i(x_i) = \begin{cases} 0, & 0 < x_i < 1, \\ x_i^{(\lambda_i - \frac{\varepsilon}{p_i})-1}, & x_i \geq 1 \end{cases} \qquad (i = 1, \cdots, m),$$

$$\widetilde{a}_n = n^{(\lambda_{m+1} - \frac{\varepsilon}{p_{m+1}})-1}, n \in \mathbf{N}.$$

If there exists a constant $k(\leq k(\lambda_{m+1}))$, such that (6.19) is valid as we replace $k(\lambda_{m+1})$ by k, then, by (6.13), we find

$$k(\lambda_{m+1}) + o(1) = I(\varepsilon)$$

$$= \varepsilon \sum_{n=1}^{\infty} \widetilde{a}_n \int_0^{\infty} \cdots \int_0^{\infty} k_\lambda(x_1, \cdots, x_m, n)$$

$$\times \prod_{j=1}^{m} \widetilde{f}_j(x_j) dx_1 \cdots dx_m$$

$$< \varepsilon k ||\widetilde{a}||_{p_{m+1},\psi} \prod_{i=1}^{m} ||\widetilde{f}_i||_{p_i,\varphi_i}$$

$$= \varepsilon k \left(1 + \sum_{n=2}^{\infty} \frac{1}{n^{1+\varepsilon}}\right)^{\frac{1}{p_{m+1}}} \prod_{i=1}^{m} \left(\frac{1}{\varepsilon}\right)^{\frac{1}{p_i}}$$

$$< \varepsilon k \left(1 + \int_1^{\infty} \frac{dy}{y^{1+\varepsilon}}\right)^{\frac{1}{p_{m+1}}} \prod_{i=1}^{m} \left(\frac{1}{\varepsilon}\right)^{\frac{1}{p_i}} = k(1+\varepsilon)^{\frac{1}{p_{m+1}}},$$

and then, $k(\lambda_{m+1}) \leq k$ ($\varepsilon \to 0^+$). Hence, $k = k(\lambda_{m+1})$ is the best constant factor of (6.19).

By the equivalency, the constant factor $k(\lambda_{m+1})$ in (6.20) is still the best possible. Otherwise, it leads to a contradiction by (6.21) that the constant factor in (6.19) is not the best possible. $\qquad\square$

Theorem 6.2. *For* $\lambda_{m+1} = \frac{\lambda}{2}$, *replacing* $k_\lambda(x_1, \cdots, x_m, y) \, y^{\widetilde{\lambda}_{m+1}-1}$ *by*

$$k_\lambda(yx_1, \cdots, yx_m, 1) \, y^{\widetilde{\lambda}_{m+1}-1},$$

in this case, let the assumptions of Theorem 6.1 be fulfilled. Then, we have the following equivalent inequalities with the same best constant factor $k(\lambda_{m+1})$:

$$\sum_{n=1}^{\infty} a_n \int_0^{\infty} \cdots \int_0^{\infty} k_\lambda(nx_1, \cdots, nx_m, 1) \prod_{j=1}^{m} f_j(x_j) dx_1 \cdots dx_m$$

$$< k(\lambda_{m+1}) \|a\|_{p_{m+1}, \psi} \prod_{i=1}^{m} \|f_i\|_{p_i, \varphi_i}, \tag{6.22}$$

and

$$\left\{ \sum_{n=1}^{\infty} n^{\frac{p\lambda}{2}-1} \left(\int_0^{\infty} \cdots \int_0^{\infty} k_\lambda(nx_1, \cdots, nx_m, 1) \right. \right.$$

$$\left. \left. \times \prod_{j=1}^{m} f_j(x_j) dx_1 \cdots dx_m \right)^p \right\}^{\frac{1}{p}} < k(\lambda_{m+1}) \prod_{i=1}^{m} \|f_i\|_{p_i, \varphi_i}. \tag{6.23}$$

Proof. We only prove that the constant factor in (6.22) is the best possible. The other parts of the proof are omitted.

For $0 < \varepsilon < \delta_0 \min_{1 \le j \le m+1}\{p_j, p\}$, we set $\widetilde{f}_i(x_i)$ and $\widetilde{a}_n$ as follows:

$$\widetilde{f}_i(x_i) = \begin{cases} x_i^{(\lambda_i + \frac{\varepsilon}{p_i})-1}, & 0 < x_i \le 1, \\ 0, & x_i > 1 \end{cases} \qquad (i = 1, \cdots, m),$$

$$\widetilde{a}_n = n^{(\frac{\lambda}{2} - \frac{\varepsilon}{p_{m+1}})-1}, n \in \mathbf{N}.$$

If there exists a constant $k(\le k(\lambda_{m+1}))$, such that (6.22) is valid as we replace $k(\lambda_{m+1})$ by k, then, by (6.15), we have

$$k(\lambda_{m+1}) + o(1) = \widetilde{I}(\varepsilon)$$

$$= \varepsilon \sum_{n=1}^{\infty} \widetilde{a}_n \int_0^{\infty} \cdots \int_0^{\infty} k_\lambda(nx, \cdots, nx_m, 1) \prod_{j=1}^{m} \widetilde{f}_j(x_j) dx_1 \cdots dx_m$$

$$< \varepsilon k \|\widetilde{a}\|_{p_{m+1}, \psi} \prod_{i=1}^{m} \|\widetilde{f}_i\|_{p_i, \varphi_i}$$

$$= k \left(1 + \sum_{n=2}^{\infty} \frac{1}{n^{1+\varepsilon}} \right)^{\frac{1}{p_{m+1}}} \prod_{i=1}^{m} \left(\frac{1}{\varepsilon} \right)^{\frac{1}{p_i}}$$

$$< \varepsilon k \left(1 + \int_1^{\infty} \frac{dy}{y^{1+\varepsilon}} \right)^{\frac{1}{p_{m+1}}} \prod_{i=1}^{m} \left(\frac{1}{\varepsilon} \right)^{\frac{1}{p_i}} = k(1+\varepsilon)^{\frac{1}{p_{m+1}}},$$

and then, $k(\lambda_{m+1}) \leq k(\varepsilon \to 0^+)$. Hence, $k = k(\lambda_{m+1})$ is the best possible constant factor of (6.22). $\qquad \square$

Remark 6.1. With the assumptions of Theorem 6.1, we define a first kind of multiple half-discrete Hilbert-type operator with the homogeneous kernel

$$T : \prod_{i=1}^{m} L_{p_i, \varphi_i}(\mathbf{R}_+) \to l_{p_{m+1}, \psi^{1-p}}$$

as follows:

For any $f = (f_1, \cdots, f_m) \in \prod_{i=1}^{m} L_{p_i, \varphi_i}(\mathbf{R}_+)$, there exists a Tf, satisfying

$$Tf(n) = \int_0^\infty \cdots \int_0^\infty k_\lambda(x_1, \cdots, x_m, n) \prod_{j=1}^{m} f_j(x_j) dx_1 \cdots dx_m \qquad (n \in \mathbf{N}).$$
(6.24)

Then, by (6.20), it follows that

$$||Tf||_{p, \psi^{1-p}} < k(\lambda_{m+1}) \prod_{i=1}^{m} ||f_i||_{p_i, \varphi_i},$$

and then, $Tf \in l_{p, \psi^{1-p}}$. Hence, T is a bounded linear operator with $||T|| \leq k(\lambda_{m+1})$.

Since the constant factor in (6.20) is the best possible, we have

$$||T|| = \sup_{f(\neq \theta) \in \prod_{i=1}^{m} L_{p_i, \varphi_i}(\mathbf{R}_+)} \frac{||Tf||_{p, \psi^{1-p}}}{\prod_{i=1}^{m} ||f_i||_{p_i, \varphi_i}} = k(\lambda_{m+1}).$$
(6.25)

With the assumptions of Theorem 6.2, we define a first kind of multiple half-discrete Hilbert-type operator with the non-homogeneous kernel

$$T_1 : \prod_{i=1}^{m} L_{p_i, \varphi_i}(\mathbf{R}_+) \to l_{p_{m+1}, \psi^{1-p}}$$

as follows:

For any $f = (f_1, \cdots, f_m) \in \prod_{i=1}^{m} L_{p_i, \varphi_i}(\mathbf{R}_+)$, there exists a $T_1 f$, satisfying

$$T_1 f(n) = \int_0^\infty \cdots \int_0^\infty k_\lambda(nx_1, \cdots, nx_m, 1) \prod_{j=1}^{m} f_j(x_j) dx_1 \cdots dx_m \ (n \in \mathbf{N}).$$
(6.26)

Then, by (6.23), it follows that

$$||T_1 f||_{p, \psi^{1-p}} < k(\lambda_{m+1}) \prod_{i=1}^{m} ||f_i||_{p_i, \varphi_i},$$

and then, $T_1 f \in l_{p,\psi^{1-p}}$. Hence, T_1 is a bounded linear operator with $||T_1|| \leq k(\lambda_{m+1})$.

Since the constant factor in (6.23) is the best possible, we have

$$||T_1|| = \sup_{f(\neq\theta)\in\prod_{i=1}^{m} L_{p_i,\varphi_i}(\mathbf{R}_+)} \frac{||T_1 f||_{p,\psi^{1-p}}}{\prod_{i=1}^{m}||f_i||_{p_i,\varphi_i}} = k(\lambda_{m+1}). \tag{6.27}$$

6.2.4 *Some Kinds of Reverse Inequalities*

For $\varphi_1(x_1) = x_1^{p_1(1-\lambda_1)-1}$,

$$\theta_1(x_1) = \frac{1}{k(\lambda_{m+1})} \int_0^{1/x_1} u_{m+1}^{\lambda_{m+1}-1} \left[\int_0^\infty \cdots \int_0^\infty k_\lambda(1, u_2, \cdots, u_{m+1}) \right.$$
$$\left. \times \prod_{j=2}^{m} u_j^{\lambda_j-1} du_2 \cdots du_m \right] du_{m+1},$$

and for $\lambda_{m+1} = \frac{\lambda}{2}$,

$$\widetilde{\theta}_1(x_1) = \frac{1}{k(\lambda_{m+1})} \int_0^{x_1} u_1^{\lambda_1-1} \left[\int_0^\infty \cdots \int_0^\infty k_\lambda(u_1, \cdots, u_m, 1) \right.$$
$$\left. \times \prod_{j=2}^{m} u_j^{\lambda_j-1} du_2 \cdots du_m \right] du_1.$$

We also set two functions

$$\Phi_1(x_1) = (1 - \theta_1(x_1))\varphi_1(x_1),$$

and

$$\widetilde{\Phi}_1(x_1) = (1 - \widetilde{\theta}_1(x_1))\varphi_1(x_1).$$

For $p_i < 1 (p_i \neq 0)$, the spaces $L_{p_i,\varphi_i}(\mathbf{R}_+)$ and $l_{p_{m+1},\psi}$ with $||f_i||_{p_i,\varphi_i}$ and $||a||_{p_{m+1},\psi}$ are not normed spaces. But we still use them as the formal symbols in the following:

Theorem 6.3. *Suppose that* $0 < p_1 < 1$, $p_i < 0$ $(i = 2, \cdots, m+1)$, *there exists a constant* $\delta_0 > 0$, *such that, for any* $\widetilde{\lambda}_i \in (\lambda_i - \delta_0, \lambda_i + \delta_0)$ $(i = 1, \cdots, m+1)$, $\sum_{i=1}^{m+1} \widetilde{\lambda}_i = \lambda$,

$$k(\widetilde{\lambda}_{m+1}) = \int_0^\infty \cdots \int_0^\infty k_\lambda(u_1, \cdots, u_m, 1) \prod_{j=1}^{m} u_j^{\widetilde{\lambda}_j-1} du_1 \cdots du_m \in \mathbf{R}_+,$$

and $k_\lambda(x_1, \cdots, x_m, y)\, y^{\tilde{\lambda}_{m+1}-1}$ *is decreasing with respect to* $y \in \mathbf{R}_+$ *and strictly decreasing in an interval* $I \subset (1, \infty)$. *There exist constants* $\alpha, L > 0$, *such that (6.9) is satisfied for* $i = 1$, *and*

$$A_1(u_{m+1}) \le L\, u_{m+1}^{\alpha - \lambda_{m+1}} \quad (u_{m+1} \in \mathbf{R}_+).$$

If $f_1(x_1) \ge 0$, $f_1 \in L_{p_1, \Phi_1}(\mathbf{R}_+)$, $||f_1||_{p_1, \Phi_1} > 0$, $f_i(x_i) \ge 0$, $f_i \in L_{p_i, \varphi_i}(\mathbf{R}_+)$, $||f_i||_{p_i, \varphi_i} > 0$ $(i = 2, \cdots, m)$, $a_n \ge 0, a = \{a_n\}_{n=1}^\infty \in l_{p_{m+1}, \psi}$, $||a||_{p_{m+1}, \psi} > 0$, *then, we have the following equivalent inequalities with the same best possible constant factor* $k(\lambda_{m+1})$:

$$I = \sum_{n=1}^\infty a_n \int_0^\infty \cdots \int_0^\infty k_\lambda(x_1, \cdots, x_m, n) \prod_{j=1}^m f_j(x_j) dx_1 \cdots dx_m$$

$$> k(\lambda_{m+1}) ||a||_{p_{m+1}, \psi} ||f_1||_{p_1, \Phi_1} \prod_{i=2}^m ||f_i||_{p_i, \varphi_i}, \qquad (6.28)$$

and

$$J = \left\{ \sum_{n=1}^\infty n^{p\lambda_{m+1}-1} \left(\int_0^\infty \cdots \int_0^\infty k_\lambda(x_1, \cdots, x_m, n) \right. \right.$$

$$\left. \left. \times \prod_{j=1}^m f_j(x_j) dx_1 \cdots dx_m \right)^p \right\}^{\frac{1}{p}} > k(\lambda_{m+1}) ||f_1||_{p_1, \Phi_1} \prod_{i=2}^m ||f_i||_{p_i, \varphi_i}.$$

$$(6.29)$$

Proof.   By the reverse of (6.17), since

$$\varpi_1(x_1) > k(\lambda_{m+1})(1 - \theta_1(x_1)), \text{ and } \varpi_i(x_i) < k(\lambda_{m+1}),$$

in view of the assumptions made, we have (6.29).

By Hölder's inequality, we obtain

$$I = \sum_{n=1}^\infty \left[(\psi(n))^{\frac{-1}{p_{m+1}}} \int_0^\infty \cdots \int_0^\infty k_\lambda(x_1, \cdots, x_m, n) \right.$$

$$\left. \times \prod_{j=1}^m f_j(x_j) dx_1 \cdots dx_m \right] [(\psi(n))^{\frac{1}{p_{m+1}}} a_n] \ge J ||a||_{p_{m+1}, \psi}. \quad (6.30)$$

Then, by (6.29), we have inequality (6.28).

Assuming that (6.28) is valid, setting a_n as follows:

$$a_n = n^{p\lambda_{m+1}-1} \left(\int_0^\infty \cdots \int_0^\infty k_\lambda(x_1, \cdots, x_m, n) \right.$$

$$\left. \times \prod_{j=1}^m f_j(x_j) dx_1 \cdots dx_m \right)^{p-1} \quad (n \in \mathbf{N}),$$

then, we find $J^{p-1} = ||a||_{p_{m+1}, \psi}$. By the reverse of (6.17), we have $J > 0$. If $J = \infty$, then (6.29) is trivially valid; if $J < \infty$, then by (6.28), it follows that

$$||a||_{p_{m+1}, \psi}^{p_{m+1}} = J^p = I > k(\lambda_{m+1})||a||_{p_{m+1}, \psi}||f_1||_{p_1, \Phi_1} \prod_{i=2}^{m} ||f_i||_{p_i, \varphi_i}, \text{that is,}$$

$$||a||_{p_{m+1}, \psi}^{p_{m+1}-1} = J > k(\lambda_{m+1})||f_1||_{p_1, \Phi_1} \prod_{i=2}^{m} ||f_i||_{p_i, \varphi_i}.$$

Thus, inequality (6.29) follows.

Hence, (6.28) and (6.29) are equivalent.

For $0 < \varepsilon < \delta_0 \min_{1 \le j \le m+1}\{|p_j|, p\}$, we set $\widetilde{f}_i(x_i)$ and $\widetilde{a}_n$ as follows:

$$\widetilde{f}_i(x_i) = \begin{cases} 0, & 0 < x_i < 1, \\ x_i^{(\lambda_i - \frac{\varepsilon}{p_i})-1}, & x_i \ge 1 \end{cases} \quad (i = 1, \cdots, m),$$

$$\widetilde{a}_n = n^{(\lambda_{m+1} - \frac{\varepsilon}{p_{m+1}})-1}, \quad n \in \mathbf{N}.$$

If there exists a constant $k(\ge k(\lambda_{m+1}))$ such that (6.28) is valid, as we replace $k(\lambda_{m+1})$ by k, then, by (6.13) and Lemma 6.4, we have

$$k(\lambda_{m+1}) + o(1) = I(\varepsilon)$$

$$= \varepsilon \sum_{n=1}^{\infty} \widetilde{a}_n \int_0^{\infty} \cdots \int_0^{\infty} k_\lambda(x_1, \cdots, x_m, n) \prod_{j=1}^{m} \widetilde{f}_j(x_j)dx_1 \cdots dx_m$$

$$> \varepsilon k ||\widetilde{a}||_{p_{m+1}, \psi}||f_1||_{p_1, \Phi_1} \prod_{i=2}^{m} ||\widetilde{f}_i||_{p_i, \varphi_i}$$

$$= \varepsilon k \left(\sum_{n=1}^{\infty} \frac{1}{n^{1+\varepsilon}}\right)^{\frac{1}{p_{m+1}}} \left[\int_1^{\infty} \left(1 - O\left(\frac{1}{x_1^\alpha}\right)\right) \frac{dx}{x^{1+\varepsilon}}\right]^{\frac{1}{p_1}} \prod_{i=2}^{m} \left(\frac{1}{\varepsilon}\right)^{\frac{1}{p_i}}$$

$$> \varepsilon k \left(1 + \int_1^{\infty} \frac{dy}{y^{1+\varepsilon}}\right)^{\frac{1}{p_{m+1}}} \left(\frac{1}{\varepsilon} - O(1)\right)^{\frac{1}{p_1}} \prod_{i=2}^{m} \left(\frac{1}{\varepsilon}\right)^{\frac{1}{p_i}}$$

$$= k(\varepsilon + 1)^{\frac{1}{p_{m+1}}}(1 - \varepsilon O(1))^{\frac{1}{p_1}},$$

and then, $k(\lambda_{m+1}) \ge k(\varepsilon \to 0^+)$. Hence, $k = k(\lambda_{m+1})$ is the best possible constant factor of (6.28).

By the equivalency, the constant factor $k(\lambda_{m+1})$ in (6.29) is the best possible. Otherwise, it leads to a contradiction by (6.30) that the constant factor in (6.28) is not the best possible. $\qquad \square$

Theorem 6.4. *Suppose that* $0 < p_1 < 1$, $p_i < 0 (i = 2, \cdots, m+1)$, $\lambda_{m+1} = \frac{\lambda}{2}$, *there exists a constant* $\delta_0 > 0$ *such that for any* $\widetilde{\lambda}_i \in (\lambda_i - \delta_0, \lambda_i + \delta_0)(i = 1, \cdots, m+1)$, $\sum_{i=1}^{m+1} \widetilde{\lambda}_i = \lambda$,

$$k(\widetilde{\lambda}_{m+1}) = \int_0^\infty \cdots \int_0^\infty k_\lambda(u_1, \cdots, u_m, 1) \prod_{j=1}^m u_j^{\widetilde{\lambda}_j - 1} du_1 \cdots du_m \in \mathbf{R}_+,$$

and $k_\lambda(yx_1, \cdots, yx_m, 1) y^{\widetilde{\lambda}_{m+1}-1}$ *is decreasing with respect to* $y \in \mathbf{R}_+$ *and strictly decreasing in an interval* $I \subset (1, \infty)$. *There exist constants* α, $L > 0$, *such that (6.12) is satisfied for* $i = 1, \cdots, m$, *and*

$$B_1(u_i) \le L \, u_i^{\alpha - \lambda_i} \quad (u_i \in \mathbf{R}_+) \quad (i = 1, \cdots, m).$$

If $f_1(x_1) \ge 0$, $f_1 \in L_{p_1, \widetilde{\Phi}_1}(\mathbf{R}_+)$, $||f_1||_{p_1, \widetilde{\Phi}_1} > 0$, $f_i(x_i) \ge 0$, $f_i \in L_{p_i, \varphi_i}(\mathbf{R}_+)$, $||f_i||_{p_i, \varphi_i} > 0$ $(i = 2, \cdots, m)$, $a_n \ge 0$, $a = \{a_n\}_{n=1}^\infty \in l_{p_{m+1}, \psi}$, $||a||_{p_{m+1}, \psi} > 0$, *then, we have the following equivalent inequalities with the same best possible constant factor* $k(\lambda_{m+1})$:

$$\sum_{n=1}^\infty a_n \int_0^\infty \cdots \int_0^\infty k_\lambda(nx_1, \cdots, nx_m, 1) \prod_{j=1}^m f_j(x_j) dx_1 \cdots dx_m$$

$$> k(\lambda_{m+1}) ||a||_{p_{m+1}, \psi} ||f_1||_{p_1, \widetilde{\Phi}_1} \prod_{i=2}^m ||f_i||_{p_i, \varphi_i}, \tag{6.31}$$

and

$$\left\{ \sum_{n=1}^\infty n^{\frac{p\lambda}{2} - 1} \left(\int_0^\infty \cdots \int_0^\infty k_\lambda(nx_1, \cdots, nx_m, 1) \right.\right.$$

$$\left.\left. \times \prod_{j=1}^m f_j(x_j) dx_1 \cdots dx_m \right)^p \right\}^{\frac{1}{p}}$$

$$> k(\lambda_{m+1}) ||f_1||_{p_1, \widetilde{\Phi}_1} \prod_{i=2}^m ||f_i||_{p_i, \varphi_i}. \tag{6.32}$$

Proof. We only prove that the constant factor in (6.31) is the best possible. The other parts of the proof are omitted.

For $0 < \varepsilon < \delta_0 \min_{1 \le j \le m+1} \{|p_j|, p\}$, we set $\widetilde{f}_i(x_i)$ and $\widetilde{a}_n$ as follows:

$$\widetilde{f}_i(x_i) = \begin{cases} x_i^{(\lambda_i + \frac{\varepsilon}{p_i})-1}, & 0 < x_i \le 1, \\ 0, & x_i > 1 \end{cases} \quad (i = 1, \cdots, m),$$

$$\widetilde{a}_n = n^{(\frac{\lambda}{2} - \frac{\varepsilon}{p_{m+1}})-1}, \quad n \in \mathbf{N}.$$

If there exists a constant $k(\geq k(\lambda_{m+1}))$ such that (6.31) is valid as we replace $k(\lambda_{m+1})$ by k, then, by (6.15), we have

$$
k(\lambda_{m+1}) + o(1) = \widetilde{I}(\varepsilon)
$$

$$
= \varepsilon \sum_{n=1}^{\infty} \widetilde{a}_n \int_0^{\infty} \cdots \int_0^{\infty} k_\lambda(nx, \cdots, nx_m, 1) \prod_{j=1}^{m} \widetilde{f}_j(x_j) dx_1 \cdots dx_m
$$

$$
> \varepsilon k ||\widetilde{a}||_{p_{m+1}, \psi} ||\widetilde{f}_1||_{p_1, \widetilde{\Phi}_1} \prod_{i=2}^{m} ||\widetilde{f}_i||_{p_i, \varphi_i}
$$

$$
= \varepsilon k \left(\sum_{n=1}^{\infty} \frac{1}{n^{1+\varepsilon}} \right)^{\frac{1}{p_{m+1}}} \left[\int_0^1 (1 - O(x_1^\alpha)) \frac{dx}{x^{1-\varepsilon}} \right]^{\frac{1}{p_1}} \prod_{i=1}^{m} \left(\frac{1}{\varepsilon} \right)^{\frac{1}{p_i}}
$$

$$
> \varepsilon k \left(1 + \int_1^{\infty} \frac{dy}{y^{1+\varepsilon}} \right)^{\frac{1}{p_{m+1}}} \left(\frac{1}{\varepsilon} - O(1) \right)^{\frac{1}{p_1}} \prod_{i=2}^{m} \left(\frac{1}{\varepsilon} \right)^{\frac{1}{p_i}}
$$

$$
= k(\varepsilon + 1)^{\frac{1}{p_{m+1}}} (1 + \varepsilon O(1))^{\frac{1}{p_1}},
$$

and then, $k(\lambda_{m+1}) \geq k(\varepsilon \to 0^+)$. Hence, $k = k(\lambda_{m+1})$ is the best value of (6.31). $\qquad \square$

Similarly, we still have

Theorem 6.5. *Suppose that $p_i < 0$ ($i = 1, \cdots, m$), $0 < p_{m+1} < 1$, there exists a constant $\delta_0 > 0$, such that, for any $\widetilde{\lambda}_i \in (\lambda_i - \delta_0, \lambda_i + \delta_0)(i = 1, \cdots, m+1)$, $\sum_{i=1}^{m+1} \widetilde{\lambda}_i = \lambda$,*

$$
k(\widetilde{\lambda}_{m+1}) = \int_0^{\infty} \cdots \int_0^{\infty} k_\lambda(u_1, \cdots, u_m, 1) \prod_{j=1}^{m} u_j^{\widetilde{\lambda}_j - 1} du_1 \cdots du_m \in \mathbf{R}_+,
$$

and $k_\lambda(x_1, \cdots, x_m, y) \, y^{\widetilde{\lambda}_{m+1}-1}$ is strictly decreasing with respect to $y \in \mathbf{R}_+$ and strictly decreasing in an interval $I \subset (1, \infty)$. If $f_i(x_i), a_n \geq 0, f_i \in L_{p_i, \varphi_i}(\mathbf{R}_+)$ ($i = 1, \cdots, m$), $a = \{a_n\}_{n=1}^{\infty} \in l_{p_{m+1}, \psi}, ||f_i||_{p_i, \varphi_i} > 0$, $||a||_{p_{m+1}, \psi} > 0$, then, we have the following equivalent reverse inequalities with the same best possible constant factor $k(\lambda_{m+1})$:

$$
\sum_{n=1}^{\infty} a_n \int_0^{\infty} \cdots \int_0^{\infty} k_\lambda(x_1, \cdots, x_m, n) \prod_{j=1}^{m} f_j(x_j) dx_1 \cdots dx_m
$$

$$
> k(\lambda_{m+1}) ||a||_{p_{m+1}, \psi} \prod_{i=1}^{m} ||f_i||_{p_i, \varphi_i}, \tag{6.33}
$$

and

$$\left\{ \sum_{n=1}^{\infty} n^{p\lambda_{m+1}-1} \left(\int_0^{\infty} \cdots \int_0^{\infty} k_\lambda(x_1, \cdots, x_m, n) \right. \right.$$

$$\left. \left. \times \prod_{j=1}^{m} f_j(x_j) dx_1 \cdots dx_m \right)^p \right\}^{\frac{1}{p}}$$

$$> k(\lambda_{m+1}) \prod_{i=1}^{m} ||f_i||_{p_i, \varphi_i}. \tag{6.34}$$

Theorem 6.6. *Suppose that* $p_i < 0 (i = 1, \cdots, m), 0 < p_{m+1} < 1, \lambda_{m+1} = \frac{\lambda}{2}$, *there exists a constant* $\delta_0 > 0$ *such that for any* $\widetilde{\lambda}_i \in (\lambda_i - \delta_0, \lambda_i + \delta_0)$ $(i = 1, \cdots, m + 1)$, $\sum_{i=1}^{m+1} \widetilde{\lambda}_i = \lambda$,

$$k(\widetilde{\lambda}_{m+1}) = \int_0^{\infty} \cdots \int_0^{\infty} k_\lambda(u_1, \cdots, u_m, 1) \prod_{j=1}^{m} u_j^{\widetilde{\lambda}_j - 1} du_1 \cdots du_m \in \mathbf{R}_+,$$

and $k_\lambda(yx_1, \cdots, yx_m, 1) y^{\widetilde{\lambda}_{m+1}-1}$ *is decreasing with respect to* $y \in \mathbf{R}_+$ *and strictly decreasing in an interval* $I \subset (1, \infty)$. *There exist constants* $\alpha, L > 0$, *such that (6.12) is satisfied for* $i = 1, \cdots, m$, *and*

$$B_1(u_i) \le Lu_i^{\alpha - \lambda_i} \ (u_i \in \mathbf{R}_+; \quad i = 1, \cdots, m).$$

If $f_i(x_i), a_n \ge 0, f_i \in L_{p_i, \varphi_i}(\mathbf{R}_+)$ $(i = 1, \cdots, m)$, $a = \{a_n\}_{n=1}^{\infty} \in l_{p_{m+1}, \psi}$, $||f_i||_{p_i, \varphi_i} > 0$, $||a||_{p_{m+1}, \psi} > 0$, *then, we have the following equivalent inequalities with the same best possible constant factor* $k(\lambda_{m+1})$:

$$\sum_{n=1}^{\infty} a_n \int_0^{\infty} \cdots \int_0^{\infty} k_\lambda(nx_1, \cdots, nx_m, 1) \prod_{j=1}^{m} f_j(x_j) dx_1 \cdots dx_m$$

$$> k(\lambda_{m+1}) ||a||_{p_{m+1}, \psi} \prod_{i=1}^{m} ||f_i||_{p_i, \varphi_i}, \tag{6.35}$$

$$\left\{ \sum_{n=1}^{\infty} n^{\frac{p\lambda}{2}-1} \left(\int_0^{\infty} \cdots \int_0^{\infty} k_\lambda(nx_1, \cdots, nx_m, 1) \right. \right.$$

$$\left. \left. \times \prod_{j=1}^{m} f_j(x_j) dx_1 \cdots dx_m \right)^p \right\}^{\frac{1}{p}} > k(\lambda_{m+1}) \prod_{i=1}^{m} ||f_i||_{p_i, \varphi_i}. \tag{6.36}$$

6.3 Second Kind of Multiple Hilbert-type Inequalities

6.3.1 *Lemmas Related to the Weight Functions*

Definition 6.3. If $n_i \in \mathbf{N}(i = 1, \cdots, m)$, $x_{m+1} \in \mathbf{R}_+$, we define weight functions $\omega_i(n_i)$ and $\varpi_{m+1}(x_{m+1})$ as follows:

$$\omega_i(n_i) = n_i^{\lambda_i} \int_0^\infty x_{m+1}^{\lambda_{m+1}-1} \sum_{n_m=1}^\infty \cdots \sum_{n_{i+1}=1}^\infty \sum_{n_{i-1}=1}^\infty$$

$$\cdots \sum_{n_1=1}^\infty k_\lambda(n_1, \cdots, n_m, x_{m+1}) \prod_{j=1(j \neq i)}^m n_j^{\lambda_j-1} dx_{m+1}, \quad (6.37)$$

$$\varpi_{m+1}(x_{m+1}) = x_{m+1}^{\lambda_{m+1}} \sum_{n_m=1}^\infty \cdots \sum_{n_1=1}^\infty k_\lambda(n_1, \cdots, n_m, x_{m+1}) \prod_{j=1}^m n_j^{\lambda_j-1}. \quad (6.38)$$

In particular, for $\lambda_{m+1} = \frac{\lambda}{2}$, replacing the kernel $k_\lambda(n_1, \cdots, n_m, x_{m+1})$ by $k_\lambda(x_{m+1}n_1, \cdots, x_{m+1}n_m, 1)$ in (6.37) and (6.38), we can still define another weight coefficients $w_i(n_i)(i = 1, \cdots, m)$ and $\widetilde{w}_{m+1}(x_{m+1})$ as follows:

$$w_i(n_i) = n_i^{\lambda_i} \int_0^\infty x_{m+1}^{\frac{\lambda}{2}-1} \sum_{n_m=1}^\infty \cdots \sum_{n_{i+1}=1}^\infty \sum_{n_{i-1}=1}^\infty$$

$$\cdots \sum_{n_1=1}^\infty k_\lambda(x_{m+1}n_1, \cdots, x_{m+1}n_m, 1) \prod_{j=1(j \neq i)}^m n_j^{\lambda_j-1} dx_{m+1},$$

$$\widetilde{w}_{m+1}(x_{m+1}) = x_{m+1}^{\frac{\lambda}{2}} \sum_{n_m=1}^\infty \cdots \sum_{n_1=1}^\infty k_\lambda(x_{m+1}n_1, \cdots, x_{m+1}n_m, 1) \prod_{j=1}^m n_j^{\lambda_j-1}.$$

Lemma 6.10. *Let the assumptions of Definition 6.3 be fulfilled and additionally, let*

$$k(\lambda_{m+1}) = \int_0^\infty \cdots \int_0^\infty k_\lambda(u_1, \cdots, u_m, 1) \prod_{j=1}^m u_j^{\lambda_j-1} du_1 \cdots du_m \in \mathbf{R}_+,$$

$$(6.39)$$

$k_\lambda(y_1, \cdots, y_m, x_{m+1})y_j^{\lambda_j-1}$ *be decreasing with respect to* $y_j \in \mathbf{R}_+$ *and strictly decreasing in an interval* $I_j \subset (1, \infty)$ $(j = 1, \cdots, m)$. *Then, we have*

(i)

$$0 < k(\lambda_{m+1})(1 - \theta_{m+1}(x_{m+1})) < \varpi_{m+1}(x_{m+1}) < k(\lambda_{m+1}), \quad (6.40)$$

where

$$\theta_{m+1}(x_{m+1}) = 1 - \frac{1}{k(\lambda_{m+1})} \int_{1/x_{m+1}}^{\infty} \cdots \int_{1/x_{m+1}}^{\infty} k_{\lambda}(u_1, \cdots, u_m, 1)$$
$$\times \prod_{j=1}^{m} u_j^{\lambda_j - 1} du_1 \cdots du_m > 0; \quad (6.41)$$

(ii) for $m = 1$, $\omega_1(n_1) = k(\lambda_2)$; for $m \geq 2$, $i = 1, \cdots, m$, we have

$$0 < k(\lambda_{m+1})(1 - \theta_i(n_i)) < \omega_i(n_i) < k(\lambda_{m+1}) \quad (n_i \in \mathbf{N}), \quad (6.42)$$

where

$$\theta_i(n_i) = 1 - \frac{1}{k(\lambda_{m+1})} \int_{\frac{1}{n_i}}^{\infty} \cdots \int_{\frac{1}{n_i}}^{\infty} k_{\lambda}(u_1, \cdots, u_{i-1}, 1, u_{i+1}, \cdots, u_{m+1})$$
$$\times \prod_{j=1(j \neq i)}^{m+1} u_j^{\lambda_j - 1} du_1 \cdots du_{i-1} du_{i+1} \cdots du_{m+1} > 0. \quad (6.43)$$

Moreover, if there exist constants α, $L > 0$, such that for $i = 1, \cdots,$ $m + 1$,

$$A_i(u_k) = \int_0^{\infty} \cdots \int_0^{\infty} k_{\lambda}(u_1, \cdots, u_{i-1}, 1, u_{i+1}, \cdots, u_{m+1})$$
$$\times \prod_{j=1(j \neq i,k)}^{m+1} u_j^{\lambda_j - 1} du_1 \cdots du_{i-1} du_{i+1} \cdots du_{k-1} du_{k+1} \cdots du_{m+1}$$
$$\leq L u_k^{\alpha - \lambda_k} (u_k \in \mathbf{R}_+), \quad (6.44)$$

then, we have

$$\theta_{m+1}(x_{m+1}) = O\left(\frac{1}{x_{m+1}^{\alpha}}\right), \quad (x_{m+1} > 0),$$

and

$$\theta_i(n_i) = O\left(\frac{1}{n_i^{\alpha}}\right) \quad (n \in \mathbf{N}; i = 1, \cdots, m).$$

Proof.   (i) By the decreasing property, it follows that

$$\varpi_{m+1}(x_{m+1}) < x_{m+1}^{\lambda_{m+1}} \int_0^{\infty} \cdots \int_0^{\infty} k_{\lambda}(x_1, \cdots, x_m, x_{m+1})$$
$$\times \prod_{j=1}^{m} x_j^{\lambda_j - 1} dx_1 \cdots dx_m.$$

Setting $x_j = x_{m+1}u_j (j = 1, \cdots, m)$ in the above integral, by (6.39), we find

$$\varpi_{m+1}(x_{m+1}) < x_{m+1}^{\lambda_{m+1}} x_{m+1}^m \int_0^\infty \cdots \int_0^\infty k_\lambda(x_{m+1}u_1, \cdots, x_{m+1}u_m, x_{m+1})$$

$$\times \prod_{j=1}^m (x_{m+1}u_j)^{\lambda_j-1} du_1 \cdots du_m = k(\lambda_{m+1}).$$

By the decreasing property, we still have

$$\varpi_{m+1}(x_{m+1}) > x_{m+1}^{\lambda_{m+1}} \int_1^\infty \cdots \int_1^\infty k_\lambda(x_1, \cdots, x_m, x_{m+1})$$

$$\times \prod_{j=1}^m x_j^{\lambda_j-1} dx_1 \cdots dx_m$$

$$= \int_{\frac{1}{x_{m+1}}}^\infty \cdots \int_{\frac{1}{x_{m+1}}}^\infty k_\lambda(u_1, \cdots, u_m, 1) \prod_{j=1}^m u_j^{\lambda_j-1} du_1 \cdots du_m$$

$$= k(\lambda_{m+1})(1 - \theta_{m+1}(x_{m+1})) > 0,$$

where $\theta_{m+1}(x_{m+1})$ is defined by (6.41).

(ii) For $m = 1$, we find

$$\omega_1(n_1) = n_1^{\lambda_1} \int_0^\infty k_\lambda(n_1, x_2) x_2^{\lambda_2-1} dx_2$$

$$= \int_0^\infty k_\lambda(u_1, 1) u_1^{\lambda_1-1} du_1 = k(\lambda_2);$$

for $m \geq 2$, $i = 1, \cdots, m$, by the decreasing property and Lemma 6.2, we have

$$\omega_i(n_i) < n_i^{\lambda_i} \int_0^\infty \cdots \int_0^\infty k_\lambda(x_1, \cdots, x_{i-1}, n_i, x_{i+1}, \cdots, x_m, x_{m+1})$$

$$\times \prod_{j=1(j\neq i)}^{m+1} x_j^{\lambda_j-1} dx_1 \cdots dx_{i-1} dx_{i+1} \cdots dx_{m+1}, \ u_j = x_j/n_i (j \neq i)$$

$$= H(i) = k(\lambda_{m+1}),$$

$$\omega_i(n_i) > n_i^{\lambda_i} \int_1^\infty \cdots \int_1^\infty k_\lambda(x_1, \cdots, x_{i-1}, n_i, x_{i+1}, \cdots, x_m, x_{m+1})$$

$$\times \prod_{j=1(j\neq i)}^{m+1} x_j^{\lambda_j-1} dx_1 \cdots dx_{i-1} dx_{i+1} \cdots dx_{m+1}$$

$$= \int_{1/n_i}^{\infty} \cdots \int_{1/n_i}^{\infty} k_\lambda(u_1, \cdots, u_{i-1}, 1, u_{i+1}, \cdots, u_{m+1})$$

$$\times \prod_{j=1(j \neq i)}^{m+1} u_j^{\lambda_j - 1} du_1 \cdots du_{i-1} du_{i+1} \cdots du_{m+1}$$

$$= k(\lambda_{m+1})(1 - \theta_i(n_i)) > 0,$$

where $\theta_i(n_i)$ is given by (6.43).

Moreover, we have

$$0 < \theta_{m+1}(x_{m+1}) = \frac{1}{k(\lambda_{m+1})} \left[k(\lambda_{m+1}) \right.$$

$$\left. - \int_{1/x_{m+1}}^{\infty} \cdots \int_{1/x_{m+1}}^{\infty} k_\lambda(u_1, \cdots, u_m, 1) \prod_{j=1}^{m} u_j^{\lambda_j - 1} du_1 \cdots du_m \right]$$

$$\leq \frac{1}{k(\lambda_{m+1})} \sum_{k=1}^{m} \int_0^{1/x_{m+1}} u_k^{\lambda_k - 1} A_{m+1}(u_k) du_k,$$

where

$$A_{m+1}(u_k) = \int_0^{\infty} \cdots \int_0^{\infty} k_\lambda(u_1, \cdots, u_m, 1)$$

$$\times \prod_{j=1(j \neq k)}^{m} u_j^{\lambda_j - 1} du_1 \cdots du_{k-1} du_{k+1} \cdots du_m.$$

By (6.44) (for $i = m+1$), it follows that

$$0 < \theta_{m+1}(x_{m+1}) \leq \frac{L}{k(\lambda_{m+1})} \sum_{k=1}^{m} \int_0^{1/x_{m+1}} u_k^{\lambda_k - 1} u_k^{\alpha - \lambda_k} du_k$$

$$= \frac{mL}{\alpha k(\lambda_{m+1}) x_{m+1}^{\alpha}},$$

and then

$$\theta_{m+1}(x_{m+1}) = O\left(\frac{1}{x_{m+1}^{\alpha}} \right) \qquad (x_{m+1} > 0).$$

By (6.43), (6.44) and the same way, we have $\theta_i(n_i) = O\left(\frac{1}{n_i^{\alpha}} \right)$ $(n \in \mathbf{N}; i = 1, \cdots, m)$. $\qquad\square$

Lemma 6.11. *Let the assumptions of Definition 6.3 be fulfilled and additionally, let $\lambda_{m+1} = \frac{\lambda}{2}$, $k(\lambda_{m+1}) \in \mathbf{R}_+$, and*

$$k_\lambda(x_{m+1} y_1, \cdots, x_{m+1} y_m, 1) y_j^{\lambda_{m+1} - 1}$$

be decreasing with respect to $y_j \in \mathbf{R}_+$ and strictly decreasing in an interval $I_j \subset (1, \infty)$ $(j = 1, \cdots, m)$. Then

(i) we have

$$0 < k(\lambda_{m+1})(1 - \widetilde{\theta}_{m+1}(x_{m+1})) < \widetilde{w}_{m+1}(x_{m+1}) < k(\lambda_{m+1}), \qquad (6.45)$$

where

$$\widetilde{\theta}_{m+1}(x_{m+1}) = 1 - \frac{1}{k(\lambda_{m+1})} \int_{x_{m+1}}^{\infty} \cdots \int_{x_{m+1}}^{\infty} k_\lambda(u_1, \cdots, u_m, 1)$$
$$\times \prod_{j=1}^{m} u_j^{\lambda_j - 1} du_1 \cdots du_m > 0; \quad (6.46)$$

(ii) for $m = 1$, $w_1(n_1) = k(\lambda_2)$; for $m \geq 2$, $i = 1, \cdots, m$, we have

$$0 < k(\lambda_{m+1})(1 - \widetilde{\theta}_i(n_i)) < w_i(n_i) < k(\lambda_{m+1}) \ (n_i \in \mathbf{N}), \qquad (6.47)$$

where

$$\widetilde{\theta}_i(n_i) = 1 - \frac{1}{k(\lambda_{m+1})} \int_{\frac{1}{n_i}}^{\infty} \cdots \int_{\frac{1}{n_i}}^{\infty} k_\lambda(u_1, \cdots, u_{i-1}, 1, u_{i+1}, \cdots, u_{m+1})$$
$$\times \prod_{j=1(j\neq i)}^{m+1} u_j^{\lambda_j - 1} du_1 \cdots du_{i-1} du_{i+1} \cdots du_{m+1} > 0. \qquad (6.48)$$

Moreover, if there exist constants $\alpha, L > 0$, such that for $i = 1, \cdots, m + 1$, inequality (6.44) follows. Thus,

$$A_i(u_k) \leq L u_k^{\alpha - \lambda_k} \ (u_k \in \mathbf{R}_+),$$

then, we have $\widetilde{\theta}_{m+1}(x_{m+1}) = O(x_{m+1}^{\alpha}) \ (x_{m+1} > 0)$, and

$$\widetilde{\theta}_i(n_i) = O(n_i^{-\alpha}) \ (n \in \mathbf{N}; i = 1, \cdots, m).$$

Proof. (i) By the decreasing property, it follows that

$$\widetilde{w}_{m+1}(x_{m+1}) < x_{m+1}^{\frac{\lambda}{2}}$$
$$\times \int_0^{\infty} \cdots \int_0^{\infty} k_\lambda(x_{m+1}x_1, \cdots, x_{m+1}x_m, 1) \prod_{j=1}^{m} x_j^{\lambda_j - 1} dx_1 \cdots dx_m.$$

Setting $u_j = x_{m+1}x_j$ $(j = 1, \cdots, m)$ in the above integral, we find

$$\widetilde{w}_{m+1}(x_{m+1}) < \int_0^\infty \cdots \int_0^\infty k_\lambda(u_1, \cdots, u_m, 1)$$

$$\times \prod_{j=1}^m u_j^{\lambda_j - 1} du_1 \cdots du_m = k(\lambda_{m+1}),$$

$$\widetilde{w}_{m+1}(x_{m+1}) > x_{m+1}^{\frac{\lambda}{2}} \int_1^\infty \cdots \int_1^\infty k_\lambda(x_{m+1}x_1, \cdots, x_{m+1}x_m, 1)$$

$$\times \prod_{j=1}^m x_j^{\lambda_j - 1} dx_1 \cdots dx_m$$

$$= \int_{x_{m+1}}^\infty \cdots \int_{x_{m+1}}^\infty k_\lambda(u_1, \cdots, u_m, 1) \prod_{j=1}^m u_j^{\lambda_j - 1} du_1 \cdots du_m$$

$$= k(\lambda_{m+1})(1 - \widetilde{\theta}_{m+1}(x_{m+1})) > 0,$$

where $\widetilde{\theta}_{m+1}(x_{m+1})$ is given by (6.46).

(ii) For $m = 1$, since $\lambda_2 = \frac{\lambda}{2} = \lambda_1$, we find

$$w_1(n_1) = n_1^{\lambda_1} \int_0^\infty k_\lambda(x_2 n_1, 1) x_2^{\lambda_2 - 1} dx_2$$

$$= \int_0^\infty k_\lambda(u_1, 1) u_1^{\lambda_1 - 1} du_1 = k(\lambda_2);$$

for $m \geq 2$, $i = 1, \cdots, m$, by the decreasing property, we find

$$w_i(n_i) < n_i^{\lambda_i} \int_0^\infty \cdots \int_0^\infty w_i(n_i) < n_i^{\lambda_i} \int_0^\infty \cdots \int_0^\infty$$

$$\times k_\lambda(x_{m+1}x_1, \cdots, x_{m+1}x_{i-1}, x_{m+1}n_i, x_{m+1}x_{i+1}, \cdots, x_{m+1}x_m, 1)$$

$$\times \prod_{j=1(j \neq i)}^{m+1} x_j^{\lambda_j - 1} dx_1 \cdots dx_{i-1} dx_{i+1} \cdots dx_{m+1}$$

$$= \int_0^\infty \cdots \int_0^\infty k_\lambda(u_1, \cdots u_m, 1) \prod_{j=1}^m u_j^{\lambda_j - 1} du_1 \cdots du_m = k(\lambda_{m+1}),$$

$$w_i(n_i) > n_i^{\lambda_i} \int_1^\infty \cdots \int_1^\infty$$

$$\times k_\lambda(x_{m+1}x_1, \cdots, x_{m+1}x_{i-1}, x_{m+1}n_i, x_{m+1}x_{i+1}, \cdots, x_{m+1}x_m, 1)$$

$$\times \prod_{j=1(j \neq i)}^{m+1} x_j^{\lambda_j - 1} dx_1 \cdots dx_{i-1} dx_{i+1} \cdots dx_{m+1}. \qquad (6.49)$$

Setting $u_j = x_j/n_i$ $(j \neq i, m+1)$, $u_{m+1} = (x_{m+1}n_i)^{-1}$ in (6.49), it follows that

$$w_i(n_i) > \int_{1/n_i}^{\infty} \cdots \int_{1/n_i}^{\infty} k_\lambda(u_1, \cdots, u_{i-1}, 1, u_{i+1}, \cdots, u_{m+1})$$

$$\times \prod_{j=1(j\neq i)}^{m+1} u_j^{\lambda_j-1} du_1 \cdots du_{i-1} du_{i+1} \cdots du_{m+1}$$

$$= k(\lambda_{m+1})(1 - \widetilde{\theta}_i(x_i)) > 0,$$

where $\widetilde{\theta}_i(x_i)$ is given by (6.48).

Moreover, by inequality (6.44), we have

$$0 < \widetilde{\theta}_{m+1}(x_{m+1}) = \frac{1}{k(\lambda_{m+1})} \left[k(\lambda_{m+1}) \right.$$

$$\left. - \int_{x_{m+1}}^{\infty} \cdots \int_{x_{m+1}}^{\infty} k_\lambda(u_1, \cdots, u_m, 1) \prod_{j=1}^{m} u_j^{\lambda_j-1} du_1 \cdots du_m \right]$$

$$\leq \frac{1}{k(\lambda_{m+1})} \sum_{k=1}^{m} \int_0^{x_{m+1}} u_k^{\lambda_k-1} A_{m+1}(u_k) du_k,$$

where

$$A_{m+1}(u_k) = \int_0^{\infty} \cdots \int_0^{\infty} k_\lambda(u_1, \cdots, u_m, 1)$$

$$\times \prod_{j=1(j\neq k)}^{m} u_j^{\lambda_j-1} du_1 \cdots du_{k-1} du_{k+1} \cdots du_m.$$

By inequality (6.44) (for $i = m+1$), it follows that

$$0 < \widetilde{\theta}_{m+1}(x_{m+1})$$

$$\leq \frac{L}{k(\lambda_{m+1})} \sum_{k=1}^{m} \int_0^{x_{m+1}} u_k^{\lambda_k-1} u_k^{\alpha-\lambda_k} du_k = \frac{mLx_{m+1}^{\alpha}}{\alpha k(\lambda_{m+1})},$$

and then

$$\widetilde{\theta}_{m+1}(x_{m+1}) = O\left(x_{m+1}^{\alpha}\right) \ (x_{m+1} > 0).$$

By (6.48), (6.44) and the same way, we have $\widetilde{\theta}_i(n_i) = O(n_i^{-\alpha})$ $(n \in \mathbf{N}; i = 1, \cdots, m)$. □

Lemma 6.12. *Suppose that there exists a constant $\delta_0 > 0$, such that for any $\widetilde{\lambda}_i \in (\lambda_i - \delta_0, \lambda_i + \delta_0)(i = 1, \cdots, m+1)$, $\sum_{i=1}^{m+1} \widetilde{\lambda}_i = \lambda$, $k(\widetilde{\lambda}_{m+1}) \in \mathbf{R}_+$. If*

$$k_\lambda(y_1, \cdots, y_m, x_{m+1})y_j^{\widetilde{\lambda}_{m+1}-1}$$

is decreasing with respect to $y_j \in \mathbf{R}_+$ $(j = 1, \cdots, m)$, then, for $0 < \varepsilon < \delta_0 \min_{1 \le j \le m+1}\{|p_j|, |p|\}$, we have

$$I_1(\varepsilon) = \varepsilon \int_1^\infty x_{m+1}^{\lambda_{m+1}-\frac{\varepsilon}{p_{m+1}}-1} \sum_{n_m=1}^\infty \cdots \sum_{n_1=1}^\infty k_\lambda(n_1, \cdots, n_m, x_{m+1})$$
$$\times \prod_{j=1}^m n_j^{(\lambda_j-\frac{\varepsilon}{p_j})-1} dx_{m+1} = k(\lambda_{m+1}) + o(1) \quad (\varepsilon \to 0^+). \quad (6.50)$$

Proof. By the decreasing property, we find

$$I_1(\varepsilon) \ge \varepsilon \int_1^\infty x_{m+1}^{\lambda_{m+1}-\frac{\varepsilon}{p_{m+1}}-1} \int_1^\infty \cdots \int_1^\infty k_\lambda(x_1, \cdots, x_m, x_{m+1})$$
$$\times \prod_{j=1}^m x_j^{(\lambda_j-\frac{\varepsilon}{p_j})-1} dx_1 \cdots dx_m dx_{m+1}.$$

In view of the proof of Lemma 6.6, we have

$$I_1(\varepsilon) \ge k\left(\lambda_{m+1} + \frac{\varepsilon}{p}\right) - \varepsilon O(1).$$

Since we have

$$I_1(\varepsilon) \le \varepsilon \int_1^\infty x_{m+1}^{-\varepsilon-1}\left[x_{m+1}^{\lambda_{m+1}+\varepsilon-\frac{\varepsilon}{p_{m+1}}} \int_0^\infty \cdots \int_0^\infty k_\lambda(x_1, \cdots, x_m, x_{m+1})\right.$$
$$\left. \times \prod_{j=1}^m x_j^{(\lambda_j-\frac{\varepsilon}{p_j})-1} dx_1 \cdots dx_m\right] dx_{m+1}$$
$$= \varepsilon \int_1^\infty x_{m+1}^{-\varepsilon-1} k\left(\lambda_{m+1} + \frac{\varepsilon}{p}\right) dx_{m+1} = k\left(\lambda_{m+1} + \frac{\varepsilon}{p}\right),$$

then, by (6.4), we have (6.50). $\qquad \square$

Lemma 6.13. *Suppose that* $\lambda_{m+1} = \frac{\lambda}{2}$, *and there exists a constant* $\delta_0 > 0$, *such that, for any* $\widetilde{\lambda}_i \in (\lambda_i - \delta_0, \lambda_i + \delta_0)(i = 1, \cdots, m + 1)$, $\sum_{i=1}^{m+1} \widetilde{\lambda}_i = \lambda$, $k(\widetilde{\lambda}_{m+1}) \in \mathbf{R}_+$. *If*

$$k_\lambda(x_{m+1}y_1, \cdots, x_{m+1}y_m, 1) y_j^{\widetilde{\lambda}_{m+1}-1}$$

is decreasing with respect to $y_j \in \mathbf{R}_+$ $(j = 1, \cdots, m)$, *then, for*

$$0 < \varepsilon < \delta_0 \min_{1 \le j \le m+1}\{|p_j|, |p|\},$$

we have

$$\tilde{I}_1(\varepsilon) = \varepsilon \int_0^1 x_{m+1}^{\frac{\lambda}{2}+\frac{\varepsilon}{p_{m+1}}-1}$$

$$\times \sum_{n_m=1}^{\infty} \cdots \sum_{n_1=1}^{\infty} k_\lambda(x_{m+1}n_1, \cdots, x_{m+1}n_m, 1)$$

$$\times \prod_{j=1}^{m} n_j^{(\lambda_j-\frac{\varepsilon}{p_j})-1} dx_{m+1} = k(\lambda_{m+1}) + o(1)(\varepsilon \to 0^+). \quad (6.51)$$

Proof. By the decreasing property, we find

$$\tilde{I}_1(\varepsilon) \geq \varepsilon \int_0^1 x_{m+1}^{(\frac{\lambda}{2}+\frac{\varepsilon}{p_{m+1}})-1} \left[\int_1^{\infty} \cdots \int_1^{\infty} k_\lambda(x_{m+1}x_1, \cdots, x_{m+1}x_m, 1) \right.$$

$$\left. \times \prod_{j=1}^{m} x_j^{(\lambda_j-\frac{\varepsilon}{p_j})-1} dx_1 \cdots dx_m \right] dx_{m+1}, \quad (u_j = x_{m+1}x_j (j=1, \cdots, m))$$

$$= \varepsilon \int_0^1 x_{m+1}^{\varepsilon-1} \left[\int_{x_{m+1}}^{\infty} \cdots \int_{x_{m+1}}^{\infty} k_\lambda(u_1, \cdots, u_m, 1) \right.$$

$$\left. \times \prod_{j=1}^{m} u_j^{(\lambda_j-\frac{\varepsilon}{p_j})-1} du_1 \cdots du_m \right] dx_{m+1}$$

$$\geq \varepsilon \int_0^1 x_{m+1}^{\varepsilon-1} \left[\int_0^{\infty} \cdots \int_0^{\infty} k_\lambda(u_1, \cdots, u_m, 1) \right.$$

$$\left. \times \prod_{j=1}^{m} u_j^{(\lambda_j-\frac{\varepsilon}{p_j})-1} du_1 \cdots du_m \right] dx_{m+1} - \varepsilon \sum_{i=1}^{m} \int_0^1 x_{m+1}^{-1} B_i(x_{m+1}) dx_{m+1}$$

$$= k\left(\lambda_{m+1}+\frac{\varepsilon}{p}\right) - \varepsilon \sum_{i=1}^{m} \int_0^1 x_{m+1}^{-1} B_i(x_{m+1}) dx_{m+1}, \quad (6.52)$$

where

$$B_i(x_{m+1}) = \int_0^{\infty} \cdots \int_0^{\infty} \int_0^{x_{m+1}} \int_0^{\infty} \cdots \int_0^{\infty} k_\lambda(u_1, \cdots, u_m, 1)$$

$$\times \prod_{j=1}^{m} u_j^{(\lambda_j-\frac{\varepsilon}{p_j})-1} du_1 \cdots du_{i-1} du_i du_{i+1} \cdots du_m \quad (1 \leq i \leq m).$$

For $i = 1, \cdots, m$, without loss of generality, we estimate

$\int_1^\infty x_{m+1}^{-1} B_m(x_{m+1})\, dx_{m+1}$ as follows:

$$0 < \int_0^1 x_{m+1}^{-1} B_m(x_{m+1}) dx_{m+1}$$

$$= \int_0^1 x_{m+1}^{-1} \left[\int_0^{x_{m+1}} \int_0^\infty \cdots \int_0^\infty k_\lambda(u_1, \cdots, u_m, 1) \right.$$

$$\left. \times \prod_{j=1}^m u_j^{(\lambda_j - \frac{\varepsilon}{p_j})-1} du_1 \cdots du_{m-1} du_m \right] dx_{m+1}$$

$$= \int_0^1 \left(\int_{u_m}^1 x_{m+1}^{-1} dx_{m+1} \right) \int_0^\infty \cdots \int_0^\infty k_\lambda(u_1, \cdots, u_m, 1)$$

$$\times \prod_{j=1}^m u_j^{(\lambda_j - \frac{\varepsilon}{p_j})-1} du_1 \cdots du_{m-1} du_m$$

$$= \int_0^1 (-\ln u_m) \int_0^\infty \cdots \int_0^\infty k_\lambda(u_1, \cdots, u_m, 1)$$

$$\times \prod_{j=1}^m u_j^{(\lambda_j - \frac{\varepsilon}{p_j})-1} du_1 \cdots du_{m-1} du_m.$$

Setting $\alpha > 0$ such that

$$\max \left\{ \alpha + \frac{\varepsilon}{|p_m|}, \alpha + \frac{\varepsilon}{|p|} \right\} < \delta_0,$$

since $\lim_{u_m \to 0^+} u_m^\alpha (-\ln u_m) = 0$, there exists a constant $M_m > 0$, such that

$$0 < u_m^\alpha (-\ln u_m) \leq M_m (u_m \in (0, 1]).$$

We find

$$0 < \int_0^1 x_{m+1}^{-1} B_m(x_{m+1}) \, dx_{m+1}$$

$$\leq M_m \int_0^\infty \int_0^\infty \cdots \int_0^\infty k_\lambda(u_1, \cdots, u_m, 1)$$

$$\times \prod_{j=1}^{m-1} u_j^{(\lambda_j - \frac{\varepsilon}{p_j})-1} u_m^{(\lambda_m - \frac{\alpha p_m + \varepsilon}{p_m})-1} du_1 \cdots du_{m-1} du_m$$

$$= k\left(\lambda_{m+1} - \alpha + \frac{\varepsilon}{p} \right) < \infty.$$

Hence, by (6.52), we have

$$\tilde{I}_1(\varepsilon) \geq k\left(\lambda_{m+1} + \frac{\varepsilon}{p} \right) - \varepsilon O(1). \tag{6.53}$$

We still have

$$\widetilde{I}_1(\varepsilon) \leq \varepsilon \int_0^1 x_{m+1}^{\frac{\lambda}{2}+\frac{\varepsilon}{p_{m+1}}-1} \left[\int_0^\infty \cdots \int_0^\infty k_\lambda(x_{m+1}x_1, \cdots, x_{m+1}x_m, 1) \right.$$

$$\left. \times \prod_{j=1}^m x_j^{(\lambda_j-\frac{\varepsilon}{p_j})-1} dx_1 \cdots dx_m \right] dx_{m+1}, \ u_j = x_{m+1}x_j (j=1, \cdots, m)$$

$$= \int_0^\infty \cdots \int_0^\infty k_\lambda(u_1, \cdots, u_m, 1)$$

$$\times \prod_{j=1}^m u_j^{(\lambda_j-\frac{\varepsilon}{p_j})-1} du_1 \cdots du_m$$

$$= k \left(\lambda_{m+1} + \frac{\varepsilon}{p} \right).$$

Then, by (6.53) and (6.4), we have (6.51). □

6.3.2 *Two Preliminary Inequalities*

Lemma 6.14. *If $a_{n_j}^{(j)} \geq 0$ $(n_j \in \mathbf{N}; j = 1, \cdots, m)$, then*

(i) *for $p_i > 1(i = 1, \cdots, m+1)$, we have the following inequality:*

$$J_1 = \left\{ \int_0^\infty \frac{x_{m+1}^{p\lambda_{m+1}-1}}{[\omega_{m+1}(x_{m+1})]^{p-1}} \left(\sum_{n_m=1}^\infty \cdots \sum_{n_1=1}^\infty k_\lambda(n_1, \cdots, n_m, x_{m+1}) \right. \right.$$

$$\left. \left. \times \prod_{j=1}^m a_{n_j}^{(j)} \right)^p dx_{m+1} \right\}^{\frac{1}{p}}$$

$$\leq \prod_{i=1}^m \left\{ \sum_{n_i=1}^\infty \omega_i(n_i) n_i^{p_i(1-\lambda_i)-1} (a_{n_i}^{(i)})^{p_i} \right\}^{\frac{1}{p_i}} ; \tag{6.54}$$

(ii) *for $0 < p_1 < 1, p_i < 0$ $(i = 2, \cdots, m+1)$, or for $p_i < 0$ $(i = 1, \cdots, m)$, $0 < p_{m+1} < 1$, we have the reverse of (6.54).*

Proof. (i) For $p_i > 1$ $(i = 1, \cdots, m+1)$, by Hölder's inequality, (6.37) and (6.38), we have

$$\left(\sum_{n_m=1}^\infty \cdots \sum_{n_1=1}^\infty k_\lambda(n_1, \cdots, n_m, x_{m+1}) \prod_{j=1}^m a_{n_j}^{(j)} \right)^p$$

$$= \left\{ \sum_{n_m=1}^\infty \cdots \sum_{n_1=1}^\infty k_\lambda(n_1, \cdots, n_m, x_{m+1}) \right.$$

$$\times \prod_{i=1}^{m} \left[n_i^{(\lambda_i-1)(1-p_i)} x_{m+1}^{\lambda_{m+1}-1} \prod_{j=1(j\neq i)}^{m} n_j^{\lambda_j-1} \right]^{\frac{1}{p_i}} a_{n_i}^{(i)}$$

$$\times \left[x_{m+1}^{(\lambda_{m+1}-1)(1-p_{m+1})} \prod_{j=1}^{m} n_j^{\lambda_j-1} \right]^{\frac{1}{p_{m+1}}} \Bigg\}^{p}$$

$$\leq \sum_{n_m=1}^{\infty} \cdots \sum_{n_1=1}^{\infty} k_\lambda(n_1,\cdots,n_m,x_{m+1})$$

$$\times \prod_{i=1}^{m} \left[n_i^{(\lambda_i-1)(1-p_i)} x_{m+1}^{\lambda_{m+1}-1} \prod_{j=1(j\neq i)}^{m} n_j^{\lambda_j-1} \right]^{\frac{p}{p_i}} (a_{n_i}^{(i)})^p \Bigg\{ x_{m+1}^{p_{m+1}(1-\lambda_{m+1})-1} $$

$$\times \left[x_{m+1}^{\lambda_{m+1}} \sum_{n_m=1}^{\infty} \cdots \sum_{n_1=1}^{\infty} k_\lambda(n_1,\cdots,n_m,x_{m+1}) \prod_{j=1}^{m} n_j^{\lambda_j-1} \right]^{\frac{p}{p_{m+1}}} \Bigg\}$$

$$= \frac{[\omega_{m+1}(x_{m+1})]^{p-1}}{x_{m+1}^{p\lambda_{m+1}-1}} \Bigg\{ \sum_{n_m=1}^{\infty} \cdots \sum_{n_1=1}^{\infty} k_\lambda(n_1,\cdots,n_m,x_{m+1})$$

$$\times \prod_{i=1}^{m} \left[n_i^{(\lambda_i-1)(1-p_i)} x_{m+1}^{\lambda_{m+1}-1} \prod_{j=1(j\neq i)}^{m} n_j^{\lambda_j-1} \right]^{\frac{p}{p_i}} (a_{n_i}^{(i)})^p \Bigg\}.$$

Then, by the Lebesgue term by term integration theorem, we obtain

$$J \leq \Bigg\{ \int_0^{\infty} \sum_{n_m=1}^{\infty} \cdots \sum_{n_1=1}^{\infty} k_\lambda(n_1,\cdots,n_m,x_{m+1})$$

$$\times \prod_{i=1}^{m} \left[n_i^{(\lambda_i-1)(1-p_i)} x_{m+1}^{\lambda_{m+1}-1} \prod_{j=1(j\neq i)}^{m} n_j^{\lambda_j-1} \right]^{\frac{p}{p_i}} (a_{n_i}^{(i)})^p dx_m \Bigg\}^{\frac{1}{p}}$$

$$= \Bigg\{ \sum_{n_m=1}^{\infty} \cdots \sum_{n_1=1}^{\infty} \int_0^{\infty} k_\lambda(n_1,\cdots,n_m,x_{m+1})$$

$$\times \prod_{i=1}^{m} \left[n_i^{(\lambda_i-1)(1-p_i)} x_{m+1}^{\lambda_{m+1}-1} \prod_{j=1(j\neq i)}^{m} n_j^{\lambda_j-1} \right]^{\frac{p}{p_i}} dx_m (a_{n_i}^{(i)})^p \Bigg\}^{\frac{1}{p}}.$$

Since $\sum_{i=1}^m \frac{1}{(p_i/p)} = 1$, still by Hölder's inequality, it follows that

$$J_1 \le \prod_{i=1}^m \left\{ \sum_{n_i=1}^\infty \left[\sum_{n_m=1}^\infty \cdots \sum_{n_{i+1}=1}^\infty \sum_{n_{i-1}=1}^\infty \cdots \sum_{n_1=1}^\infty \int_0^\infty k_\lambda(n_1,\cdots,n_m,x_{m+1}) \right. \right.$$

$$\left. \left. \times n_i^{(\lambda_i-1)(1-p_i)} x_{m+1}^{\lambda_{m+1}-1} \prod_{j=1(j\ne i)}^m n_j^{\lambda_j-1} dx_m \right] (a_{n_i}^{(i)})^{p_i} \right\}^{\frac{1}{p_i}}$$

$$= \prod_{i=1}^m \left\{ \sum_{n_i=1}^\infty \omega_i(n_i) n_i^{p_i(1-\lambda_i)-1} (a_{n_i}^{(i)})^{p_i} \right\}^{\frac{1}{p_i}}.$$

Hence, inequality (6.54) follows.

(ii) For $0 < p_1 < 1$, $p_i < 0$ $(i = 2,\cdots,m+1)$, or for $p_i < 0$ $(i = 1,\cdots,m)$, $0 < p_{m+1} < 1$, in view of the assumptions and by the same way, we obtain the reverse of (6.54). □

In particular, for $\lambda_{m+1} = \frac{\lambda}{2}$, we still have

Lemma 6.15. *If $\lambda_{m+1} = \frac{\lambda}{2}$, $a_{n_j}^{(j)} \ge 0$ $(n_j \in \mathbf{N}; j = 1,\cdots,m)$, then*

(i) for $p_i > 1$ $(i = 1,\cdots,m+1)$, we have the following inequality:

$$J_2 = \left\{ \int_0^\infty \frac{x_{m+1}^{\frac{p\lambda}{2}-1}}{[\varpi_{m+1}(x_{m+1})]^{p-1}} \right.$$

$$\left. \times \left(\sum_{n_m=1}^\infty \cdots \sum_{n_1=1}^\infty k_\lambda(x_{m+1}n_1,\cdots,x_{m+1}n_m,1) \prod_{j=1}^m a_{n_j}^{(j)} \right)^p dx_{m+1} \right\}^{\frac{1}{p}}$$

$$\le \prod_{i=1}^m \left\{ \sum_{n_i=1}^\infty \varpi_i(n_i) n_i^{p_i(1-\lambda_i)-1} (a_{n_i}^{(i)})^{p_i} \right\}^{\frac{1}{p_i}}; \quad (6.55)$$

(ii) for $0 < p_1 < 1$, $p_i < 0$ $(i = 2,\cdots,m+1)$, or for $p_i < 0$ $(i = 1,\cdots,m)$, $0 < p_{m+1} < 1$, we have the reverse of (6.55).

6.3.3 *Main Results and Operator Expressions*

In the following, we set the functions:

$$\psi_i(n_i) = n_i^{p_i(1-\lambda_i)-1} \ (n_i \in \mathbf{N}; i = 1,\cdots,m),$$

and $\varphi(x_{m+1}) = x_{m+1}^{p_{m+1}(1-\lambda_{m+1})-1}(x_{m+1} \in \mathbf{R}_+)$. The spaces $L_{p_{m+1},\varphi}(\mathbf{R}_+)$ and l_{p_i,ψ_i} with the norms $||f||_{p_{m+1},\varphi}$ and $||a||_{p_i,\psi_i}$ are defined in Chapter 3.

Theorem 6.7. *Suppose that $p_i > 1$ $(i = 1, \cdots, m + 1)$, there exists a constant $\delta_0 > 0$ such that for any $\widetilde{\lambda}_i \in (\lambda_i - \delta_0, \lambda_i + \delta_0)$ $(i = 1, \cdots, m+1)$, $\sum_{i=1}^{m+1} \widetilde{\lambda}_i = \lambda$,*

$$k(\widetilde{\lambda}_{m+1}) = \int_0^\infty \cdots \int_0^\infty k_\lambda(u_1, \cdots, u_m, 1) \prod_{j=1}^m u_j^{\widetilde{\lambda}_j - 1} du_1 \cdots du_m \in \mathbf{R}_+,$$

and $k_\lambda(y_1, \cdots, y_m, x_{m+1}) \, y_j^{\lambda_j - 1}$ is decreasing with respect to $y_j \in \mathbf{R}_+$ and strictly decreasing in an interval $I_j \subset (1, \infty)(j = 1, \cdots, m)$. If $f(x_{m+1}) \geq 0$, $f \in L_{p_{m+1}, \varphi}(\mathbf{R}_+), \|f\|_{p_{m+1}, \varphi} > 0, a_{n_i}^{(i)} \geq 0, a^{(i)} = \{a_{n_i}^{(i)}\}_{n_i=1}^\infty \in l_{p_i, \psi_i}, \|a^{(i)}\|_{p_i, \psi_i} > 0$ $(i = 1, \cdots, m)$, then, we have the following equivalent inequalities:

$$I = \int_0^\infty f(x_{m+1}) \sum_{n_m=1}^\infty \cdots \sum_{n_1=1}^\infty k_\lambda(n_1, \cdots, n_m, x_{m+1}) \prod_{j=1}^m a_{n_j}^{(j)} dx_{m+1}$$

$$< k(\lambda_{m+1}) \|f\|_{p_{m+1}, \varphi} \prod_{i=1}^m \|a^{(i)}\|_{p_i, \psi_i}, \tag{6.56}$$

$$J = \left\{ \int_0^\infty x_{m+1}^{p\lambda_{m+1}-1} \left(\sum_{n_m=1}^\infty \cdots \sum_{n_1=1}^\infty k_\lambda(n_1, \cdots, n_m, x_{m+1}) \right.\right.$$

$$\left.\left. \times \prod_{j=1}^m a_{n_j}^{(j)} \right)^p dx_{m+1} \right\}^{\frac{1}{p}} < k(\lambda_{m+1}) \prod_{i=1}^m \|a^{(i)}\|_{p_i, \psi_i}, \tag{6.57}$$

where the constant factor $k(\lambda_{m+1})$ in the above inequalities is the best possible.

Proof.   By (6.40), (6.42), and (6.54) and the assumptions, we have (6.57). By Hölder's inequality, we obtain

$$I = \int_0^\infty \left[x_{m+1}^{\frac{-1}{p}+\lambda_{m+1}} \sum_{n_m=1}^\infty \cdots \sum_{n_1=1}^\infty k_\lambda(n_1, \cdots, n_m, x_{m+1}) \prod_{j=1}^m a_{n_j}^{(j)} \right]$$

$$\times [x_{m+1}^{\frac{1}{p}-\lambda_{m+1}} f(x_{m+1})] dx_{m+1} \leq J \|f\|_{p_{m+1}, \varphi}. \tag{6.58}$$

Then, by (6.57), we have inequality (6.56).

   Assuming that (6.56) is valid, setting

$$f(x_{m+1}) = x_{m+1}^{p\lambda_{m+1}-1} \left(\sum_{n_m=1}^\infty \cdots \sum_{n_1=1}^\infty k_\lambda(n_1, \cdots, n_m, x_{m+1}) \prod_{j=1}^m a_{n_j}^{(j)} \right)^{p-1},$$

then, we find $J^{p-1} = ||f||_{p_{m+1}, \varphi}$. By (6.54), we have $J < \infty$. If $J = 0$, then (6.57) is trivially valid; if $J > 0$, then, by (6.56), it follows that

$$||f||_{p_{m+1}, \varphi}^{p_{m+1}} = J^p = I < k(\lambda_{m+1})||f||_{p_{m+1}, \varphi} \prod_{i=1}^{m} ||a^{(i)}||_{p_i, \psi_i}, \quad \text{that is,}$$

$$||f||_{p_{m+1}, \varphi}^{p_{m+1}-1} = J < k(\lambda_{m+1}) \prod_{i=1}^{m} ||a^{(i)}||_{p_i, \psi_i},$$

and then, inequality (6.57) follows.

Hence (6.56) and (6.57) are equivalent.

For $0 < \varepsilon < \delta_0 \min_{1 \le j \le m+1}\{p_j, p\}$, we set $\widetilde{f}(x_{m+1})$ and $\widetilde{a}_{n_i}^{(i)}$ as follows:

$$\widetilde{f}(x_{m+1}) = \begin{cases} 0, & 0 < x_{m+1} < 1, \\ x_{m+1}^{(\lambda_{m+1} - \frac{\varepsilon}{p_{m+1}})-1}, & x_{m+1} \ge 1, \end{cases}$$

$$\widetilde{a}_{n_i}^{(i)} = n_i^{(\lambda_i - \frac{\varepsilon}{p_i})-1}, \quad n_i \in \mathbf{N} \quad (i = 1, \cdots, m).$$

If there exists a constant $k(\le k(\lambda_{m+1}))$ such that (6.56) is valid as we replace $k(\lambda_{m+1})$ by k, then, in particular, by (6.50), we have

$$k(\lambda_{m+1}) + o(1) = I_1(\varepsilon)$$

$$= \varepsilon \int_0^\infty \widetilde{f}(x_{m+1}) \sum_{n_m=1}^\infty \cdots \sum_{n_1=1}^\infty k_\lambda(n_1, \cdots, n_m, x_{m+1}) \prod_{j=1}^{m} \widetilde{a}_{n_j}^{(j)} dx_{m+1}$$

$$< \varepsilon k ||\widetilde{f}||_{p_{m+1}, \varphi} \prod_{i=1}^{m} ||\widetilde{a}^{(i)}||_{p_i, \psi_i} = \varepsilon k \left(\frac{1}{\varepsilon}\right)^{\frac{1}{p_{m+1}}} \prod_{i=1}^{m} \left(1 + \sum_{n_i=1}^\infty \frac{1}{n_i^{1+\varepsilon}}\right)^{\frac{1}{p_i}}$$

$$< \varepsilon k \left(\frac{1}{\varepsilon}\right)^{\frac{1}{p_{m+1}}} \prod_{i=1}^{m} \left(1 + \int_1^\infty \frac{dy}{y^{1+\varepsilon}}\right)^{\frac{1}{p_i}} = k \prod_{i=1}^{m} (1+\varepsilon)^{\frac{1}{p_i}},$$

and then, $k(\lambda_{m+1}) \le k$ $(\varepsilon \to 0^+)$. Hence $k = k(\lambda_{m+1})$ is the best possible constant factor of inequality (6.56).

By the equivalency, the constant factor $k(\lambda_{m+1})$ in (6.57) is still the best possible. Otherwise, it leads to a contradiction by (6.58) that the constant factor in (6.56) is not the best possible. $\qquad \square$

Theorem 6.8. *Suppose that $p_i > 1$, $\lambda_{m+1} = \frac{\lambda}{2}$, there exists a constant $\delta_0 > 0$ such that, for any $\widetilde{\lambda}_i \in (\lambda_i - \delta_0, \lambda_i + \delta_0)$ $(i = 1, \cdots, m+1)$, $\sum_{i=1}^{m+1} \widetilde{\lambda}_i = \lambda$,*

$$k(\widetilde{\lambda}_{m+1}) = \int_0^\infty \cdots \int_0^\infty k_\lambda(u_1, \cdots, u_m, 1) \prod_{j=1}^{m} u_j^{\widetilde{\lambda}_j - 1} du_1 \cdots du_m \in \mathbf{R}_+,$$

and $k_\lambda(x_{m+1}y_1, \cdots, x_{m+1}y_m, 1) \, y_j^{\lambda_j-1}$ *is decreasing with respect to* $y_j \in$ $\mathbf{R}_+$ *and strictly decreasing in an interval* $I_j \subset (1, \infty)$ $(j = 1, \cdots, m)$. *If* $f(x_{m+1}) \geq 0$, $f \in L_{p_{m+1}, \varphi}(\mathbf{R}_+)$, $\|f\|_{p_{m+1}, \varphi} > 0$, $a_{n_i}^{(i)} \geq 0$, $a^{(i)} =$ $\{a_{n_i}^{(i)}\}_{n_i=1}^\infty \in l_{p_i, \psi_i}$, $\|a^{(i)}\|_{p_i, \psi_i} > 0 (i = 1, \cdots, m)$, *then, we have the following equivalent inequalities with the same best constant factor* $k(\lambda_{m+1})$:

$$\int_0^\infty f(x_{m+1}) \sum_{n_m=1}^\infty \cdots \sum_{n_1=1}^\infty k_\lambda(x_{m+1}n_1, \cdots, x_{m+1}n_m, 1)$$

$$\times \prod_{j=1}^m a_{n_j}^{(j)} dx_{m+1}$$

$$< k(\lambda_{m+1})\|f\|_{p_{m+1}, \varphi} \prod_{i=1}^m \|a^{(i)}\|_{p_i, \psi_i}, \tag{6.59}$$

and

$$\left\{ \int_0^\infty x_{m+1}^{\frac{p\lambda}{2}-1} \left(\sum_{n_m=1}^\infty \cdots \sum_{n_1=1}^\infty k_\lambda(x_{m+1}n_1, \cdots, x_{m+1}n_m, 1) \right. \right.$$

$$\left. \left. \times \prod_{j=1}^m a_{n_j}^{(j)} \right)^p dx_{m+1} \right\}^{\frac{1}{p}}$$

$$< k(\lambda_{m+1}) \prod_{i=1}^m \|a^{(i)}\|_{p_i, \psi_i}. \tag{6.60}$$

Proof. We only prove that the constant factor in (6.59) is the best possible. The other parts of the proof are omitted.

For $0 < \varepsilon < \delta_0 \min_{1 \leq j \leq m+1} \{p_j, p\}$, we set $\widetilde{f}(x_{m+1})$ and $\widetilde{a}_{n_i}^{(i)}$ as follows:

$$\widetilde{f}(x_{m+1}) = \begin{cases} x_{m+1}^{(\frac{\lambda}{2} + \frac{\varepsilon}{p_{m+1}})-1}, & 0 < x_{m+1} \leq 1 \\ 0, & x_{m+1} > 1 \end{cases},$$

$$\widetilde{a}_{n_i}^{(i)} = n_i^{\lambda_i - \frac{\varepsilon}{p_i} - 1}, \quad n_i \in \mathbf{N}(i = 1, \cdots, m).$$

If there exists a constant $k(\leq k(\lambda_{m+1}))$, such that (6.59) is valid as we replace $k(\lambda_{m+1})$ by k, then, by (6.51), we have

$$k(\lambda_{m+1}) + o(1) = \widetilde{I}_1(\varepsilon)$$

$$= \varepsilon \int_0^\infty \widetilde{f}(x_{m+1}) \sum_{n_m=1}^\infty \cdots \sum_{n_1=1}^\infty k_\lambda(x_{m+1}n_1, \cdots, x_{m+1}n_m, 1) \prod_{j=1}^m \widetilde{a}_{n_j}^{(j)} dx_{m+1}$$

$$< \varepsilon k ||\widetilde{f}||_{p_{m+1},\varphi} \prod_{i=1}^{m} ||\widetilde{a}^{(i)}||_{p_i,\psi_i} = \varepsilon k (\frac{1}{\varepsilon})^{\frac{1}{p_{m+1}}} \prod_{i=1}^{m} \left(\sum_{n_i=1}^{\infty} \frac{1}{n_i^{1+\varepsilon}} \right)^{\frac{1}{p_i}}$$

$$< \varepsilon k \left(\frac{1}{\varepsilon}\right)^{\frac{1}{p_{m+1}}} \prod_{i=1}^{m} \left(1 + \int_{1}^{\infty} \frac{dy}{y^{1+\varepsilon}} \right)^{\frac{1}{p_i}} = k \prod_{i=1}^{m} (1+\varepsilon)^{\frac{1}{p_i}},$$

and then, $k(\lambda_{m+1}) \leq k(\varepsilon \to 0^+)$. Hence, $k = k(\lambda_{m+1})$ is the best possible value of (6.59). □

Remark 6.2. With the assumptions of Theorem 6.7, we define a second kind of multiple half-discrete Hilbert-type operator with the homogeneous kernel $\widetilde{T} : \prod_{i=1}^{m} l_{p_i,\psi_i} \to L_{p,\varphi^{1-p}}(\mathbf{R}_+)$ as follows:

For any $a = (a^{(1)}, \cdots, a^{(m)}) \in \prod_{i=1}^{m} l_{p_i,\psi_i}$, there exists a $\widetilde{T}a$, satisfying

$$\widetilde{T}a(x_{m+1}) = \sum_{n_m=1}^{\infty} \cdots \sum_{n_1=1}^{\infty} k_\lambda(n_1, \cdots, n_m, x_{m+1}) \prod_{j=1}^{m} a_{n_j}^{(j)} \quad (x_{m+1} \in \mathbf{R}_+).$$
(6.61)

Then, by (6.57), we have

$$||\widetilde{T}a||_{p,\varphi^{1-p}} < k(\lambda_{m+1}) \prod_{i=1}^{m} ||a^{(i)}||_{p_i,\psi_i},$$

and then, $\widetilde{T}a \in L_{p,\varphi^{1-p}}(\mathbf{R}_+)$. Hence, $\widetilde{T}$ is a bounded linear operator with $||\widetilde{T}|| \leq k(\lambda_{m+1})$. Since the constant factor in (6.57) is the best possible, we have

$$||\widetilde{T}|| = \sup_{a(\neq\theta)\in\prod_{i=1}^{m} l_{p_i,\psi_i}} \frac{||\widetilde{T}a||_{p,\varphi^{1-p}}}{\prod_{i=1}^{m} ||a^{(i)}||_{p_i,\psi_i}} = k(\lambda_{m+1}). \qquad (6.62)$$

With the assumptions of Theorem 6.8, we define a second kind of multiple half-discrete Hilbert-type operator with the non-homogeneous kernel $\widetilde{T}_1 : \prod_{i=1}^{m} l_{p_i,\psi_i} \to L_{p,\varphi^{1-p}}(\mathbf{R}_+)$ as follows:

For any $a = (a^{(1)}, \cdots, a^{(m)}) \in \prod_{i=1}^{m} l_{p_i,\psi_i}$, there exists a $\widetilde{T}_1 a$, satisfying

$$\widetilde{T}_1 a(x_{m+1}) = \sum_{n_m=1}^{\infty} \cdots \sum_{n_1=1}^{\infty} k_\lambda(x_{m+1}n_1, \cdots, x_{m+1}n_m, 1) \prod_{j=1}^{m} a_{n_j}^{(j)}$$

$$(x_{m+1} \in \mathbf{R}_+). \qquad (6.63)$$

Then, by (6.60), we have

$$||\widetilde{T}_1 a||_{p,\varphi^{1-p}} < k(\lambda_{m+1}) \prod_{i=1}^{m} ||a^{(i)}||_{p_i,\psi_i},$$

and then, $\widetilde{T}_1 a \in L_{p,\varphi^{1-p}}(\mathbf{R}_+)$. Hence, $\widetilde{T}_1$ is a bounded linear operator with $||\widetilde{T}_1|| \leq k(\lambda_{m+1})$. Since the constant factor in (6.60) is the best possible, we have

$$||\widetilde{T}_1|| = \sup_{a(\neq\theta)\in\prod_{i=1}^m l_{p_i,\psi_i}} \frac{||\widetilde{T}_1 a||_{p,\varphi^{1-p}}}{\prod_{i=1}^m ||a^{(i)}||_{p_i,\psi_i}} = k(\lambda_{m+1}). \tag{6.64}$$

6.3.4   *Some Kinds of Reverse Inequalities*

For $\psi_1(n_1) = n_1^{p_1(1-\lambda_1)-1}$ and $\varphi(x_{m+1}) = x_{m+1}^{p_{m+1}(1-\lambda_{m+1})-1}$, we set functions

$$\Psi_1(n_1) = (1 - \theta_1(n_1))\psi_1(n_1),$$
$$\widetilde{\Psi}_1(n_1) = (1 - \widetilde{\theta}_1(n_1))\psi_1(n_1),$$
$$\Phi(x_{m+1}) = (1 - \theta_{m+1}(x_{m+1}))\varphi(x_{m+1}),$$

and

$$\widetilde{\Phi}(x_{m+1}) = (1 - \widetilde{\theta}_{m+1}(x_{m+1}))\varphi(x_{m+1}),$$

where $\theta_1(n_1)$, $\widetilde{\theta}_1(n_1)$, $\theta_{m+1}(x_{m+1})$ and $\widetilde{\theta}_{m+1}(x_{m+1})$ are given by (6.43), (6.48), (6.41) and (6.46).

For $p_i < 1 (p_i \neq 0)$, the spaces l_{p_i,ψ_i} and $l_{p_{m+1}\varphi}(\mathbf{R}_+)$ with $||a^{(i)}||_{p_i,\psi_i}$ and $||f||_{p_{m+1},\varphi}$ are not normed spaces. But we still use them as the formal symbols in the following:

Theorem 6.9. *Suppose that* $0 < p_1 < 1$, $p_i < 0$ $(i = 2, \cdots, m + 1)$, *there exists a constant* $\delta_0 > 0$, *such that for any* $\widetilde{\lambda}_i \in (\lambda_i - \delta_0, \lambda_i + \delta_0)$ $(i = 1, \cdots, m + 1)$, $\sum_{i=1}^{m+1} \widetilde{\lambda}_i = \lambda$,

$$k(\widetilde{\lambda}_{m+1}) = \int_0^\infty \cdots \int_0^\infty k_\lambda(u_1, \cdots, u_m, 1) \prod_{j=1}^m u_j^{\widetilde{\lambda}_j-1} du_1 \cdots du_m \in \mathbf{R}_+,$$

and $k_\lambda(y_1, \cdots, y_m, x_{m+1}) y_j^{\lambda_j-1}$ *is decreasing with respect to* $y_j \in \mathbf{R}_+$ *and strictly decreasing in an interval* $I_j \subset (1, \infty)$ $(j = 1, \cdots, m)$. *There exist constants* α, $L > 0$ *such that (6.44) is satisfied for* $i = 1$, *and*

$$A_1(u_k) \leq L u_k^{\alpha-\lambda_1} \qquad (u_k \in \mathbf{R}_+; i = 1, \cdots, m + 1).$$

If $f(x_{m+1}) \geq 0$, $f \in L_{p_{m+1},\varphi}(\mathbf{R}_+)$, $||f||_{p_{m+1},\varphi} > 0$, $a_{n_1}^{(1)} \geq 0$, $a^{(1)} = \{a_{n_1}^{(1)}\}_{n_1=1}^\infty \in l_{p_1,\Psi_1}$, $||a^{(1)}||_{p_1,\Psi_1} > 0$, $a_{n_i}^{(i)} \geq 0$, $a^{(i)} = \{a_{n_i}^{(i)}\}_{n_i=1}^\infty \in l_{p_i,\psi_i}$,

$||a^{(i)}||_{p_i,\psi_i} > 0$ $(i = 2, \cdots, m)$, then, we have the following equivalent inequalities with the same best possible constant factor $k(\lambda_{m+1})$:

$$I = \int_0^\infty f(x_{m+1}) \sum_{n_m=1}^\infty \cdots \sum_{n_1=1}^\infty k_\lambda(n_1, \cdots, n_m, x_{m+1}) \prod_{j=1}^m a_{n_j}^{(j)} dx_{m+1}$$

$$> k(\lambda_{m+1})||f||_{p_{m+1},\varphi}||a^{(1)}||_{p_1,\Psi_1} \prod_{i=2}^m ||a^{(i)}||_{p_i,\psi_i}, \qquad (6.65)$$

and

$$J = \left\{ \int_0^\infty x_{m+1}^{p\lambda_{m+1}-1} \right.$$

$$\left. \times \left(\sum_{n_m=1}^\infty \cdots \sum_{n_1=1}^\infty k_\lambda(n_1, \cdots, n_m, x_{m+1}) \prod_{j=1}^m a_{n_j}^{(j)} \right)^p dx_{m+1} \right\}^{\frac{1}{p}}$$

$$> k(\lambda_{m+1})||a^{(1)}||_{p_1,\Psi_1} \prod_{i=2}^m ||a^{(i)}||_{p_i,\psi_i}. \qquad (6.66)$$

Proof. By (6.40), (6.42), the reverse of (6.54) and the assumption, we have (6.66).

By the reverse Hölder's inequality, we obtain

$$I = \int_0^\infty \left[x_{m+1}^{\frac{-1}{p}+\lambda_{m+1}} \sum_{n_m=1}^\infty \cdots \sum_{n_1=1}^\infty k_\lambda(n_1, \cdots, n_m, x_{m+1}) \prod_{j=1}^m a_{n_j}^{(j)} \right]$$

$$\times [x_{m+1}^{\frac{1}{p}-\lambda_{m+1}} f(x_{m+1})] dx_{m+1} \geq J||f||_{p_{m+1},\varphi}. \qquad (6.67)$$

Then, by (6.66), we have (6.65).

Assuming that (6.65) is valid, and setting $f(x_{m+1})$ as follows:

$$f(x_{m+1}) = x_{m+1}^{p\lambda_{m+1}-1} \left(\sum_{n_m=1}^\infty \cdots \sum_{n_1=1}^\infty k_\lambda(n_1, \cdots, n_m, x_{m+1}) \prod_{j=1}^m a_{n_j}^{(j)} \right)^{p-1},$$

then, we find $J^{p-1} = ||f||_{p_{m+1},\varphi}$. By the reverse of (6.54), we have $J > 0$. If $J = \infty$, then, (6.66) is trivially valid; if $J < \infty$, then, by (6.65), it follows that

$$||f||_{p_{m+1},\varphi}^{p_{m+1}} = J^p = I$$

$$> k(\lambda_{m+1})||f||_{p_{m+1},\varphi}||a^{(1)}||_{p_1,\Psi_1} \prod_{i=2}^m ||a^{(i)}||_{p_i,\psi_i}, \quad \text{that is}$$

$$||f||_{p_{m+1},\varphi}^{p_{m+1}-1} = J > k(\lambda_{m+1})||a^{(1)}||_{p_1,\Psi_1} \prod_{i=2}^m ||a^{(i)}||_{p_i,\psi_i},$$

and inequality (6.66) follows.

Hence, (6.65) and (6.66) are equivalent.

For $0 < \varepsilon < \delta_0 \min_{1 \le j \le m+1}\{|p_j|, p\}$, we set $\widetilde{f}(x_{m+1})$ and $\widetilde{a}_{n_i}^{(i)}$ as follows:

$$\widetilde{f}(x_{m+1}) = \begin{cases} 0, & 0 < x_{m+1} < 1, \\ x_{m+1}^{\left(\lambda_{m+1} - \frac{\varepsilon}{p_{m+1}}\right) - 1}, & x_{m+1} \ge 1, \end{cases}$$

$$\widetilde{a}_{n_i}^{(i)} = n_i^{\left(\lambda_i - \frac{\varepsilon}{p_i}\right) - 1}, \qquad n_i \in \mathbf{N} (i = 1, \cdots, m).$$

If there exists a constant $k(\ge k(\lambda_{m+1}))$, such that (6.65) is valid as we replace $k(\lambda_{m+1})$ by k, then, in particular, by (6.50) and Lemma 6.4, we have

$$k(\lambda_{m+1}) + o(1) = I_1(\varepsilon)$$

$$= \varepsilon \int_0^\infty \widetilde{f}(x_{m+1}) \sum_{n_m=1}^\infty \cdots \sum_{n_1=1}^\infty k_\lambda(n_1, \cdots, n_m, x_{m+1}) \prod_{j=1}^m \widetilde{a}_{n_j}^{(j)} dx_{m+1}$$

$$> \varepsilon k \|\widetilde{f}\|_{p_{m+1}, \varphi} \|\widetilde{a}^{(1)}\|_{p_1, \Psi_1} \prod_{i=2}^m \|\widetilde{a}^{(i)}\|_{p_i, \psi_i}$$

$$= \varepsilon k \left(\frac{1}{\varepsilon}\right)^{\frac{1}{p_{m+1}}} \left[\sum_{n_1=1}^\infty \left(1 - O\left(\frac{1}{n_1^\alpha}\right)\right) \frac{1}{n_1^{1+\varepsilon}}\right]^{\frac{1}{p_1}} \prod_{i=2}^m \left(\sum_{n_i=1}^\infty \frac{1}{n_i^{1+\varepsilon}}\right)^{\frac{1}{p_i}}$$

$$= \varepsilon k \left(\frac{1}{\varepsilon}\right)^{\frac{1}{p_{m+1}}} \left(\sum_{n_1=1}^\infty \frac{1}{n_1^{1+\varepsilon}} - O(1)\right)^{\frac{1}{p_1}} \prod_{i=2}^m \left(1 + \sum_{n_i=2}^\infty \frac{1}{n_i^{1+\varepsilon}}\right)^{\frac{1}{p_i}}$$

$$> \varepsilon k \left(\frac{1}{\varepsilon}\right)^{\frac{1}{p_{m+1}}} \left(\int_1^\infty \frac{dy}{y^{1+\varepsilon}} - O(1)\right)^{\frac{1}{p_1}} \prod_{i=2}^m \left(1 + \int_1^\infty \frac{dx_i}{x_i^{1+\varepsilon}}\right)^{\frac{1}{p_i}}$$

$$= k(1 - \varepsilon O(1))^{\frac{1}{p_1}} \prod_{i=2}^m (\varepsilon + 1)^{\frac{1}{p_i}},$$

and then, $k(\lambda_{m+1}) \ge k$ $(\varepsilon \to 0^+)$. Hence, $k = k(\lambda_{m+1})$ is the best constant factor of inequality (6.65).

By the equivalency, the constant factor $k(\lambda_{m+1})$ in (6.66) is still the best possible. Otherwise, it leads to a contradiction by (6.37) that the constant factor in (6.65) is not the best possible. □

Theorem 6.10. *Suppose that* $0 < p_1 < 1$, $p_i < 0$ $(i = 2, \cdots, m + 1)$, $\lambda_{m+1} = \frac{\lambda}{2}$, *there exists a constant* $\delta_0 > 0$, *such that for any* $\widetilde{\lambda}_i \in (\lambda_i - \delta_0, \lambda_i + \delta_0)$ $(i = 1, \cdots, m + 1)$, $\sum_{i=1}^{m+1} \widetilde{\lambda}_i = \lambda$,

$$k(\widetilde{\lambda}_{m+1}) = \int_0^\infty \cdots \int_0^\infty k_\lambda(u_1, \cdots, u_m, 1) \prod_{j=1}^m u_j^{\widetilde{\lambda}_j - 1} du_1 \cdots du_m \in \mathbf{R}_+,$$

and $k_\lambda(x_{m+1}y_1, \cdots, x_{m+1}y_m, 1)y_j^{\lambda_j-1}$ is decreasing with respect to $y_j \in \mathbf{R}_+$ and strictly decreasing in an interval $I_j \subset (1, \infty)(j = 1, \cdots, m)$. If there exist constants $\alpha, L > 0$, such that (6.44) is satisfied, and
$$A_i(u_k) \le Lu_k^{\alpha-\lambda_k} \qquad (u_k \in \mathbf{R}_+; i = 1, \cdots, m+1),$$
then, we have the following equivalent inequalities with the same best possible constant factor $k(\lambda_{m+1})$:

$$\int_0^\infty f(x_{m+1}) \sum_{n_m=1}^\infty \cdots \sum_{n_1=1}^\infty k_\lambda(x_{m+1}n_1, \cdots, x_{m+1}n_m, 1) \prod_{j=1}^m a_{n_j}^{(j)} dx_{m+1}$$

$$> k(\lambda_{m+1})||f||_{p_{m+1},\varphi} ||a^{(1)}||_{p_1, \Psi_1} \prod_{i=2}^m ||a^{(i)}||_{p_i, \psi_i}, \qquad (6.68)$$

and

$$\left\{ \int_0^\infty x_{m+1}^{\frac{p\lambda}{2}-1} \left(\sum_{n_m=1}^\infty \cdots \sum_{n_1=1}^\infty k_\lambda(x_{m+1}n_1, \cdots, x_{m+1}n_m, 1) \right. \right.$$

$$\left. \left. \times \prod_{j=1}^m a_{n_j}^{(j)} \right)^p dx_{m+1} \right\}^{\frac{1}{p}}$$

$$> k(\lambda_{m+1})||a^{(1)}||_{p_1, \Psi_1} \prod_{i=2}^m ||a^{(i)}||_{p_i, \psi_i}. \qquad (6.69)$$

Proof. We only prove that the constant factor in (6.68) is the best possible. Other parts of the proof are omitted.

For $0 < \varepsilon < \delta_0 \min_{1 \le j \le m+1}\{|p_j|, p\}$, we set $\widetilde{f}(x_{m+1})$ and $\widetilde{a}_{n_i}^{(i)}$ as follows:

$$\widetilde{f}(x_{m+1}) = \begin{cases} 0, & 0 < x_{m+1} < 1, \\ x_{m+1}^{\left(\lambda_{m+1}-\frac{\varepsilon}{p_{m+1}}\right)-1}, & x_{m+1} \ge 1, \end{cases}$$

$$\widetilde{a}_{n_i}^{(i)} = n_i^{\left(\lambda_i-\frac{\varepsilon}{p_i}\right)-1}, \qquad n_i \in \mathbf{N}(i = 1, \cdots, m).$$

If there exists a constant $k(\ge k(\lambda_{m+1}))$, such that (6.68) is valid as we replace $k(\lambda_{m+1})$ by k, then, in particular, by (6.51), we have

$$k(\lambda_{m+1}) + o(1) = \widetilde{I}_1(\varepsilon)$$

$$= \varepsilon \int_0^\infty \widetilde{f}(x_{m+1}) \sum_{n_m=1}^\infty \cdots \sum_{n_1=1}^\infty k_\lambda(x_{m+1}n_1, \cdots, x_{m+1}n_m, 1) \prod_{j=1}^m \widetilde{a}_{n_j}^{(j)} dx_{m+1}$$

$$> \varepsilon k ||\widetilde{f}||_{p_{m+1},\varphi} ||\widetilde{a}^{(1)}||_{p_1, \Psi_1} \prod_{i=2}^m ||\widetilde{a}^{(i)}||_{p_i, \psi_i}$$

$$= \varepsilon k \left(\frac{1}{\varepsilon}\right)^{\frac{1}{p_{m+1}}} \left[\sum_{n_1=1}^\infty \left(1 - O\left(\frac{1}{n_1^\alpha}\right)\right) \frac{1}{n_1^{1+\varepsilon}} \right]^{\frac{1}{p_1}} \prod_{i=2}^m \left(\sum_{n_i=1}^\infty \frac{1}{n_i^{1+\varepsilon}} \right)^{\frac{1}{p_i}}$$

$$= \varepsilon k \left(\frac{1}{\varepsilon} \right)^{\frac{1}{p_{m+1}}} \left(\sum_{n_1=1}^{\infty} \frac{1}{n_1^{1+\varepsilon}} - O(1) \right)^{\frac{1}{p_1}} \prod_{i=2}^{m} \left(1 + \sum_{n_i=2}^{\infty} \frac{1}{n_i^{1+\varepsilon}} \right)^{\frac{1}{p_i}}$$

$$> \varepsilon k \left(\frac{1}{\varepsilon} \right)^{\frac{1}{p_{m+1}}} \left(\int_1^{\infty} \frac{dy}{y^{1+\varepsilon}} - O(1) \right)^{\frac{1}{p_1}} \prod_{i=2}^{m} \left(1 + \int_1^{\infty} \frac{dx_i}{x_i^{1+\varepsilon}} \right)^{\frac{1}{p_i}}$$

$$= k \left(1 - \varepsilon O(1) \right)^{\frac{1}{p_1}} \prod_{i=2}^{m} (\varepsilon + 1)^{\frac{1}{p_i}},$$

and then, $k(\lambda_{m+1}) \geq k(\varepsilon \to 0^+)$. Hence, $k = k(\lambda_{m+1})$ is the best constant factor of (6.68). $\qquad \square$

Similarly, we still have:

Theorem 6.11. *Suppose that $p_i < 0 (i = 1, \cdots, m)$, $0 < p_{m+1} < 1$, there exists a constant $\delta_0 > 0$ such that for any $\widetilde{\lambda}_i \in (\lambda_i - \delta_0, \lambda_i + \delta_0)$ $(i = 1, \cdots, m+1)$, $\sum_{i=1}^{m+1} \widetilde{\lambda}_i = \lambda$,*

$$k(\widetilde{\lambda}_{m+1}) = \int_0^{\infty} \cdots \int_0^{\infty} k_{\lambda}(u_1, \cdots, u_m, 1) \prod_{j=1}^{m} u_j^{\widetilde{\lambda}_j - 1} du_1 \cdots du_m \in \mathbf{R}_+,$$

and $k_{\lambda}(y_1, \cdots, y_m, x_{m+1}) y_j^{\lambda_j - 1}$ is decreasing with respect to $y_j \in \mathbf{R}_+$ and strictly decreasing in an interval $I_j \subset (1, \infty)$ $(j = 1, \cdots, m)$. If $f(x_{m+1}) \geq 0, f \in L_{p_{m+1}, \Phi}(\mathbf{R}_+)$, $||f||_{p_{m+1}, \Phi} > 0, a_{n_i}^{(i)} \geq 0$, $a^{(i)} = \{a_{n_i}^{(i)}\}_{n_i=1}^{\infty} \in l_{p_i, \psi_i}$, $||a^{(i)}||_{p_i, \psi_i} > 0$ $(i = 1, \cdots, m)$, then, we have the following equivalent reverse inequalities with the same best possible constant factor $k(\lambda_{m+1})$:

$$\int_0^{\infty} f(x_{m+1}) \sum_{n_m=1}^{\infty} \cdots \sum_{n_1=1}^{\infty} k_{\lambda}(n_1, \cdots, n_m, x_{m+1}) \prod_{j=1}^{m} a_{n_j}^{(j)} dx_{m+1}$$

$$> k(\lambda_{m+1}) ||f||_{p_{m+1}, \Phi} \prod_{i=1}^{m} ||a^{(i)}||_{p_i, \psi_i}, \qquad (6.70)$$

and

$$\left\{ \int_0^{\infty} \frac{x_{m+1}^{p\lambda_{m+1} - 1}}{[1 - \theta(x_{m+1})]^{p-1}} \left(\sum_{n_m=1}^{\infty} \cdots \sum_{n_1=1}^{\infty} k_{\lambda}(n_1, \cdots, n_m, x_{m+1}) \right. \right.$$

$$\left. \left. \times \prod_{j=1}^{m} a_{n_j}^{(j)} \right)^p dx_{m+1} \right\}^{\frac{1}{p}}$$

$$> k(\lambda_{m+1}) \prod_{i=1}^{m} ||a^{(i)}||_{p_i, \psi_i}. \qquad (6.71)$$

Theorem 6.12. *Suppose that $p_i < 0 (i = 1, \cdots, m)$, $0 < p_{m+1} \leq 1$, $\lambda_{m+1} = \frac{\lambda}{2}$, there exists a constant $\delta_0 > 0$ such that for any $\widetilde{\lambda}_i \in (\lambda_i - \delta_0, \lambda_i + \delta_0)$ $(i = 1, \cdots, m+1)$, $\sum_{i=1}^{m+1} \widetilde{\lambda}_i = \lambda$,*

$$k(\widetilde{\lambda}_{m+1}) = \int_0^\infty \cdots \int_0^\infty k_\lambda(u_1, \cdots, u_m, 1) \prod_{j=1}^m u_j^{\widetilde{\lambda}_j - 1} du_1 \cdots du_m \in \mathbf{R}_+,$$

and $k_\lambda(x_{m+1}y_1, \cdots, x_{m+1}y_m, 1) y_j^{\lambda_j - 1}$ is decreasing with respect to $y_j \in \mathbf{R}_+$ and strictly decreasing in an interval $I_j \subset (1, \infty) (j = 1, \cdots, m)$. If $f(x_{m+1}) \geq 0$, $f \in L_{p_{m+1}, \widetilde{\Phi}}(\mathbf{R}_+)$, $||f||_{p_{m+1}, \widetilde{\Phi}} > 0$, then, we have the following equivalent inequalities with the same best constant factor $k(\lambda_{m+1})$:

$$\int_0^\infty f(x_{m+1}) \sum_{n_m=1}^\infty \cdots \sum_{n_1=1}^\infty k_\lambda(x_{m+1}n_1, \cdots, x_{m+1}n_m, 1) \prod_{j=1}^m a_{n_j}^{(j)} dx_{m+1}$$

$$> k(\lambda_{m+1}) ||f||_{p_{m+1}, \widetilde{\Phi}} \prod_{i=1}^m ||a^{(i)}||_{p_i, \psi_i}, \qquad (6.72)$$

and

$$\left\{ \int_0^\infty \frac{x_{m+1}^{\frac{p\lambda}{2} - 1}}{[1 - \widetilde{\theta}(x_{m+1})]^{p-1}} \left(\sum_{n_m=1}^\infty \cdots \sum_{n_1=1}^\infty (x_{m+1}n_1, \cdots, x_{m+1}n_m, 1) \right. \right.$$

$$\left. \left. \times \prod_{j=1}^m a_{n_j}^{(j)} \right)^p dx_{m+1} \right\}^{\frac{1}{p}}$$

$$> k(\lambda_{m+1}) \prod_{i=1}^m ||a^{(i)}||_{p_i, \psi_i}. \qquad (6.73)$$

6.4 Some Examples with the Particular Kernels

In the following, we still assume that $m \in \mathbf{N}$, $p_i, p \in \mathbf{R}\backslash\{0, 1\}$, $\lambda_i \in \mathbf{R}(i = 1, \cdots, m+1)$, $\sum_{i=1}^{m+1} \lambda_i = \lambda$, $\frac{1}{p} = \sum_{i=1}^m \frac{1}{p_i} = 1 - \frac{1}{p_{m+1}}$. In the following Corollaries 6.1 to 6.21, we set

$$\varphi_i(x_i) = x_i^{p_i(1-\lambda_i)-1} (x_i \in \mathbf{R}_+; i = 1, \cdots, m),$$

$$\psi(n) = n^{p_{m+1}(1-\lambda_{m+1})-1} (n \in \mathbf{N}),$$

$\Phi_1(x_1) = (1 - \theta_1(x_1))\varphi_1(x_1)$ and $\widetilde{\Phi}_1(x_1) = (1 - \widetilde{\theta}_1(x_1))\varphi_1(x_1)$, where

$$\theta_1(x_1) = \frac{1}{k(\lambda_{m+1})} \int_0^{1/x_1} u_{m+1}^{\lambda_{m+1}-1} \left[\int_0^\infty \cdots \int_0^\infty k_\lambda(1, u_2, \cdots, u_{m+1}) \right.$$

$$\left. \times \prod_{j=2}^m u_j^{\lambda_j - 1} du_2 \cdots du_m \right] du_{m+1},$$

and

$$\widetilde{\theta}_1(x_1) = \frac{1}{k(\lambda_{m+1})} \int_0^{x_1} u_1^{\lambda_1-1} \left[\int_0^\infty \cdots \int_0^\infty k_\lambda(u_1,\cdots,u_m,1) \right.$$
$$\left. \times \prod_{j=2}^m u_j^{\lambda_j-1} du_2 \cdots du_m \right] du_1.$$

In the following Corollaries 6.4 to 6.24, we set

$$\psi_i(n_i) = n_i^{p_i(1-\lambda_i)-1} \qquad (n_i \in \mathbf{N}; i = 1, \cdots, m),$$
$$\varphi(x_{m+1}) = x_{m+1}^{p_{m+1}(1-\lambda_{m+1})-1} \qquad (x_{m+1} \in \mathbf{R}_+),$$
$$\Phi(x_{m+1}) = (1 - \theta_{m+1}(x_{m+1}))\varphi(x_{m+1}) \qquad (x_{m+1} \in \mathbf{R}_+),$$
$$\widetilde{\Phi}(x_{m+1}) = (1 - \widetilde{\theta}_{m+1}(x_{m+1}))\varphi(x_{m+1}) \qquad (x_{m+1} \in \mathbf{R}_+),$$

$\Psi(n_1) = (1 - \theta_1(n_1))\psi(n_1)$ and $\widetilde{\Psi}(n_1) = (1 - \widetilde{\theta}_1(n_1))\psi(n_1)$, where

$$\theta_{m+1}(x_{m+1}) = 1 - \frac{1}{k(\lambda_{m+1})} \int_{1/x_{m+1}}^\infty \cdots \int_{1/x_{m+1}}^\infty k_\lambda(u_1,\cdots,u_m,1)$$
$$\times \prod_{j=1}^m u_j^{\lambda_j-1} du_1 \cdots du_m,$$

and

$$\widetilde{\theta}_{m+1}(x_{m+1}) = 1 - \frac{1}{k(\lambda_{m+1})} \int_{x_{m+1}}^\infty \cdots \int_{x_{m+1}}^\infty k_\lambda(u_1,\cdots,u_m,1)$$
$$\times \prod_{j=1}^m u_j^{\lambda_j-1} du_1 \cdots du_m.$$

6.4.1   *The Case of* $k_\lambda(x_1,\cdots,x_m,x_{m+1}) = \frac{1}{(\sum_{i=1}^{m+1} x_i)^\lambda}$

Lemma 6.16. *If* λ, $\lambda_i > 0$ $(i = 1,\cdots,m+1)$, *and*

$$k_\lambda(x_1,\cdots,x_{m+1}) = \frac{1}{(\sum_{i=1}^{m+1} x_i)^\lambda},$$

then, we have

$$k(\lambda_{m+1}) = \int_0^\infty \cdots \int_0^\infty \frac{\prod_{j=1}^m u_j^{\lambda_j-1}}{(\sum_{j=1}^m u_j + 1)^\lambda} du_1 \cdots du_m$$
$$= \frac{1}{\Gamma(\lambda)} \prod_{i=1}^{m+1} \Gamma(\lambda_i). \tag{6.74}$$

Proof. For $M > 0$, we set

$$D_M = \{(u_1, \cdots, u_m)|u_i > 0, \ \sum_{i=1}^{m} \left(\frac{u_i}{M}\right) \leq 1\}.$$

In view of the assumptions with reference (see Yang [131], (9.18)), we have

$$\int \cdots \int_{D_M} \psi\left(\sum_{i=1}^{m} \left(\frac{u_i}{M}\right)\right) \prod_{j=1}^{m} u_j^{\lambda_j - 1} du_1 \cdots du_m$$

$$= \frac{M^{\lambda - \lambda_{m+1}}}{\Gamma(\lambda - \lambda_{m+1})} \prod_{i=1}^{m} \Gamma(\lambda_i) \int_0^1 \psi(u) u^{\lambda - \lambda_{m+1} - 1} du. \quad (6.75)$$

Setting $\psi(u) = \frac{1}{(Mu+1)^\lambda}$, by (6.75), we find

$$k(\lambda_{m+1}) = \lim_{M \to \infty} \int \cdots \int_{D_M} \frac{\prod_{j=1}^{m} u_j^{\lambda_j - 1}}{[M \sum_{i=1}^{m} (\frac{u_i}{M}) + 1]^\lambda} du_1 \cdots du_m$$

$$= \lim_{M \to \infty} \frac{M^{\lambda - \lambda_{m+1}}}{\Gamma(\lambda - \lambda_{m+1})} \prod_{i=1}^{m} \Gamma(\lambda_i) \int_0^1 \frac{u^{\lambda - \lambda_{m+1} - 1}}{(Mu + 1)^\lambda} du, \ (v = Mu)$$

$$= \frac{1}{\Gamma(\lambda - \lambda_{m+1})} \prod_{i=1}^{m} \Gamma(\lambda_i) \int_0^\infty \frac{v^{\lambda - \lambda_{m+1} - 1}}{(v + 1)^\lambda} dv$$

$$= \frac{1}{\Gamma(\lambda - \lambda_{m+1})} \prod_{i=1}^{m} \Gamma(\lambda_i) \left[\frac{\Gamma(\lambda - \lambda_{m+1})\Gamma(\lambda_{m+1})}{\Gamma(\lambda)}\right]$$

$$= \frac{1}{\Gamma(\lambda)} \prod_{i=1}^{m+1} \Gamma(\lambda_i),$$

then, inequality (6.74) follows. $\qquad \square$

With the assumptions of Lemma 6.16, if $\lambda_{m+1} < 1$, then, we set $\delta_0 = \min_{1 \leq i \leq m+1}\{\lambda_i, 1 - \lambda_{m+1}\} > 0$. For any $\tilde{\lambda}_i \in (\lambda_i - \delta_0, \lambda_i + \delta_0)$ ($i = 1, \cdots, m + 1$), $\sum_{i=1}^{m+1} \tilde{\lambda}_i = \lambda$, it follows that $\tilde{\lambda}_i > 0$ ($i = 1, \cdots, m + 1$), $\tilde{\lambda}_{m+1} < \lambda_{m+1} + \delta_0 \leq \lambda_{m+1} + 1 - \lambda_{m+1} = 1$,

$$k(\tilde{\lambda}_{m+1}) = \int_0^\infty \cdots \int_0^\infty \frac{\prod_{j=1}^{m} u_j^{\tilde{\lambda}_j - 1}}{(\sum_{j=1}^{m} u_j + 1)^\lambda} du_1 \cdots du_m$$

$$= \frac{1}{\Gamma(\lambda)} \prod_{i=1}^{m+1} \Gamma(\tilde{\lambda}_i) \in \mathbf{R}_+,$$

and both expressions $\frac{1}{(x_1 + \cdots + x_m + y)^\lambda} y^{\tilde{\lambda}_{m+1} - 1}$ and $\frac{1}{(yx_1 + \cdots + yx_m + 1)^\lambda} y^{\tilde{\lambda}_{m+1} - 1}$ are strictly decreasing with respect to $y \in \mathbf{R}_+$. Then, for

$$k_\lambda(x_1, \cdots, x_m, x_{m+1}) = \frac{1}{(\sum_{i=1}^{m+1} x_i)^\lambda}$$

in Theorems 6.1 and 6.2, it follows that

Corollary 6.1. *Suppose that $p_i > 1$, $\lambda_i > 0$ ($i = 1, \cdots, m+1$), $\lambda_{m+1} < 1$. If $f_i(x_i) \geq 0$, $f_i \in L_{p_i, \varphi_i}(\mathbf{R}_+)$, $||f_i||_{p_i, \varphi_i} > 0$ ($i = 1, \cdots, m$), $a_n \geq 0$, $a = \{a_n\}_{n=1}^{\infty} \in l_{p_{m+1}, \psi}$, $||a||_{p_{m+1}, \psi} > 0$, then*

(i) *we have the following equivalent inequalities with the best possible constant factor $\frac{1}{\Gamma(\lambda)} \prod_{i=1}^{m+1} \Gamma(\lambda_i)$:*

$$\sum_{n=1}^{\infty} a_n \int_0^{\infty} \cdots \int_0^{\infty} \frac{\prod_{j=1}^m f_j(x_j)}{(\sum_{j=1}^m x_j + n)^{\lambda}} dx_1 \cdots dx_m$$
$$< \frac{\Gamma(\lambda_{m+1})}{\Gamma(\lambda)} \left(\prod_{i=1}^m \Gamma(\lambda_i) ||f_i||_{p_i, \varphi_i} \right) ||a||_{p_{m+1}, \psi}, \quad (6.76)$$

and

$$\left\{ \sum_{n=1}^{\infty} n^{p\lambda_{m+1}-1} \left[\int_0^{\infty} \cdots \int_0^{\infty} \frac{\prod_{j=1}^m f_j(x_j) dx_1 \cdots dx_m}{(\sum_{j=1}^m x_j + n)^{\lambda}} \right]^p \right\}^{\frac{1}{p}}$$
$$< \frac{\Gamma(\lambda_{m+1})}{\Gamma(\lambda)} \prod_{i=1}^m \Gamma(\lambda_i) ||f_i||_{p_i, \varphi_i}; \quad (6.77)$$

(ii) *for $\lambda_{m+1} = \frac{\lambda}{2}(< 1)$, we have the following equivalent inequalities with the same best possible constant factor $\frac{\Gamma(\lambda/2)}{\Gamma(\lambda)} \prod_{i=1}^m \Gamma(\lambda_i)$:*

$$\sum_{n=1}^{\infty} a_n \int_0^{\infty} \cdots \int_0^{\infty} \frac{\prod_{j=1}^m f_j(x_j)}{(\sum_{j=1}^m nx_j + 1)^{\lambda}} dx_1 \cdots dx_m$$
$$< \frac{\Gamma(\lambda/2)}{\Gamma(\lambda)} \left(\prod_{i=1}^m \Gamma(\lambda_i) ||f_i||_{p_i, \varphi_i} \right) ||a||_{p_{m+1}, \psi}, \quad (6.78)$$

and

$$\left\{ \sum_{n=1}^{\infty} n^{\frac{p\lambda}{2}-1} \left[\int_0^{\infty} \cdots \int_0^{\infty} \frac{\prod_{j=1}^m f_j(x_j) dx_1 \cdots dx_m}{(\sum_{j=1}^m nx_j + 1)^{\lambda}} \right]^p \right\}^{\frac{1}{p}}$$
$$< \frac{\Gamma(\lambda/2)}{\Gamma(\lambda)} \prod_{i=1}^m \Gamma(\lambda_i) ||f_i||_{p_i, \varphi_i}. \quad (6.79)$$

In view of (6.9) and (6.12), for

$$k_{\lambda}(x_1, \cdots, x_m, x_{m+1}) = \frac{1}{(\sum_{i=1}^{m+1} x_i)^{\lambda}},$$

we find, for $m \geq 2$,

$$A_1(u_{m+1}) = \int_0^\infty \cdots \int_0^\infty \frac{\prod_{j=2}^m u_j^{\lambda_j - 1}}{(1 + \sum_{j=2}^{m+1} u_j)^\lambda} \, du_2 \cdots du_m$$

$$\leq \int_0^\infty \cdots \int_0^\infty \frac{\prod_{j=2}^m u_j^{\lambda_j - 1} du_2 \cdots du_m}{(1 + \sum_{j=2}^m u_j)^{\lambda + \alpha - \lambda_{m+1}} u_{m+1}^{\lambda_{m+1} - \alpha}}$$

$$= \frac{\Gamma(\lambda_1 + \alpha) \prod_{i=2}^m \Gamma(\lambda_i)}{\Gamma(\lambda + \alpha - \lambda_{m+1})} (u_{m+1}^{\alpha - \lambda_{m+1}}) \quad (0 < \alpha < \lambda_{m+1}),$$

and

$$B_1(u_i) = \int_0^\infty \cdots \int_0^\infty \frac{\prod_{j=1(j \neq i)}^m u_j^{\lambda_j - 1}}{(\sum_{j=1}^m u_j + 1)^\lambda} du_1 \cdots du_{i-1} du_{i+1} \cdots du_m$$

$$\leq \int_0^\infty \cdots \int_0^\infty \frac{\prod_{j=1(j \neq i)}^m u_j^{\lambda_j - 1} du_1 \cdots du_{i-1} du_{i+1} \cdots du_m}{(\sum_{j=1(j \neq i)}^m u_j + 1)^{\lambda + \alpha - \lambda_i} u_i^{\lambda_i - \alpha}}$$

$$= \frac{\Gamma(\lambda_{m+1} + \alpha)}{\Gamma(\lambda + \alpha - \lambda_i)} \prod_{j=1(j \neq i)}^m \Gamma(\lambda_j)(u_i^{\alpha - \lambda_i}) \quad (0 < \alpha < \lambda_i);$$

for $m = 1$, we can still obtain the same results.

By Theorems 6.3 and 6.4, it follows that

Corollary 6.2. *Suppose that* $0 < p_1 < 1$, $p_i < 0$ $(i = 2, \cdots, m + 1)$, $\lambda_i > 0$ $(i = 1, \cdots, m + 1)$, $\lambda_{m+1} < 1$. *If* $f_1(x_1) \geq 0$, $f_1 \in L_{p_1, \Phi_1}(\mathbf{R}_+)$, $||f_1||_{p_1, \Phi_1} > 0$, $f_i(x_i) \geq 0$, $f_i \in L_{p_i, \varphi_i}(\mathbf{R}_+)$, $||f_i||_{p_i, \varphi_i} > 0$ $(i = 2, \cdots, m)$, $a_n \geq 0$, $a = \{a_n\}_{n=1}^\infty \in l_{p_{m+1}, \psi}$, $||a||_{p_{m+1}, \psi} > 0$, *then*

(i) *we have the following equivalent inequalities with the same best possible constant factor* $\frac{1}{\Gamma(\lambda)} \prod_{i=1}^{m+1} \Gamma(\lambda_i)$:

$$\sum_{n=1}^\infty a_n \int_0^\infty \cdots \int_0^\infty \frac{\prod_{j=1}^m f_j(x_j)}{(\sum_{j=1}^m x_j + n)^\lambda} dx_1 \cdots dx_m$$

$$> \frac{1}{\Gamma(\lambda)} \prod_{i=1}^{m+1} \Gamma(\lambda_i) ||a||_{p_{m+1}, \psi} ||f_1||_{p_1, \Phi_1} \prod_{i=2}^m ||f_i||_{p_i, \varphi_i}, \quad (6.80)$$

and

$$\left\{ \sum_{n=1}^\infty n^{p\lambda_{m+1} - 1} \left[\int_0^\infty \cdots \int_0^\infty \frac{\prod_{j=1}^m f_j(x_j)}{(\sum_{j=1}^m x_j + n)^\lambda} dx_1 \cdots dx_m \right]^p \right\}^{\frac{1}{p}}$$

$$> \frac{1}{\Gamma(\lambda)} \prod_{i=1}^{m+1} \Gamma(\lambda_i) ||f_1||_{p_1, \Phi_1} \prod_{i=2}^m ||f_i||_{p_i, \varphi_i}; \quad (6.81)$$

(ii) *for* $\lambda_{m+1} = \frac{\lambda}{2}(< 1), f_1 \in L_{p_1,\widetilde{\Phi}_1}(\mathbf{R}_+), \|f_1\|_{p_1,\widetilde{\Phi}_1} > 0,$ *we have the following equivalent inequalities with the same best possible constant factor* $\frac{\Gamma(\lambda/2)}{\Gamma(\lambda)} \prod_{i=1}^{m} \Gamma(\lambda_i)$:

$$\sum_{n=1}^{\infty} a_n \int_0^{\infty} \cdots \int_0^{\infty} \frac{\prod_{j=1}^{m} f_j(x_j)}{(\sum_{j=1}^{m} nx_j + 1)^{\lambda}} dx_1 \cdots dx_m$$
$$> \frac{\Gamma(\lambda/2)}{\Gamma(\lambda)} \left(\prod_{i=1}^{m} \Gamma(\lambda_i) \right) \|a\|_{p_{m+1},\psi} \|f_1\|_{p_1,\widetilde{\Phi}_1} \prod_{i=2}^{m} \|f_i\|_{p_i,\varphi_i},$$
$$(6.82)$$

and

$$\left\{ \sum_{n=1}^{\infty} n^{\frac{p\lambda}{2}-1} \left[\int_0^{\infty} \cdots \int_0^{\infty} \frac{\prod_{j=1}^{m} f_j(x_j)}{(\sum_{j=1}^{m} nx_j + 1)^{\lambda}} dx_1 \cdots dx_m \right]^p \right\}^{\frac{1}{p}}$$
$$> \frac{\Gamma(\lambda/2)}{\Gamma(\lambda)} \left(\prod_{i=1}^{m} \Gamma(\lambda_i) \right) \|f_1\|_{p_1,\widetilde{\Phi}_1} \prod_{i=2}^{m} \|f_i\|_{p_i,\varphi_i}. \quad (6.83)$$

By Theorems 6.5 and 6.6, we still have

Corollary 6.3. *Suppose that* $\lambda_i > 0, p_i < 0$ $(i = 1, \cdots, m), 0 < p_{m+1} < 1,$ $0 < \lambda_{m+1} < 1.$ *If* $f_i(x_i), a_n \geq 0, f_i \in L_{p_i,\varphi_i}(\mathbf{R}_+), \|f_i\|_{p_i,\varphi_i} > 0 (i = 1, \cdots, m), a = \{a_n\}_{n=1}^{\infty} \in l_{p_{m+1},\psi}, \|a\|_{p_{m+1},\psi} > 0,$ *then*

(i) *we have the following equivalent reverse inequalities with the best possible constant factor* $\frac{1}{\Gamma(\lambda)} \prod_{i=1}^{m+1} \Gamma(\lambda_i)$:

$$\sum_{n=1}^{\infty} a_n \int_0^{\infty} \cdots \int_0^{\infty} \frac{\prod_{j=1}^{m} f_j(x_j)}{(\sum_{j=1}^{m} x_j + n)^{\lambda}} dx_1 \cdots dx_m$$
$$> \frac{\Gamma(\lambda_{m+1})}{\Gamma(\lambda)} \left(\prod_{i=1}^{m} \Gamma(\lambda_i) \|f_i\|_{p_i,\varphi_i} \right) \|a\|_{p_{m+1},\psi}, \quad (6.84)$$

and

$$\left\{ \sum_{n=1}^{\infty} n^{p\lambda_{m+1}-1} \left[\int_0^{\infty} \cdots \int_0^{\infty} \frac{\prod_{j=1}^{m} f_j(x_j)}{(\sum_{j=1}^{m} x_j + n)^{\lambda}} dx_1 \cdots dx_m \right]^p \right\}^{\frac{1}{p}}$$
$$> \frac{\Gamma(\lambda_{m+1})}{\Gamma(\lambda)} \prod_{i=1}^{m} \Gamma(\lambda_i) \|f_i\|_{p_i,\varphi_i}; \quad (6.85)$$

(ii) *for* $\lambda_{m+1} = \frac{\lambda}{2} < 1$, *we have the following equivalent inequalities with the same best possible constant factor* $\frac{\Gamma(\lambda/2)}{\Gamma(\lambda)} \prod_{i=1}^{m} \Gamma(\lambda_i)$:

$$\sum_{n=1}^{\infty} a_n \int_0^{\infty} \cdots \int_0^{\infty} \frac{\prod_{j=1}^m f_j(x_j)}{(\sum_{j=1}^m nx_j + 1)^{\lambda}} dx_1 \cdots dx_m$$
$$> \frac{\Gamma(\lambda/2)}{\Gamma(\lambda)} \left(\prod_{i=1}^m \Gamma(\lambda_i) \|f_i\|_{p_i, \varphi_i} \right) \|a\|_{p_{m+1}, \psi}, \quad (6.86)$$

and

$$\left\{ \sum_{n=1}^{\infty} n^{\frac{p\lambda}{2}-1} \left[\int_0^{\infty} \cdots \int_0^{\infty} \frac{\prod_{j=1}^m f_j(x_j)}{(\sum_{j=1}^m nx_j + 1)^{\lambda}} dx_1 \cdots dx_m \right]^p \right\}^{\frac{1}{p}}$$
$$> \frac{\Gamma(\lambda/2)}{\Gamma(\lambda)} \prod_{i=1}^m \Gamma(\lambda_i) \|f_i\|_{p_i, \varphi_i}. \quad (6.87)$$

By Theorems 6.7 and 6.8, we have

Corollary 6.4. *Suppose that* $p_i > 1$, $0 < \lambda_i < 1$ $(i = 1, \cdots, m)$, $\lambda_{m+1} > 0$. *If* $f(x_{m+1}) \geq 0$, $f \in L_{p_{m+1}, \varphi}(\mathbf{R}_+)$, $\|f\|_{p_{m+1}, \varphi} > 0$, $a_{n_i}^{(i)} \geq 0, a^{(i)} = \{a_{n_i}^{(i)}\}_{n_i=1}^{\infty} \in l_{p_i, \psi_i}$, $\|a^{(i)}\|_{p_i, \psi_i} > 0$ $(i = 1, \cdots, m)$, *then*

(i) *we have the following equivalent inequalities with the best possible constant factor* $\frac{1}{\Gamma(\lambda)} \prod_{i=1}^{m+1} \Gamma(\lambda_i)$:

$$\int_0^{\infty} f(x_{m+1}) \sum_{n_m=1}^{\infty} \cdots \sum_{n_1=1}^{\infty} \frac{\prod_{j=1}^m a_{n_j}^{(j)}}{(\sum_{j=1}^m n_j + x_{m+1})^{\lambda}} dx_{m+1}$$
$$< \frac{\Gamma(\lambda_{m+1})}{\Gamma(\lambda)} \left(\prod_{i=1}^m \Gamma(\lambda_i) \|a^{(i)}\|_{p_i, \psi_i} \right) \|f\|_{p_{m+1}, \varphi}, \quad (6.88)$$

and

$$\left\{ \int_0^{\infty} x_{m+1}^{p\lambda_{m+1}-1} \left[\sum_{n_m=1}^{\infty} \cdots \sum_{n_1=1}^{\infty} \frac{\prod_{j=1}^m a_{n_j}^{(j)}}{(\sum_{j=1}^m n_j + x_{m+1})^{\lambda}} \right]^p dx_{m+1} \right\}^{\frac{1}{p}}$$
$$< \frac{\Gamma(\lambda_{m+1})}{\Gamma(\lambda)} \prod_{i=1}^m \Gamma(\lambda_i) \|a^{(i)}\|_{p_i, \psi_i}; \quad (6.89)$$

(ii) *for* $\lambda_{m+1} = \frac{\lambda}{2}$, *we have the following equivalent inequalities with the best possible constant factor* $\frac{\Gamma(\lambda/2)}{\Gamma(\lambda)} \prod_{i=1}^{m} \Gamma(\lambda_i)$:

$$\int_0^\infty f(x_{m+1}) \sum_{n_m=1}^\infty \cdots \sum_{n_1=1}^\infty \frac{\prod_{j=1}^m a_{n_j}^{(j)}}{(\sum_{j=1}^m x_{m+1} n_j + 1)^\lambda} dx_{m+1}$$

$$< \frac{\Gamma(\lambda/2)}{\Gamma(\lambda)} \left(\prod_{i=1}^m \Gamma(\lambda_i) \|a^{(i)}\|_{p_i, \psi_i} \right) \|f\|_{p_{m+1}, \varphi}, \qquad (6.90)$$

and

$$\left\{ \int_0^\infty x_{m+1}^{\frac{p\lambda}{2}-1} \left[\sum_{n_m=1}^\infty \cdots \sum_{n_1=1}^\infty \frac{\prod_{j=1}^m a_{n_j}^{(j)}}{(\sum_{j=1}^m x_{m+1} n_j + 1)^\lambda} \right]^p dx_{m+1} \right\}^{\frac{1}{p}}$$

$$< \frac{\Gamma(\lambda/2)}{\Gamma(\lambda)} \prod_{i=1}^m \Gamma(\lambda_i) \|a^{(i)}\|_{p_i, \psi_i}. \qquad (6.91)$$

In (6.44), for

$$k_\lambda(x_1, \cdots, x_m, x_{m+1}) = \frac{1}{(\sum_{i=1}^{m+1} x_i)^\lambda},$$

we find, for $m \geq 2$,

$$A_i(u_k) = \int_0^\infty \cdots \int_0^\infty \frac{1}{(\sum_{j=1(j\neq i)}^{m+1} u_j + 1)^\lambda}$$

$$\times \prod_{j=1(j\neq i,k)}^{m+1} u_j^{\lambda_j - 1} du_1 \cdots du_{i-1} du_{i+1} \cdots du_{k-1} du_{k+1} \cdots du_{m+1}$$

$$\leq \int_0^\infty \cdots \int_0^\infty \frac{\prod_{j=1(j\neq i,k)}^{m+1} u_j^{\lambda_j - 1} du_1 \cdots du_{i-1} du_{i+1} \cdots du_{k-1} du_{k+1} \cdots du_{m+1}}{(\sum_{j=1(j\neq i,k)}^{m+1} u_j + 1)^{\lambda+\alpha-\lambda_k} u_k^{\lambda_k - \alpha}}$$

$$= \frac{\Gamma(\lambda_i + \alpha)}{\Gamma(\lambda + \alpha - \lambda_k)} \prod_{j=1(j\neq i,k)}^{m+1} \Gamma(\lambda_j)(u_k^{\alpha-\lambda_k})(0 < \alpha < \lambda_k);$$

for $m = 1$, we can still find the same result.

If $0 < \lambda_i < 1(i = 1, \cdots, m)$, $\lambda_{m+1} > 0$, then, we set $\delta_0 = \min_{1\leq i\leq m}\{\lambda_{m+1}, \lambda_i, 1 - \lambda_i\} > 0$. For any $\widetilde{\lambda}_i \in (\lambda_i - \delta_0, \lambda_i + \delta_0)(i = 1, \cdots, m + 1)$, $\sum_{i=1}^{m+1} \widetilde{\lambda}_i = \lambda$, we find that $\widetilde{\lambda}_{m+1} > 0$, $0 < \widetilde{\lambda}_i < \lambda_i + \delta_0 \leq 1(i = 1, \cdots, m)$,

$$k(\widetilde{\lambda}_{m+1}) = \frac{1}{\Gamma(\lambda)} \prod_{i=1}^{m+1} \Gamma(\widetilde{\lambda}_i) \in \mathbf{R}_+,$$

and both $\dfrac{1}{(y_1+\cdots+y_m+x_{m+1})^\lambda}y_j^{\widetilde{\lambda}_j-1}$ and $\dfrac{1}{(x_{m+1}y_1+\cdots+x_{m+1}y_m+1)^\lambda}y_j^{\widetilde{\lambda}_j-1}$ are strictly decreasing with respect to $y_j \in \mathbf{R}(j=1,\cdots,m)$.

Then, by Theorems 6.9 and 6.10, it follows that

Corollary 6.5. *Suppose that* $0 < p_1 < 1$, $p_i < 0(i = 2, \cdots, m+1)$, $0 < \lambda_i < 1(i = 1, \cdots, m)$, $\lambda_{m+1} > 0$. *If* $f(x_{m+1}) \geq 0$, $f \in L_{p_{m+1},\varphi}(\mathbf{R}_+)$, $||f||_{p_{m+1},\varphi} > 0$, $a_{n_1}^{(1)} \geq 0$, $a^{(1)} = \{a_{n_1}^{(1)}\}_{n_1=1}^\infty \in l_{p_1,\Psi_1}$, $||a^{(1)}||_{p_1,\Psi_1} > 0$, $a_{n_i}^{(i)} \geq 0$, $a^{(i)} = \{a_{n_i}^{(i)}\}_{n_i=1}^\infty \in l_{p_i,\psi_i}$, $||a^{(i)}||_{p_i,\psi_i} > 0$ $(i = 2, \cdots, m)$, *then*

(i) *we have the following equivalent inequalities with the same best possible constant factor* $\frac{1}{\Gamma(\lambda)}\prod_{i=1}^{m+1}\Gamma(\lambda_i)$:

$$\int_0^\infty f(x_{m+1}) \sum_{n_m=1}^\infty \cdots \sum_{n_1=1}^\infty \frac{\prod_{j=1}^m a_{n_j}^{(j)}}{(\sum_{j=1}^m n_j + x_{m+1})^\lambda}dx_{m+1}$$
$$> \frac{1}{\Gamma(\lambda)}\left(\prod_{i=1}^{m+1}\Gamma(\lambda_i)\right)||f||_{p_{m+1},\varphi}||a^{(1)}||_{p_1,\Psi_1}\prod_{i=2}^m||a^{(i)}||_{p_i,\psi_i}, \quad (6.92)$$

and

$$\left\{\int_0^\infty x_{m+1}^{p\lambda_{m+1}-1}\left[\sum_{n_m=1}^\infty \cdots \sum_{n_1=1}^\infty \frac{\prod_{j=1}^m a_{n_j}^{(j)}}{(\sum_{j=1}^m n_j + x_{m+1})^\lambda}\right]^p dx_{m+1}\right\}^{\frac{1}{p}}$$
$$> \frac{1}{\Gamma(\lambda)}\left(\prod_{i=1}^{m+1}\Gamma(\lambda_i)\right)||a^{(1)}||_{p_1,\Psi_1}\prod_{i=2}^m||a^{(i)}||_{p_i,\psi_i}; \quad (6.93)$$

(ii) *for* $\lambda_{m+1} = \frac{\lambda}{2}$, *we have the following equivalent inequalities with the same best possible constant factor* $\frac{\Gamma(\lambda/2)}{\Gamma(\lambda)}\prod_{i=1}^m\Gamma(\lambda_i)$:

$$\int_0^\infty f(x_{m+1}) \sum_{n_m=1}^\infty \cdots \sum_{n_1=1}^\infty \frac{\prod_{j=1}^m a_{n_j}^{(j)}}{(\sum_{j=1}^m x_{m+1}n_j + 1)^\lambda}dx_{m+1}$$
$$> \frac{\Gamma(\lambda/2)}{\Gamma(\lambda)}\left(\prod_{i=1}^m\Gamma(\lambda_i)\right)||f||_{p_{m+1},\varphi}||a^{(1)}||_{p_1,\Psi_1}\prod_{i=2}^m||a^{(i)}||_{p_i,\psi_i}, \quad (6.94)$$

and

$$\left\{\int_0^\infty x_{m+1}^{\frac{p\lambda}{2}-1}\left[\sum_{n_m=1}^\infty \cdots \sum_{n_1=1}^\infty \frac{\prod_{j=1}^m a_{n_j}^{(j)}}{(\sum_{j=1}^m x_{m+1}n_j + 1)^\lambda}\right]^p dx_{m+1}\right\}^{\frac{1}{p}}$$
$$> \frac{\Gamma(\lambda/2)}{\Gamma(\lambda)}\left(\prod_{i=1}^m\Gamma(\lambda_i)\right)||a^{(1)}||_{p_1,\Psi_1}\prod_{i=2}^m||a^{(i)}||_{p_i,\psi_i}. \quad (6.95)$$

By Theorems 6.11 and 6.12, it follows that

Corollary 6.6. *Suppose that $p_i < 0, 0 < \lambda_i < 1 (i = 1, \cdots, m), \lambda_{m+1} > 0, 0 < p_{m+1} < 1$. If $f(x_{m+1}) \geq 0, f \in L_{p_{m+1}, \Phi}(\mathbf{R}_+), ||f||_{p_{m+1}, \Phi} > 0, a_{n_i}^{(i)} \geq 0, a^{(i)} = \{a_{n_i}^{(i)}\}_{n_i=1}^{\infty} \in l_{p_i, \psi_i}, ||a^{(i)}||_{p_i, \psi_i} > 0 \ (i = 1, \cdots, m)$, then*

(i) *we have the following equivalent reverse inequalities with the same best possible constant factor $\frac{1}{\Gamma(\lambda)} \prod_{i=1}^{m+1} \Gamma(\lambda_i)$:*

$$\int_0^\infty f(x_{m+1}) \sum_{n_m=1}^\infty \cdots \sum_{n_1=1}^\infty \frac{\prod_{j=1}^m a_{n_j}^{(j)}}{(\sum_{j=1}^m n_j + x_{m+1})^\lambda} dx_{m+1}$$

$$> \frac{\Gamma(\lambda_{m+1})}{\Gamma(\lambda)} \left(\prod_{i=1}^m \Gamma(\lambda_i) ||a^{(i)}||_{p_i, \psi_i} \right) ||f||_{p_{m+1}, \Phi}, \quad (6.96)$$

and

$$\left\{ \int_0^\infty \frac{x_{m+1}^{p\lambda_{m+1}-1}}{[1 - \theta_{m+1}(x_{m+1})]^{p-1}} \right.$$

$$\left. \times \left[\sum_{n_m=1}^\infty \cdots \sum_{n_1=1}^\infty \frac{\prod_{j=1}^m a_{n_j}^{(j)}}{(\sum_{j=1}^m n_j + x_{m+1})^\lambda} \right]^p dx_{m+1} \right\}^{\frac{1}{p}}$$

$$> \frac{\Gamma(\lambda_{m+1})}{\Gamma(\lambda)} \prod_{i=1}^m \Gamma(\lambda_i) ||a^{(i)}||_{p_i, \psi_i}; \quad (6.97)$$

(ii) *for $\lambda_{m+1} = \frac{\lambda}{2}, f(x_{m+1}) \geq 0, f \in L_{p_{m+1}, \widetilde{\Phi}}(\mathbf{R}_+), ||f||_{p_{m+1}, \widetilde{\Phi}} > 0$, we have the following equivalent inequalities with the same best possible constant factor $\frac{\Gamma(\lambda/2)}{\Gamma(\lambda)} \prod_{i=1}^m \Gamma(\lambda_i)$:*

$$\int_0^\infty f(x_{m+1}) \sum_{n_m=1}^\infty \cdots \sum_{n_1=1}^\infty \frac{\prod_{j=1}^m a_{n_j}^{(j)}}{(\sum_{j=1}^m x_{m+1} n_j + 1)^\lambda} dx_{m+1}$$

$$> \frac{\Gamma(\lambda/2)}{\Gamma(\lambda)} \left(\prod_{i=1}^m \Gamma(\lambda_i) ||a^{(i)}||_{p_i, \psi_i} \right) ||f||_{p_{m+1}, \widetilde{\Phi}}, \quad (6.98)$$

and

$$\left\{ \int_0^\infty \frac{x_{m+1}^{\frac{p\lambda}{2}-1}}{[1 - \widetilde{\theta}_{m+1}(x_{m+1})]^{p-1}} \right.$$

$$\left. \times \left[\sum_{n_m=1}^\infty \cdots \sum_{n_1=1}^\infty \frac{\prod_{j=1}^m a_{n_j}^{(j)}}{(\sum_{j=1}^m x_{m+1} n_j + 1)^\lambda} \right]^p dx_{m+1} \right\}^{\frac{1}{p}}$$

$$> \frac{\Gamma(\lambda/2)}{\Gamma(\lambda)} \prod_{i=1}^m \Gamma(\lambda_i) ||a^{(i)}||_{p_i, \psi_i}. \quad (6.99)$$

Remark 6.3. (i) If, in (6.24), we set

$$k_\lambda(x_1, \cdots, x_m, n) = \frac{1}{(\sum_{j=1}^{\infty} x_j + n)^\lambda},$$

then, in view of Corollary 6.1 and equality (6.25), we have

$$||T|| = k(\lambda_{m+1}) = \frac{1}{\Gamma(\lambda)} \prod_{i=1}^{m+1} \Gamma(\lambda_i).$$

If, in (6.26), for $\lambda_{m+1} = \frac{\lambda}{2}(< 1)$, we set

$$k_\lambda(nx_1, \cdots, nx_m, 1) = \frac{1}{(\sum_{j=1}^{\infty} nx_j + 1)^\lambda},$$

then, in view of Corollary 6.4 and equality (6.27), we have

$$||T_1|| = k(\lambda_{m+1}) = \frac{\Gamma(\lambda/2)}{\Gamma(\lambda)} \prod_{i=1}^{m} \Gamma(\lambda_i).$$

(ii) If, in (6.61), we set

$$k_\lambda(n_1, \cdots, n_m, x_{m+1}) = \frac{1}{(\sum_{j=1}^{\infty} n_j + x_{m+1})^\lambda},$$

then, in view of Corollary 6.4 and equality (6.62), we have

$$||\widetilde{T}|| = k(\lambda_{m+1}) = \frac{1}{\Gamma(\lambda)} \prod_{i=1}^{m+1} \Gamma(\lambda_i).$$

If, in (6.63), for $\lambda_{m+1} = \frac{\lambda}{2}$, we set

$$k_\lambda(x_{m+1}n_1, \cdots, x_{m+1}n_m, 1) = \frac{1}{(\sum_{j=1}^{\infty} x_{m+1}n_j + 1)^\lambda},$$

then, in view of Corollary 6.4 and (6.64), we have

$$||\widetilde{T}_1|| = k(\lambda_{m+1}) = \frac{\Gamma(\lambda/2)}{\Gamma(\lambda)} \prod_{i=1}^{m} \Gamma(\lambda_i).$$

6.4.2 The Case of $k_\lambda(x_1, \cdots, x_{m+1}) = \prod_{k=1}^{s} \frac{1}{\sum_{i=1}^{m} x_i^{\lambda/s} + c_k x_{m+1}^{\lambda/s}}$

Lemma 6.17. *If* $s \in \mathbf{N}$, $0 < c_1 < \cdots < c_s$, λ, $\lambda_i > 0 (i = 1, \cdots, m+1)$, *and*

$$k_\lambda(x_1, \cdots, x_{m+1}) = \prod_{k=1}^{s} \frac{1}{\sum_{i=1}^{m} x_i^{\lambda/s} + c_k x_{m+1}^{\lambda/s}},$$

then, we have

$$k(\lambda_{m+1}) = \int_0^\infty \cdots \int_0^\infty \frac{\prod_{j=1}^m u_j^{\lambda_j - 1}}{\prod_{k=1}^s (\sum_{i=1}^m u_i^{\lambda/s} + c_k)} du_1 \cdots du_m$$

$$= \beta(\lambda_{m+1}) = \frac{\prod_{i=1}^m \Gamma(\frac{s}{\lambda}\lambda_i)}{\Gamma(\frac{s}{\lambda}(\lambda - \lambda_{m+1}))} \frac{(\frac{s}{\lambda})^m \pi}{\sin[\frac{\pi s}{\lambda}(\lambda - \lambda_{m+1})]}$$

$$\times \sum_{k=1}^s c_k^{\frac{s}{\lambda}(\lambda - \lambda_{m+1}) - 1} \prod_{j=1(j\neq k)}^s \frac{1}{c_j - c_k}. \quad (6.100)$$

In particular, for $\lambda_{m+1} = \frac{\lambda}{2}$,

$$\beta\left(\frac{\lambda}{2}\right) = \frac{\prod_{i=1}^m \Gamma(\frac{s}{\lambda}\lambda_i)}{\Gamma(\frac{s}{2})} \frac{(\frac{s}{\lambda})^m \pi}{\sin(\frac{\pi s}{2})} \sum_{k=1}^s c_k^{\frac{s}{2} - 1} \prod_{j=1(j\neq k)}^s \frac{1}{c_j - c_k}. \quad (6.101)$$

Proof.   For $M > 0$, setting

$$D_M = \{(u_1, \cdots, u_m) | u_i > 0, \quad \sum_{i=1}^m \left(\frac{u_i}{M}\right) \leq 1\},$$

and $\psi(u) = \prod_{k=1}^s \frac{1}{Mu+c_k}$, by (6.75) and (3.12), we have

$$k(\lambda_{m+1}) = \left(\frac{s}{\lambda}\right)^m \int_0^\infty \cdots \int_0^\infty \frac{\prod_{j=1}^m v_j^{\frac{s}{\lambda}\lambda_j - 1} dv_1 \cdots dv_m}{\prod_{k=1}^s (\sum_{i=1}^m v_i + c_k)}, \quad (v_i = u_i^{\lambda/s})$$

$$= \left(\frac{s}{\lambda}\right)^m \lim_{M\to\infty} \int \cdots \int_{D_M} \frac{\prod_{j=1}^m v_j^{\frac{s}{\lambda}\lambda_j - 1}}{\prod_{k=1}^s [M\sum_{i=1}^m (\frac{v_i}{M}) + c_k]} dv_1 \cdots dv_m$$

$$= \left(\frac{s}{\lambda}\right)^m \lim_{M\to\infty} \frac{M^{\frac{s}{\lambda}(\lambda - \lambda_{m+1})} \prod_{i=1}^m \Gamma(\frac{s}{\lambda}\lambda_i)}{\Gamma(\frac{s}{\lambda}(\lambda - \lambda_{m+1}))} \int_0^1 \frac{u^{\frac{s}{\lambda}(\lambda - \lambda_{m+1}) - 1} du}{\prod_{k=1}^s (Mu + c_k)},$$

$$(v = Mu)$$

$$= \left(\frac{s}{\lambda}\right)^m \frac{\prod_{i=1}^m \Gamma(\frac{s}{\lambda}\lambda_i)}{\Gamma(\frac{s}{\lambda}(\lambda - \lambda_{m+1}))} \int_0^\infty \frac{v^{\frac{s}{\lambda}(\lambda - \lambda_{m+1}) - 1}}{\prod_{k=1}^s (v + c_k)} dv$$

$$= \frac{\prod_{i=1}^m \Gamma(\frac{s}{\lambda}\lambda_i)}{\Gamma(\frac{s}{\lambda}(\lambda - \lambda_{m+1}))} \frac{(\frac{s}{\lambda})^m \pi}{\sin[\frac{\pi s}{\lambda}(\lambda - \lambda_{m+1})]}$$

$$\times \sum_{k=1}^s c_k^{\frac{s}{\lambda}(\lambda - \lambda_{m+1}) - 1} \prod_{j=1(j\neq k)}^s \frac{1}{c_j - c_k}$$

$$= \beta(\lambda_{m+1}),$$

and then (6.74) follows. $\square$

With the assumptions of Lemma 6.16, if $\lambda_{m+1} < 1$, then, we set $\delta_0 = \min_{1 \leq i \leq m+1}\{\lambda_i, 1 - \lambda_{m+1}\} > 0$. For any $\widetilde{\lambda}_i \in (\lambda_i - \delta_0, \lambda_i + \delta_0)$ $(i = 1, \cdots, m+1)$, $\sum_{i=1}^{m+1} \widetilde{\lambda}_i = \lambda$, it follows that $\widetilde{\lambda}_i > 0$ $(i = 1, \cdots, m+1)$, $\widetilde{\lambda}_{m+1} < \lambda_{m+1} + \delta_0 \leq \lambda_{m+1} + 1 - \lambda_{m+1} = 1$,

$$
k(\widetilde{\lambda}_{m+1}) = \beta(\widetilde{\lambda}_{m+1}) = \int_0^\infty \cdots \int_0^\infty \frac{\prod_{j=1}^m u_j^{\widetilde{\lambda}_j - 1} du_1 \cdots du_m}{\prod_{k=1}^s (\sum_{i=1}^m u_i^{\lambda/s} + c_k)}
$$

$$
= \frac{\prod_{i=1}^m \Gamma(\frac{s}{\lambda}\widetilde{\lambda}_i)}{\Gamma(\frac{s}{\lambda}(\lambda - \widetilde{\lambda}_{m+1}))} \frac{(\frac{s}{\lambda})^m \pi}{\sin[\frac{\pi s}{\lambda}(\lambda - \widetilde{\lambda}_{m+1})]}
$$

$$
\times \sum_{k=1}^s c_k^{\frac{s}{\lambda}(\lambda - \widetilde{\lambda}_{m+1}) - 1} \prod_{j=1 (j \neq k)}^s \frac{1}{c_j - c_k} \in \mathbf{R}_+,
$$

and both

$$
\prod_{k=1}^s \frac{1}{\sum_{i=1}^m x_i^{\lambda/s} + c_k y^{\lambda/s}} y^{\widetilde{\lambda}_{m+1} - 1} \quad \text{and} \quad \prod_{k=1}^s \frac{1}{\sum_{i=1}^m (y x_i)^{\lambda/s} + c_k} y^{\widetilde{\lambda}_{m+1} - 1}
$$

are strictly decreasing with respect to $y \in \mathbf{R}_+$. Then, for

$$
k_\lambda(x_1, \cdots, x_m, x_{m+1}) = \prod_{k=1}^s \frac{1}{\sum_{i=1}^m x_i^{\lambda/s} + c_k x_{m+1}^{\lambda/s}}
$$

in Theorems 6.1 and 6.2, it follows that

Corollary 6.7. *Suppose that $s \in \mathbf{N}$, $0 < c_1 < \cdots < c_s$, $p_i > 1, \lambda_i > 0$ $(i = 1, \cdots, m+1)$, $\lambda_{m+1} < 1$. If $f_i(x_i) \geq 0$, $f_i \in L_{p_i, \varphi_i}(\mathbf{R}_+)$, $\|f_i\|_{p_i, \varphi_i} > 0 (i = 1, \cdots, m)$, $a_n \geq 0$, $a = \{a_n\}_{n=1}^\infty \in l_{p_{m+1}, \psi}$, $\|a\|_{p_{m+1}, \psi} > 0$, then*

(i) *we have the following equivalent inequalities with the best possible constant factor $\beta(\lambda_{m+1})$:*

$$
\sum_{n=1}^\infty a_n \int_0^\infty \cdots \int_0^\infty \frac{\prod_{j=1}^m f_j(x_j)}{\prod_{k=1}^s (\sum_{i=1}^m x_i^{\lambda/s} + c_k n^{\lambda/s})} dx_1 \cdots dx_m
$$

$$
< \beta(\lambda_{m+1}) \left(\prod_{i=1}^m \|f_i\|_{p_i, \varphi_i} \right) \|a\|_{p_{m+1}, \psi}, \qquad (6.102)
$$

and

$$
\left\{ \sum_{n=1}^\infty n^{p\lambda_{m+1} - 1} \left[\int_0^\infty \cdots \int_0^\infty \frac{\prod_{j=1}^m f_j(x_j) dx_1 \cdots dx_m}{\prod_{k=1}^s (\sum_{i=1}^m x_i^{\lambda/s} + c_k n^{\lambda/s})} \right]^p \right\}^{\frac{1}{p}}
$$

$$
< \beta(\lambda_{m+1}) \prod_{i=1}^m \|f_i\|_{p_i, \varphi_i}; \qquad (6.103)
$$

(ii) *for* $\lambda_{m+1} = \frac{\lambda}{2} < 1$, *we have the following equivalent inequalities with the same best possible constant factor* $\beta(\frac{\lambda}{2})$:

$$\sum_{n=1}^{\infty} a_n \int_0^{\infty} \cdots \int_0^{\infty} \frac{\prod_{j=1}^m f_j(x_j)}{\prod_{k=1}^s [\sum_{i=1}^m (nx_i)^{\lambda/s} + c_k]} dx_1 \cdots dx_m$$

$$< \beta\left(\frac{\lambda}{2}\right) \left(\prod_{i=1}^m \|f_i\|_{p_i, \varphi_i}\right) \|a\|_{p_{m+1}, \psi}, \qquad (6.104)$$

and

$$\left\{ \sum_{n=1}^{\infty} n^{\frac{p\lambda}{2}-1} \left[\int_0^{\infty} \cdots \int_0^{\infty} \frac{\prod_{j=1}^m f_j(x_j) dx_1 \cdots dx_m}{\prod_{k=1}^s [\sum_{i=1}^m (nx_i)^{\lambda/s} + c_k]} \right]^p \right\}^{\frac{1}{p}}$$

$$< \beta\left(\frac{\lambda}{2}\right) \prod_{i=1}^m \|f_i\|_{p_i, \varphi_i}. \qquad (6.105)$$

In (6.9) and (6.12), for

$$k_\lambda(x_1, \cdots, x_m, x_{m+1}) = \prod_{k=1}^s \frac{1}{\sum_{i=1}^m x_i^{\lambda/s} + c_k x_{m+1}^{\lambda/s}},$$

we find, for $m \geq 2$,

$$A_1(u_{m+1}) = \int_0^{\infty} \cdots \int_0^{\infty} \frac{\prod_{j=2}^m u_j^{\lambda_j - 1}}{\prod_{k=1}^s (1 + \sum_{i=2}^m u_i^{\lambda/s} + c_k u_{m+1}^{\lambda/s})} du_2 \cdots du_m$$

$$\leq \int_0^{\infty} \cdots \int_0^{\infty} \frac{\prod_{j=2}^m u_j^{\lambda_j - 1}}{(1 + \sum_{i=2}^m u_i^{\lambda/s} + c_1 u_{m+1}^{\lambda/s})^s} du_2 \cdots du_m$$

$$\leq \int_0^{\infty} \cdots \int_0^{\infty} \frac{\prod_{j=2}^m u_j^{\lambda_j - 1} du_2 \cdots du_m}{c_1^{\frac{s}{\lambda}(\lambda_{m+1} - \alpha)} (1 + \sum_{i=2}^m u_i^{\lambda/s})^{s - \frac{s}{\lambda}(\lambda_{m+1} - \alpha)}}$$

$$\times \left(\frac{1}{u_{m+1}^{\lambda_{m+1} - \alpha}}\right)$$

$$= A u_{m+1}^{\alpha - \lambda_{m+1}} \quad (0 < \alpha < \lambda_{m+1}),$$

where

$$A = \int_0^{\infty} \cdots \int_0^{\infty} \frac{\prod_{j=2}^m u_j^{\lambda_j - 1} du_2 \cdots du_m}{c_1^{\frac{s}{\lambda}(\lambda_{m+1} - \alpha)} (1 + \sum_{i=2}^m u_i^{\lambda/s})^{s - \frac{s}{\lambda}(\lambda_{m+1} - \alpha)}}, \quad (v_i = u_i^{\lambda/s})$$

$$= \frac{(\frac{s}{\lambda})^{m-2}}{c_1^{\frac{s}{\lambda}(\lambda_{m+1} - \alpha)}} \int_0^{\infty} \cdots \int_0^{\infty} \frac{\prod_{j=2}^m v_j^{\frac{s}{\lambda}\lambda_j - 1} dv_2 \cdots dv_m}{(1 + \sum_{i=2}^m v_i)^{s - \frac{s}{\lambda}(\lambda_{m+1} - \alpha)}}$$

$$= \frac{(\frac{s}{\lambda})^{m-2}}{c_1^{\frac{s}{\lambda}(\lambda_{m+1} - \alpha)}} \frac{\Gamma\left(s - \frac{s}{\lambda}(\lambda - \lambda_1 - \alpha)\right)}{\Gamma(s - \frac{s}{\lambda}(\lambda_{m+1} - \alpha))} \prod_{i=2}^m \Gamma(\frac{s\lambda_i}{\lambda}) < \infty,$$

and

$$B_1(u_i) = \int_0^\infty \cdots \int_0^\infty \frac{\prod_{j=1(j\neq i)}^m u_j^{\lambda_j-1} du_1 \cdots du_{i-1}du_{i+1} \cdots du_m}{\prod_{k=1}^s (\sum_{i=1}^m u_i^{\lambda/s} + c_k)}$$

$$\leq \int_0^\infty \cdots \int_0^\infty \frac{\prod_{j=1(j\neq i)}^m u_j^{\lambda_j-1} du_1 \cdots du_{i-1}du_{i+1} \cdots du_m}{(\sum_{i=1}^m u_i^{\lambda/s} + c_1)^s}$$

$$\leq \int_0^\infty \cdots \int_0^\infty \frac{\prod_{j=1(j\neq i)}^m u_j^{\lambda_j-1} du_1 \cdots du_{i-1}du_{i+1} \cdots du_m}{(\sum_{i=1}^m u_i^{\lambda/s} + c_1)^{s-\frac{s}{\lambda}(\lambda_i-\alpha)}}$$

$$\times \left(\frac{1}{u_i^{\lambda_i-\alpha}} \right)$$

$$\leq \int_0^\infty \cdots \int_0^\infty \frac{\prod_{j=1(j\neq i)}^m u_j^{\lambda_j-1} du_1 \cdots du_{i-1}du_{i+1} \cdots du_m}{(\sum_{j=1(j\neq i)}^m u_j^{\lambda/s} + c_1)^{s-\frac{s}{\lambda}(\lambda_i-\alpha)} u_i^{\lambda_i-\alpha}}$$

$$= B u_i^{\alpha-\lambda_i} \qquad (0 < \alpha < \lambda_i),$$

where

$$B = \int_0^\infty \cdots \int_0^\infty \frac{\prod_{j=1(j\neq i)}^m u_j^{\lambda_j-1} du_1 \cdots du_{i-1}du_{i+1} \cdots du_m}{(\sum_{j=1(j\neq i)}^m u_j^{\lambda/s} + c_1)^{s-\frac{s}{\lambda}(\lambda_i-\alpha)}}, \quad (v_j = u_j^{\lambda/s}/c_1).$$

$$= \frac{\left(\frac{s}{\lambda}\right)^{m-1}}{c_1^{s-\frac{s}{\lambda}(\lambda-\lambda_{m+1}-\alpha)}}$$

$$\times \int_0^\infty \cdots \int_0^\infty \frac{\prod_{j=1(j\neq i)}^m v_j^{\frac{s}{\lambda}\lambda_j-1} dv_1 \cdots dv_{i-1}dv_{i+1} \cdots dv_m}{(\sum_{j=1(j\neq i)}^m v_j + 1)^{s-\frac{s}{\lambda}(\lambda_i-\alpha)}}$$

$$= \frac{\left(\frac{s}{\lambda}\right)^{m-1} \Gamma(s - \frac{s}{\lambda}(\lambda - \lambda_{m+1} - \alpha)) \prod_{j=1(j\neq i)}^m \Gamma(\frac{s}{\lambda}\lambda_j)}{c_1^{s-\frac{s}{\lambda}(\lambda-\lambda_{m+1}-\alpha)} \Gamma(s - \frac{s}{\lambda}(\lambda_i - \alpha))} < \infty.$$

For $m = 1$, we can still find some similar results.

By Theorems 6.3 and 6.4, it follows that

Corollary 6.8. *Suppose that $s \in \mathbf{N}, 0 < c_1 < \cdots < c_s, 0 < p_1 < 1, p_i < 0(i = 2, \cdots, m+1), \lambda_i > 0 \ (i = 1, \cdots, m+1), \ \lambda_{m+1} < 1$. If $f_1(x_1) \geq 0$, $f_1 \in L_{p_1,\Phi_1}(\mathbf{R}_+), \|f_1\|_{p_1,\Phi_1} > 0, f_i(x_i) \geq 0, \ f_i \in L_{p_i,\varphi_i}(\mathbf{R}_+), \|f_i\|_{p_i,\varphi_i} > 0 \ (i = 2, \cdots, m), \ a_n \geq 0, \ a = \{a_n\}_{n=1}^\infty \in l_{p_{m+1},\psi}, \|a\|_{p_{m+1},\psi} > 0$, then*

(i) we have the following equivalent inequalities with the same best possible

constant factor $\beta(\lambda_{m+1})$:

$$\sum_{n=1}^{\infty} a_n \int_0^{\infty} \cdots \int_0^{\infty} \frac{\prod_{j=1}^m f_j(x_j)}{\prod_{k=1}^s (\sum_{i=1}^m x_i^{\lambda/s} + c_k n^{\lambda/s})} dx_1 \cdots dx_m$$

$$> \beta(\lambda_{m+1}) \|a\|_{p_{m+1},\psi} \|f_1\|_{p_1,\Phi_1} \prod_{i=2}^m \|f_i\|_{p_i,\varphi_i}, \qquad (6.106)$$

and

$$\left\{ \sum_{n=1}^{\infty} n^{p\lambda_{m+1}-1} \left[\int_0^{\infty} \cdots \int_0^{\infty} \frac{\prod_{j=1}^m f_j(x_j) dx_1 \cdots dx_m}{\prod_{k=1}^s (\sum_{i=1}^m x_i^{\lambda/s} + c_k n^{\lambda/s})} \right]^p \right\}^{\frac{1}{p}}$$

$$> \beta(\lambda_{m+1}) \|f_1\|_{p_1,\Phi_1} \prod_{i=2}^m \|f_i\|_{p_i,\varphi_i}; \qquad (6.107)$$

(ii) *for* $\lambda_{m+1} = \frac{\lambda}{2} < 1, f_1 \in L_{p_1,\tilde{\Phi}_1}(\mathbf{R}_+), \|f_1\|_{p_1,\tilde{\Phi}_1} > 0,$ *we have the following equivalent inequalities with the same best possible constant factor* $\beta(\frac{\lambda}{2})$:

$$\sum_{n=1}^{\infty} a_n \int_0^{\infty} \cdots \int_0^{\infty} \frac{\prod_{j=1}^m f_j(x_j)}{\prod_{k=1}^s [\sum_{i=1}^m (nx_i)^{\lambda/s} + c_k]} dx_1 \cdots dx_m$$

$$> \beta\left(\frac{\lambda}{2}\right) \|a\|_{p_{m+1},\psi} \|f_1\|_{p_1,\tilde{\Phi}_1} \prod_{i=2}^m \|f_i\|_{p_i,\varphi_i}, \qquad (6.108)$$

and

$$\left\{ \sum_{n=1}^{\infty} n^{\frac{p\lambda}{2}-1} \left[\int_0^{\infty} \cdots \int_0^{\infty} \frac{\prod_{j=1}^m f_j(x_j)}{\prod_{k=1}^s [\sum_{i=1}^m (nx_i)^{\lambda/s} + c_k]} dx_1 \cdots dx_m \right]^p \right\}^{\frac{1}{p}}$$

$$> \beta\left(\frac{\lambda}{2}\right) \|f_1\|_{p_1,\tilde{\Phi}_1} \prod_{i=2}^m \|f_i\|_{p_i,\varphi_i}. \qquad (6.109)$$

By Theorems 6.5 and 6.6, we still have

Corollary 6.9. *Suppose that* $s \in \mathbf{N}, 0 < c_1 < \cdots < c_s, \lambda_i > 0,$ $p_i < 0 (i = 1, \cdots, m), 0 < p_{m+1} < 1, 0 < \lambda_{m+1} < 1.$ *If* $f_i(x_i), a_n \geq 0,$ $f_i \in L_{p_i,\varphi_i}(\mathbf{R}_+), \|f_i\|_{p_i,\varphi_i} > 0 \ (i = 1, \cdots, m), a = \{a_n\}_{n=1}^{\infty} \in l_{p_{m+1},\psi},$ $\|a\|_{p_{m+1},\psi} > 0,$ *then*

(i) *we have the following equivalent reverse inequalities with the best possible constant factor* $\beta(\lambda_{m+1})$:

$$\sum_{n=1}^{\infty} a_n \int_0^{\infty} \cdots \int_0^{\infty} \frac{\prod_{j=1}^m f_j(x_j)}{\prod_{k=1}^s (\sum_{i=1}^m x_i^{\lambda/s} + c_k n^{\lambda/s})} dx_1 \cdots dx_m$$

$$> \beta(\lambda_{m+1}) \left(\prod_{i=1}^m \|f_i\|_{p_i,\varphi_i} \right) \|a\|_{p_{m+1},\psi}, \qquad (6.110)$$

and

$$\left\{ \sum_{n=1}^{\infty} n^{p\lambda_{m+1}-1} \left[\int_0^{\infty} \cdots \int_0^{\infty} \frac{\prod_{j=1}^m f_j(x_j) dx_1 \cdots dx_m}{\prod_{k=1}^s (\sum_{i=1}^m x_i^{\lambda/s} + c_k n^{\lambda/s})} \right]^p \right\}^{\frac{1}{p}}$$

$$> \beta(\lambda_{m+1}) \prod_{i=1}^m \|f_i\|_{p_i, \varphi_i}; \tag{6.111}$$

(ii) *for* $\lambda_{m+1} = \frac{\lambda}{2} (< 1)$, *we have the following equivalent inequalities with the same best possible constant factor* $\beta(\frac{\lambda}{2})$:

$$\sum_{n=1}^{\infty} a_n \int_0^{\infty} \cdots \int_0^{\infty} \frac{\prod_{j=1}^m f_j(x_j)}{\prod_{k=1}^s [\sum_{i=1}^m (nx_i)^{\lambda/s} + c_k]} dx_1 \cdots dx_m$$

$$> \beta\left(\frac{\lambda}{2}\right) \left(\prod_{i=1}^m \|f_i\|_{p_i, \varphi_i} \right) \|a\|_{p_{m+1}, \psi}, \tag{6.112}$$

and

$$\left\{ \sum_{n=1}^{\infty} n^{\frac{p\lambda}{2}-1} \left[\int_0^{\infty} \cdots \int_0^{\infty} \frac{\prod_{j=1}^m f_j(x_j) dx_1 \cdots dx_m}{\prod_{k=1}^s [\sum_{i=1}^m (nx_i)^{\lambda/s} + c_k]} \right]^p \right\}^{\frac{1}{p}}$$

$$> \beta\left(\frac{\lambda}{2}\right) \prod_{i=1}^m \|f_i\|_{p_i, \varphi_i}. \tag{6.113}$$

By Theorems 6.7 and 6.8, we have

Corollary 6.10. *Suppose that* $s \in \mathbf{N}$, $0 < c_1 < \cdots < c_s$, $p_i > 1$, $0 < \lambda_i < 1 (i = 1, \cdots, m)$, $\lambda_{m+1} > 0$. *If* $f(x_{m+1}) \geq 0$, $f \in L_{p_{m+1}, \varphi}(\mathbf{R}_+)$, $\|f\|_{p_{m+1}, \varphi} > 0$, $a_{n_i}^{(i)} \geq 0$, $a^{(i)} = \{a_{n_i}^{(i)}\}_{n_i=1}^{\infty} \in l_{p_i, \psi_i}$, $\|a^{(i)}\|_{p_i, \psi_i} > 0 (i = 1, \cdots, m)$, *then*

(i) *we have the following equivalent inequalities with the best possible constant factor* $\beta(\lambda_{m+1})$:

$$\int_0^{\infty} f(x_{m+1}) \sum_{n_m=1}^{\infty} \cdots \sum_{n_1=1}^{\infty} \frac{\prod_{j=1}^m a_{n_j}^{(j)}}{\prod_{k=1}^s (\sum_{i=1}^m n_i^{\lambda/s} + c_k x_{m+1}^{\lambda/s})} dx_{m+1}$$

$$< \beta(\lambda_{m+1}) \left(\prod_{i=1}^m \|a^{(i)}\|_{p_i, \psi_i} \right) \|f\|_{p_{m+1}, \varphi}, \tag{6.114}$$

and

$$\left\{ \int_0^{\infty} x_{m+1}^{p\lambda_{m+1}-1} \left[\sum_{n_m=1}^{\infty} \cdots \sum_{n_1=1}^{\infty} \frac{\prod_{j=1}^m a_{n_j}^{(j)}}{\prod_{k=1}^s (\sum_{i=1}^m n_i^{\lambda/s} + c_k x_{m+1}^{\lambda/s})} \right]^p dx_{m+1} \right\}^{\frac{1}{p}}$$

$$< \beta(\lambda_{m+1}) \prod_{i=1}^m \|a^{(i)}\|_{p_i, \psi_i}; \tag{6.115}$$

(ii) *for* $\lambda_{m+1} = \frac{\lambda}{2}$, *we have the following equivalent inequalities with the best possible constant factor* $\beta(\frac{\lambda}{2})$:

$$\int_0^\infty f(x_{m+1}) \sum_{n_m=1}^\infty \cdots \sum_{n_1=1}^\infty \frac{\prod_{j=1}^m a_{n_j}^{(j)}}{\prod_{k=1}^s [\sum_{i=1}^m (x_{m+1} n_i)^{\lambda/s} + c_k]} dx_{m+1}$$

$$< \beta\left(\frac{\lambda}{2}\right) \left(\prod_{i=1}^m \|a^{(i)}\|_{p_i,\psi_i}\right) \|f\|_{p_{m+1},\varphi}, \qquad (6.116)$$

and

$$\left\{\int_0^\infty x_{m+1}^{\frac{p\lambda}{2}-1} \left[\sum_{n_m=1}^\infty \cdots \sum_{n_1=1}^\infty \frac{\prod_{j=1}^m a_{n_j}^{(j)}}{\prod_{k=1}^s [\sum_{i=1}^m (x_{m+1} n_i)^{\lambda/s} + c_k]}\right]^p dx_{m+1}\right\}^{\frac{1}{p}}$$

$$< \beta\left(\frac{\lambda}{2}\right) \prod_{i=1}^m \|a^{(i)}\|_{p_i,\psi_i}. \qquad (6.117)$$

In (6.44), for

$$k_\lambda(x_1,\cdots,x_m,x_{m+1}) = \frac{1}{\prod_{k=1}^s (\sum_{i=1}^m x_i^{\lambda/s} + c_k x_{m+1}^{\lambda/s})},$$

we find, for $k \neq m+1 (m \geq 2)$,

$$A_i(u_k) = \int_0^\infty \cdots \int_0^\infty \frac{\prod_{j=1(j\neq i,k)}^{m+1} u_j^{\lambda_j-1}}{\prod_{l=1}^s (1 + \sum_{j=1(j\neq i)}^m u_j^{\lambda/s} + c_l u_{m+1}^{\lambda/s})}$$

$$\times du_1 \cdots du_{i-1} du_{i+1} \cdots du_{k-1} du_{k+1} \cdots du_{m+1}$$

$$\leq \int_0^\infty \cdots \int_0^\infty \frac{\prod_{j=1(j\neq i,k)}^{m+1} u_j^{\lambda_j-1} du_1 \cdots du_{i-1} du_{i+1} \cdots du_{k-1} du_{k+1} \cdots du_{m+1}}{(1 + \sum_{j=1(j\neq i,k)}^m u_j^{\lambda/s} + c_1 u_{m+1}^{\lambda/s})^{s-\frac{s}{\lambda}(\lambda_k-\alpha)} u_k^{\lambda_k-\alpha}}$$

$$= C u_k^{\alpha-\lambda_k} \qquad (0 < \alpha < \lambda_k),$$

where

$$C = \int_0^\infty \cdots \int_0^\infty \frac{\prod_{j=1(j\neq i,k)}^{m+1} u_j^{\lambda_j-1} du_1 \cdots du_{i-1} du_{i+1} \cdots du_{k-1} du_{k+1} \cdots du_{m+1}}{(1 + \sum_{j=1(j\neq i,k)}^m u_j^{\lambda/s} + c_1 u_{m+1}^{\lambda/s})^{s-\frac{s}{\lambda}(\lambda_k-\alpha)}}.$$

Setting $v_j = u_j^{\lambda/s}(j \neq i, k, m+1)$, $v_{m+1} = c_1 u_{m+1}^{\lambda/s}$ in the above integral, we obtain

$$
\begin{aligned}
C &= \frac{(\frac{s}{\lambda})^{m-1}}{c_1^{s/\lambda}} \int_0^\infty \cdots \int_0^\infty \frac{\prod_{j=1(j\neq i,k)}^{m+1} v_j^{\frac{s}{\lambda}\lambda_j-1}}{(1+\sum_{j=1(j\neq i,k)}^{m+1} v_j)^{s-\frac{s}{\lambda}(\lambda_k-\alpha)}} \\
&\qquad \times dv_1 \cdots dv_{i-1} dv_{i+1} \cdots dv_{k-1} dv_{k+1} \cdots dv_{m+1} \\
&= \frac{(\frac{s}{\lambda})^{m-1}}{c_1^{s/\lambda}} \frac{\Gamma(s-\frac{s}{\lambda}(\lambda-\lambda_i-\alpha))}{\Gamma(s-\frac{s}{\lambda}(\lambda_k-\alpha))} \prod_{j=1(j\neq i,k)}^{m+1} \Gamma(\tfrac{s}{\lambda}\lambda_j) < \infty.
\end{aligned}
$$

Similarly, we can find that

$$
A_i(u_{m+1}) \leq C_1 u_{m+1}^{\alpha-\lambda_{m+1}} \qquad (0 < \alpha < \lambda_{m+1}, m \geq 2),
$$

and the same result in the case of $m = 1$.

If $0 < \lambda_i < 1(i = 1, \cdots, m)$, $\lambda_{m+1} > 0$, then, we set $\delta_0 = \min_{1\leq i\leq m}\{\lambda_{m+1}, \lambda_i, 1-\lambda_i\} > 0$. For any $\widetilde{\lambda}_i \in (\lambda_i - \delta_0, \lambda_i + \delta_0)(i = 1, \cdots, m+1)$, $\sum_{i=1}^{m+1} \widetilde{\lambda}_i = \lambda$, we find that $\widetilde{\lambda}_{m+1} > 0$, $0 < \widetilde{\lambda}_i < \lambda_i + \delta_0 \leq 1(i = 1, \cdots, m)$, $k(\widetilde{\lambda}_{m+1}) = \beta(\widetilde{\lambda}_{m+1}) \in \mathbf{R}_+$, and both $\prod_{k=1}^s \frac{1}{\sum_{i=1}^m y_i^{\lambda/s}+c_k x_{m+1}^{\lambda/s}} y_j^{\widetilde{\lambda}_j-1}$ and $\prod_{k=1}^s \frac{1}{\sum_{i=1}^m (x_{m+1}y_i)^{\lambda/s}+c_k} y_j^{\widetilde{\lambda}_j-1}$ are strictly decreasing with respect to $y_j \in \mathbf{R}$ $(j = 1, \cdots, m)$.

Then, by Theorems 6.9 and 6.10, it follows that

Corollary 6.11. *Suppose that* $s \in \mathbf{N}$, $0 < c_1 < \cdots < c_s$, $0 < p_1 < 1$, $p_i < 0(i = 2, \cdots, m+1)$, $0 < \lambda_i < 1(i = 1, \cdots, m)$, $\lambda_{m+1} > 0$. *If* $f(x_{m+1}) \geq 0$, $f \in L_{p_{m+1},\varphi}(\mathbf{R}_+)$, $\|f\|_{p_{m+1},\varphi} > 0$, $a_{n_1}^{(1)} \geq 0$, $a^{(1)} = \{a_{n_1}^{(1)}\}_{n_1=1}^\infty \in l_{p_1,\Psi_1}$, $\|a^{(1)}\|_{p_1,\Psi_1} > 0$, $a_{n_i}^{(i)} \geq 0$, $a^{(i)} = \{a_{n_i}^{(i)}\}_{n_i=1}^\infty \in l_{p_i,\psi_i}$, $\|a^{(i)}\|_{p_i,\psi_i} > 0(i = 2, \cdots, m)$, *then*

(i) *we have the following equivalent inequalities with the same best possible constant factor* $\beta(\lambda_{m+1})$:

$$
\int_0^\infty f(x_{m+1}) \sum_{n_m=1}^\infty \cdots \sum_{n_1=1}^\infty \frac{\prod_{j=1}^m a_{n_j}^{(j)}}{\prod_{k=1}^s(\sum_{i=1}^m n_i^{\lambda/s}+c_k x_{m+1}^{\lambda/s})} dx_{m+1}
$$
$$
> \beta(\lambda_{m+1})\|f\|_{p_{m+1},\varphi}\|a^{(1)}\|_{p_1,\Psi_1}\prod_{i=2}^m \|a^{(i)}\|_{p_i,\psi_i}, \qquad (6.118)
$$

and

$$
\left\{\int_0^\infty x_{m+1}^{p\lambda_{m+1}-1}\left[\sum_{n_m=1}^\infty \cdots \sum_{n_1=1}^\infty \frac{\prod_{j=1}^m a_{n_j}^{(j)}}{\prod_{k=1}^s(\sum_{i=1}^m n_i^{\lambda/s}+c_k x_{m+1}^{\lambda/s})}\right]^p dx_{m+1}\right\}^{\frac{1}{p}}
$$
$$
> \beta(\lambda_{m+1})\|a^{(1)}\|_{p_1,\Psi_1}\prod_{i=2}^m \|a^{(i)}\|_{p_i,\psi_i}; \qquad (6.119)
$$

(ii) *for* $\lambda_{m+1} = \frac{\lambda}{2}$, *we have the following equivalent inequalities with the same best possible constant factor* $\beta(\frac{\lambda}{2})$:

$$\int_0^\infty f(x_{m+1}) \sum_{n_m=1}^\infty \cdots \sum_{n_1=1}^\infty \frac{\prod_{j=1}^m a_{n_j}^{(j)}}{\prod_{k=1}^s [\sum_{i=1}^m (x_{m+1} n_i)^{\lambda/s} + c_k]} \, dx_{m+1}$$

$$> \beta\left(\frac{\lambda}{2}\right) ||f||_{p_{m+1},\varphi} ||a^{(1)}||_{p_1,\Psi_1} \prod_{i=2}^m ||a^{(i)}||_{p_i,\psi_i}, \qquad (6.120)$$

and

$$\left\{ \int_0^\infty x_{m+1}^{\frac{p\lambda}{2}-1} \left[\sum_{n_m=1}^\infty \cdots \sum_{n_1=1}^\infty \frac{\prod_{j=1}^m a_{n_j}^{(j)}}{\prod_{k=1}^s [\sum_{i=1}^m (x_{m+1} n_i)^{\lambda/s} + c_k]} \right]^p dx_{m+1} \right\}^{\frac{1}{p}}$$

$$> \beta\left(\frac{\lambda}{2}\right) ||a^{(1)}||_{p_1,\Psi_1} \prod_{i=2}^m ||a^{(i)}||_{p_i,\psi_i}. \qquad (6.121)$$

By Theorems 6.10 and 6.12, it follows that

Corollary 6.12. *Suppose that* $s \in \mathbf{N}$, $0 < c_1 < \cdots < c_s$, $p_i < 0$, $0 < \lambda_i < 1 (i = 1, \cdots, m)$, $\lambda_{m+1} > 0$, $0 < p_{m+1} < 1$. *If* $f(x_{m+1}) \geq 0$, $f \in L_{p_{m+1},\Phi}(\mathbf{R}_+)$, $||f||_{p_{m+1},\Phi} > 0$, $a_{n_i}^{(i)} \geq 0$, $a^{(i)} = \{a_{n_i}^{(i)}\}_{n_i=1}^\infty \in l_{p_i,\psi_i}$, $||a^{(i)}||_{p_i,\psi_i} > 0 (i = 1, \cdots, m)$, *then*

(i) *we have the following equivalent reverse inequalities with the same best possible constant factor* $\beta(\lambda_{m+1})$:

$$\int_0^\infty f(x_{m+1}) \sum_{n_m=1}^\infty \cdots \sum_{n_1=1}^\infty \frac{\prod_{j=1}^m a_{n_j}^{(j)}}{\prod_{k=1}^s (\sum_{i=1}^m n_i^{\lambda/s} + c_k x_{m+1}^{\lambda/s})} \, dx_{m+1}$$

$$> \beta(\lambda_{m+1}) \left(\prod_{i=1}^m ||a^{(i)}||_{p_i,\psi_i} \right) ||f||_{p_{m+1},\Phi}, \qquad (6.122)$$

and

$$\left\{ \int_0^\infty \frac{x_{m+1}^{p\lambda_{m+1}-1}}{[1 - \theta_{m+1}(x_{m+1})]^{p-1}} \right.$$

$$\times \left. \left[\sum_{n_m=1}^\infty \cdots \sum_{n_1=1}^\infty \frac{\prod_{j=1}^m a_{n_j}^{(j)}}{\prod_{k=1}^s (\sum_{i=1}^m n_i^{\lambda/s} + c_k x_{m+1}^{\lambda/s})} \right]^p dx_{m+1} \right\}^{\frac{1}{p}}$$

$$> \beta(\lambda_{m+1}) \prod_{i=1}^m ||a^{(i)}||_{p_i,\psi_i}; \qquad (6.123)$$

(ii) *for* $\lambda_{m+1} = \frac{\lambda}{2}$, $f(x_{m+1}) \geq 0$, $f \in L_{p_{m+1}, \widetilde{\Phi}}(\mathbf{R}_+)$, $||f||_{p_{m+1}, \widetilde{\Phi}} > 0$, *we have the following equivalent inequalities with the same best possible constant factor* $\beta(\frac{\lambda}{2})$:

$$\int_0^\infty f(x_{m+1}) \sum_{n_m=1}^\infty \cdots \sum_{n_1=1}^\infty \frac{\prod_{j=1}^m a_{n_j}^{(j)}}{\prod_{k=1}^s [\sum_{i=1}^m (x_{m+1}n_i)^{\lambda/s} + c_k]} \, dx_{m+1}$$

$$> \beta\left(\frac{\lambda}{2}\right) \left(\prod_{i=1}^m ||a^{(i)}||_{p_i, \psi_i}\right) ||f||_{p_{m+1}, \widetilde{\Phi}}, \tag{6.124}$$

and

$$\left\{ \int_0^\infty \frac{x_{m+1}^{\frac{p\lambda}{2}-1}}{[1 - \widetilde{\theta}_{m+1}(x_{m+1})]^{p-1}} \right.$$

$$\left. \times \left[\sum_{n_m=1}^\infty \cdots \sum_{n_1=1}^\infty \frac{\prod_{j=1}^m a_{n_j}^{(j)}}{\prod_{k=1}^s [\sum_{i=1}^m (x_{m+1}n_i)^{\lambda/s} + c_k]} \right]^p dx_{m+1} \right\}^{\frac{1}{p}}$$

$$> \beta\left(\frac{\lambda}{2}\right) \prod_{i=1}^m ||a^{(i)}||_{p_i, \psi_i}. \tag{6.125}$$

Remark 6.4. (i) If in (6.24), we set

$$k_\lambda(x_1, \cdots, x_m, n) = \frac{1}{\prod_{k=1}^s (\sum_{i=1}^m n_i^{\lambda/s} + c_k x_{m+1}^{\lambda/s})},$$

then, in view of Corollary 6.1 and equality (6.25), we have

$$||T|| = k(\lambda_{m+1}) = \beta(\lambda_{m+1}).$$

If, in (6.26), for $\lambda_{m+1} = \frac{\lambda}{2} < 1$, we set

$$k_\lambda(nx_1, \cdots, nx_m, 1) = \frac{1}{\prod_{k=1}^s [\sum_{i=1}^m (x_{m+1}n_i)^{\lambda/s} + c_k]},$$

then, in view of Corollary 6.1 and equality (6.27), we have $||T_1|| = \beta(\frac{\lambda}{2})$.

(ii) If, in (6.61), we set

$$k_\lambda(n_1, \cdots, n_m, x_{m+1}) = \frac{1}{\prod_{k=1}^s (\sum_{i=1}^m n_i^{\lambda/s} + c_k x_{m+1}^{\lambda/s})},$$

then, in view of Corollary 6.4 and (6.62), we have

$$||\widetilde{T}|| = k(\lambda_{m+1}) = \beta(\lambda_{m+1}).$$

If, in (6.63), for $\lambda_{m+1} = \frac{\lambda}{2}$, we set

$$k_\lambda(x_{m+1}n_1, \cdots, x_{m+1}n_m, 1) = \frac{1}{\prod_{k=1}^s [\sum_{i=1}^m (x_{m+1}n_i)^{\lambda/s} + c_k]},$$

then, in view of Corollary 6.4 and (6.64), we have $||\widetilde{T}_1|| = \beta(\frac{\lambda}{2})$.

6.4.3   The Case of
$$k_\lambda(x_1, \cdots, x_m, x_{m+1}) = \frac{1}{(\max_{1 \le i \le m+1}\{x_i\})^\lambda}$$

Lemma 6.18. *If* $\lambda_i > 0$ $(i = 1, \cdots, m+1)$,

$$k_\lambda(x_1, \cdots, x_{m+1}) = \frac{1}{(\max_{1 \le i \le m+1}\{x_i\})^\lambda},$$

then, we have

$$k(\lambda_{m+1}) = \int_0^\infty \cdots \int_0^\infty \frac{\prod_{j=1}^m u_j^{\lambda_j - 1}}{(\max_{1 \le j \le m}\{u_j, 1\})^\lambda} du_1 \cdots du_m$$

$$= \lambda \prod_{i=1}^{m+1} \frac{1}{\lambda_i}. \tag{6.126}$$

Proof.   We prove (6.126) by mathematical induction (see Yang [128]).
For $m = 1$, we find

$$k(\lambda_2) = \int_0^\infty \frac{u_1^{\lambda_1 - 1}}{(\max\{u_1, 1\})^\lambda} du_1$$

$$= \int_0^1 u_1^{\lambda_1 - 1} du_1 + \int_1^\infty \frac{u_1^{\lambda_1 - 1}}{u_1^\lambda} du_1$$

$$= \frac{1}{\lambda_1} + \frac{1}{\lambda - \lambda_1} = \frac{\lambda}{\lambda_1 \lambda_2},$$

and then (6.126) follows.
Assuming that (6.126) is valid for $m = n - 1$, then, for $m = n$, we have

$$k(\lambda_{n+1}) = \int_0^\infty \cdots \int_0^\infty \frac{\prod_{j=1}^n u_j^{\lambda_j - 1}}{(\max_{1 \le j \le n}\{u_j, 1\})^\lambda} du_1 \cdots du_n$$

$$= \int_0^\infty \cdots \int_0^\infty \left[\int_0^\infty \frac{u_n^{\lambda_n - 1}}{(\max_{1 \le j \le n}\{u_j, 1\})^\lambda} du_n \right]$$

$$\times \prod_{j=1}^{n-1} u_j^{\lambda_j - 1} du_1 \cdots du_{n-1}.$$

Since, we find

$$\int_0^\infty \frac{u_n^{\lambda_n - 1}}{(\max_{1 \le j \le n}\{u_j, 1\})^\lambda} du_n$$

$$= \int_0^{\max_{1 \le j \le n-1}\{u_j, 1\}} \frac{u_n^{\lambda_n - 1}}{(\max_{1 \le j \le n}\{u_j, 1\})^\lambda} du_n$$

$$+ \int_{\max_{1 \le j \le n-1}\{u_j, 1\}}^{\infty} \frac{u_n^{\lambda_n-1}}{(\max_{1 \le j \le n}\{u_j, 1\})^{\lambda}} \, du_n$$

$$= \int_0^{\max_{1 \le j \le n-1}\{u_j, 1\}} \frac{u_n^{\lambda_n-1}}{(\max_{1 \le j \le n-1}\{u_j, 1\})^{\lambda}} \, du_n$$

$$+ \int_{\max_{1 \le j \le n-1}\{u_j, 1\}}^{\infty} \frac{u_n^{\lambda_n-1}}{u_n^{\lambda}} \, du_n$$

$$= \frac{(\max_{1 \le j \le n-1}\{u_j, 1\})^{\lambda_n}}{\lambda_n (\max_{1 \le j \le n-1}\{u_j, 1\})^{\lambda}} + \frac{(\max_{1 \le j \le n-1}\{u_j, 1\})^{\lambda_n-\lambda}}{\lambda - \lambda_n}$$

$$= \frac{\lambda}{\lambda_n(\lambda - \lambda_n)} \frac{1}{(\max_{1 \le j \le n-1}\{u_j, 1\})^{\lambda-\lambda_n}},$$

then, by the assumption of $m = n - 1$, it follows that

$$k(\lambda_{n+1}) = \frac{\lambda}{\lambda_n(\lambda - \lambda_n)} \int_0^{\infty} \cdots \int_0^{\infty} \frac{\prod_{j=1}^{n-1} u_j^{\lambda_j-1} du_1 \cdots du_{n-1}}{(\max_{1 \le j \le n-1}\{u_j, 1\})^{\lambda-\lambda_n}}$$

$$= \frac{\lambda(\lambda - \lambda_n)}{\lambda_n(\lambda - \lambda_n)} \prod_{i=1(i \ne n)}^{n+1} \frac{1}{\lambda_i} = \lambda \prod_{i=1}^{n+1} \frac{1}{\lambda_i},$$

and then (6.99) follows. $\square$

With the assumptions of Lemma 6.17, if $\lambda_{m+1} < 1$, then we set $\delta_0 = \min_{1 \le i \le m+1}\{\lambda_i, 1-\lambda_{m+1}\} > 0$. For any $\tilde{\lambda}_i \in (\lambda_i-\delta_0, \lambda_i+\delta_0)(i = 1, \cdots, m+1)$, $\sum_{i=1}^{m+1} \tilde{\lambda}_i = \lambda$, it follows that $\tilde{\lambda}_i > 0(i = 1, \cdots, m+1)$, $\tilde{\lambda}_{m+1} < \lambda_{m+1} + \delta_0 \le \lambda_{m+1} + 1 - \lambda_{m+1} = 1$,

$$k(\tilde{\lambda}_{m+1}) = \int_0^{\infty} \cdots \int_0^{\infty} \frac{\prod_{j=1}^{m} u_j^{\tilde{\lambda}_j-1}}{(\max_{1 \le j \le m}\{u_j, 1\})^{\lambda}} \, du_1 \cdots du_m$$

$$= \lambda \prod_{i=1}^{m+1} \frac{1}{\tilde{\lambda}_i} \in \mathbf{R}_+,$$

and both $\frac{1}{(\max_{1 \le i \le m}\{x_i, y\})^{\lambda}} y^{\tilde{\lambda}_{m+1}-1}$ and $\frac{1}{(\max_{1 \le i \le m}\{yx_i, 1\})^{\lambda}} y^{\tilde{\lambda}_{m+1}-1}$ are strictly decreasing with respect to $y \in \mathbf{R}_+$. Then for

$$k_{\lambda}(x_1, \cdots, x_m, x_{m+1}) = \frac{1}{(\max_{1 \le i \le m+1}\{x_i\})^{\lambda}}$$

in Theorems 6.1 and 6.2, it follows that

Corollary 6.13. *Suppose that* $p_i > 1, \lambda_i > 0(i = 1, \cdots, m+1)$, $\lambda_{m+1} < 1$. *If* $f_i(x_i) \ge 0$, $f_i \in L_{p_i, \varphi_i}(\mathbf{R}_+)$, $\|f_i\|_{p_i, \varphi_i} > 0(i = 1, \cdots, m)$, $a_n \ge 0$, $a = \{a_n\}_{n=1}^{\infty} \in l_{p_{m+1}, \psi}$, $\|a\|_{p_{m+1}, \psi} > 0$, *then*

(i) *we have the following equivalent inequalities with the best possible constant factor* $\lambda \prod_{i=1}^{m+1} \frac{1}{\lambda_i}$:

$$\sum_{n=1}^{\infty} a_n \int_0^{\infty} \cdots \int_0^{\infty} \frac{\prod_{j=1}^{m} f_j(x_j)}{(\max_{1 \le i \le m}\{x_i, n\})^{\lambda}} dx_1 \cdots dx_m$$

$$< \frac{\lambda}{\lambda_{m+1}} \left(\prod_{i=1}^{m} \frac{1}{\lambda_i} ||f_i||_{p_i, \varphi_i} \right) ||a||_{p_{m+1}, \psi}, \qquad (6.127)$$

and

$$\left\{ \sum_{n=1}^{\infty} n^{p\lambda_{m+1}-1} \left[\int_0^{\infty} \cdots \int_0^{\infty} \frac{\prod_{j=1}^{m} f_j(x_j) dx_1 \cdots dx_m}{(\max_{1 \le i \le m}\{x_i, n\})^{\lambda}} \right]^p \right\}^{\frac{1}{p}}$$

$$< \frac{\lambda}{\lambda_{m+1}} \prod_{i=1}^{m} \frac{1}{\lambda_i} ||f_i||_{p_i, \varphi_i}; \qquad (6.128)$$

(ii) *for* $\lambda_{m+1} = \frac{\lambda}{2}(< 1)$, *we have the following equivalent inequalities with the same best constant factor* $2 \prod_{i=1}^{m} \frac{1}{\lambda_i}$:

$$\sum_{n=1}^{\infty} a_n \int_0^{\infty} \cdots \int_0^{\infty} \frac{\prod_{j=1}^{m} f_j(x_j)}{(\max_{1 \le i \le m}\{nx_i, 1\})^{\lambda}} dx_1 \cdots dx_m$$

$$< 2 \left(\prod_{i=1}^{m} \frac{1}{\lambda_i} ||f_i||_{p_i, \varphi_i} \right) ||a||_{p_{m+1}, \psi}, \qquad (6.129)$$

and

$$\left\{ \sum_{n=1}^{\infty} n^{\frac{p\lambda}{2}-1} \left[\int_0^{\infty} \cdots \int_0^{\infty} \frac{\prod_{j=1}^{m} f_j(x_j) dx_1 \cdots dx_m}{(\max_{1 \le i \le m}\{nx_i, 1\})^{\lambda}} \right]^p \right\}^{\frac{1}{p}}$$

$$< 2 \prod_{i=1}^{m} \frac{1}{\lambda_i} ||f_i||_{p_i, \varphi_i}. \qquad (6.130)$$

In (6.9) and (6.12), for

$$k_{\lambda}(x_1, \cdots, x_m, x_{m+1}) = \frac{1}{(\max_{1 \le i \le m+1}\{x_i\})^{\lambda}},$$

we find, for $m \ge 2$,

$$A_1(u_{m+1}) = \int_0^{\infty} \cdots \int_0^{\infty} \frac{\prod_{j=2}^{m} u_j^{\lambda_j - 1}}{(\max_{2 \le j \le m+1}\{1, u_j\})^{\lambda}} du_2 \cdots du_m$$

$$\le \int_0^{\infty} \cdots \int_0^{\infty} \frac{\prod_{j=2}^{m} u_j^{\lambda_j - 1} du_2 \cdots du_m}{(\max_{2 \le j \le m}\{1, u_j\})^{\lambda + \alpha - \lambda_{m+1}} u_{m+1}^{\lambda_{m+1} - \alpha}}$$

$$= \left(\frac{\lambda + \alpha - \lambda_{m+1}}{\lambda_1 + \alpha} \right) \prod_{i=2}^{m} \frac{1}{\lambda_i} (u_{m+1}^{\alpha - \lambda_{m+1}}) \quad (0 < \alpha < \lambda_{m+1}),$$

and

$$B_1(u_i) = \int_0^\infty \cdots \int_0^\infty \frac{\prod_{j=1(j\neq i)}^m u_j^{\lambda_j-1} du_1 \cdots du_{i-1} du_{i+1} \cdots du_m}{(\max_{1\le j\le m}\{1,u_j\})^\lambda}$$

$$\le \int_0^\infty \cdots \int_0^\infty \frac{\prod_{j=1(j\neq i)}^m u_j^{\lambda_j-1} du_1 \cdots du_{i-1} du_{i+1} \cdots du_m}{(\max_{1\le j\neq i\le m}\{1,u_j\})^{\lambda+\alpha-\lambda_i} u_i^{\lambda_i-\alpha}}$$

$$= \left(\frac{\lambda+\alpha-\lambda_i}{\lambda_{m+1}+\alpha}\right) \prod_{j=1(j\neq i)}^m \frac{1}{\lambda_j}(u_i^{\alpha-\lambda_i}) \quad (0 < \alpha < \lambda_i).$$

For $m = 1$, we can still find the same results.

By Theorems 6.3 and 6.4, it follows that

Corollary 6.14. *Suppose that* $0 < p_1 < 1$, $p_i < 0$ $(i = 2, \cdots, m+1)$, $\lambda_i > 0 (i = 1, \cdots, m+1)$, $\lambda_{m+1} < 1$. *If* $f_1(x_1) \ge 0$, $f_1 \in L_{p_1,\Phi_1}(\mathbf{R}_+)$, $\|f_1\|_{p_1,\Phi_1} > 0$, $f_i(x_i) \ge 0$, $f_i \in L_{p_i,\varphi_i}(\mathbf{R}_+)$, $\|f_i\|_{p_i,\varphi_i} > 0 (i = 2, \cdots, m)$, $a_n \ge 0$, $a = \{a_n\}_{n=1}^\infty \in l_{p_{m+1},\psi}$, $\|a\|_{p_{m+1},\psi} > 0$, *then*

(i) *we have the following equivalent inequalities with the same best possible constant factor* $\lambda \prod_{i=1}^{m+1}\frac{1}{\lambda_i}$:

$$\sum_{n=1}^\infty a_n \int_0^\infty \cdots \int_0^\infty \frac{\prod_{j=1}^m f_j(x_j)}{(\max_{1\le i\le m}\{x_i,n\})^\lambda} dx_1 \cdots dx_m$$

$$> \lambda \prod_{i=1}^{m+1}\frac{1}{\lambda_i}\|a\|_{p_{m+1},\psi}\|f_1\|_{p_1,\Phi_1}\prod_{i=2}^m \|f_i\|_{p_i,\varphi_i}, \quad (6.131)$$

and

$$\left\{\sum_{n=1}^\infty n^{p\lambda_{m+1}-1}\left[\int_0^\infty \cdots \int_0^\infty \frac{\prod_{j=1}^m f_j(x_j)dx_1 \cdots dx_m}{(\max_{1\le i\le m}\{x_i,n\})^\lambda}\right]^p\right\}^{\frac{1}{p}}$$

$$> \lambda \prod_{i=1}^{m+1}\frac{1}{\lambda_i}\|f_1\|_{p_1,\Phi_1}\prod_{i=2}^m \|f_i\|_{p_i,\varphi_i}; \quad (6.132)$$

(ii) *for* $\lambda_{m+1} = \frac{\lambda}{2} (< 1)$, $f_1 \in L_{p_1,\tilde{\Phi}_1}(\mathbf{R}_+)$, $\|f_1\|_{p_1,\tilde{\Phi}_1} > 0$, *we have the following equivalent inequalities with the same best possible constant factor* $2 \prod_{i=1}^m \frac{1}{\lambda_i}$:

$$\sum_{n=1}^\infty a_n \int_0^\infty \cdots \int_0^\infty \frac{\prod_{j=1}^m f_j(x_j)}{(\max_{1\le i\le m}\{nx_i,1\})^\lambda}dx_1 \cdots dx_m$$

$$> 2\left(\prod_{i=1}^m \frac{1}{\lambda_i}\right)\|a\|_{p_{m+1},\psi}\|f_1\|_{p_1,\tilde{\Phi}_1}\prod_{i=2}^m \|f_i\|_{p_i,\varphi_i}, \quad (6.133)$$

and

$$\left\{\sum_{n=1}^{\infty} n^{p\frac{\lambda}{2}-1}\left[\int_0^{\infty}\cdots\int_0^{\infty}\frac{\prod_{j=1}^m f_j(x_j)dx_1\cdots dx_m}{(\max_{1\le i\le m}\{nx_i,1\})^{\lambda}}\right]^p\right\}^{\frac{1}{p}}$$

$$> 2\left(\prod_{i=1}^m\frac{1}{\lambda_i}\right)||f_1||_{p_1,\widetilde{\Phi}_1}\prod_{i=2}^m||f_i||_{p_i,\varphi_i}. \tag{6.134}$$

By Theorems 6.5 and 6.6, we still have

Corollary 6.15. *Suppose that $p_i < 0$, $\lambda_i > 0 (i = 1, \cdots, m)$, $0 < p_{m+1} < 1$, $0 < \lambda_{m+1} < 1$. If $f_i(x_i)$, $a_n \ge 0$, $f_i \in L_{p_i,\varphi_i}(\mathbf{R}_+)$, $||f_i||_{p_i,\varphi_i} > 0 (i = 1, \cdots, m)$, $a = \{a_n\}_{n=1}^{\infty} \in l_{p_{m+1},\psi}$, $||a||_{p_{m+1},\psi} > 0$, then*

(i) *we have the following equivalent reverse inequalities with the same best possible constant factor $\lambda\prod_{i=1}^{m+1}\frac{1}{\lambda_i}$:*

$$\sum_{n=1}^{\infty} a_n\int_0^{\infty}\cdots\int_0^{\infty}\frac{\prod_{j=1}^m f_j(x_j)}{(\max_{1\le i\le m}\{x_i,n\})^{\lambda}}\,dx_1\cdots dx_m$$

$$> \frac{\lambda}{\lambda_{m+1}}\left(\prod_{i=1}^m\frac{1}{\lambda_i}||f_i||_{p_i,\varphi_i}\right)||a||_{p_{m+1},\psi}, \tag{6.135}$$

and

$$\left\{\sum_{n=1}^{\infty} n^{p\lambda_{m+1}-1}\left[\int_0^{\infty}\cdots\int_0^{\infty}\frac{\prod_{j=1}^m f_j(x_j)dx_1\cdots dx_m}{(\max_{1\le i\le m}\{x_i,n\})^{\lambda}}\right]^p\right\}^{\frac{1}{p}}$$

$$> \frac{\lambda}{\lambda_{m+1}}\prod_{i=1}^m\frac{1}{\lambda_i}||f_i||_{p_i,\varphi_i}; \tag{6.136}$$

(ii) *for $\lambda_{m+1} = \frac{\lambda}{2} (< 1)$, we have the following equivalent inequalities with the same best possible constant factor $2\prod_{i=1}^m\frac{1}{\lambda_i}$:*

$$\sum_{n=1}^{\infty} a_n\int_0^{\infty}\cdots\int_0^{\infty}\frac{\prod_{j=1}^m f_j(x_j)}{(\max_{1\le i\le m}\{nx_i,1\})^{\lambda}}\,dx_1\cdots dx_m$$

$$> 2\left(\prod_{i=1}^m\frac{1}{\lambda_i}||f_i||_{p_i,\varphi_i}\right)||a||_{p_{m+1},\psi}, \tag{6.137}$$

and

$$\left\{\sum_{n=1}^{\infty} n^{\frac{p\lambda}{2}-1}\left[\int_0^{\infty}\cdots\int_0^{\infty}\frac{\prod_{j=1}^m f_j(x_j)dx_1\cdots dx_m}{(\max_{1\le i\le m}\{nx_i,1\})^{\lambda}}\right]^p\right\}^{\frac{1}{p}}$$

$$> 2\prod_{i=1}^m\frac{1}{\lambda_i}||f_i||_{p_i,\varphi_i}. \tag{6.138}$$

By Theorems 6.7 and 6.8, we have

Corollary 6.16. *Suppose that* $p_i > 1, 0 < \lambda_i < 1(i = 1, \cdots, m), \lambda_{m+1} > 0$. *If* $f(x_{m+1}) \geq 0$, $f \in L_{p_{m+1},\varphi}(\mathbf{R}_+)$, $||f||_{p_{m+1},\varphi} > 0$, $a_{n_i}^{(i)} \geq 0$, $a^{(i)} = \{a_{n_i}^{(i)}\}_{n_i=1}^{\infty} \in l_{p_i,\psi_i}, ||a^{(i)}||_{p_i,\psi_i} > 0(i = 1, \cdots, m)$, *then*

(i) we have the following equivalent inequalities with the best constant factor $\lambda \prod_{i=1}^{m+1} \frac{1}{\lambda_i}$:

$$\int_0^\infty f(x_{m+1}) \sum_{n_m=1}^{\infty} \cdots \sum_{n_1=1}^{\infty} \frac{\prod_{j=1}^m a_{n_j}^{(j)}}{(\max_{1\leq i\leq m}\{n_i, x_{m+1}\})^\lambda} dx_{m+1}$$

$$< \frac{\lambda}{\lambda_{m+1}} \left(\prod_{i=1}^m \frac{1}{\lambda_i} ||a^{(i)}||_{p_i,\psi_i} \right) ||f||_{p_{m+1},\varphi}, \qquad (6.139)$$

and

$$\left\{ \int_0^\infty x_{m+1}^{p\lambda_{m+1}-1} \left[\sum_{n_m=1}^{\infty} \cdots \sum_{n_1=1}^{\infty} \frac{\prod_{j=1}^m a_{n_j}^{(j)}}{(\max_{1\leq i\leq m}\{n_i, x_{m+1}\})^\lambda} \right]^p dx_{m+1} \right\}^{\frac{1}{p}}$$

$$< \frac{\lambda}{\lambda_{m+1}} \prod_{i=1}^m \frac{1}{\lambda_i} ||a^{(i)}||_{p_i,\psi_i}; \qquad (6.140)$$

(ii) for $\lambda_{m+1} = \frac{\lambda}{2}$, *we have the following equivalent inequalities with the best possible constant factor* $2 \prod_{i=1}^m \frac{1}{\lambda_i}$:

$$\int_0^\infty f(x_{m+1}) \sum_{n_m=1}^{\infty} \cdots \sum_{n_1=1}^{\infty} \frac{\prod_{j=1}^m a_{n_j}^{(j)}}{\max_{1\leq i\leq m}\{x_{m+1}n_i, 1\})^\lambda} dx_{m+1}$$

$$< 2 \left(\prod_{i=1}^m \frac{1}{\lambda_i} ||a^{(i)}||_{p_i,\psi_i} \right) ||f||_{p_{m+1},\varphi}, \qquad (6.141)$$

and

$$\left\{ \int_0^\infty x_{m+1}^{\frac{p\lambda}{2}-1} \left[\sum_{n_m=1}^{\infty} \cdots \sum_{n_1=1}^{\infty} \frac{\prod_{j=1}^m a_{n_j}^{(j)}}{\max_{1\leq i\leq m}\{x_{m+1}n_i, 1\})^\lambda} \right]^p dx_{m+1} \right\}^{\frac{1}{p}}$$

$$< 2 \prod_{i=1}^m \frac{1}{\lambda_i} ||a^{(i)}||_{p_i,\psi_i}. \qquad (6.142)$$

In (6.44), for

$$k_\lambda(x_1, \cdots, x_m, x_{m+1}) = \frac{1}{(\max_{1\leq i\leq m+1}\{x_i\})^\lambda},$$

we find, for $m \geq 2$,

$$A_i(u_k) = \int_0^\infty \cdots \int_0^\infty \frac{\prod_{j=1(j\neq i,k)}^{m+1} u_j^{\lambda_j-1}}{(\max_{1\leq j\neq i\leq m+1}\{u_j,1\})^\lambda}$$

$$\times \, du_1 \cdots du_{i-1}du_{i+1}\cdots du_{k-1}du_{k+1}\cdots du_{m+1}$$

$$\leq \int_0^\infty \cdots \int_0^\infty \frac{\prod_{j=1(j\neq i,k)}^{m+1} u_j^{\lambda_j-1} du_1 \cdots du_{i-1}du_{i+1}\cdots du_{k-1}du_{k+1}\cdots du_{m+1}}{(\max_{1\leq j\neq i,k\leq m+1}\{u_j,1\})^{\lambda+\alpha-\lambda_k} u_k^{\lambda_k-\alpha}}$$

$$= \left(\frac{\lambda+\alpha-\lambda_k}{\lambda_i+\alpha}\right)\prod_{j=1(j\neq i,k)}^{m+1} \frac{1}{\lambda_j}(u_k^{\alpha-\lambda_k}) \, (0 < \alpha < \lambda_k).$$

For $m = 1$, we can still find the similar result.

If $0 < \lambda_i < 1(i = 1,\cdots,m)$, $\lambda_{m+1} > 0$, then we set $\delta_0 = \min_{1\leq i\leq m}\{\lambda_{m+1}, \lambda_i, 1-\lambda_i\} > 0$. For any $\widetilde{\lambda}_i \in (\lambda_i - \delta_0, \lambda_i + \delta_0)(i = 1,\cdots,m+1)$, $\sum_{i=1}^{m+1}\widetilde{\lambda}_i = \lambda$, we find $\widetilde{\lambda}_{m+1} > 0$, $0 < \widetilde{\lambda}_i < 1(i = 1,\cdots,m)$,

$$k(\widetilde{\lambda}_{m+1}) = \lambda \prod_{i=1}^{m+1} \frac{1}{\widetilde{\lambda}_i} \in \mathbf{R}_+$$

and both $\frac{1}{(\max_{1\leq i\leq m}\{y_i,x_{m+1}\})^\lambda} y_j^{\widetilde{\lambda}_j-1}$ and $\frac{1}{(\max_{1\leq i\leq m}\{x_{m+1}y_i,1\})^\lambda} y_j^{\widetilde{\lambda}_j-1}$ are strictly decreasing with respect to $y_j \in \mathbf{R}_+(j = 1,\cdots,m)$.

Then by Theorems 6.9 and 6.10, it follows that

Corollary 6.17. *Suppose that* $0 < p_1 < 1$, $p_i < 0(i = 2,\cdots,m+1)$, $0 < \lambda_i < 1(i = 1,\cdots,m)$, $\lambda_{m+1} > 0$. *If* $f(x_{m+1}) \geq 0$, $f \in L_{p_{m+1},\varphi}(\mathbf{R}_+)$, $||f||_{p_{m+1},\varphi} > 0, a_{n_1}^{(1)} \geq 0$, $a^{(1)} = \{a_{n_1}^{(1)}\}_{n_1=1}^\infty \in l_{p_1,\Psi_1}$, $||a^{(1)}||_{p_1,\Psi_1} > 0$, $a_{n_i}^{(i)} \geq 0$, $a^{(i)} = \{a_{n_i}^{(i)}\}_{n_i=1}^\infty \in l_{p_i,\psi_i}$, $||a^{(i)}||_{p_i,\psi_i} > 0(i = 2,\cdots,m)$, *then*

(i) *we have the following equivalent inequalities with the same best possible constant factor* $\lambda \prod_{i=1}^{m+1} \frac{1}{\lambda_i}$:

$$\int_0^\infty f(x_{m+1}) \sum_{n_m=1}^\infty \cdots \sum_{n_1=1}^\infty \frac{\prod_{j=1}^m a_{n_j}^{(j)}}{(\max_{1\leq i\leq m}\{n_i,x_{m+1}\})^\lambda} \, dx_{m+1}$$

$$> \lambda\left(\prod_{i=1}^{m+1}\frac{1}{\lambda_i}\right)||f||_{p_{m+1},\varphi}||a^{(1)}||_{p_1,\Psi_1}\prod_{i=2}^m ||a^{(i)}||_{p_i,\psi_i}, \quad (6.143)$$

and

$$\left\{ \int_0^\infty x_{m+1}^{p\lambda_{m+1}-1} \left[\sum_{n_m=1}^\infty \cdots \sum_{n_1=1}^\infty \frac{\prod_{j=1}^m a_{n_j}^{(j)}}{(\max_{1\le i\le m}\{n_i, x_{m+1}\})^\lambda} \right]^p dx_{m+1} \right\}^{\frac{1}{p}}$$

$$> \lambda \prod_{i=1}^{m+1} \frac{1}{\lambda_i} ||a^{(1)}||_{p_1, \Psi_1} \prod_{i=2}^m ||a^{(i)}||_{p_i, \psi_i}; \qquad (6.144)$$

(ii) *for,* $\lambda_{m+1} = \frac{\lambda}{2}$, *we have the following equivalent inequalities with the same best possible constant factor* $2 \prod_{i=1}^m \frac{1}{\lambda_i}$:

$$\int_0^\infty f(x_{m+1}) \sum_{n_m=1}^\infty \cdots \sum_{n_1=1}^\infty \frac{\prod_{j=1}^m a_{n_j}^{(j)}}{(\max_{1\le i\le m}\{x_{m+1}n_i, 1\})^\lambda} dx_{m+1}$$

$$> \frac{2}{\lambda_1} \left(\prod_{i=2}^m \frac{1}{\lambda_i} ||a^{(i)}||_{p_i, \psi_i} \right) ||f||_{p_{m+1}, \varphi} ||a^{(1)}||_{p_1, \Psi_1}, \quad (6.145)$$

and

$$\left\{ \int_0^\infty x_{m+1}^{\frac{p\lambda}{2}-1} \left[\sum_{n_m=1}^\infty \cdots \sum_{n_1=1}^\infty \frac{\prod_{j=1}^m a_{n_j}^{(j)}}{(\max_{1\le i\le m}\{x_{m+1}n_i, 1\})^\lambda} \right]^p dx_{m+1} \right\}^{\frac{1}{p}}$$

$$> \frac{2}{\lambda_1} \left(\prod_{i=2}^m \frac{1}{\lambda_i} ||a^{(i)}||_{p_i, \psi_i} \right) ||a^{(1)}||_{p_1, \Psi_1}. \qquad (6.146)$$

By Theorems 6.11 and 6.12, it follows that

Corollary 6.18. *Suppose that* $p_i < 0$, $0 < \lambda_i < 1 (i = 1, \cdots, m)$, $\lambda_{m+1} > 0$, $0 < p_{m+1} < 1$. *If* $f(x_{m+1}) \ge 0$, $f \in L_{p_{m+1}, \Phi}(\mathbf{R}_+)$, $||f||_{p_{m+1}, \Phi} > 0$, $a_{n_i}^{(i)} \ge 0$, $a^{(i)} = \{a_{n_i}^{(i)}\}_{n_i=1}^\infty \in l_{p_i, \psi_i}$, $||a^{(i)}||_{p_i, \psi_i} > 0 (i = 1, \cdots, m)$, *then*

(i) *we have the following equivalent reverse inequalities with the same best possible constant factor* $\lambda \prod_{i=1}^{m+1} \frac{1}{\lambda_i}$:

$$\int_0^\infty f(x_{m+1}) \sum_{n_m=1}^\infty \cdots \sum_{n_1=1}^\infty \frac{\prod_{j=1}^m a_{n_j}^{(j)}}{(\max_{1\le i\le m}\{n_i, x_{m+1}\})^\lambda} dx_{m+1}$$

$$> \frac{\lambda}{\lambda_{m+1}} \left(\prod_{i=1}^m \frac{1}{\lambda_i} ||a^{(i)}||_{p_i, \psi_i} \right) ||f||_{p_{m+1}, \Phi}, \qquad (6.147)$$

and

$$\left\{ \int_0^\infty \frac{x_{m+1}^{p\lambda_{m+1}-1}}{[1-\theta_{m+1}(x_{m+1})]^{p-1}} \right.$$

$$\left. \times \left[\sum_{n_m=1}^\infty \cdots \sum_{n_1=1}^\infty \frac{\prod_{j=1}^m a_{n_j}^{(j)}}{(\max_{1\le i\le m}\{n_i,x_{m+1}\})^\lambda} \right]^p dx_{m+1} \right\}^{\frac{1}{p}}$$

$$> \frac{\lambda}{\lambda_{m+1}} \left(\prod_{i=1}^m \frac{1}{\lambda_i} ||a^{(i)}||_{p_i,\psi_i} \right); \qquad (6.148)$$

(ii) *for,* $\lambda_{m+1} = \frac{\lambda}{2}$, $f(x_{m+1}) \ge 0$, $f \in L_{p_{m+1},\widetilde{\Phi}}(\mathbf{R}_+)$, $||f||_{p_{m+1},\widetilde{\Phi}} > 0$, *we have the following equivalent inequalities with the same best possible constant factor* $2\prod_{i=1}^m \frac{1}{\lambda_i}$:

$$\int_0^\infty f(x_{m+1}) \sum_{n_m=1}^\infty \cdots \sum_{n_1=1}^\infty \frac{\prod_{j=1}^m a_{n_j}^{(j)} dx_{m+1}}{(\max_{1\le i\le m}\{x_{m+1}n_i,1\})^\lambda}$$

$$> 2 \left(\prod_{i=1}^m \frac{1}{\lambda_i} ||a^{(i)}||_{p_i,\psi_i} \right) ||f||_{p_{m+1},\widetilde{\Phi}}, \qquad (6.149)$$

and

$$\left\{ \int_0^\infty \frac{x_{m+1}^{\frac{p\lambda}{2}-1}}{[1-\widetilde{\theta}_{m+1}(x_{m+1})]^{p-1}} \right.$$

$$\left. \times \left[\sum_{n_m=1}^\infty \cdots \sum_{n_1=1}^\infty \frac{\prod_{j=1}^m a_{n_j}^{(j)}}{(\max_{1\le i\le m}\{x_{m+1}n_i,1\})^\lambda} \right]^p dx_{m+1} \right\}^{\frac{1}{p}}$$

$$> 2\prod_{i=1}^m \frac{1}{\lambda_i} ||a^{(i)}||_{p_i,\psi_i}. \qquad (6.150)$$

Remark 6.5. (i) If in (6.24), we set

$$k_\lambda(x_1,\cdots,x_m,n) = \frac{1}{(\max_{1\le i\le m}\{x_i,n\})^\lambda},$$

then, in view of Corollary 6.12 and (6.25), we have

$$||T|| = k(\lambda_{m+1}) = \lambda \prod_{i=1}^{m+1} \frac{1}{\lambda_i};$$

if, in (6.26), for $\lambda_{m+1} = \frac{\lambda}{2} < 1$, we set

$$k_\lambda(nx_1,\cdots,nx_m,1) = \frac{1}{(\max_{1\le i\le m}\{nx_i,1\})^\lambda},$$

then, in view of Corollary 6.13 and (6.27), we have

$$\|T_1\| = k(\lambda_{m+1}) = 2 \prod_{i=1}^{m} \frac{1}{\lambda_i}.$$

(ii) If, in (6.61), we set

$$k_\lambda(n_1, \cdots, n_m, x_{m+1}) = \frac{1}{(\max_{1 \leq i \leq m}\{n_i, x_{m+1}\})^\lambda},$$

then, in view of Corollary 6.16 and (6.62), we have

$$\|\widetilde{T}\| = k(\lambda_{m+1}) = \lambda \prod_{i=1}^{m+1} \frac{1}{\lambda_i};$$

if, in (6.63), for $\lambda_{m+1} = \frac{\lambda}{2}$, we set

$$k_\lambda(x_{m+1}n_1, \cdots, x_{m+1}n_m, 1) = \frac{1}{(\max_{1 \leq i \leq m}\{x_{m+1}n_i, 1\})^\lambda},$$

then, in view of Corollary 6.16 and (6.64), we have

$$\|\widetilde{T}_1\| = k(\lambda_{m+1}) = 2 \prod_{i=1}^{m} \frac{1}{\lambda_i}.$$

6.4.4 The Case of
$$k_\lambda(x_1, \cdots, x_m, x_{m+1}) = \frac{1}{(\min_{1 \leq i \leq m+1}\{x_i\})^\lambda}$$

Lemma 6.19. *If λ_i, $\lambda < 0$ ($i = 1, \cdots, m+1$),*

$$k_\lambda(x_1, \cdots, x_{m+1}) = \frac{1}{(\min_{1 \leq i \leq m+1}\{x_i\})^\lambda},$$

then, we have

$$k(\lambda_{m+1}) = \int_0^\infty \cdots \int_0^\infty \frac{\prod_{j=1}^{m} u_j^{\lambda_j - 1}}{(\min_{1 \leq j \leq m+1}\{u_j, 1\})^\lambda} \, du_1 \cdots du_m$$

$$= (-1)^m \lambda \prod_{i=1}^{m+1} \frac{1}{\lambda_i}. \tag{6.151}$$

Proof. We prove (6.151) by mathematical induction.

For $m = 1$, we find

$$k(\lambda_2) = \int_0^\infty \frac{u_1^{\lambda_1 - 1}}{(\min\{u_1, 1\})^\lambda} du_1$$

$$= \int_0^1 \frac{u_1^{\lambda_1 - 1}}{u_1^\lambda} du_1 + \int_1^\infty u_1^{\lambda_1 - 1} du_1$$

$$= \frac{1}{-\lambda + \lambda_1} + \frac{1}{-\lambda_1} = \frac{-\lambda}{\lambda_1 \lambda_2},$$

and then, (6.151) follows.

Assuming that (6.151) is valid for $m = n - 1$, then, for $m = n$, we have

$$k(\lambda_{n+1}) = \int_0^\infty \cdots \int_0^\infty \frac{\prod_{j=1}^n u_j^{\lambda_j - 1}}{(\min_{1 \leq j \leq n}\{u_j, 1\})^\lambda} \, du_1 \cdots du_n$$

$$= \int_0^\infty \cdots \int_0^\infty \left[\int_0^\infty \frac{u_n^{\lambda_n - 1}}{(\min_{1 \leq j \leq n}\{u_j, 1\})^\lambda} du_n \right]$$

$$\times \prod_{j=1}^{n-1} u_j^{\lambda_j - 1} \, du_1 \cdots du_{n-1}.$$

Since, we find

$$\int_0^\infty \frac{u_n^{\lambda_n - 1}}{(\min_{1 \leq j \leq n}\{u_j, 1\})^\lambda} \, du_n$$

$$= \int_0^{\min_{1 \leq j \leq n-1}\{u_j, 1\}} \frac{u_n^{\lambda_n - 1}}{(\min_{1 \leq j \leq n}\{u_j, 1\})^\lambda} \, du_n$$

$$+ \int_{\min_{1 \leq j \leq n-1}\{u_j, 1\}}^\infty \frac{u_n^{\lambda_n - 1}}{(\min_{1 \leq j \leq n}\{u_j, 1\})^\lambda} \, du_n$$

$$= \int_0^{\min_{1 \leq j \leq n-1}\{u_j, 1\}} \frac{u_n^{\lambda_n - 1}}{u_n^\lambda} \, du_n$$

$$+ \int_{\min_{1 \leq j \leq n-1}\{u_j, 1\}}^\infty \frac{u_n^{\lambda_n - 1}}{(\min_{1 \leq j \leq n}\{u_j, 1\})^\lambda} \, du_n$$

$$= \frac{(\min_{1 \leq j \leq n-1}\{u_j, 1\})^{\lambda_n - \lambda}}{\lambda_n - \lambda} + \frac{-(\min_{1 \leq j \leq n-1}\{u_j, 1\})^{\lambda_n}}{\lambda_n (\min_{1 \leq j \leq n-1}\{u_j, 1\})^\lambda}$$

$$= \frac{\lambda}{\lambda_n(\lambda_n - \lambda)} \frac{1}{(\min_{1 \leq j \leq n-1}\{u_j, 1\})^{\lambda - \lambda_n}},$$

then, by the assumption of $m = n - 1$, it follows that

$$k(\lambda_{n+1}) = \frac{\lambda}{\lambda_n(\lambda_n - \lambda)} \int_0^\infty \cdots \int_0^\infty \frac{\prod_{j=1}^{n-1} u_j^{\lambda_j - 1} du_1 \cdots du_{n-1}}{(\min_{1 \leq j \leq n-1}\{u_j, 1\})^{\lambda - \lambda_n}}$$

$$= \frac{\lambda(\lambda - \lambda_n)(-1)^{n-1}}{\lambda_n(\lambda_n - \lambda)} \prod_{i=1(i \neq n)}^{n+1} \frac{1}{\lambda_i} = (-1)^n \lambda \prod_{i=1}^{n+1} \frac{1}{\lambda_i},$$

and then, (6.151) follows. $\square$

With the assumptions of Lemma 6.18, if $\lambda_{m+1} < \lambda + 1$, then we set $\delta_0 = \min_{1 \leq i \leq m+1}\{-\lambda_i, \lambda + 1 - \lambda_{m+1}\} > 0$. For any $\tilde{\lambda}_i \in (\lambda_i - \delta_0, \lambda_i +$

$\delta_0)(i = 1, \cdots, m+1)$, $\sum_{i=1}^{m+1} \widetilde{\lambda}_i = \lambda$, it follows that $\widetilde{\lambda}_i < \lambda_i + \delta_0 \leq 0 (i = 1, \cdots, m+1)$, $\widetilde{\lambda}_{m+1} < \lambda_{m+1} + \delta_0 \leq \lambda + 1$,

$$k(\widetilde{\lambda}_{m+1}) = \int_0^\infty \cdots \int_0^\infty \frac{\prod_{j=1}^m u_j^{\widetilde{\lambda}_j - 1}}{(\min_{1 \leq j \leq n-1}\{u_j, 1\})^\lambda} \, du_1 \cdots du_m$$

$$= (-1)^m \lambda \prod_{i=1}^{m+1} \frac{1}{\widetilde{\lambda}_i} \in \mathbf{R}_+,$$

and both $\frac{1}{(\min_{1 \leq i \leq m}\{x_i, y\})^\lambda} y^{\widetilde{\lambda}_{m+1}-1}$, and $\frac{1}{(\min_{1 \leq i \leq m}\{yx_i, 1\})^\lambda} y^{\widetilde{\lambda}_{m+1}-1}$ are strictly decreasing with respect to $y \in \mathbf{R}_+$. Then, for

$$k_\lambda(x_1, \cdots, x_m, x_{m+1}) = \frac{1}{(\min_{1 \leq i \leq m+1}\{x_i\})^\lambda} \quad (\lambda < 0)$$

in Theorems 6.1 and 6.2, it follows that

Corollary 6.19. *Suppose that* $p_i > 1$, $\lambda_i < 0 (i = 1, \cdots, m+1)$, $\lambda_{m+1} < \lambda + 1$. *If* $f_i(x_i) \geq 0$, $f_i \in L_{p_i, \varphi_i}(\mathbf{R}_+)$, $||f_i||_{p_i, \varphi_i} > 0 (i = 1, \cdots, m)$, $a_n \geq 0$, $a = \{a_n\}_{n=1}^\infty \in l_{p_{m+1}, \psi}$, $||a||_{p_{m+1}, \psi} > 0$, *then*

(i) *we have the following equivalent inequalities with the best possible constant factor* $(-1)^m \lambda \prod_{i=1}^{m+1} \frac{1}{\widetilde{\lambda}_i}$:

$$\sum_{n=1}^\infty a_n \int_0^\infty \cdots \int_0^\infty \frac{\prod_{j=1}^m f_j(x_j)}{(\min_{1 \leq i \leq m}\{x_i, n\})^\lambda} \, dx_1 \cdots dx_m$$

$$< \frac{(-1)^m \lambda}{\lambda_{m+1}} \left(\prod_{i=1}^m \frac{1}{\widetilde{\lambda}_i} ||f_i||_{p_i, \varphi_i} \right) ||a||_{p_{m+1}, \psi}, \quad (6.152)$$

and

$$\left\{ \sum_{n=1}^\infty n^{p\lambda_{m+1}-1} \left[\int_0^\infty \cdots \int_0^\infty \frac{\prod_{j=1}^m f_j(x_j) dx_1 \cdots dx_m}{(\min_{1 \leq i \leq m}\{x_i, n\})^\lambda} \right]^p \right\}^{\frac{1}{p}}$$

$$< \frac{(-1)^m \lambda}{\lambda_{m+1}} \prod_{i=1}^m \frac{1}{\widetilde{\lambda}_i} ||f_i||_{p_i, \varphi_i}; \quad (6.153)$$

(ii) *for,* $\lambda_{m+1} = \frac{\lambda}{2}(-\frac{1}{2} < \lambda < 0)$, *we have the following equivalent inequalities with the same best possible constant factor,* $2(-1)^m \prod_{i=1}^m \frac{1}{\widetilde{\lambda}_i}$:

$$\sum_{n=1}^\infty a_n \int_0^\infty \cdots \int_0^\infty \frac{\prod_{j=1}^m f_j(x_j)}{(\min_{1 \leq i \leq m}\{nx_i, 1\})^\lambda} \, dx_1 \cdots dx_m$$

$$< 2(-1)^m \left(\prod_{i=1}^m \frac{1}{\widetilde{\lambda}_i} ||f_i||_{p_i, \varphi_i} \right) ||a||_{p_{m+1}, \psi}, \quad (6.154)$$

and

$$\left\{\sum_{n=1}^{\infty} n^{\frac{p\lambda}{2}-1}\left[\int_0^{\infty}\cdots\int_0^{\infty}\frac{\prod_{j=1}^m f_j(x_j)dx_1\cdots dx_m}{(\min_{1\leq i\leq m}\{nx_i,1\})^{\lambda}}\right]^p\right\}^{\frac{1}{p}}$$

$$< 2(-1)^m\prod_{i=1}^m\frac{1}{\lambda_i}||f_i||_{p_i,\varphi_i}. \qquad (6.155)$$

In (6.9) and (6.12), for

$$k_{\lambda}(x_1,\cdots,x_m,x_{m+1})=\frac{1}{(\min_{1\leq i\leq m+1}\{x_i\})^{\lambda}}\quad(\lambda<0),$$

we find, for $m \geq 2$,

$$A_1(u_{m+1})=\int_0^{\infty}\cdots\int_0^{\infty}\frac{\prod_{j=2}^m u_j^{\lambda_j-1}}{(\min_{2\leq j\leq m+1}\{1,u_j\})^{\lambda}}\,du_2\cdots du_m$$

$$\leq\int_0^{\infty}\cdots\int_0^{\infty}\frac{\prod_{j=2}^m u_j^{\lambda_j-1}du_2\cdots du_m}{(\min_{2\leq j\leq m+1}\{1,u_j\})^{\lambda+\alpha-\lambda_{m+1}}u_{m+1}^{\lambda_{m+1}-\alpha}}$$

$$=\frac{\lambda+\alpha-\lambda_{m+1}}{\lambda_1+\alpha}\,(-1)^{m-1}\prod_{i=2}^m\frac{1}{\lambda_i}(u_{m+1}^{\alpha-\lambda_{m+1}}),$$

where, $0<\alpha<\min\{-\lambda_1,\lambda_{m+1}-\lambda\}$,

$$B_1(u_i)=\int_0^{\infty}\cdots\int_0^{\infty}\frac{\prod_{j=1(j\neq i)}^m u_j^{\lambda_j-1}du_1\cdots du_{i-1}du_{i+1}\cdots du_m}{(\min_{1\leq j\leq m}\{1,u_j\})^{\lambda}}$$

$$\leq\int_0^{\infty}\cdots\int_0^{\infty}\frac{\prod_{j=1(j\neq i)}^m u_j^{\lambda_j-1}du_1\cdots du_{i-1}du_{i+1}\cdots du_m}{(\min_{1\leq j\neq i\leq m}\{1,u_j\})^{\lambda+\alpha-\lambda_i}u_i^{\lambda_i-\alpha}}$$

$$=\frac{\lambda+\alpha-\lambda_i}{\lambda_{m+1}+\alpha}\,(-1)^{m-1}\prod_{j=1(j\neq i)}^m\frac{1}{\lambda_j}(u_i^{\alpha-\lambda_i}),$$

where $0<\alpha<\min\{\lambda_i-\lambda,-\lambda_{m+1}\}$.

For $m=1$, we can obtain some similar results.

By Theorems 6.3 and 6.4, it follows that

Corollary 6.20. *Suppose that* $0<p_1<1$, $p_i<0(i=2,\cdots,m+1)$, $\lambda_i<0(i=1,\cdots,m+1)$, $\lambda_{m+1}<\lambda+1$. *If* $f_1(x_1)\geq 0$, $f_1\in L_{p_1,\Phi_1}(\mathbf{R}_+)$, $||f_1||_{p_1,\Phi_1}>0$, $f_i(x_i)\geq 0$, $f_i\in L_{p_i,\varphi_i}(\mathbf{R}_+)$, $||f_i||_{p_i,\varphi_i}>0(i=2,\cdots,m)$, $a_n\geq 0$, $a=\{a_n\}_{n=1}^{\infty}\in l_{p_{m+1},\psi}$, $||a||_{p_{m+1},\psi}>0$, *then*

(i) *we have the following equivalent inequalities with the same best possible constant factor* $(-1)^m \lambda \prod_{i=1}^{m+1} \frac{1}{\lambda_i}$:

$$\sum_{n=1}^{\infty} a_n \int_0^{\infty} \cdots \int_0^{\infty} \frac{\prod_{j=1}^m f_j(x_j)}{(\min_{1 \le i \le m}\{x_i, n\})^{\lambda}} \, dx_1 \cdots dx_m$$

$$> (-1)^m \lambda \prod_{i=1}^{m+1} \frac{1}{\lambda_i} ||a||_{p_{m+1}, \psi} ||f_1||_{p_1, \Phi_1} \prod_{i=2}^m ||f_i||_{p_i, \varphi_i}, \quad (6.156)$$

and

$$\left\{ \sum_{n=1}^{\infty} n^{p\lambda_{m+1}-1} \left[\int_0^{\infty} \cdots \int_0^{\infty} \frac{\prod_{j=1}^m f_j(x_j) \, dx_1 \cdots dx_m}{(\min_{1 \le i \le m}\{x_i, n\})^{\lambda}} \right]^p \right\}^{\frac{1}{p}}$$

$$> (-1)^m \lambda \prod_{i=1}^{m+1} \frac{1}{\lambda_i} ||f_1||_{p_1, \Phi_1} \prod_{i=2}^m ||f_i||_{p_i, \varphi_i}; \quad (6.157)$$

(ii) *for* $\lambda_{m+1} = \frac{\lambda}{2}(-\frac{1}{2} < \lambda < 0), f_1 \in L_{p_1, \widetilde{\Phi}_1}(\mathbf{R}_+), ||f_1||_{p_1, \widetilde{\Phi}_1} > 0,$ *we have the following equivalent inequalities with the best possible constant factor* $2(-1)^m \prod_{i=1}^m \frac{1}{\lambda_i}$:

$$\sum_{n=1}^{\infty} a_n \int_0^{\infty} \cdots \int_0^{\infty} \frac{\prod_{j=1}^m f_j(x_j)}{(\min_{1 \le i \le m}\{nx_i, 1\})^{\lambda}} \, dx_1 \cdots dx_m$$

$$> 2(-1)^m \left(\prod_{i=1}^m \frac{1}{\lambda_i} \right) ||a||_{p_{m+1}, \psi} ||f_1||_{p_1, \widetilde{\Phi}_1} \prod_{i=2}^m ||f_i||_{p_i, \varphi_i}, \quad (6.158)$$

and

$$\left\{ \sum_{n=1}^{\infty} n^{p\frac{\lambda}{2}-1} \left[\int_0^{\infty} \cdots \int_0^{\infty} \frac{\prod_{j=1}^m f_j(x_j) dx_1 \cdots dx_m}{(\min_{1 \le i \le m}\{nx_i, 1\})^{\lambda}} \right]^p \right\}^{\frac{1}{p}}$$

$$> 2(-1)^m \left(\prod_{i=1}^m \frac{1}{\lambda_i} \right) ||f_1||_{p_1, \widetilde{\Phi}_1} \prod_{i=2}^m ||f_i||_{p_i, \varphi_i}. \quad (6.159)$$

By Theorems 6.5 and 6.6, we still have

Corollary 6.21. *Suppose that* $p_i < 0, \lambda_i < 0(i = 1, \cdots, m), 0 < p_{m+1} < 1, \lambda_{m+1} < \min\{0, \lambda+1\}$. *If* $f_i(x_i), a_n \ge 0, f_i \in L_{p_i, \varphi_i}(\mathbf{R}_+) \ (i = 1, \cdots, m), a = \{a_n\}_{n=1}^{\infty} \in l_{p_{m+1}, \psi}, ||f_i||_{p_i, \varphi_i} > 0, ||a||_{p_{m+1}, \psi} > 0$. *then*

(i) *we have the following equivalent reverse inequalities with the same best possible constant factor* $(-1)^m \lambda \prod_{i=1}^{m+1} \frac{1}{\lambda_i}$:

$$\sum_{n=1}^{\infty} a_n \int_0^{\infty} \cdots \int_0^{\infty} \frac{\prod_{j=1}^m f_j(x_j)}{(\min_{1 \le i \le m}\{x_i, n\})^{\lambda}} dx_1 \cdots dx_m$$

$$> \frac{(-1)^m \lambda}{\lambda_{m+1}} \left(\prod_{i=1}^m \frac{1}{\lambda_i} ||f_i||_{p_i, \varphi_i} \right) ||a||_{p_{m+1}, \psi}, \quad (6.160)$$

and

$$\left\{ \sum_{n=1}^{\infty} n^{p\lambda_{m+1}-1} \left[\int_0^{\infty} \cdots \int_0^{\infty} \frac{\prod_{j=1}^m f_j(x_j)dx_1 \cdots dx_m}{(\min_{1\le i\le m}\{x_i,n\})^{\lambda}} \right]^p \right\}^{\frac{1}{p}}$$

$$> \frac{(-1)^m \lambda}{\lambda_{m+1}} \prod_{i=1}^m \frac{1}{\lambda_i} ||f_i||_{p_i,\varphi_i}; \qquad (6.161)$$

(ii) *for* $\lambda_{m+1} = \frac{\lambda}{2}$ $(-\frac{1}{2} < \lambda < 0)$, *we have the following equivalent inequalities with the same best constant factor* $2(-1)^m \prod_{i=1}^m \frac{1}{\lambda_i}$:

$$\sum_{n=1}^{\infty} a_n \int_0^{\infty} \cdots \int_0^{\infty} \frac{\prod_{j=1}^m f_j(x_j)}{(\min_{1\le i\le m}\{nx_i,1\})^{\lambda}} \, dx_1 \cdots dx_m$$

$$> 2(-1)^m \left(\prod_{i=1}^m \frac{1}{\lambda_i} ||f_i||_{p_i,\varphi_i} \right) ||a||_{p_{m+1},\psi}, \qquad (6.162)$$

and

$$\left\{ \sum_{n=1}^{\infty} n^{\frac{p\lambda}{2}-1} \left[\int_0^{\infty} \cdots \int_0^{\infty} \frac{\prod_{j=1}^m f_j(x_j)dx_1 \cdots dx_m}{(\min_{1\le i\le m}\{nx_i,1\})^{\lambda}} \right]^p \right\}^{\frac{1}{p}}$$

$$> 2(-1)^m \prod_{i=1}^m \frac{1}{\lambda_i} ||f_i||_{p_i,\varphi_i}. \qquad (6.163)$$

By Theorems 6.7 and 6.8, we have the following

Corollary 6.22. *Suppose that* $p_i > 1, \lambda_i < 0(i = 1, \cdots, m), \lambda_{m+1} < \min\{0, \lambda + 1\}$. *If* $f(x_{m+1}) \ge 0$, $f \in L_{p_{m+1},\varphi}(\mathbf{R}_+)$, $||f||_{p_{m+1},\varphi} > 0$, $a_{n_i}^{(i)} \ge 0$, $a^{(i)} = \{a_{n_i}^{(i)}\}_{n_i=1}^{\infty} \in l_{p_i,\psi_i}, ||a^{(i)}||_{p_i,\psi_i} > 0(i = 1, \cdots, m)$, *then*

(i) *we have the following equivalent inequalities with the best possible constant factor* $(-1)^m \lambda \prod_{i=1}^{m+1} \frac{1}{\lambda_i}$:

$$\int_0^{\infty} f(x_{m+1}) \sum_{n_m=1}^{\infty} \cdots \sum_{n_1=1}^{\infty} \frac{\prod_{j=1}^m a_{n_j}^{(j)}}{(\min_{1\le i\le m}\{n_i, x_{m+1}\})^{\lambda}} \, dx_{m+1}$$

$$< \frac{(-1)^m \lambda}{\lambda_{m+1}} \left(\prod_{i=1}^m \frac{1}{\lambda_i} ||a^{(i)}||_{p_i,\psi_i} \right) ||f||_{p_{m+1},\varphi}, \qquad (6.164)$$

and

$$\left\{ \int_0^{\infty} x_{m+1}^{p\lambda_{m+1}-1} \left[\sum_{n_m=1}^{\infty} \cdots \sum_{n_1=1}^{\infty} \frac{\prod_{j=1}^m a_{n_j}^{(j)}}{(\min_{1\le i\le m}\{n_i, x_{m+1}\})^{\lambda}} \right]^p dx_{m+1} \right\}^{\frac{1}{p}}$$

$$< \frac{(-1)^m \lambda}{\lambda_{m+1}} \prod_{i=1}^m \frac{1}{\lambda_i} ||a^{(i)}||_{p_i,\psi_i}; \qquad (6.165)$$

(ii) *for $\lambda_{m+1} = \frac{\lambda}{2}$ $(-\frac{1}{2} < \lambda < 0)$, we have the following equivalent inequalities with the best possible constant factor $2(-1)^m \prod_{i=1}^{m} \frac{1}{\lambda_i}$:*

$$\int_0^\infty f(x_{m+1}) \sum_{n_m=1}^{\infty} \cdots \sum_{n_1=1}^{\infty} \frac{\prod_{j=1}^{m} a_{n_j}^{(j)}}{(\min_{1\le i\le m}\{n_i x_{m+1}, 1\})^\lambda} \, dx_{m+1}$$

$$< 2(-1)^m \left(\prod_{i=1}^{m} \frac{1}{\lambda_i} \|a^{(i)}\|_{p_i,\psi_i} \right) \|f\|_{p_{m+1},\varphi}, \qquad (6.166)$$

and

$$\left\{ \int_0^\infty x_{m+1}^{\frac{p\lambda}{2}-1} \left[\sum_{n_m=1}^{\infty} \cdots \sum_{n_1=1}^{\infty} \frac{\prod_{j=1}^{m} a_{n_j}^{(j)}}{(\min_{1\le i\le m}\{n_i x_{m+1}, 1\})^\lambda} \right]^p dx_{m+1} \right\}^{\frac{1}{p}}$$

$$< 2(-1)^m \prod_{i=1}^{m} \frac{1}{\lambda_i} \|a^{(i)}\|_{p_i,\psi_i}. \qquad (6.167)$$

In (6.44), for

$$k_\lambda(x_1, \cdots, x_m, x_{m+1}) = \frac{1}{(\min_{1\le i\le m+1}\{x_i\})^\lambda} \quad (\lambda < 0)$$

we find, for $m \ge 2$,

$$A_i(u_k) = \int_0^\infty \cdots \int_0^\infty \frac{1}{(\min_{1\le j\ne i\le m+1}\{u_j, 1\})^\lambda}$$

$$\times \prod_{j=1(j\ne i,k)}^{m+1} u_j^{\lambda_j-1} du_1 \cdots du_{i-1} du_{i+1} \cdots du_{k-1} du_{k+1} \cdots du_{m+1}$$

and

$$\le \int_0^\infty \cdots \int_0^\infty \frac{\prod_{j=1(j\ne i,k)}^{m+1} u_j^{\lambda_j-1}}{(\min_{1\le j\ne i,k\le m+1}\{u_j, 1\})^{\lambda+\alpha-\lambda_k} u_k^{\lambda_k-\alpha}}$$

$$\times du_1 \cdots du_{i-1} du_{i+1} \cdots du_{k-1} du_{k+1} \cdots du_{m+1}$$

$$= \frac{\lambda+\alpha-\lambda_k}{\lambda_i+\alpha}(-1)^{m-1} \prod_{j=1(j\ne i,k)}^{m+1} \frac{1}{\lambda_j}(u_k^{\alpha-\lambda_k}),$$

where $0 < \alpha < \min\{\lambda_k - \lambda, -\lambda_i\}$. For $m = 1$, we can obtain a similar result.

If $\lambda_i < \min\{0, 1 - \lambda\}(i = 1, \cdots, m)$, $\lambda_{m+1} < 0$, then, we set $\delta_0 = \min_{1\le i\le m}\{-\lambda_{m+1}, -\lambda_i, 1-\lambda-\lambda_i\} > 0$. For any $\widetilde{\lambda}_i \in (\lambda_i - \delta_0, \lambda_i + \delta_0)$ $(i = 1, \cdots, m+1)$, $\sum_{i=1}^{m+1} \widetilde{\lambda}_i = \lambda$, we find $\widetilde{\lambda}_{m+1} < 0$, $\widetilde{\lambda}_i < \min\{0, 1 - \lambda\}$ $(i = 1, \cdots, m)$,

$$k(\widetilde{\lambda}_{m+1}) = (-1)^m \lambda \prod_{i=1}^{m+1} \frac{1}{\widetilde{\lambda}_i} \in \mathbf{R}_+$$

and both $\dfrac{1}{(\min_{1\le i\le m} y_i, x_{m+1})^\lambda} y^{\tilde\lambda_j - 1}$ and $\dfrac{1}{(\min_{1\le i\le m}\{x_{m+1}y_i, 1\})^\lambda} y^{\tilde\lambda_j - 1}$ are strictly decreasing with respect to $y_j \in \mathbf{R}_+ (j = 1, \cdots, m)$.

Then by Theorems 6.9 and 6.10, it follows that

Corollary 6.23. *Suppose that $0 < p_1 < 1$, $p_i < 0 (i = 2, \cdots, m + 1)$, $\lambda_i < \min\{0, 1+\lambda\}\ (i = 1, \cdots, m)$, $\lambda_{m+1} < 0$. If $f(x_{m+1}) \ge 0$, $f \in L_{p_{m+1}, \varphi}(\mathbf{R}_+)$, $\|f\|_{p_{m+1},\varphi} > 0$, $a_{n_1}^{(1)} \ge 0$, $a^{(1)} = \{a_{n_1}^{(1)}\}_{n_1=1}^{\infty} \in l_{p_1, \Psi_1}$, $\|a^{(1)}\|_{p_1, \Psi_1} > 0$, $a_{n_i}^{(i)} \ge 0$, $a^{(i)} = \{a_{n_i}^{(i)}\}_{n_i=1}^{\infty} \in l_{p_i, \psi_i}$, $\|a^{(i)}\|_{p_i, \psi_i} > 0 (i = 2, \cdots, m)$, then*

(i) *we have the following equivalent inequalities with the same best possible constant factor $(-1)^m \lambda \prod_{i=1}^{m+1} \frac{1}{\lambda_i}$:*

$$\int_0^\infty f(x_{m+1}) \sum_{n_m=1}^\infty \cdots \sum_{n_1=1}^\infty \frac{\prod_{j=1}^m a_{n_j}^{(j)}}{(\min_{1\le i\le m}\{n_i, x_{m+1}\})^\lambda} dx_{m+1}$$
$$> (-1)^m \lambda \left(\prod_{i=1}^{m+1} \frac{1}{\lambda_i}\right) \|f\|_{p_{m+1},\varphi} \|a^{(1)}\|_{p_1, \Psi_1} \prod_{i=2}^m \|a^{(i)}\|_{p_i, \psi_i}, \quad (6.168)$$

and

$$\left\{\int_0^\infty x_{m+1}^{p\lambda_{m+1}-1} \left[\sum_{n_m=1}^\infty \cdots \sum_{n_1=1}^\infty \frac{\prod_{j=1}^m a_{n_j}^{(j)}}{(\min_{1\le i\le m} n_i, x_{m+1})^\lambda}\right]^p dx_{m+1}\right\}^{\frac{1}{p}}$$
$$> (-1)^m \lambda \prod_{i=1}^{m+1} \frac{1}{\lambda_i} \|a^{(1)}\|_{p_1, \Psi_1} \prod_{i=2}^m \|a^{(i)}\|_{p_i, \psi_i}; \quad (6.169)$$

(ii) *for, $\lambda_{m+1} = \frac{\lambda}{2}$, we have the following equivalent inequalities with the same best possible constant factor $2(-1)^m \prod_{i=1}^m \frac{1}{\lambda_i}$:*

$$\int_0^\infty f(x_{m+1}) \sum_{n_m=1}^\infty \cdots \sum_{n_1=1}^\infty \frac{\prod_{j=1}^m a_{n_j}^{(j)}}{(\min_{1\le i\le m}\{x_{m+1}n_i, 1\})^\lambda} dx_{m+1}$$
$$> \frac{2(-1)^m}{\lambda_1} \left(\prod_{i=2}^m \frac{1}{\lambda_i} \|a^{(i)}\|_{p_i, \psi_i}\right) \|f\|_{p_{m+1},\varphi} \|a^{(1)}\|_{p_1, \Psi_1}, \quad (6.170)$$

and

$$\left\{\int_0^\infty x_{m+1}^{\frac{p\lambda}{2}-1} \left[\sum_{n_m=1}^\infty \cdots \sum_{n_1=1}^\infty \frac{\prod_{j=1}^m a_{n_j}^{(j)}}{(\min_{1\le i\le m}\{x_{m+1}n_i, 1\})^\lambda}\right]^p dx_{m+1}\right\}^{\frac{1}{p}}$$
$$> \frac{2(-1)^m}{\lambda_1} \left(\prod_{i=2}^m \frac{1}{\lambda_i} \|a^{(i)}\|_{p_i, \psi_i}\right) \|a^{(1)}\|_{p_1, \Psi_1}. \quad (6.171)$$

By Theorems 6.11 and 6.12, it follows that

Corollary 6.24. *Suppose that $p_i < 0$, $\lambda_i < \min\{0, 1+\lambda\}(i = 1, \cdots, m)$, $\lambda_{m+1} < 0$, $0 < p_{m+1} < 1$. If $f(x_{m+1}) \geq 0$, $f \in L_{p_{m+1}, \Phi}(\mathbf{R}_+)$, $\|f\|_{p_{m+1}, \Phi} > 0$, $a_{n_i}^{(i)} \geq 0$, $a^{(i)} = \{a_{n_i}^{(i)}\}_{n_i=1}^{\infty} \in l_{p_i, \psi_i}$, $\|a^{(i)}\|_{p_i, \psi_i} > 0(i = 1, \cdots, m)$, then*

(i) *we have the following equivalent reverse inequalities with the same best possible constant factor $(-1)^m \lambda \prod_{i=1}^{m+1} \frac{1}{\lambda_i}$:*

$$\int_0^\infty f(x_{m+1}) \sum_{n_m=1}^\infty \cdots \sum_{n_1=1}^\infty \frac{\prod_{j=1}^m a_{n_j}^{(j)}}{(\min_{1 \leq i \leq m}\{n_i, x_{m+1}\})^\lambda} \, dx_{m+1}$$
$$> \frac{(-1)^m \lambda}{\lambda_{m+1}} \left(\prod_{i=1}^m \frac{1}{\lambda_i} \|a^{(i)}\|_{p_i, \psi_i} \right) \|f\|_{p_{m+1}, \Phi}, \quad (6.172)$$

and

$$\left\{ \int_0^\infty \frac{x_{m+1}^{p\lambda_{m+1}-1}}{[1 - \theta_{m+1}(x_{m+1})]^{p-1}} \right.$$
$$\times \left[\sum_{n_m=1}^\infty \cdots \sum_{n_1=1}^\infty \frac{\prod_{j=1}^m a_{n_j}^{(j)}}{(\min_{1 \leq i \leq m}\{n_i, x_{m+1}\})^\lambda} \right]^p dx_{m+1} \right\}^{\frac{1}{p}}$$
$$> \frac{(-1)^m \lambda}{\lambda_{m+1}} \left(\prod_{i=1}^m \frac{1}{\lambda_i} \|a^{(i)}\|_{p_i, \psi_i} \right); \quad (6.173)$$

(ii) *for, $\lambda_{m+1} = \frac{\lambda}{2}$, $f(x_{m+1}) \geq 0, f \in L_{p_{m+1}, \widetilde{\Phi}}(\mathbf{R}_+)$, $\|f\|_{p_{m+1}, \widetilde{\Phi}} > 0$, we have the following equivalent inequalities with the same best possible constant factor $2(-1)^m \prod_{i=1}^m \frac{1}{\lambda_i}$:*

$$\int_0^\infty f(x_{m+1}) \sum_{n_m=1}^\infty \cdots \sum_{n_1=1}^\infty \frac{\prod_{j=1}^m a_{n_j}^{(j)} dx_{m+1}}{(\min_{1 \leq i \leq m}\{x_{m+1}n_i, 1\})^\lambda}$$
$$> 2(-1)^m \left(\prod_{i=1}^m \frac{1}{\lambda_i} \|a^{(i)}\|_{p_i, \psi_i} \right) \|f\|_{p_{m+1}, \widetilde{\Phi}}, \quad (6.174)$$

and

$$\left\{ \int_0^\infty \frac{x_{m+1}^{\frac{p\lambda}{2}-1}}{[1 - \widetilde{\theta}_{m+1}(x_{m+1})]^{p-1}} \right.$$
$$\times \left[\sum_{n_m=1}^\infty \cdots \sum_{n_1=1}^\infty \frac{\prod_{j=1}^m a_{n_j}^{(j)}}{(\min_{1 \leq i \leq m}\{x_{m+1}n_i, 1\})^\lambda} \right]^p dx_{m+1} \right\}^{\frac{1}{p}}$$
$$> 2(-1)^m \prod_{i=1}^m \frac{1}{\lambda_i} \|a^{(i)}\|_{p_i, \psi_i}. \quad (6.175)$$

Remark 6.6. (i) If, in (6.24), we set

$$k_\lambda(x_1, \cdots, x_m, n) = \frac{1}{(\min_{1 \le i \le m}\{x_i, n\})^\lambda} \quad (\lambda < 0),$$

then, in view of Corollary 6.19 and (6.25), we have

$$\|T\| = k(\lambda_{m+1}) = (-1)^m \lambda \prod_{i=1}^{m+1} \frac{1}{\lambda_i};$$

if, in (6.26), for $\lambda_{m+1} = \frac{\lambda}{2}$, we set

$$k_\lambda(nx_1, \cdots, nx_m, 1) = \frac{1}{(\min_{1 \le i \le m}\{nx_i, 1\})^\lambda} \quad (\lambda < 0),$$

then, in view of Corollary 6.19 and (6.27), we have

$$\|T_1\| = k(\lambda_{m+1}) = 2(-1)^m \prod_{i=1}^{m} \frac{1}{\lambda_i}.$$

(ii) If, in (6.61), we set

$$k_\lambda(n_1, \cdots, n_m, x_{m+1}) = \frac{1}{(\min_{1 \le i \le m}\{n_i, x_{m+1}\})^\lambda} \quad (\lambda < 0),$$

then, in view of Corollary 6.22 and equality (6.62), we have

$$\|\widetilde{T}\| = k(\lambda_{m+1}) = (-1)^m \lambda \prod_{i=1}^{m+1} \frac{1}{\lambda_i};$$

if, in (6.63), for $\lambda_{m+1} = \frac{\lambda}{2}$, we set

$$k_\lambda(x_{m+1}n_1, \cdots, x_{m+1}n_m, 1) = \frac{1}{(\min_{1 \le i \le m}\{x_{m+1}n_i, 1\})^\lambda} \quad (\lambda < 0),$$

then, in view of Corollary 6.22 and (6.64), we have

$$\|\widetilde{T_1}\| = k(\lambda_{m+1}) = 2(-1)^m \prod_{i=1}^{m} \frac{1}{\lambda_i}.$$

Remark 6.7. Putting $m = 1$ in the theorems and corollaries of this chapter, we obtain the corresponding results of Chapters 3 and 4.

Bibliography

This bibliography is not by any means a complete one for the subject. Most of them consist of research papers and books to which reference is made in the text. Many other selected books and papers related to material of the subject have been included so that they may serve to stimulate new interest in future study and research.

[1] Bényi A. and Oh Choonghong, Best constants for certain multilinear integral operator, *Journal of Inequalities and Applications*, Vol. 2006, Art. ID28582: 1–12.

[2] Bonsall F. F., Inequalities with non-conjugate parameter, *J. Math. Oxford Ser.*, **2:2** (1951) 135–150.

[3] Brnetić I. and Pečarić J., Generalization of Hilbert's integral inequality, *Math. Ineq. & Appl.*, **7:2** (2004) 199–205.

[4] Brnetić I., Krnić M. and Pečarić J., Multiple Hilbert and Hardy-Hilbert inequalities with non-conjugate parameters, *Bull. Austral. Math. Soc.*, **71** (2005) 447–457.

[5] Carleman T., *Sur les equations integrals singulieres a noyau reel et symetrique*, Uppsala, 1923.

[6] Chen Z. Q. and Xu J. S., New extensions of Hilbert's inequality with multiple parameters, *Acta Math. Hungar.*, **117:4** (2007) 383–400.

[7] Chen Q. and Yang B. C., On a more accurate half-discrete Mulholland's inequality and an extension, *Journal of Inequalities and Applications*, **70** (2012) doi:10.1186/1029-242X-2012-70.

[8] Chen W., Jin C., Li C., and Lim J., Weighted Hardy-Littlewood-Sobolev Inequalities and Systems of Integral Equations, Discrete and Continuous Dynamical Systems, Supplement Volume: (2005) 164–172.

[9] Cheng Q. X., *Basic on real variable functions and functional analysis*, Beijing: Higher Education Press, 2003.

[10] Debnath, L. and Yang, B.C., Recent developments of Hilbert-type discrete and integral inequalities with applications, *Internat. Jour. Math. and Math. Sci.*, **2012** (2012) 1–30.

[11] Dračic B. Ban, and Pogány T. K., Discrete Hilbert type inequality with non-homogeneous kernel, *Appl. Anal. Discrete Math.*, **3:1** (2009) 88–96.

[12] Dračic B. Ban, Pečarić J., and Pogány T. K., On a discrete Hilbert type inequality with non-homogeneous kernel, *Sarajevo J. Math.*, **6:1** (2010) 23–34.

[13] Dračic B. Ban, Pečarić J., Peric I., and Pogány T. K., Discrete multiple Hilbert type inequality with non-homogeneous kernel, *J. Korean Math. Soc.*, **47:3** (2010) 537–546.

[14] Gao M. Z., A note on Hilbert double series theorem, *Hunan Mathematical Annal*, **12:1-2** (1992) 143–147.

[15] Gao M. Z., On the Hilbert inequality, *J. Anal. Appl.*, **18:4** (1999) 1117–1122.

[16] Gao M. Z., A new Hardy-Hilbert's type inequality for double series and its applications, *The Australian Journal of Mathematical Analysis and Appl.*, **3:1** (2005) Art.13: 1–10.

[17] Gao M. Z. and Hsu L. C., A survey of various refinements and generalizations of Hilbert's inequalities, *J. Math. Res. Exp.*, **25:2** (2005) 227–243.

[18] Gao M. Z. and Yang B. C., On the extended Hilbert's inequality, *Proc. Amer. Math. Soc.*, **126:3** (1998) 751–759.

[19] Gao M. Z., Jia W. J. and Gao X. M., On an improvement of Hardy-Hilbert's inequality, *J. Math.*, **26:6** (2006) 647–651.

[20] Hardy G. H., Note on a theorem of Hilbert concerning series of positive term, *Proceedings of the London Mathematical Society*, **23** (1925) 45–46.

[21] Hardy G. H., Littlewood J. E., and Pòlya G., *Inequalities*, Cambridge University Press, Cambridge, 1934.

[22] He B., On a Hilbert-type integral inequality with a homogeneous kernel in $\mathbf{R}^2$ and its equivalent form, *Journal of Inequalities and Applications*, 2012, **94** (2012) doi:10.1186/1029-242X-2012-94.

[23] He B. and Li Y. J., On several new inequalities close to Hilbert-Pachpatte's inequality, *J. Ineq. in Pure and Applied Math.*, **7:4** (2006) Art.154: 1–9.

[24] He B. and Yang B. C., On a half-discrete inequality with a general homogeneous kernel, *Journal of Inequalities and Applications*, 2012, **30** (2012) doi:10.1186/1029-242X-2012-30.

[25] He L. P., Gao M. Z. and Jia W. J., On a new strengthened Hardy-Hilbert's inequality, *J. Math. Res. Exp.*, **26:2** (2006) 276–282.

[26] He L. P., Jia W. J. and Gao M. Z., A Hardy-Hilbert's type inequality with gamma function and its applications, *Integral Transforms and Special functions*, **17:5** (2006) 355–363.

[27] He B., Qian Y. and Li Y. J., On analogues of the Hilbert's inequality, *Comm. in Math. Anal.*, **4:2** (2008) 47–53.

[28] He L. P., Yu J. M. and Gao M. Z., An extension of Hilbert's integral inequality, *Journal of Shaoguan University (Natural Science)*, **23:3** (2002) 25–30.

[29] Hong Y., All-side generalization about Hardy-Hilbert integral inequalities, *Acta Mathematica Sinica*, **44:4** (2001) 619–626.

[30] Hong Y., On Hardy-Hilbert integral inequalities with some parameters, *J. Ineq. in Pure & Applied Math.*, **6:4** (2005) Art. 92: 1–10.

[31] Hong Y., On multiple Hardy-Hilbert integral inequalities with some parameters, *Journal of Inequalities and Applications*, Vol. 2006, Art. ID 94960: 1–11.

[32] Hsu L. C. and Wang Y. J., A refinement of Hilbert's double series theorem, *J. Math. Res. Exp.*, **11:1** (1991) 143–144.

[33] Hu K., A few important inequalities, *Journal of Jianxi Teacher's College (Natural Science)*, **3:1** (1979) 1–4.

[34] Hu K., *Some problems in analysis inequalities*, Wuhan: Wuhan University Press, 2007.

[35] Huang Q. L., On a Multiple Hilbert's inequality with parameters, *Journal of Inequalities and Applications*, Volume 2010, Article ID 309319, 12 pages.

[36] Huang Q. L. and Yang B. C., On a multiple Hilbert-type integral operator and applications, *Journal of Inequalities and Applications*, Volume 2009, Article ID 192197, 13 pages.

[37] Huang Q., Yang B. and Debnath L., A multiple more accurate Hardy-Littlewood-Pòlya Inequality, *Le Matematiche* (accepted 2012).

[38] Huang Q. L. and Yang B. C., On a more accurate half-discrete Hilbert's inequality, *Journal of Inequalities and Applications*, **106** (2012), doi:10.1186/1029-242X-2012-106.

[39] Ingham A.E., A note on Hilbert's inequality, *J. London Math. Soc.*, **11** (1936) 237–240.

[40] Jia W. J., Gao M. Z. and Debnath L., Some new improvement of the Hardy -Hilbert inequality with applications, *International Journal of Pure and Applied Math.*, **11:1** (2004) 21–28.

[41] Jia W. J., Gao M. Z. and Gao X. M., On an extension of the Hardy-Hilbert theorem, *Studia Scientiarum Mathematicarum Hungarica*, **42:1** (2005) 21–35.

[42] Khotyakov M., Two proofs of the sharp Hardy-Littlewood-Sobolev inequality, Bachelor Thesis, Mathematics Department, LMU Munich (2011).

[43] Krnić M. and Pečarić J., General Hilbert's and Hardy's inequalities, *Math. Ineq. & Appl.*, **8:1** (2005) 29–51.

[44] Krnić M., Tomovski Z., and Pečarić J., Hilbert inequalities related to generalized hypergeometric functions, *Balkanica N.S.*, **22** (2008) 307–322.

[45] Krnić M., Gao M. Z., Pečarić J. and Gao X. M., On the best constant in Hilbert's inequality, *Math. Ineq. & Appl.*, **8:2** (2005) 317–329.

[46] Kuang J. C., On new extension of Hilbert's integral inequality, *J. Math. Anal. Appl.*, **235** (1999) 608–614.

[47] Kuang J. C., *Applied inequalities*, Jinan: Shandong Science Technic Press, 2004.

[48] Kuang J. C., New progress in inequality study in China, *Journal of Beijing Union University (Natural Science)*, **19:1** (2005) 29–37.

[49] Kuang J. C., *Introduction to real analysis, Hunan Education Press*, Chansha, China, 1996.

[50] Laith E. A., On some extensions of Hardy-Hilbert's inequality and applications, *Journal of Inequalities and Applications*, volume 2008, Article ID 546828, 14 pages.

[51] Knopp K., *Theory and application of infinite series*, Londen: Blackie & Son Limited, 1928.

[52] Levin V., Two remarks on Hilbert's double series theorem, *J. Indian Math. Soc.*, **11** (1937) 111–115.

[53] Li Y. J. and He B., On inequalities of Hilbert's type, *Bull. Austral. Math. Soc.*, **76** (2007) 1–13.

[54] Liu X. D. and Yang B. C., On a new Hilbert-Hardy-type integral operator and applications, *Journal of Inequalities and Applications*, Volume 2010, Article ID 812636, 10 pages.

[55] Lu Z. X., Some new inverse type Hilbert-Pachpatte inequalities, *Tamkang Journal of Mathematics*, **34:2** (2003) 155–161.

[56] Lu Z. X., On new generalizations of Hilbert's inequalities, *Tamkang Journal of Mathematics*, **35:1** (2004) 77–86.

[57] Mitrinović J. E., Pečarić J. E., and Fink A. M., *Inequalities involving functions and their integrals and derivatives*, Boston: Kluwer Acaremic Publishers, 1991.

[58] Pachpatte B. G., On some new inequalities similar to Hilbert's inequality, *J. Math. Anal. Appl.*, **226** (1998) 166–179.

[59] Pachpatte B. G., *Mathematical inequalities*, Elsevier B. V., Netherland, 2005.

[60] Pogány T. K., Hilbert's double series theorem extended to the case of non-homogeneous kernels, *J. Math. Anal. Appl.*, **342:2** (2008) 1485–1489.

[61] Pogány T. K., New class of inequalities associated with the Hilbert's double series theorem, *Appl. Math. E-Notes*, **10** (2010) 47–51.

[62] Pang C. D. and Pang C. B., *Basic on analysis number theory*, Beijing: Science Press, 1990.

[63] Pan Y. L, Wang H. T and Wang F. T., *On complex functions*, Science Press, Beijing, 2006.

[64] Qu W. L., *Combination mathematics*, Beijing: Beijing University Press, 1989.

[65] Salem S. R., Some new Hilbert type inequalities, *Kyungpook Math. J.*, **46** (2006) 19–29.

[66] Schur I., Bernerkungen sur Theorie der beschrankten Bilinearformen mit unendlich vielen veranderlichen, *Journal of Math.*, **140** (1911) 1–28.

[67] Stein E.M. and Weiss G., Fractional integrals in n-dimensional Euclidean space, *J. Math. Mech.*, **7** (1958) 503–514.

[68] Sulaiman W. T., On Hardy-Hilbert's integral inequality, *J. Ineq. in Pure & Appl. Math.*, **5:2** Art.25 (2004) 1–9.

[69] Sulaiman W. T., New ideas on Hardy-Hilbert's integral inequality (I), *Pan American Math. J.*, **15:2** (2005) 95–100.

[70] Sun B. J., Best generalization of a Hilbert type inequality, *J. Ineq. in Pure & Applied Math.*, **7:3** Art.113 (2006) 1–7.

[71] Titchmarsh E. C., *The theory of the Riemann Zeta-function*, Oxford: Clarendon Press, 1986.

[72] Tailor A. E., Lay D. C., *Introduction to functional analysis*, New York: John Wiley & Sons, 1980.

[73] Wang Z. Q. and Guo D. R., *Introduction to Special Functions*, Science Press, Beijing, 1979.

[74] Wang W. H. and Xin D. M., On a new strengthened version of a Hardy-Hilbert type inequality and applications, *J. Ineq. in Pure & Applied Math.*, **7:5** Art.180 (2006) 1–7.

[75] Wang W. H. and Yang B. C., A strengthened Hardy-Hilbert's type inequality, *The Australian Journal of Mathematical Analysis and Applications*, **3:2** (2006) Art.17: 1–7.

[76] Wang A. Z. and Yang B. C., A new Hilbert-type integral inequality in the whole plane with the non-homogeneous kernel, *Journal of Inequalities and Applications*, 2011, 2011:123, doi:10.1186/1029-242X-2011-123.

[77] Weyl H., *Singulare integral gleichungen mit besonderer berucksichtigung des fourierschen integral theorems*, Inaugeral-Dissertation, Gottingen, 1908.

[78] Wilhelm M., On the spectrum of Hilbert's matrix, *Amer J. Math.*, **72** (1950) 699–704.

[79] Xi G. W., A reverse Hardy-Hilbert-type inequality, *Journal of Inequalities and Appl.*, Vol. 2007, Art.ID79758: 1–7.

[80] Xie H. and Lu Z., Discrete Hardy-Hilbert's inequalities in $\mathbf{R}^n$, *Northeast. Math.*, **21:1** (2005) 87–94.

[81] Xie Z. T., A new Hilbert-type inequality with the kernel of 3 -homogeneous, *Journal of Jilin University (Science Edition)*, **45:3** (2007) 369–373.

[82] Xie Z. T., A Hilbert-type integral inequality with non-homogeneous kernel and withe the integral in whole plane, *Journal of Guangdong University of Education*, **31:3** (2011) 8–12.

[83] Xie Z. T., A new half-discrete Hilbert's inequality wit the homogeneous kernel of degree -4μ, *Journal of Zhanjiang Normal College*, **32:6** (2011) 13–19.

[84] Xie Z. T. and Zheng Z., A Hilbert-type inequality with parameters, *J. Xiangtan Univ. (Natural Science)*, **29:3** (2007) 24–28.

[85] Xie Z. T. and Zheng Z., A Hilbert-type integral inequality whose kernel is a homogeneous form of degree -3, *J. Math. Anal. Appl.*, **339** (2007) 324–331.

[86] Xie Z. T. and Zheng Z., A new Hilbert-type integral inequality and its reverse, *Soochow Journal of Math.*, **33:4** (2007) 751–759.

[87] Xin D. M., Best generalization of Hardy-Hilbert's inequality with multi-parameters, *J. Ineq. in Pure and Applied Math.*, **7:4** Art.153 (2006) 1–8.

[88] Xin D. M. and Yang B. C., A basic Hilbert-type inequality, *Journal of Mathematics*, **30:3** (2010) 554–560.

[89] Xin D. M. and Yang B. C., A Hilbert-type integral inequality in the whole plane with the homogeneous kernel of degree -2, *Journal of Inequalities and Applications*, Volume 2010, Article ID 401428, 11 pages.

[90] Xu J. S., Hardy-Hilbert's inequalities with two parameters, *Advances in Mathematics*, **36:2** (2007) 189–198.

[91] Xu L. Z. and Guo Y. K., Note on Hardy-Riesz's extension of Hilbert's inequality, *Chin. Quart. J. Math.*, **6:1** (1991) 75–77.

[92] Xu L. Z, Wang X. H., *Methods on mathematical analysis and examples*, Beijing: Higher Education Press, 1985.

[93] Xie Z. T., A generalization of Stirling formula, *Mathematical Practice and Cognition*, **36:6** (2006) 331–333.

[94] Xie Z. T. and Zeng Z., A new half-discrete Hilbert's inequality with the homogeneous kernel of degree -4μ, *Journal of Zhanjiang Normal College*, **32:6** (2011) 13–19.

[95] Yang B. C., A note on Hilbert's integral inequalities, *Chinese Quarterly J. Math.*, **13:4** (1998) 83–86.

[96] Yang B. C., On Hilbert's integral inequality, *J. Math. Anal. Appl.*, **220** (1998) 778–785.

[97] Yang B. C., On a strengthened version of the more accurate Hardy-Hilbert's inequality, *Acta Mathematica Sinica*, **42:6** (1999) 1103–1110.

[98] Yang B. C., A general Hardy-Hilbert's integral inequality with a best value, *Chinese Annals of Mathematics*, **21A:4** (2000) 401–408.

[99] Yang B. C., On a generalization of Hilbert's double series theorem, *Journal of Nanjing University-Mathematical Biquarterly*, **18:1** (2001) 145–151.

[100] Yang B. C., On a general Hardy-Hilbert's inequality, *Chinese Annals of Mathematics*, **23A:2** (2002) 247–254.

[101] Yang B. C., On a multiple Hardy-Hilbert's integral inequality, *Chinese Annals of Mathematics*, **24A:6** (2003) 743–750.

[102] Yang B. C., On a new inequality similar to Hardy-Hilbert's inequality, *Math. Ineq. Appl.*, **6:1** (2003) 37–44.

[103] Yang B. C., On a new Hardy-Hilbert's type inequality, *Math. Ineq. Appl.*, **7:3** (2004) 355–363.

[104] Yang B. C., On new extensions of Hilbert's inequality, *Acta Math. Hungar.*, **104:4** (2004) 291–299.

[105] Yang B. C., On an extension of Hilbert's integral inequality with some parameters, *The Australian Journal of Math. Analysis and Applications*, **1:1** Art.11 (2004) 1–8.

[106] Yang B. C., A new Hilbert-type integral inequality and its generalization, *Journal of Jilin University (Science Edition)*, **43:5** (2005) 580–584.

[107] Yang B. C., A mixed Hilbert-type inequality with a best constant factor, *International Journal of Pure and Applied Mathematics*, **20:3** (2005) 319–328.

[108] Yang B. C., On best extensions of Hardy-Hilbert's inequality with two parameters, *J. Ineq. in Pure & Applied Math.*, **6:3** (2005) Art. 81: 1–15.

[109] Yang B. C., A reverse of the Hardy-Hilbert's type inequality, *Journal of Southwest China Normal University (Natural Science)*, **30:6** (2005) 1012–1015.

[110] Yang B. C., On the way of weight function and research for Hilbert's type

integral inequalities, *Journal of Guangdong Education Institute (Natural Science)*, **25:3** (2005) 1–6.

[111] Yang B. C., On a more accurate Hardy-Hilbert's type inequality and its applications, *Acta Mathematica Sinica*, **49:3** (2006) 363–368.

[112] Yang B. C., On a relation to Hardy-Hilbert's inequality and Mulholland's inequality, *Acta Mathematica Sinica*, **49:3** (2006) 559–566.

[113] Yang B. C., On the norm of an integral operator and applications, *J. Math. Anal. Appl.*, **321** (2006) 182–192.

[114] Yang B. C., A bilinear inequality with the kernel of -2-order homogeneous, *Journal of Xiamen University (Natural Science)*, **45:6** (2006) 752–755.

[115] Yang B. C., A dual Hardy-Hilbert's inequality and generalizations, *Advances in Math.*, **35:1** (2006) 102–108.

[116] Yang B. C., On the norm of a self-adjoint operator and applications to Hilbert's type inequalities, *Bull. Belg. Math. Soc.*, **13** (2006) 577–584.

[117] Yang B. C., A new Hilbert-type inequality, *Bull. Belg. Math. Soc.*, **13** (2006) 479–487.

[118] Yang B. C., A new Hilbert-type inequality, *Journal of Shanghai Univ. (Natural Science)*, **13:3** (2007) 274–278.

[119] Yang B. C., On the norm of a self-adjoint operator and a new bilinear integral inequality, *Acta Mathematica Sinica, English Series*, **23:7** (2007) 1311–1316.

[120] Yang B. C., On the norm of a Hilbert's type linear operator and applications, *J. Math. Anal. Appl.*, **325** (2007) 529–541.

[121] Yang B. C., A Hilbert-type inequality with two pairs of conjugate exponents, *Journal of Jilin University (Science Edition)*, **45:4** (2007) 524–528.

[122] Yang B. C., On a Hilbert-type operator with a symmetric homogeneous kernel of -1-order and applications, *Journal of Inequalities and Applications*, Volume 2007, Article ID 47812, 9 pages.

[123] Yang B. C., A Hilbert-type integral inequality with the kernel of -3-order homogeneous, *Journal of Yunnam University*, **30:4** (2008) 325–330.

[124] Yang B. C., On the norm of a linear operator and its applications, *Indian Journal of Pure and Applied Mathematics*, **39:3** (2008) 237–250.

[125] Yang B. C., On the norm of a certain self-adjoint integral operator and applications to bilinear integral inequalities, *Taiwan Journal of Mathematics*, **12:2** (2008) 315–324.

[126] Yang B. C., A basic Hilbert-type integral inequality with the homogeneous kernel of -1-degree and extensions, *Journal of Guangdong Education Institute*, **28:3** (2008) 1–10.

[127] Yang B. C., On a Hilbert-type operator with a class of homogeneous kernels, *Journal of Inequalities and Applications*, Volume 2009, Article ID 572176, 9 pages.

[128] Yang B. C., *Hilbert-type integral inequalities*, Bentham Science Publishers Ltd., 2009.

[129] Yang B. C., A survey of the study of Hilbert-type inequalities with parameters, *Advances in Mathematics*, **38:3** (2009) 257–268.

[130] Yang B. C., *On the norm of operator and Hilbert-type inequalities*, Beijing: Science Press, 2009.

[131] Yang B. C., On an Application of Hilbert's inequality with multi-parameters, *Journal of Beijing Union University (Natural Sciences)*, **24:4** (1010) 78–84.

[132] Yang B. C., An Application of the reverse Hilbert's inequality, *Journal of Xinxiang University (Natural Sciences)*, **27:4** (1010) 2–5.

[133] Yang B. C., A new Hilbert-type operator and applications, *Publ. Math. Debrecen*, **76:1-2** (2010) 147–156.

[134] Yang B. C., *Discrete Hilbert-type inequalities*, Bentham Science Publishers Ltd., 2011.

[135] Yang B. C., A half-discrete Hilbert-type inequality, *Journal of Guangdong University of Education*, **31:3** (2011) 1–7.

[136] Yang B. C., A Hilbert-type integral inequality with the non-homogeneous kernel on the plane, *Journal of Guangdong University of Education*, **31:5** (2011) 5–10.

[137] Yang B. C., A new half-discrete Mulholland-type inequality with parameters, *Ann. Funct. Anal.*, **3:1** (2012) 142–150.

[138] Yang B. C. and Chen Q., A half-discrete Hilbert-type inequality with a homogeneous kernel and an extension, *Journal of Inequalities and Applications*, 2011, **124** (2011), doi:10.1186/1029-242X-2011-124.

[139] Yang B. C. and Chen Q., A half-discrete Hilbert-type inequality with the non-monotone kernel, *Journal of Jilin University (Science Edition)*, **50:2** (2012) 167–172.

[140] Yang B. C. and Debnath L., On new strengthened Hardy-Hilbert's inequality, *Internat. J. Math. & Math. Soc.*, **21:2** (1998) 403–408.

[141] Yang B. C. and Debnath L., On a new generalization of Hardy-Hilbert's inequality, *J. Math. Anal. Appl.*, **233** (1999) 484–497.

[142] Yang B. C. and Debnath L., On the extended Hardy-Hilbert's inequality, *J. Math. Anal. Appl.*, **72:2** (2002) 187–199.

[143] Yang B. C. and Debnath L., A strengthened Hardy-Hilbert's inequality, *Proceedings of the Jangjeon Mathematical Society*, **6:2** (2003) 119–124.

[144] Yang B. C. and Gao M. Z., On a best value of Hardy-Hilbert's inequality, *Advances in Math.*, **26:2** (1997) 159–164.

[145] Yang B. C. and Krnić M., Hilbert-type inequalities and related operators with homogeneous kernel of degree 0, *Mathematical Inequalities & Applications*, **13:4** (2010) 817–839.

[146] Yang B. C. and Liang H. W., A new Hilbert-type integral inequality with a parameter, *Journal of Henan University (Natural Science)*, **35:4** (2005) 4–8.

[147] Yang B. C. and Rassias T. M., On the way of weight coefficient and research for Hilbert-type inequalities, *Math. Ineq. Appl.*, **6:4** (2003) 625–658.

[148] Yang B. C. and Rassias Th. M., On a Hilbert-type integral inequality in the subinterval and its operator expression, *Banach J. Math. Anal.*, **4:2** (2010) 100–110.

[149] Yang B. C., Brnetć I., Krnić M. and Pečarić J., Generalization of Hilbert

and Hardy-Hilbert integral inequalities, *Math. Ineq. & Appl.*, **8:2** (2005) 259–272.

[150] Yang B. C., A new formula for evaluating the sum of d-th powers of the first n terms of an arithmetic sequence, *Journal of South China Normal University*, **1** (1996) 129–137.

[151] Yang B. C., The formula about the sum of powers of natural numbers relating Bernoulli numbers, *Mathematical Practice and Cognition*, **4** (1994) 52–56.

[152] Yang B. C. and Zhu Y. H., Inequalities on the Hurwitz Zeta-function restricted to the axis of positive reals, *Acta Scientiarum Naturalium Universitis Sunyatseni*, **36:3** (1997) 30–35.

[153] Yang B. C., The evaluating formulas on the convergence p-series relating Bernoulli numbers, *Journal of Guangdong Education Institute*, **3** (1992) 19–27.

[154] Yang B. C. and Li D. C., Estimation of the sum for -1-th powers of the first n terms of an arithmetic sequence, *Natural Science Journal of Hainan Teachers College*, **10:1** (1997) 19–24.

[155] Yang B. C. and Wang G. Q., Some inequalities on harmonic series. *Journal of Mathematics Study*, **29:3** (1996) 90–97.

[156] Yang B. C., Some new inequalities on step multiply, *Journal of Guangdong Education Institute*, **22:2** (2002) 1–4.

[157] Yang B. C. and Krnić M., A half-discrete Hilbert-type inequality with a general homogeneous kernel of degree o, *Journal of Mathematical Inequalities*, **6:3** (2012) 401–417.

[158] Yang B. C., On a more accurate half-discrete Hilbert's inequality, *Journal of Beijing Union University (Natural Sciences)*, **26:2** (2012) 63–68.

[159] Yang B. C., On a half-discrete Hilbert-type inequality, *Journal of Shantou University (Natural Sciences)*, **26:4** (2011) 5–10.

[160] Yang B. C., A more accurate half-discrete reverse Hilbert-type inequality, *Journal of Hunan Institute of Science and Technology (Natural Sciences)*, **24:4** (2011) 1–6.

[161] Yang B. C., On two classes of more accurate half-discrete reverse Hilbert's inequalities, *Journal of Xinxiang University (Natural Science Edition)*, **29:1** (2012) 9–14.

[162] Yang B. C., A half-discrete reverse Hilbert-type inequality with a homogeneous kernel of positive degree, *Journal of Zhanjiang Normal College*, **32:3** (2011) 5–9.

[163] Yang B. C., On a half-discrete Mulholland's inequality and its extension, *Thai Journal of Mathematics*, (2012) (to appear).

[164] Yang B. C., A half-discrete Hilbert's inequality with a non-homogeneous kernel, *Journal of Zhanjiang Normal College*, **32:6** (2011) 5–11.

[165] Yang B. C., A half-discrete Hilbert-type inequality with the non-homogeneous kernel, *Journal of Xinxiang University: Natural Science Edition*, **28:5** (2011) 385–387.

[166] Yang B. C., On a half-discrete reverse Hilbert-type inequality with a non-

homogeneous kernel, *Journal of Inner Mongolia Normal University (Natural Science Edition)*, **40:5** (2011) 433–337.

[167] Yang B. C., A half-discrete reverse Hilbert-type inequality with a non-homogeneous kernel, *Journal of Hunan Institute of Science and Technology (Natural Science)*, **24:3** (2011) 1–4.

[168] Yang B. C., A more accurate half-discrete reverse Hilbert's inequality with the non-homogeneous kernel, *Journal of Guangdong University of Education*, **32:3** (2012) 1–7.

[169] Yang B. C., On a best extension of a half-discrete Hilbert-type inequalities, *Jordan Journal of Mathematics and Statistics*, **5:4** (2012) 267–281.

[170] Yang B. C., A half-discrete Hilbert-type inequality with a non-homogeneous kernel and two variables, *Mediterranean Journal of Methematics*, 2012, doi: 10.1007/s00009-012-0213-50 online first.

[171] Zeng Z. and Xie Z. T., On a new Hilbert-type integral inequality with the integral in whole plane, *Journal of Inequalities and Applications*, Volume 2010, Article ID 256796, 8 pages.

[172] Zhang K. W., A bilinear inequality, *J. Math. Anal. Appl.*, **271** (2002) 288–296.

[173] Zhao C. J. and Debnath L., Some new type Hilbert integral inequalities, *J. Math. Anal. Appl.*, **262** (2001) 411–418.

[174] Zhong W. Y., A mixed Hilbert-type inequality and its equivalent forms, *Journal of Guangdong University of Education*, **31:5** (2011) 18–22.

[175] Zhong J. H. and Yang B. C., On an extension of a more accurate Hilbert-type inequality, *Journal of Zhejiang University (Science Edition)*, **35:2** (2008) 121–124.

[176] Zhong W. I. and Yang B. C., A best extension of Hilbert inequality involving several parameters, *Journal of Jinan University (Natural Science)*, **28:1** (2007) 20–23.

[177] Zhong W. I. and Yang B. C., A reverse Hilbert's type integral inequality with some parameters and the equivalent forms, *Pure and Applied Mathematics*, **24:2** (2008) 401–407.

[178] Zhong W. Y. and Yang B. C., On Multiple's Hardy-Hilbert integral inequality with kernel, *Journal of Inequalities and Applications*, Vol. 2007, Art.ID 27962, 17 pages, doi:10.1155/2007/27.

[179] Zhu Y. H, Yang B. C., Accurate inequalities of partial sums on a type of divergent series, *Acta Scientiarum Naturalium Universitis Sunyatseni*, **37:4** (1998) 33–37.

[180] Zhu Y. H., Yang B. C., Improvement on Euler's summation formula and some inequalities on sums of powers, *Acta Scientiarum Naturalium Universitis Sunyatseni*, **36:4** (1997) 21–26.

[181] Zhong W. Y., A mixed Hilbert-type inequality and its equivalent forms, *Journal of Guangdong University of Education*, **31:5** (2011) 18–22.

Index